Modern Experimental Design

THE WILEY BICENTENNIAL–KNOWLEDGE FOR GENERATIONS

$\mathscr{E}$ach generation has its unique needs and aspirations. When Charles Wiley first opened his small printing shop in lower Manhattan in 1807, it was a generation of boundless potential searching for an identity. And we were there, helping to define a new American literary tradition. Over half a century later, in the midst of the Second Industrial Revolution, it was a generation focused on building the future. Once again, we were there, supplying the critical scientific, technical, and engineering knowledge that helped frame the world. Throughout the 20th Century, and into the new millennium, nations began to reach out beyond their own borders and a new international community was born. Wiley was there, expanding its operations around the world to enable a global exchange of ideas, opinions, and know-how.

For 200 years, Wiley has been an integral part of each generation's journey, enabling the flow of information and understanding necessary to meet their needs and fulfill their aspirations. Today, bold new technologies are changing the way we live and learn. Wiley will be there, providing you the must-have knowledge you need to imagine new worlds, new possibilities, and new opportunities.

Generations come and go, but you can always count on Wiley to provide you the knowledge you need, when and where you need it!

WILLIAM J. PESCE
PRESIDENT AND CHIEF EXECUTIVE OFFICER

PETER BOOTH WILEY
CHAIRMAN OF THE BOARD

Modern Experimental Design

THOMAS P. RYAN

Acworth, GA

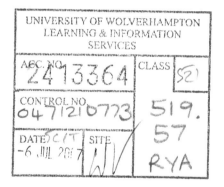

BICENTENNIAL
1807
WILEY
2007
BICENTENNIAL

WILEY-INTERSCIENCE
A JOHN WILEY & SONS, INC., PUBLICATION

Library of Congress Cataloging-in-Publication Data:

Ryan, Thomas P.
  Modern experimental design / by Thomas P. Ryan
      p.   cm.
  Includes bibliographical references and index.
  ISBN 978-0-471-21077-1

Printed in the United States of America
10 9 8 7 6 5 4 3 2 1

Contents

Preface **XV**

1 **Introduction** **1**

 1.1 Experiments All Around Us 2
 1.2 Objectives for Experimental Designs 3
 1.3 Planned Experimentation versus Use of
   Observational Data 5
 1.4 Basic Design Concepts 6
  1.4.1 Randomization 6
  1.4.2 Replication versus Repeated Measurements 7
  1.4.3 Example 8
  1.4.4 Size of an Effect That Can be Detected 11
 1.5 Terminology 12
 1.6 Steps for the Design of Experiments 13
  1.6.1 Recognition and Statement of the Problem 14
  1.6.2 Selection of Factors and Levels 14
   1.6.2.1 Choice of Factors 14
   1.6.2.2 Choice of Levels 15
 1.7 Processes Should Ideally be in a State of Statistical
   Control 18
 1.8 Types of Experimental Designs 20
 1.9 Analysis of Means 20
 1.10 Missing Data 22
 1.11 Experimental Designs and Six Sigma 22
 1.12 Quasi-Experimental Design 23
 1.13 Summary 23
   References 23
   Exercises 26

2 Completely Randomized Design   **31**

 2.1 Completely Randomized Design 31
  2.1.1 Model 32
  2.1.2 Example: One Factor, Two Levels 33
   2.1.2.1 Assumptions 33
  2.1.3 Examples: One Factor, More Than Two Levels 35
   2.1.3.1 Multiple Comparisons 36
   2.1.3.2 Unbalanced and Missing Data 39
   2.1.3.3 Computations 40
  2.1.4 Example Showing the Effect of Unequal Variances 41
 2.2 Analysis of Means 42
  2.2.1 ANOM for a Completely Randomized Design 43
   2.2.1.1 Example 44
  2.2.2 ANOM with Unequal Variances 45
   2.2.2.1 Applications 47
  2.2.3 Nonparametric ANOM 47
  2.2.4 ANOM for Attributes Data 47
 2.3 Software for Experimental Design 48
 2.4 Missing Values 48
 2.5 Summary 48
  Appendix 49
  References 49
  Exercises 51

3 Designs that Incorporate Extraneous (Blocking) Factors   **56**

 3.1 Randomized Block Design 56
  3.1.1 Assumption 57
  3.1.2 Blocking an Out-of-Control Process 60
  3.1.3 Efficiency of a Randomized Block Design 61
  3.1.4 Example 61
   3.1.4.1 Critique 63
  3.1.5 ANOM 64
 3.2 Incomplete Block Designs 65
  3.2.1 Balanced Incomplete Block Designs 65
   3.2.1.1 Analysis 66
   3.2.1.2 Recovery of Interblock Information 68
   3.2.1.3 ANOM 68
  3.2.2 Partially Balanced Incomplete Block Designs 69
   3.2.2.1 Lattice Design 70
  3.2.3 Nonparametric Analysis for Incomplete Block Designs 70
  3.2.4 Other Incomplete Block Designs 70
 3.3 Latin Square Design 71
  3.3.1 Assumptions 72
  3.3.2 Model 74

	3.3.3	Example	74
	3.3.4	Efficiency of a Latin Square Design	77
	3.3.5	Using Multiple Latin Squares	77
	3.3.6	ANOM	79
3.4		Graeco–Latin Square Design	80
	3.4.1	Model	80
	3.4.2	Degrees of Freedom Limitations on the Design Construction	81
	3.4.3	Sets of Graeco–Latin Square Designs	82
	3.4.4	Application	82
	3.4.5	ANOM	83
3.5		Youden Squares	84
	3.5.1	Model	85
	3.5.2	Lists of Youden Designs	86
	3.5.3	Using Replicated Youden Designs	86
	3.5.4	Analysis	86
3.6		Missing Values	86
3.7		Software	89
3.8		Summary	90
		References	91
		Exercises	93

4 Full Factorial Designs with Two Levels — **101**

4.1		The Nature of Factorial Designs	101
4.2		The Deleterious Effects of Interactions	106
	4.2.1	Conditional Effects	107
	4.2.1.1	Sample Sizes for Conditional Effects Estimation	113
	4.2.2	Can We "Transform Away" Interactions?	114
4.3		Effect Estimates	114
4.4		Why Not One-Factor-at-a-Time Designs?	115
4.5		ANOVA Table for Unreplicated Two-Factor Design?	116
4.6		The 2^3 Design	119
4.7		Built-in Replication	122
4.8		Multiple Readings versus Replicates	123
4.9		Reality versus Textbook Examples	124
	4.9.1	Factorial Design but not "Factorial Model"	124
4.10		Bad Data in Factorial Designs	127
	4.10.1	ANOM Display	134
4.11		Normal Probability Plot Methods	136
4.12		Missing Data in Factorial Designs	138
	4.12.1	Resulting from Bad Data	139
	4.12.2	Proposed Solutions	140
4.13		Inaccurate Levels in Factorial Designs	140
4.14		Checking for Statistical Control	141
4.15		Blocking 2^k Designs	142

4.16   The Role of Expected Mean Squares in Experimental Design   144
4.17   Hypothesis Tests with Only Random Factors in 2^k Designs?
       Avoid Them!   146
4.18   Hierarchical versus Nonhierarchical Models   147
4.19   Hard-to-Change Factors   148
       4.19.1   Software for Designs with Hard-to-Change Factors   150
4.20   Factors Not Reset   150
4.21   Detecting Dispersion Effects   150
4.22   Software   151
4.23   Summary   151
       Appendix A   Derivation of Conditional Main Effects   152
       Appendix B   Relationship Between Effect Estimates and
                    Regression Coefficients: 153
       Appendix C   Precision of the Effect Estimates   153
       Appendix D   Expected Mean Squares for the Replicated 2^2
                    Design   153
       Appendix E   Expected Mean Squares, in General   155
       References   157
       Exercises   162

5   Fractional Factorial Designs with Two Levels   169

5.1 2^{k-1} Designs   170
       5.1.1   Which Fraction?   176
       5.1.2   Effect Estimates and Regression Coefficients   177
       5.1.3   Alias Structure   177
       5.1.4   What if I Had Used the Other Fraction?   179
5.2 2^{k-2} Designs   181
       5.2.1   Basic Concepts   185
5.3   Designs with $k - p = 16$ 187
       5.3.1   Normal Probability Plot Methods when $k - p = 16$ 187
       5.3.2   Other Graphical Methods   188
5.4   Utility of Small Fractional Factorials vis-à-vis Normal
       Probability Plots   188
5.5   Design Efficiency   190
5.6   Retrieving a Lost Defining Relation   190
5.7   Minimum Aberration Designs and Minimum Confounded
       Effects Designs   192
5.8   Blocking Factorial Designs   194
       5.8.1   Blocking Fractional Factorial Designs   195
               5.8.1.1   Blocks of Size 2   200
5.9   Foldover Designs   201
       5.9.1   Semifolding   203
               5.9.1.1   Conditional Effects   208
               5.9.1.2   Semifolding a 2^{k-1} Design   210

	5.9.1.3	General Strategy?	215
	5.9.1.4	Semifolding with Software	215
5.10	John's 3/4 Designs		216
5.11	Projective Properties of 2^{k-p} Designs		219
5.12	Small Fractions and Irregular Designs		220
5.13	An Example of Sequential Experimentation		222
	5.13.1	Critique of Example	224
5.14	Inadvertent Nonorthogonality—Case Study		225
5.15	Fractional Factorial Designs for Natural Subsets of Factors		226
5.16	Relationship Between Fractional Factorials and Latin Squares		228
5.17	Alternatives to Fractional Factorials		229
	5.17.1	Designs Attributed to Genichi Taguchi	229
5.18	Missing and Bad Data		230
5.19	Plackett–Burman Designs		230
5.20	Software		230
5.21	Summary		233
	References		234
	Exercises		238

6 Designs With More Than Two Levels — **248**

6.1	3^k Designs		248
	6.1.1	Decomposing the $A*B$ Interaction	251
	6.1.2	Inference with Unreplicated 3^k Designs	252
6.2	Conditional Effects		255
6.3	3^{k-p} Designs		257
	6.3.1	Understanding 3^{k-p} Designs	259
	6.3.2	Constructing 3^{k-p} Designs	260
	6.3.3	Alias Structure	262
	6.3.4	Constructing a 3^{3-1} Design	262
	6.3.5	Need for Mixed Number of Levels	263
	6.3.6	Replication of 3^{k-p} Designs?	264
6.4	Mixed Factorials		264
	6.4.1	Constructing Mixed Factorials	265
	6.4.2	Additional Examples	266
6.5	Mixed Fractional Factorials		274
6.6	Orthogonal Arrays with Mixed Levels		275
6.7	Minimum Aberration Designs and Minimum Confounded Effects Designs		277
6.8	Four or More Levels		278
6.9	Software		280
6.10	Catalog of Designs		284
6.11	Summary		284
	References		284
	Exercises		286

7 Nested Designs   **291**

7.1   Various Examples   294
7.2   Software Shortcomings   295
       7.2.1   A Workaround   295
7.3   Staggered Nested Designs   298
7.4   Nested and Staggered Nested Designs with Factorial
       Structure   300
7.5   Estimating Variance Components   300
7.6   ANOM for Nested Designs?   302
7.7   Summary   302
       References   302
       Exercises   304

8 Robust Designs   **311**

8.1   "Taguchi Designs?"   312
8.2   Identification of Dispersion Effects   314
8.3   Designs with Noise Factors   316
8.4   Product Array, Combined Array, or Compound Array?   318
8.5   Software   320
8.6   Further Reading   322
8.7   Summary   322
       References   323
       Exercises   326

9 Split-Unit, Split-Lot, and Related Designs   **330**

9.1   Split-Unit Design   331
       9.1.1   Split-Plot Mirror Image Pairs Designs   336
       9.1.2   Split-Unit Designs in Industry   336
       9.1.3   Split-Unit Designs with Fractional Factorials   340
       9.1.4   Blocking Split-Plot Designs   342
       9.1.5   Split-Unit Plackett-Burman Designs   343
       9.1.6   Examples of Split-Plot Designs for Hard-to-Change
               Factors   343
       9.1.7   Split-Split-Plot Designs   345
9.2   Split-Lot Design   345
       9.2.1   Strip-Plot Design   346
               9.2.1.1   Applications of Strip-Block (Strip-Plot)
                         Designs   347
9.3   Commonalities and Differences Between these Designs   349
9.4   Software   350
9.5   Summary   351
       References   351
       Exercises   354

10 Response Surface Designs   **360**

10.1 Response Surface Experimentation: One Design or More Than One?   362
10.2 Which Designs?   364
10.3 Classical Response Surface Designs versus Alternatives   364
     10.3.1 Effect Estimates?   369
10.4 Method of Steepest Ascent (Descent)   370
10.5 Central Composite Designs   373
     10.5.1 CCD Variations   377
     10.5.2 Small Composite Designs   377
        10.5.2.1 Draper–Lin Designs   378
     10.5.3 Additional Applications   383
10.6 Properties of Space-Filling Designs   384
10.7 Applications of Uniform Designs   386
10.8 Box–Behnken Designs   386
     10.8.1 Application   388
10.9 Conditional Effects?   389
10.10 Other Response Surface Designs   390
     10.10.1 Hybrid Designs   390
     10.10.2 Uniform Shell Designs   393
     10.10.3 Koshal Designs   393
     10.10.4 Hoke Designs   394
10.11 Blocking Response Surface Designs   394
     10.11.1 Blocking Central Composite Designs   394
     10.11.2 Blocking Box–Behnken Designs   396
     10.11.3 Blocking Other Response Surface Designs   396
10.12 Comparison of Designs   397
10.13 Analyzing the Fitted Surface   398
     10.13.1 Characterization of Stationary Points   401
     10.13.2 Confidence Regions on Stationary Points   402
     10.13.3 Ridge Analysis   403
        10.13.3.1 Ridge Analysis with Noise Factors   404
     10.13.4 Optimum Conditions and Regions of Operability   404
10.14 Response Surface Designs for Computer Simulations   404
10.15 ANOM with Response Surface Designs?   405
10.16 Further Reading   405
10.17 The Present and Future Direction of Response Surface Designs   406
10.18 Software   406
10.19 Catalogs of Designs   408
10.20 Summary   408
     References   409
     Exercises   414

11   Repeated Measures Designs   **425**

  11.1   One Factor   426
         11.1.1   The Example in Section 2.1.2   428
  11.2   More Than One Factor   428
  11.3   Crossover Designs   429
  11.4   Designs for Carryover Effects   432
  11.5   How Many Repeated Measures?   437
  11.6   Further Reading   438
  11.7   Software   438
  11.8   Summary   439
         References   439
         Exercises   444

12   Multiple Responses   **447**

  12.1   Overlaying Contour Plots   448
  12.2   Seeking Multiple Response Optimization with Desirability
         Functions   449
         12.2.1   Weight and Importance   451
  12.3   Dual Response Optimization   452
  12.4   Designs Used with Multiple Responses   452
  12.5   Applications   453
  12.6   Multiple Response Optimization Variations   463
  12.7   The Importance of Analysis   469
  12.8   Software   469
  12.9   Summary   471
         References   472
         Exercises   474

13   Miscellaneous Design Topics   **483**

  13.1   One-Factor-at-a-Time Designs   483
  13.2   Cotter Designs   487
  13.3   Rotation Designs   488
  13.4   Screening Designs   489
         13.4.1   Plackett–Burman Designs   489
                  13.4.1.1   Projection Properties of Plackett–Burman
                             Designs   493
                  13.4.1.2   Applications   494
         13.4.2   Supersaturated Designs   498
                  13.4.2.1   Applications   499
         13.4.3   Lesser-Known Screening Designs   500
  13.5   Design of Experiments for Analytic Studies   500
  13.6   Equileverage Designs   501
         13.6.1   One Factor, Two Levels   502
         13.6.2   Are Commonly Used Designs Equileverage?   502

13.7	Optimal Designs	503
	13.7.1 Alphabetic Optimality	504
	13.7.2 Applications of Optimal Designs	507
13.8	Designs for Restricted Regions of Operability	508
13.9	Space-Filling Designs	514
	13.9.1 Uniform Designs	515
	13.9.1.1 From Raw Form to Coded Form	518
	13.9.2 Sphere-Packing Designs	518
	13.9.3 Latin Hypercube Design	519
13.10	Trend-Free Designs	521
13.11	Cost-Minimizing Designs	522
13.12	Mixture Designs	522
	13.12.1 Optimal Mixture Designs or Not?	523
	13.12.2 ANOM	523
13.13	Design of Measurement Capability Studies	523
13.14	Design of Computer Experiments	523
13.15	Design of Experiments for Categorical Response Variables	524
13.16	Weighing Designs and Calibration Designs	524
	13.16.1 Calibration Designs	525
	13.16.2 Weighing Designs	526
13.17	Designs for Assessing the Capability of a System	528
13.18	Designs for Nonlinear Models	528
13.19	Model-Robust Designs	528
13.20	Designs and Analyses for Non-normal Responses	529
13.21	Design of Microarray Experiments	529
13.22	Multi-Vari Plot	530
13.23	Evolutionary Operation	531
13.24	Software	531
13.25	Summary	532
	References	533
	Exercises	542
14	**Tying It All Together**	**544**
14.1	Training for Experimental Design Use	544
	References	545
	Exercises	546
Answers to Selected Exercises		**551**
Appendix: Statistical Tables		**565**
Author Index		**575**
Subject Index		**587**

Preface

Although there is a moderate amount of data analysis, especially in certain chapters, the emphasis in this book is on the statistical design of experiments. Such emphasis is justified by the widely held view that data from a well-designed experiment are easy to analyze. Certain types of designs are not simple, however, such as those covered in Chapters 7, 8, and 11, and the problem is compounded by the fact that some popular statistical software packages have quite limited capability for those designs.

The book would be suitable for an undergraduate one-semester course in design of experiments. For a course taught to nonstatistics majors, an instructor may wish to cover Chapters 1–4, part of Chapter 5, and then pick and choose from the other chapters in accordance with the needs of the students. The selection might include either or both of Chapters 10 and 12 and then cover sections of interest in Chapter 13.

For statistics majors, the book would be suitable for use in an advanced undergraduate course, perhaps covering Chapters 1–5, 7, 8, and much of Chapter 13. There is also enough advanced material for the book to be useful as a reference book in a graduate course taught to statistics majors, and might also be used in a graduate course for nonstatistics majors, depending on the needs and backgrounds of the students.

There is also enough material for a two-semester course, with the first course perhaps covering Chapters 1–6 and the second course covering Chapters 7–12 and 14, and parts of Chapter 13.

There is a considerable amount of material that is not covered to any extent, if at all, in other books on the subject, and some or all of this material might be used in special topics courses. These topics include conditional effects, uniform designs, and designs for restricted operating regions. (I have covered this material in an Internet course.)

A two-semester course in statistical methods should provide more than enough background for the book since the emphasis is on designs rather than statistical concepts. Matrix algebra is used in various places in the book, although it is not used extensively. Nevertheless, proficiency in the basics of matrix algebra is necessary for following some of the material.

One of the special features of the book is the emphasis on conditional effects in Chapters 4, 5, 6, and 10. This is an important topic that is not covered to any extent in most books and is addressed in very few journal articles. Another somewhat unique feature is moderate use of URLs, especially links to published articles that are available to the general public as well as article preprints and technical reports. There are other links for articles that are available to certain groups, such as members of the American Society for Quality. Some of those URLs might of course become outdated but I decided to list them since many of them, such as links to journal articles, will probably not become outdated in the near future. They make available to the reader a considerate amount of important resource material.

It is worth noting that this book does not contain catalogs of designs, as are given in some other books on the subject. Rather, the emphasis is on understanding design concepts and properties, the software that is available for generating specific designs and when to use those designs, and as stated, a moderate amount of analysis of data from experiments in which the designs are used, with extensive analysis provided in some case studies. Although there is some hand computation, the emphasis is on using appropriate software to generate output and interpret the output.

It is also worth noting that whereas there are case studies and a moderate amount of data analyses, there is not a "full" analysis of any dataset as that would include checking for outliers and influential observations, testing assumptions, and so on, which are covered in books on statistical methods. This is important but comes under the heading of data analysis rather than design and analysis of experiments. Although this book has more analysis than most books on design of experiments, it is not intended to be a handbook on data analysis.

I wish to gratefully acknowledge my editor, Steve Quigley, who motivated me to write this book, in addition to the contributions of associate editor Susanne Steitz, production editor Rosalyn Farkas, and colleagues who have made helpful comments, including Dennis Lin and Ivelisse Aviles, plus the helpful comments of three anonymous reviewers.

THOMAS P. RYAN

CHAPTER 1

Introduction

The statistical design of experiments plays a prominent role in experimentation. As George Box has stated, to see how a system functions when you have interfered with it, you have to interfere with it. That "interference" must be done in a systematic way so that the data from the experiment produce meaningful information.

The design of an experiment should be influenced by (1) the objectives of the experiment, (2) the extent to which sequential experimentation will be performed, if at all, (3) the number of factors under investigation, (4) the possible presence of identifiable and nonidentifiable extraneous factors, (5) the amount of money available for the experimentation, and (6) the purported model for modeling the response variable. Inman, Ledolter, Lenth, and Niemi (1992) stated, "Finally, it is impossible to overemphasize the importance of using a statistical model that matches the experimental design that was actually used." If we turn that statement around, we should use a design that matches a tentative model, recognizing that we won't know the model exactly.

In general, the design that is used for an experiment should be guided by these objectives. In many cases, the conditions and objectives will lead to an easy choice of a design, but this will not always be the case. Software is almost indispensable in designing experiments, although commonly used software will sometimes be inadequate, such as when there is a very large number of factors. Special-purpose software, not all of which is commercially available, will be needed in some circumstances. Various software programs are discussed throughout the book, with strong emphasis on Design-Expert®, which has certain features reminiscent of expert systems software, JMP®, and MINITAB®. (Readers intending to use the latter for designing experiments and analyzing the resultant data may be interested in Mathews (2004), although the latter is largely an introductory statistics book. Parts of the book are available online to members of the American Society for Quality (ASQ) at http://qualitypress.asq.org/chapters/H1233.pdf.) Although it is freeware, GOSSET is far more powerful than typical freeware. It is especially good for optimal designs (see Section 13.7) and runs on Unix, Linux, and Mac operating systems. Since GOSSET

Modern Experimental Design   By Thomas P. Ryan

is not as well known by experimenters, its Web site has been given here, which is http://www.research.att.com/~njas/gosset/index.html.

Design-Expert is a registered trademark of Stat-Ease, Inc. JMP is a registered trademark of SAS Institute, Inc. MINITAB is a registered trademark of MINITAB, Inc.

1.1   EXPERIMENTS ALL AROUND US

People perform experiments all of the time: workers who are new to a city want to find the shortest and/or fastest route to work, chefs experiment with new recipes, computer makers try to make better and faster computers, and so on. Improvement in processes is often the objective, as is optimality, such as finding the shortest route to work.

A pharmaceutical company that invents a new drug it believes is effective in combating a particular disease has to support its belief with the results of clinical trials, a form of experimentation. A scientist who believes he or she has made an important discovery needs to have the result supported by the results of experimentation. Although books on design of experiments did not begin to appear until well into the twentieth century, experimentation is certainly about as old as mankind.

Undoubtedly, all kinds of experiments were performed centuries ago that did not become a part of recorded history. About 100 years ago some rather extreme and bizarre experiments performed by Duncan MacDougall, MD, did become part of recorded history, however. He postulated that the human soul has measurable mass that falls within a specific range of weights. To prove this, he performed experiments on humans and dogs. In experimentation described at http://www.snopes.com/religion/soulweight.asp, Dr. MacDougall supposedly used six terminal patients and weighed them before, during, and after the process of death. The first patient lost three-fourths of an ounce and Dr. MacDougall, who apparently sought to conduct his experiments in a manner approximating the scientific method (see, e.g., Beveridge, 1960), ruled out all possible physiological explanations for the loss of weight. Since 3/4 ounce equals 21.26 grams, the result of this experimentation is believed to form the basis for the title of the movie *21 Grams* that was released in 2003 and starred Sean Penn and Naomi Watts.

To help confirm his conclusion, Dr. MacDougall decided to perform the same experiment on 15 dogs and found that the weight of the dogs did not change. As he stated, "the ideal test on dogs would be obtained in those dying from some disease that rendered them much exhausted and incapable of struggle." Unfortunately, he found that "it was not my good fortune to find dogs dying from such sickness." This prompted author Mary Roach (2003) to write "barring a local outbreak of distemper, one is forced to conclude that the good doctor calmly poisoned fifteen healthy canines for his little exercise in biological theology."

Accounts of Dr. MacDougall's experiments were published in the journal *American Medicine* and in *The New York Times*: "Soul has weight, physician thinks," March 11, 1907, p. 5, and "He weighed human soul," October 16, 1920, p. 13, with the latter published at the time of his death. MacDougall admitted that his experiments would have to be repeated many times with similar results before any conclusions could be drawn. Today his work is viewed as suffering from too small a sample size and an imprecise measuring instrument, and is viewed as nothing more than a curiosity (see, for example, http://www.theage.com.au/articles/2004/02/20/1077072838871.html.)

Although such experimentation is quite different from most types of experimentation that involve statistically designed experiments, small sample sizes and imprecise measuring instruments can undermine any experiment. Accordingly, attention is devoted in Section 1.4.3 and in other chapters on necessary minimum sample sizes for detecting significant effects in designed experiments.

More traditional experiments, many of which were performed more than 50 years ago, are in the 113 case studies of statistically designed experiments given by Bisgaard (1992).

When we consider all types of experiments that are performed, we find that certainly most experiments are not guided by statistical principles. Rather, most experimentation is undoubtedly trial-and-error experimentation. Much experimentation falls in the one-factor-at-a-time (OFAT) category, with each of two or more factors varied one at a time while the other factors are held fixed. Misleading information can easily result from such experiments, although OFAT designs can occasionally be used beneficially. These designs are discussed in Section 13.1.

1.2 OBJECTIVES FOR EXPERIMENTAL DESIGNS

The objectives for each experiment should be clearly delineated, as these objectives will dictate the construction of the designs, with sequential experimentation generally preferred. The latter is usually possible, depending upon the field of application. Bisgaard (1989) described a sequence of experiments and how, after considerable frustration, a satisfactory end result was finally achieved.

As explained by John (2003), however, sequential experimentation isn't very practical in the field of agronomy, as the agronomist must plan his or her experiment in the spring and harvest all of the data in the fall. Such obstacles to sequential experimentation do not exist in engineering applications, nor do they exist in most other fields of application. (John (2003) is recommended reading for its autobiographical content on one of the world's leading researchers in experimental design over a period of several decades.)

Following are a few desirable criteria for an experimental design:

(1) The design points should exert equal influence on the determination of the regression coefficients and effect estimates, as is the case with almost all the designs discussed in this book.

(2) The design should be able to detect the need for nonlinear terms.

(3) The design should be robust to model misspecification since all models are wrong.

(4) Designs in the early stage of the use of a sequential set of designs should be constructed with an eye toward providing appropriate information for follow-up experiments.

Box and Draper (1975) gave a list of 14 properties that a response surface design (see Chapter 10) should possess, and most of the properties are sufficiently general as to be applicable to virtually all types of designs. That list was published over 30 years ago and many advancements have occurred since then, although some properties, such as "provide data that will allow visual analysis," will stand the test of time.

Assume that a marathon runner would like to identify the training and nutritional regimens that will allow him or her to perform at an optimal level in a forthcoming race. Let Y denote the runner's race time and let μ denote what his or her theoretical average time would be over all training and nutritional regimens that he or she would consider and over all possible weather conditions. If no controllable or uncontrollable factors could be identified that would affect the runner's time, then the model for the race time would be

$$Y = \mu + \epsilon$$

with ϵ denoting a random error term that represents that the race time should vary in a random manner from the overall mean.

If this were the true model, then all attempts at discovering the factors that affect this person's race time would be unsuccessful. But we know this cannot be the correct model because, at the very least, weather conditions will have an affect. Weather conditions are, of course, uncontrollable, and so being able to identify weather conditions as an important factor would not be of great value to our runner. However, he or she would still be interested in knowing the effect of weather conditions on performance, just as a company would like to know how its products perform when customers use the products in some way other than the intended manner.

The runner would naturally prefer not to be greatly affected by weather conditions nor by the difficulty of the course, just as a toy manufacturer would not want its toys to fall apart if children are somewhat rough on them.

In experimental design applications we want to be able to identify both controllable and uncontrollable factors that affect our response variable (Y). We must face the fact, however, that we cannot expect to identify all of the relevant factors and the true model that is a function of them. As G. E. P. Box stated (e.g., Box, 1976), "All models are wrong, but some are useful." Our objective, then, is to identify a useful model, $Y = f(X_1, X_2, \ldots, X_k) + \epsilon$, with $X_1, X_2, \ldots, X_k$ having been identified as significant factors. Each factor is either *quantitative* or *qualitative*, and a useful model might contain a mixture of the two. For example, the type of breakfast that a runner eats would be a qualitative factor.

Since we will never have the correct model, we cannot expect to run a single experiment and learn all that we need to learn from that experiment. Indeed, Box (1993) quoted R. A. Fisher: "The best time to design an experiment is after you have done it." Thus, experimentation should (ideally) be sequential, with subsequent experiments designed using knowledge gained from prior experiments, and budgets should be constructed with this in mind. Opinions do vary on how much of the budget should be spent on the first experiment. Daniel (1976) recommends using 50–67 percent of the resources on the first experiment, whereas Box, Hunter, and Hunter (1978) more stringently recommend that at most 25 percent of the resources be used for the first experiment. Since sequential experimentation could easily involve more than two experiments, depending upon the overall objective(s), the latter seems preferable.

1.3 PLANNED EXPERIMENTATION VERSUS USE OF OBSERVATIONAL DATA

Many universities model college grade point average (GPA) as a function of variables such as high school GPA and aptitude test scores. As a simple example, assume that the model contains high school GPA and SAT total as the two variables. Clearly these two variables should be positively correlated. That is, if one is high the other will probably also be high. When we have two factors (i.e., variables) in an experimental design, we want to isolate the effect of each factor and also to determine if the interaction of the two factors is important (interaction is discussed and illustrated in detail in Section 4.2).

A factor can be either *quantitative or qualitative.* For a quantitative factor, inferences can be drawn regarding the expected change in the response variable per unit change in the factor, within the range of the experimentation, whereas, say, the "midpoint" between two levels of a qualitative factor, such as two cities, generally won't have any meaning. Quantitative and qualitative factors are discussed further in Section 1.6.2.2.

For the scenario just depicted, we do not have an experimental design, however. Rather, we have observational data, as we would "observe" the data that we would obtain in our sample of records from the Registrar's office. We can model observational data, but we cannot easily determine the separate effects of the factors since they will almost certainly be correlated, at least to some degree.

However, assume that we went to the Registrar's office and listed 25 combinations of the two variables that we wanted, and the student's college GPA was recorded for each combination. Since the values of the two variables are thus fixed, could we call this planned experimentation? No, it is still observational data. Furthermore, it would be nonrandom data if we wanted our "design" to have good properties, as we would, for example, be trying to make the two variables appear to be uncorrelated (i.e., an orthogonal design), which are actually highly correlated. So the results that were produced would probably be extremely misleading.

Returning to the runner example, let's say that our runner uses two nutritional supplement approaches (heavy and moderate), and two training regimes (intense and less intense). He wants to isolate the effects of these two factors, and he will use a prescribed course and record his running time. Assume that he is to make four runs and for two of these runs he uses a heavy supplement approach and an intense training regime, and for the other two he uses a moderate supplement approach and a less intense training regime.

Would the data obtained from this experiment be useful? No, this would be a classic example of how *not* to design an experiment. If the running time decreased when the intensity of the training regimen increased, was the decrease in running time due to the training regimen change or was it due to the increase in supplementation? In statistical parlance, these two effects are completely *confounded* and cannot be separated. (The terms *confounding* and *partial confounding* are discussed and illustrated in Section 5.1.)

Obviously the correct way to design the experiment if four runs are to be used is to use all four combinations of the two factors. Then we could identify the effects of each factor separately, as will be seen in Section 4.1 when we return to this example.

1.4 BASIC DESIGN CONCEPTS

Assume that a math teacher in an elementary school has too many students in her class one particular semester, so her class will be split and she will teach each of the two classes. She has been considering a new approach to teaching certain math concepts, and this unexpected turn of events gives her an opportunity to test the new approach against the standard approach. She will split the 40 students (20 boys and 20 girls) into two classes, and she wonders how she should perform the split so that the results of her experiment will be valid.

One obvious possibility would be to have the boys in one class and the girls in the other class. In addition to being rather unorthodox, this could create a *lurking variable* (i.e., an extraneous factor) that could undermine the results since it has been conjectured for decades that boys may take to math better than do girls. What if the split were performed alphabetically? Some people believe that there is a correlation between intelligence and the closeness to the beginning of the alphabet of the first letter in the person's last name. Although this is probably more folklore than fact, why take a chance? The safest approach would obviously be to use some random number device to assign the students to the two classes. That is, *randomization* is used. (Although this would likely create different numbers of boys and girls in each class if the 40 students were randomly divided between the two classes, the imbalance would probably be slight and not of any real concern.)

1.4.1 Randomization

IMPORTANT POINT

Randomization should be used *whenever possible and practical* so as to eliminate or at least reduce the possibility of confounding effects that could render an experiment practically useless.

Randomization is, loosely speaking, the random assignment of factor levels to experimental units. Ideally, the randomization method described by Atkinson and Bailey (2001) should be used whenever possible, although it is doubtful that hardly any experimenters actually use it. Specifically, they state, "In a completely randomised design the treatments, with their given replications, are first assigned to the experimental units systematically, and then a permutation is chosen at random from the $n!$ permutations of the experimental units (p. 57)." This is preferable to assigning the treatments (i.e., factor levels) at random to the experimental units, because a random assignment if performed sequentially will result, for example, in the last factor level being assigned to the last available experimental unit, which is clearly not a random assignment. The randomization method espoused by Atkinson and Bailey (2001) avoids these types of problems. Of course we could accomplish the same thing by, assuming t treatments, randomly selecting one of the $t!$ orderings, and then randomly selecting one of the $n!$ permutations of the experimental units and elementwise combining the juxtaposed lists.

Randomization is an important part of design of experiments because it reduces the chances of extraneous factors undermining the results, as illustrated in the preceding section. Czitrom (2003, p. 25) stated, "The results of many semiconductor experiments have been compromised by lack of randomization in the assignment of the wafers in a lot (experimental units) to experimental conditions."

Notice the words "whenever possible and practical" in italics in the Important Point, however, as randomization should not automatically be used.

In particular, randomization is not always possible, and this is especially true in regard to a randomized run order, as it will often not be possible to change factor levels at will and use certain combinations of factor levels. If randomization is not performed, however, and the results are unexpected, it may be almost impossible to quantitatively assess the effect of any distortion caused by the failure to randomize. This is an important consideration.

There are various detailed discussions of randomization in the literature, perhaps the best of which is Box (1990). The position taken by the author, which is entirely reasonable, is that randomization should be used if it only slightly complicates the experiment; it should not be used if it more than slightly complicates the experiment, but there is a strong belief that process stability has been achieved and is likely to continue during the experiment; and the experiment should not be run at all if the process is so unstable that the results would be unreliable without randomization but randomization is not practical.

The issue of process stability and its importance is discussed further in Section 1.7.

Undoubtedly there are instances, although probably rare, when the use of randomization in the form of randomly ordering the runs can cause problems. John (2003) gave an example of the random ordering of runs for an experiment with a 2^4 design (covered in Chapter 4) that created a problem. Specifically, the machinery broke down after the first week so that only 8 of the 16 runs could be made. Quoting John (2003), "It would have been so much better if we had not randomized the order. If only we had made the first eight points be one of the two resolution IV half replicates. We could have also chosen the next four points to make a twelve-point fraction of resolution V, and, then, if all was going well, complete the full factorial." (These designs are covered in Chapters 4 and 5.)

1.4.2 Replication versus Repeated Measurements

Another important concept is *replication*, and the importance of this (and the importance of doing it properly) can be illustrated as follows.

IMPORTANT POINT

Replication should be used whenever possible so as to provide an estimate of the standard deviation of the experimental error. It is important to distinguish between replicates and multiple readings. To replicate an experiment is to start from scratch and repeat an entire experiment, not to simply take more readings at each factor-level condition without resetting factor levels and doing the other things necessary to have a true replicated experiment.

The distinction between replication and multiple readings is an important one, as values of the response variable Y that result from replication can be used to estimate σ_ϵ^2, the variance of the error term for the model that is used. (Multiple readings, however, may lead to underestimation of σ_ϵ^2 because the multiple readings might be misleadingly similar.) Values of Y that result from experiments that do not meet all the requirements of a replicated experiment may have variation due to extraneous factors, which would cause σ_ϵ^2 to be overestimated, with the consequence that significant factors may be erroneously declared not significant. For the moment we will simply note that many experiments are performed that are really not true replicated experiments, and indeed the fraction of such experiments that are presumed to be replicated experiments is undoubtedly quite high. One example of such an experiment is the lead extraction from paint experiment described in Ryan (2004), which although being "close" to a replicated experiment (and assumed to be such) wasn't quite that because the specimens could not be ground down to an exact particle size, with the size of the specimen expected to influence the difficulty in grinding to the exact desired particle size. Thus, the experimental material was not quite identical between replicates, or even within replicates. Undoubtedly, occurrences of this type are very common in experimentation.

One decision that must be made when an experiment is replicated is whether or not "replications" should be isolated as a factor. If replications are to extend over a period of time and the replicated observations can be expected to differ over time, then replications should be treated as a factor.

1.4.3   Example

Let's think back a century or more when there were many rural areas, and schools in such areas might have some very small classes. Consider the extreme case where the teacher has only two students; so one student receives one method of instruction and the other student receives the other method of instruction. Then there will be two test scores, one for each method.

We could see which score is larger, but could we draw any meaningful conclusion from this? Obviously we cannot do so. We would have no estimate of the variability of the test scores for each method, and without a measure of variability we cannot make a meaningful comparison.

Now consider the other extreme and assume that we start with 600 students so that 300 will be in each class (of course many college classes are indeed of this size, and larger).

What do we gain by having such a large *sample size*? Quite frankly, we may gain something that we don't want. We are in essence testing the hypothesis that the average score will be the same for the two methods, if the process of selecting a set of students and splitting the group into two equal groups were continued a very large number of times. The larger the sample sizes, the more likely we are to conclude that there is a difference in the true means (say, μ_1 for the standard method and μ_2 for the new method), although the actual difference might be quite small and not of any practical significance. (The determination of an appropriate sample size has been described by some as a way of equating statistical significance with practical significance.)

From a practical standpoint we *know* that the true means are almost certainly different. If we record the means to the nearest hundredth of a point (e.g., 87.45), is there much chance the means could be the same? Of course not. If we rounded the means to the nearest integer, there would be a reasonable chance of equality, but then we would not be using the actual means.

The point to be made is that in some ways *hypothesis testing* is somewhat of a mindless exercise that has been criticized by many, although certain types of hypothesis tests, such as testing for a normal distribution and hoping that we don't see a great departure from normality, do make sense and are necessary. See, for example, Nester (1996) and the references cited therein regarding hypothesis testing.

A decision must be reached in some manner, however, so the teacher would have to decide the smallest value of $\mu_2 - \mu_1$ that he or she would consider to be of practical significance. Let Δ denote this difference, so that the alternative hypothesis is H_a : $\mu_2 - \mu_1 > \Delta$. If the standard method has been used for many semesters, a reasonably good estimate of σ_1, the standard deviation of scores for that method, is presumably available. If we assume $\sigma_1 \doteq \sigma_2$ (probably not an unrealistic assumption for this scenario), then following Wheeler (1974), using a significance level of $\alpha = .05$ and a probability of .90 of detecting a difference of at least Δ, we might determine the total sample size, n, as

$$n = \left(\frac{4r\sigma}{\Delta}\right)^2 \tag{1.1}$$

with r denoting the number of levels, 2 in this case, of the factor "teaching method." Thus, for example, if $\sigma_1 = \sigma_2 = \sigma = 15/8 = 1.875$ and the teacher selects $\Delta = 3$, then $n = 25$ students so use 26 in order to have 13 in each of the two classes.

Equation (1.1), although appealing because of its simplicity and for that reason has probably been used considerably and has been mentioned in various literature articles (e.g., Lucas, 1994), is an omnibus formula that does not reduce to the exact expression when $r = 2$. Furthermore, Bowman and Kastenbaum (1974) pointed out that Eq. (1.1) resulted from incorrectly applying the charts of Pearson and Hartley (1972). More specifically, Bowman and Kastenbaum (1974) stated that Eq. (1.1) is based on the false assumption that values of φ are constant, with $\varphi = [\delta^2/(v_1 + 1)]^{1/2}$, with δ^2 denoting the noncentrality parameter and $\nu_1 + 1$ denoting the number of levels of a factor.

An important general point made by Wheeler (1974) is that when an effect is not significant, the experimenter should state that if the factor has an effect, it is less than approximately Δ. Clearly this is preferable to stating that the factor has no effect, which is the same as saying that $\Delta = 0$, a statement that would not be warranted.

The appropriate expression for the number of observations to be used in *each* of $r = 2$ groups is given in many introductory statistics books and is

$$\frac{n}{2} = \frac{(z_\alpha + z_\beta)^2 (\sigma_1^2 + \sigma_2^2)}{\Delta^2} \tag{1.2}$$

Using this formula produces

$$\frac{n}{2} = \frac{(1.645 + 1.28)^2 (1.875^2 + 1.875^2)}{3^2}$$

$$= 6.68$$

with $1.645 = z_{.05}$ and $1.28 = z_{.10}$ being the standard normal variates corresponding to $\alpha = .05$ and the power of the test of .90, respectively. Thus, 7 students would be used in each class rather than 13, which is the result from the use of Eq. (1.1).

There are various other methods available for determining sample sizes in designed experiments, such as the more complicated iterative procedure given by Dean and Voss (1999, p. 50). The utility of Eq. (1.1) of course lies in its simplicity, although its approximate nature should be kept in mind and variations of it will be needed for certain types of designs, with some variations given by Wheeler (1974). If the test averages for the two classes, denoted by $\bar{y}_1$ and $\bar{y}_2$, respectively, are 79.2 and 75.8, then

$$z = \frac{\bar{y}_1 - \bar{y}_2}{\sqrt{2\sigma^2/n}}$$

$$= \frac{75.8 - 79.2}{\sqrt{2(15/8)^2/13}}$$

$$= -4.62$$

Since, assuming (approximate) normality for the statistic z, $P(z < -4.62 \,|\, \mu_1 = \mu_2) = 1.9 \times 10^{-6}$, we would conclude that there is a significant difference between the two teaching methods.

Notice that this computation is based on the assumption that σ_1 and σ_2 were known and that $\sigma_1 = \sigma_2$. Generally we want to test assumptions, so it would be advisable to use the data to test the assumption that the two variances are equal. (Of course the standard deviations will be equal if the variances are equal but the proposed tests are for testing the equality of the variances.) Preferably, we should use a test that is not sensitive to the assumption of normality, and tests such as those given by Layard

(1973) and Levene (1960) are therefore recommended, in addition to the Brown and Forsythe (1974) modification of Levene's test. (The latter is used in Section 2.1.2.1.1.)

We should also test the assumption of normality of each of the two populations. This can be done graphically by using normal probability plots (see, e.g., Section 4.9) and/or by using numerical tests. Preferably, the two types of tests should be used together.

1.4.4 Size of an Effect That Can Be Detected

It is useful to turn Eq. (1.2) and similar formulas around and solve for Δ. Doing so produces

$$\Delta = \frac{(z_\alpha + z_\beta)2\sigma}{\sqrt{n}} \tag{1.3}$$

assuming $\sigma_1 = \sigma_2 = \sigma$. For the example in Section 1.4.3, we thus have

$$\Delta = \frac{2.925(2\sigma)}{\sqrt{n}}$$
$$= \frac{5.850\sigma}{\sqrt{n}}$$

With $n = 14$, the smallest difference that can be detected with a probability of .90 and a significance level of $\alpha = .05$ is $1.56\sigma = 1.56(1.875) = 2.925$, which is slightly less than 3 because the sample size was rounded up to the next integer. (We should keep in mind that Eq. (1.3) is for a one-sided test.)

We will return to Eq. (1.3) and related formulas in subsequent chapters when we consider the magnitude of effects that can be detected with factorial designs (covered in Chapter 4) and other designs, especially small factorial designs, because it is important to know the magnitude of effect sizes that can be detected. This is something that is often overlooked. Indeed, Wheeler (1974, p. 200) stated, "The omission of such statements (crude though the numbers in them may be) is a major shortcoming of many statistical analyses."

There are various Java applets that can determine n, or Δ for a given value of n; perhaps the best known of these is the one that is due to Russ Lenth, which is found at http://www.stat.uiowa.edu/~rlenth/Power/index.html. Entering $n = 9$, $\sigma = 1.87$, and $\Delta = 3$, results in a power of .8896. There will not be exact agreement between the results obtained using this applet and the results using the previously stated equations, however, because the latter are based on the use of z, whereas that is not one of the options when the applet is used. Instead, these numbers result when the use of a t-test is assumed.

Software can of course also be used to compute power, and Design-Expert can be used for this purpose for any specific design.

In addition to these applets and software, Lynch (1993) gave tables for use in determining the minimum detectable effects in two-level fractional factorial designs (i.e., 2^{k-p} designs), which are covered in Chapter 5. These tables were computed using the noncentrality parameter of the t-distribution, which was given as

$$\lambda = \left(\frac{\Delta}{\sigma}\right)\frac{\sqrt{n}}{2}$$

with n denoting the total number of runs in the experiment and the test statistic given by

$$t = \frac{\text{Effect estimate}}{2(s_p/\sqrt{n})}$$

with s_p denoting the square root of the pooled estimate of σ^2, and s_p^2 given by

$$s_p^2 = \frac{(n_1 - 1)s_1^2 + (n_2 - 1)s_2^2}{n_1 + n_2 - 2}$$

with s_1^2 and s_2^2 denoting the sample variances for the first and second levels of the factor, respectively, and n_1 and n_2 denoting the corresponding sample sizes, with $n_1 + n_2 = n$.

The results essentially show that 2^{k-p} designs with $2^{k-p} < 16$ (i.e., 8-point designs) have poor detection properties. This is discussed in more detail in Section 5.1.

A general method for computing power for a variety of designs, including many that are given in this book, was given by Oehlert and Whitcomb (2001).

1.5 TERMINOLOGY

The terms *randomization* and *replication* were used in Sections 1.4.1 and 1.4.2, respectively. There are other terms that will be used frequently in subsequent chapters. In the example involving the math teacher, which was given in Section 1.4.3, the students are the *experimental units* to whom the two *treatments* (i.e., the two methods of teaching the class) are applied.

In that experiment the possibility of having all the girls in one class and all the boys in the other class was mentioned—and quickly dismissed. If the experiment had been conducted in this manner, this would be an example of an experiment in which factors are confounded. That is, we would estimate the gender effect—if we were interested in doing so—by taking the difference of the average of the girls' scores on the first test and the average of the boys' scores on that test. But this is exactly the same way that we would estimate the teaching method effect. Thus, one number would estimate two effects; so we would say that the effects are *confounded*. Obviously we would want to avoid confounding the two effects if we believe that they both may be statistically significant. Therefore, confounding, which is essentially unavoidable in most experiments, due to cost considerations when more than a few factors are involved, must be used judiciously. This is discussed in detail in later chapters, especially Chapter 5.

We should bear in mind, however, that true randomization will often not be possible. That is, whereas it would seem to be easy to randomly assign the girls and the boys to the two classes, physical variables can present insurmountable obstacles. For example, it might be impossible or at least impractical to make frequent temperature changes to ensure randomness, and certain combinations of temperature and other variables might not even be possible. Joiner and Campbell (1976) were among the first to discuss such problems in the statistics literature, and the reader is also referred to papers by Ganju and Lucas (1997, 1999), Bailey (1987), and Youden (1972). Hard-to-change factors and debarred observations are discussed in Sections 4.19 and 13.8, respectively.

Another important concept is *blocking*. It is often said that an experimenter should randomize over factors that can be controlled and block factors that cannot be controlled. For example, suppose that a farmer wants to conduct an experiment to compare two types of feed because he is interested in finding the best feed that will allow him to fatten up his pigs for market as quickly as possible. How quickly the pigs gain weight on a given feed is likely to depend somewhat on the litter that they are from, so it would be logical to use litter as a blocking variable. The use of blocking variables is illustrated in Chapter 3.

1.6 STEPS FOR THE DESIGN OF EXPERIMENTS

The steps that one should follow in designing and conducting experiments can be laid out in a very general way, although a cookbook set of steps cannot (or at least should not) be provided. This is because the specifics of procedures to follow will vary somewhat from setting to setting. Nevertheless, Bisgaard (1999) provided a template that is appropriate for factorial experiments (Chapters 4 and 5) that are to be used in a sequential manner. The starting point of course would be a statement of the reason for the experiment(s) and the objective(s). This should be followed by a list of the factors to be studied and the levels of each, a statement of the response variable(s) and how the measurements will be conducted. Bisgaard (1999) went on to list eight subsequent steps that are mostly nonstatistical and include using a flowchart and data collection sheets, planning for problems, assigning responsibilities to team members, and so on. One recommendation that is not generally mentioned in conjunction with designing experiments is the idea of using a Gantt chart that shows the steps to be followed and the associated dates, with a small pilot experiment included. (There is much information on Gantt charts on the Internet. Informative sources are the brief tutorial at http://www.me.umn.edu/courses/me4054/assignments/gantt.html and at the web site devoted to Gantt charts: http://www.ganttchart.com.)

Coleman and Montgomery (1993) also gave a thorough discussion of considerations that should be made in designing an experiment. They list seven steps that should be performed sequentially: (1) recognition of and statement of the problem, (2) choice of factors and levels, (3) selection of the response variable(s), (4) choice of experimental design, (5) conduction of the experiment, (6) data analysis, and (7) conclusions and recommendations. See also Van Matre and Diamond (1996–1997) for a discussion of the importance of teamwork in designing and carrying out experiments, as well as Bisgaard (1989), who gave an interesting example of team problem solving.

We will consider the first two steps in some detail and will provide some additional insight.

1.6.1   Recognition and Statement of the Problem

Montgomery (1996) points out that the problem statement is often too broad. The problem should be specific enough and the conditions under which the experiment will be performed should be understood so that an appropriate design for the experiment can be selected.

1.6.2   Selection of Factors and Levels

This issue has been addressed by Hahn (1977, 1984) and Cox (1958), in addition to Coleman and Montgomery (1993), with a more recent and more extensive discussion given by Czitrom (2003). We will review their recommendations and provide some additional insight.

1.6.2.1   Choice of Factors

The factors that are studied in the initial stages of sequential experimentation are those that are believed to be important. The set can be reduced in later stages, so it is better to start with a large set than with a small set that may not include some important factors. Of course if an experimenter knew which factors were important, then the number of stages normally used in experimentation could be reduced, but such prior knowledge is generally unavailable. Sometimes a factor may not be recognized as important simply because its level isn't changed. Indeed, Myers and Montgomery (1995, p. 636) stated, "Often we have found that a variable was not realized to be important simply because it had never been changed."

See also Barton (1997), who presented a method for identifying and classifying seemingly important factors before an experiment is run. In particular, Barton uses the term *intermediate variable* in referring to variables that may be related to the dependent value, but the values of intermediate variables are determined by the values of independent variables. An example of this was given. This is an important point because the factors used in an experimental design must be true independent variables so that the levels of the factors can be selected without having to worry about what levels will be used for the other factors. There will frequently be restrictions on combinations of factor levels that translate into restriction regions of operability (Section 13.8) and debarred observations (also Section 13.8), but that is quite different from having *every* possible level of a factor dependent upon the levels of other factors, as is the case for an intermediate variable. There is a class of designs, namely, supersaturated designs (Section 13.4.2) for which the sets of factor levels used in the design will be slightly correlated, but this is a consequence of the design (and specifically because the number of design points is less than the number of factors), not because of a deterministic relationship between any of the factors.

1.6.2.2 Choice of Levels

The experimenter must also address the question of how many levels to use and how the levels are to be selected. If only two or three levels are likely to be used in the immediate future, then those levels should be used in the experiment and the inference that is drawn will apply only to those levels. If, however, there is interest in a range of possible levels for the factors of interest, then those levels should be randomly selected from that range. In the first case the factor(s) would be classified as *fixed*; in the second case the factor(s) would be classified as *random*. The distinction between a fixed factor and a random factor is important, as will be seen in succeeding chapters.

Briefly, if a factor were fixed, we would be interested in testing the equality of the population means for those levels. If a factor were random, we would be interested in testing whether or not the variance of the effects of the factor levels is zero for the range of levels of interest. In designs involving more than one factor, the classification of factors as fixed or random determines how certain hypothesis tests are performed. So this distinction is quite important.

In the initial stages of an experimental investigation, the first objective is to determine which factors are important. This necessitates the use of a *screening design* if there are more than a few factors that are being considered. Screening designs usually have two levels, this being necessary because p^k can be quite large when k is not small and when $p > 2$, with k denoting the number of factors and p denoting the number of levels for each factor, with p usually the same for each factor. Thus, p^k is the number of observations unless the experiment is replicated.

The number of levels can be increased in the secondary stages of a sequential investigation, after the relevant factors have been identified; this is illustrated in later chapters, including Chapter 10, where it would logically be done since response surface designs generally do not have a large number of factors, but the important factors must be defined first.

Now assume that $k = 1$. For qualitative factors, the selection of levels is generally not a major issue, and indeed the word "levels" in the case of alternatives to a standard process that would be compared to the standard process in an experiment simply denotes the number of such alternatives to be examined in an experiment plus one (the standard process).

For quantitative factors, the selection of factor levels can be critical. Assume that we have one factor with two levels, with the factor denoted by X and X_1 and X_2 denoting the low level and the high level, respectively.

How should the levels be chosen? We obviously want to detect a significant effect of the factor if it exists. We have to exercise some care in defining what we mean by "effect of the factor" as well as variability of the response. In particular, we cannot speak of the latter independent of the levels of a factor.

For the sake of illustration, assume that the response variable has a normal distribution over values of a factor that has not been controlled but has freely varied within

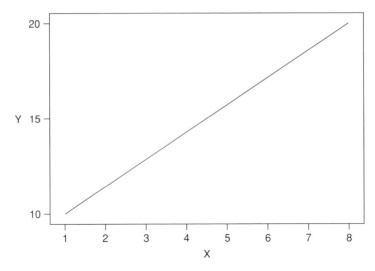

Figure 1.1   Relationship between Y and X.

a wide range. Now assume that the response variable ranges from $\mu - \sigma$ to $\mu + \sigma$ when the factor of interest is within a specific range, and the endpoints of the range are to be used in an experiment to determine if the factor is significant.

Of course we don't know what the range of the response will be—which is why we conduct experiments—but what should happen if we conduct an experiment with the endpoints of the specific range of the factor used as the two levels in an experiment, with n observations used at each level? Let's say that the two values of the factor are fixed within the aforementioned range, and assume that the conditional distribution of $Y|X_1 \sim \text{Normal}(\mu_1, (\sigma^*)^2)$ and $Y|X_2 \sim \text{Normal}(\mu_2, (\sigma^*)^2)$. That is, the conditional distributions have a common variance, $(\sigma^*)^2$, and of course different means.

The questions to be addressed are as follows: (1) is the difference between μ_1 and μ_2 of practical significance, and if so, (2) are X_1 and X_2 far enough apart, relative to σ^*, to allow the difference to be detected with a high probability? If the expected change is only σ^* and the spread from X_1 to X_2 is the largest spread that could be tolerated in, say, a manufacturing setting, then there is not much point in performing experimentation with the factor X. These are technical questions, the answers to which are beyond the intended level of this chapter.

At this point a simple graphical answer may suffice. Consider Figure 1.1.

Assume that 1–8 on the X-axis denote possible levels of X, all of which are feasible, with the line representing the relationship between Y and X, which we would expect to eventually become nonlinear if the line were extended far enough in each direction. Assume that an experiment is conducted with levels corresponding to $X = 2$ and $X = 3$. The difference between the corresponding Y-values in the graph is approximately 1.5, which is thus the difference between the conditional expectations $E(Y|X_1)$ and $E(Y|X_2)$. Let the estimators of these conditional means be denoted by

$\overline{Y}_1$ and $\overline{Y}_2$, respectively. Assume that the conditional variances are also equal and are denoted by $(\sigma^*)^2$, in accordance with the previous notation, and for the sake of illustration we will assume that this is known. Since $\text{Var}(\overline{Y}_1 - \overline{Y}_2) = (\sigma^*)^2(2/n)$, assuming an equal number of observations, n, at each value of X, we might ask whether $E(\overline{Y}_1 - \overline{Y}_2)$ exceeds $2(\sigma^*)\sqrt{(2/n)}$ (i.e, whether the expected value of the test statistic exceeds 2), as a value of the test statistic much greater than 2 would certainly suggest a difference in the conditional means for the two values of X.

Obviously the further apart the X-values are, the more likely this condition is to be met, as is apparent from Figure 1.1. Realize, of course, that spreading out the X-values will increase the spread of the Y-values, which will increase the denominator of the test statistic. Furthermore, this apparent desiderata regarding the spread of the X-values is offset by the fact that the wider the ranges of factors, the more likely interactions between factors are to be judged significant. We will see in Section 4.2 and subsequent sections how significant interactions complicate analyses.

Of course relationships such as the one depicted in Figure 1.1 will in general be unknown. Czitrom (2003, p. 14) states, "In the absence of other engineering considerations, attempt to give each factor, a priori, an "equal opportunity" to influence the result, or set the factors at, say, 10% above and below a value of interest such as the current operating conditions." Of course, feasible factor levels will frequently serve as boundaries for level settings in experimentation.

Although we speak of the variability of the response variable for a given factor, our real interest is in examining the variability of the response between factor levels, as the extent of that variability determines whether or not the factor is significant. Accordingly, the levels must be far enough apart to allow the response variable the opportunity to vary enough between levels to result in the factor being judged significant.

At times we will discuss important nonstatistical considerations that should be made in designing experiments. Detailed discussions of steps (nonstatistical steps) that are followed for an actual experiment are generally not given in the literature; an exception is Vivacqua and de Pinho (2004), which relates to the subject matter in Chapter 9 and is discussed there. We might view the selection of factors and factor levels as being almost nonstatistical considerations, but certainly very important considerations. As Czitrom (2003) stated. "...detailed guidelines for selecting factors and factor levels, which are critical inputs to an experiment, are seldom presented systematically."

The discussion to this point has been restricted to placement of the levels of factors. *If* only a single experiment were to be run, which of course is usually not a good idea, the *number* of such levels would be dictated by our belief regarding the order of a likely model for the response. Any such prior beliefs are of secondary importance when experimentation is sequential, however (as preferred), and an absence of such prior beliefs would be one way to motivate sequential experimentation.

Three levels are of course needed to detect curvature, and three levels might be a substitute for four levels if a 3-level factor had the same extreme levels as would be used if four levels were employed. Five levels are involved in central composite designs, which are response surface designs (Chapter 10), and although Czitrom

(2003, p. 16) mentions an experiment in which 20 levels were used to capture a sinusoidal behavior in the response variable, we would rarely want to use more than five levels. Generally the levels that are used should be equally spaced. Sometimes the levels of factors won't be known and will have to be approximated. In such cases the levels will almost certainly be unequally spaced.

1.7 PROCESSES SHOULD IDEALLY BE IN A STATE OF STATISTICAL CONTROL

Although this is a book on design of experiments, and not on process control, the importance of maintaining processes in a state of statistical control when designed experiments are performed cannot be overemphasized. In particular, process drift could cause a new setting for a process variable to appear to produce significantly worse results than the standard setting.

Simply stated, when designed experimentation is performed, controllable factors that could influence the value of the response variable should be kept in a state of statistical control. Ideally, the mean of the response variable should not be affected by controllable extraneous factors during an experiment, but can this idealized state be achieved? Some would say that it cannot be achieved. Box (1990) stated, in responding to the question of whether statistical control can be achieved: "The truth is that we all live in a nonstationary world; a world in which external factors never stay still. Indeed the idea of stationarity—of a stable world in which, without our intervention, things stay put over time—is a purely conceptual one." If we accept this, then we should strive for processes being in a state of near-statistical control. That might not seem particularly difficult to achieve, but Bisgaard (1994) stated that technicians often do not have very good control of their processes during the early stages of experimentation.

Does this mean that control should not be sought during experimentation? Various industrial personnel, who have learned the hard way what the consequences of out-of-control processes can be, would respond with a resounding "No!."

An example of how an out-of-control process can lead to misleading results was given in Ryan (2000, p. 361), using hypothetical data. Specifically, assume that an experiment is being conducted using two levels of a factor (e.g., the standard temperature level versus an experimental level). If a variable that is not under study suddenly goes out of statistical control (generally defined as no longer being within three standard deviations of its mean), and this causes a sharp change in the response variable, this change could erroneously be attributed to a change in the level of the factor under study if that change occurred near the point at which the extraneous variable went out of control. This problem can happen frequently in industrial and other experiments if care is not exercised.

If factors (variables) not part of a study are out of control, how can this be detected? One suggested approach is to make runs at the standard operating conditions at the beginning and end of an experiment and compare the results (see Croarkin and Tobias, 2002, Section 5.2.2 of the *NIST/SEMATECH e-Handbook of Statistical Methods:*

http://www.itl.nist.gov/div898/handbook/pri/section2/pri22.htm). If the results differ greatly, one or more of the variables not in the study are probably out of control, although of course the variance of the difference of the two numbers is $2\sigma^2$ when/if the process(es) is in control, so the difference between the numbers may not be small when a state of statistical control does exist if process variability is relatively large but stable.

Unfortunately, this placement of check runs is seldom done, however, as additional design points are generally not strategically placed among the experimental runs. *This is a major weakness in the use of experimental designs.*

Of course, process control techniques should be applied routinely to variables that could easily go out of control, regardless of whether experiments that involve those variables are being performed. Unfortunately, the need to have processes in control when experiments are performed is generally not recognized in the statistical literature. Notable exceptions are the experiment described in Jiang, Turnbull, and Clark (1999) and case study #2 in Inman et al. (1992). Clark (1999) stated the importance of statistical control about as well as it can be stated in discussing the importance of the tooling process being in control for a resistance welding operation experiment that was undertaken: "Control of the tooling proves to be as important as parameter evaluation is in achieving optimum nugget size in resistance welding. Otherwise you are trying to make a prediction about an unpredictable event. Without stability, you have nothing." See also the discussion about the need for processes being in control in Arviddson and Gremyr (2003).

The use of statistical process control (SPC) methods to ensure that processes are in control when experimentation is performed is very important and so we will at times reiterate the importance of this in subsequent chapters.

Let's assume that a process cannot be brought under a state of statistical control. What should be done? It might seem as though experiments simply should not be run under such conditions, but Box, Bisgaard, and Fung (1990, p. 190) recommend that processes be brought into the best state of control possible, and experimentation be performed in such a way that blocks are formed to represent intervals of time when the process appears to be in control. One potential impediment to the implementation of such an approach, however, is that it may be difficult to determine when changes have occurred, unless the changes are large. There are, however, methods that have been developed for trying to determine change points.

The safest approach would be to strive for tight control of processes so that the results of experiments can be interpreted unambiguously. Since statistical control methods should be used anyway, at least for critical processes, this should not entail any extra work. Furthermore, if these statistical process control methods are in force, check runs as described earlier in this section should be unnecessary, although experimenters might still want to use them if experimentation at the standard operating conditions is inexpensive.

The SPC methods that are employed include control charts in addition to more sophisticated and more efficient methods, including cumulative sum (CUSUM) and exponentially weighted moving average (EWMA) techniques. These and other SPC methods are covered extensively in Ryan (2000) and other books on the subject.

1.8 TYPES OF EXPERIMENTAL DESIGNS

There are many types of experimental designs, some of which have been in use for over 70 years. Experimental design has it roots in agriculture and much of the pioneering work was performed at the Rothamsted Experiment Station in England. Although the use of experimental designs has broadened into many areas, including such diverse fields as manufacturing and medicine, some of the terminology of the distant past remains, such as the word "plot" in the names of designs (e.g., a split-plot design) even though there will usually not be a physical "plot," such as a plot of land, in an experiment. The basic objective remains unchanged, however: to determine the factors that are influencing the variability in the response values and to use whichever experimental design will best provide this insight for a given amount of resources.

Many different experimental designs are presented in the succeeding chapters. These include designs for a single factor, with and without blocking, in addition to designs for multiple factors, including a large number of factors. Various realities of design construction must be faced, including the fact that not all combinations of factor levels are possible, and restrictions on randomization are also encountered very frequently. There is as yet no general solution to the problem of design construction when not all factor-level combinations are possible, and only during the past 10 years have the consequences of restricted randomization been discussed extensively in the literature.

The world of experimental design is thus quite different from what it was several decades ago, and contrary to what some might say, there are still important research contributions in certain areas of design that are emerging and more research that is needed.

New methods of analysis are also needed, and conditional effects analyses are presented herein for the first time as a general method of analysis. Established methods that are typically not used with experimental designs must also be considered. This includes Analysis of Means (ANOM), which is presented in the next section and is used in the following chapters in conjunction with various designs.

1.9 ANALYSIS OF MEANS

Almost 50 years ago, Ellis Ott, the founding chairman of the statistics department at Rutgers University, needed a better way to communicate with the engineers for whom he performed consulting work, as they found Analysis of Variance (ANOVA) not to be very intuitive. And indeed it isn't intuitive since squared quantities are involved, thus resulting in the unit of measurement being "squared" in the numerical analysis. Although there is squared inches, the square of most units doesn't make any sense, such as squared temperature and squared yield.

So Ott sought a method that did not cause the unit of measurement to be lost, and consequently invented Analysis of Means (ANOM), developing it in Ott (1958), and introducing it into the literature in Ott (1967); the first edition of Ott's text (1975) contains many illustrative examples. The most recent edition of Ott's book is Ott, Schilling, and Neubauer (2005), but it should be noted that there are people who consider the first edition of Ott's book to be the best version. Indeed, the graphical display of interactions in factorial and fractional factorial designs (those designs are covered herein in Chapters 4 and 5, respectively) has become distorted relative to the way that they were displayed originally by Ott.

The only book on ANOM is Nelson, Wludyka, and Copeland (2005). Other recent books that have a moderate amount of space devoted to ANOM include Ryan (2000) and Nelson, Coffin, and Copeland (2003), with the display of interactions in Ryan (2000) and in this book being the same as in Ott (1975), except for the minor difference that the latter used slanted lines instead of vertical lines.

Ott's 1967 paper appeared in the Walter Shewhart Memorial issue of *Industrial Quality Control*, and it is fitting that the January 1983 issue of the *Journal of Quality Technology* (which was previously named *Industrial Quality Control*) contained articles only on ANOM, with the purpose of serving as a tribute to Ellis Ott upon his passing.

The state of the art has advanced during the past two decades, primarily because of the research contributions of the late Peter Nelson and his students. This research has allowed ANOM to be applied to data from various types of designs with fixed factors. (One limitation is the restriction to fixed factors; recall the discussion of fixed and random factors in Section 1.6.2.2.)

To many people, the term "analysis of means" undoubtedly conjures up notions of analysis of variance (ANOVA), however, which is also concerned with the analysis of means, and which is much better known and more widely used than ANOM. The latter is apt to have more appeal to engineers and other industrial personnel than does ANOVA, however, since ANOM is inherently a graphical procedure and is somewhat similar to a control chart. The importance of using good analytic graphical procedures cannot be overemphasized, and the similarity to a control chart is also a plus since control charts or related procedures should be used in conjunction with experimental design, anyway, as was emphasized in Section 1.7.

Readers familiar with ANOVA will recall that with that method the experimenter concludes that either all of the means are equal or at least one of the means differs from the others. If the latter is concluded, then a multiple comparison procedure, of which there are many, is generally used to determine which means differ significantly.

With ANOM, however, the user will see whether or not one or more means differ from the average of all the means. Thus, what is being tested is different for the two procedures, so the results will not necessarily agree. In particular, when $k - 1$ sample averages are bunched tightly together but the kth sample average (i.e., the other one) differs considerably from the $k - 1$ averages, the F-value in ANOVA would likely be relatively small (thus indicating that the population means are equal), whereas the difference would probably be detected using ANOM. Conversely, if the differences between adjacent sample averages are both sizable and similar, the (likely) difference in the population means is more apt to be detected with ANOVA than with ANOM.

One procedure need not be used to the exclusion of the other, however. As Ott (1967) indicates, ANOM can be used either alone or as a supplement to ANOVA.

Since the methods produce similar results, it is not surprising that the assumptions for each method are the same. That is, the variances must be the same within each level of the factor (and within each cell, in general, when there is more than one factor), and the populations for each level (cell) must be normally distributed. Of course, these assumptions are almost impossible to check with a small amount of data per level (cell), so fortunately the normality assumption is not crucial and the equal variances assumption is also not a problem for equal numbers of observations per level (cell) unless the variances differ greatly.

When the one or more factors all have two levels, ANOM is simply a "graphical t-test" and the results will then agree with the results obtained using either the t-test or ANOVA. Since two-level factors predominate in practice and good graphical procedures are always useful, ANOM is an important technique for analyzing data from designed experiments. Accordingly, it will be used and discussed in subsequent chapters.

Almost all of the ANOM methods are based on the assumption of a normal distribution for the plotted statistic, or the adequacy of the normal approximation to the binomial or Poisson distribution. An exception is the nonparametric procedure of Bakir (1989).

In general, the use of graphical methods in analyzing data from designed experiments is quite important. See Barton (1999) for many illustrative examples as well as standard and novel graphical methods.

1.10   MISSING DATA

Missing data is really an analysis problem, not a design problem, but since inferences are drawn from the data that are collected when a designed experiment is performed, the problem (which will often occur, such as when an experimental run is botched) must be addressed. Accordingly, it is discussed in subsequent chapters, such as Section 2.1.3.2.

1.11   EXPERIMENTAL DESIGNS AND SIX SIGMA

Because the Six Sigma phenomenon is still going strong after several years, there should probably be at least brief mention of the role of design of experiments in Six Sigma, a topic that is covered more fully in the context of robust designs in Chapter 8. For those readers unfamiliar with the term "Six Sigma," it refers to a collection of tools (and even a company way of life), both statistical and nonstatistical that are used to improve the quality of products and processes. Experimental design plays a key role in this because the factors that influence product and process quality must be identified. Readers interested in how design of experiments fits into Six Sigma may want to read Goh (2002).

1.12 QUASI-EXPERIMENTAL DESIGN

One topic that is *not* covered in this book is *quasi-experimental design*. The term is used to refer to experimentation in which there is no attempt to adhere to the tenets of experimental designs. For example, there is generally no randomization, and studies are often longitudinal with the time aspect playing a prominent role. In some fields, such as the social and behavioral sciences, it isn't possible to conduct experiments in accordance with the principles of experimental design. Quasi-experimental designs can be very useful, especially in the field of education. Readers interested in quasi-experimentation are referred to Campbell and Stanley (2005), Shadish (2001), Shadish, Cook, and Campbell (2001), Trochim (1986), Cook and Campbell (1979), and Manly (1992).

1.13 SUMMARY

Experimental design is used in a very long list of application areas, including areas where we might not expect to see it being used, such as marketing, and in areas where we might want to see greater use, such as engineering. The proper use of experimental designs requires considerable thought, however, and various obstacles discussed in later chapters, such as hard-to-change factors and restricted regions of operability, make optimal or at least judicious use of experimental designs a challenging task.

The minimal size of an effect that an experimenter wants to detect will often be used to determine the number of observations and thus the size of the experiment that will be employed.

It is very important that processes that will impact an experiment be in at least a reasonable state of statistical control when an experiment is performed. The consequences of not doing so can be severe, with the results possibly being very misleading. If experimenters are not used to checking for control, results that run counter to engineering knowledge would likely result in a repeated experiment, with the results possibly being quite different from the first experiment if processes are out of control in a way that is different from what occurred when the first experiment was run.

Appropriate analysis of data from designed experiments is also important and experimenters have certain options, including ANOM as a possible supplement to ANOVA or a replacement for it for fixed factors. The effect that restricted randomization, if present, has on analyses must also be considered. This is considered in Section 4.19 in the context of hard-to-change factors.

REFERENCES

Arviddson, M. and I. Gremyr (2003). Deliberate choices of restrictions in complete randomization. *Quality and Reliability Engineering International*, **19**, 87–99.

Atkinson, A. C. and R. A. Bailey (2001). One hundred years of the design of experiments on and off the pages of *Biometrika*. *Biometrika*, **88**, 53–97.

Bailey, R. A. (1987). Restricted randomization: A practical example. *Journal of the American Statistical Association*, **82**, 712–719.

Bakir, S. T. (1989). Analysis of means using ranks. *Communications in Statistics—Simulation and Computation*, **18**(2), 757–776.

Barton, R. R. (1997). Pre-experiment planning for designed experiments: Graphical methods. *Journal of Quality Technology*, **29**(3), 307–316.

Barton, R. R. (1999). *Graphical Methods for the Design of Experiments*. New York: Springer-Verlag.

Beveridge, W. I. (1960). *Art of Scientific Investigation*. New York: Vantage Press.

Bisgaard, S. (1989). The quality detective: A case study. *Philosophical Transactions of the Royal Society* **A327**, 499–511. (This is also available as Report No. 32 of the Center for Productivity and Quality Improvement, University of Wisconsin-Madison and can be downloaded at http://www.engr.wisc.edu/centers/cqpi/reports/pdfs/r032.pdf.)

Bisgaard, S. (1992). The early years of designed experiments in industry: Case study references and some historical anecdotes. *Quality Engineering*, **4**(4), 547–562.

Bisgaard, S. (1994). Blocking generators for small 2^{k-p} designs. *Journal of Quality Technology*, **26**(4), 288–296.

Bisgaard, S. (1999). Proposals: A mechanism for achieving better experiments. *Quality Engineering*, **11**(4), 645–649.

Bowman, K. O. and M. A. Kastenbaum (1974). Potential pitfalls of portable power. *Technometrics*, **16**(3), 349–352.

Box, G. E. P. (1976). Science and statistics. *Journal of the American Statistical Association*, **71**, 791–799.

Box, G. E. P. (1990). Must we randomize our experiment? *Quality Engineering*, **2**, 497–502.

Box, G. E. P. (1993). George's Column. *Quality Engineering*, **5**(2), 321–330.

Box, G. E. P., S. Bisgaard, and C. Fung (1990). *Designing Industrial Experiments*. Madison, WI: BBBF Books.

Box, G. E. P. and N. R. Draper (1975). Robust designs. *Biometrika*, **62**, 347–352.

Box, G. E. P., W. G. Hunter, and J. S. Hunter (1978). *Statistics for Experimenters*. New York: Wiley.

Brown, M. B. and A. B. Forsythe (1974). Robust tests for equality of variances. *Journal of the American Statistical Association*, **69**, 364–367.

Buckner, J., B. L. Chin, and J. Henri (1997). Prometrix RS35e gauge study in five two-level factors and one three-level factor. In *Statistical Case Studies for Industrial Process Improvement*, Chapter 2 (V. Czitrom and P. D. Spagon, eds.). Philadelphia: Society for Industrial and Applied Mathematics, and Alexandria, VA: American Statistical Association.

Campbell, D. T. and J. C. Stanley (2005). *Experimental and Quasi-Experimental Designs for Research*. Boston: Houghton Mifflin Company.

Clark, J. B. (1999). Response surface modeling for resistance welding. In *Annual Quality Congress Proceedings*, American Society for Quality, Milwaukee, WI.

Coleman, D. E. and D. C. Montgomery (1993). A systematic approach to planning for a designed industrial experiment. *Technometrics*, **35**(1), 1–12; discussion:13–27.

Cook, T. D. and D. T. Campbell (1979). *Quasi-Experimentation: Design and Analysis Issues*. Boston: Houghton Mifflin Company.

Cox, D. R. (1958). *Planning of Experiments*. New York: Wiley.

Croarkin, C. and P. Tobias, eds. (2002). *NIST/SEMATECH e-Handbook of Statistical Methods* (http://www.itl.nist.gov/div898/handbook), a joint effort of the National Institute of Standards and Technology and International SEMATECH.

Czitrom, V. (2003). Guidelines for selecting factors and factor levels for an industrial designed experiment. In *Handbook of Statistics*, Vol. 22: *Statistics in Industry* (R. Khattree and C. R. Rao, eds.). Amsterdam: Elsevier Science B. V.

Daniel, C. (1976). *Applications of Statistics to Industrial Experimentation.* New York: Wiley.

Dean, A. and D. Voss (1999). *Design and Analysis of Experiments.* New York: Springer-Verlag.

Ganju, J. and J. M. Lucas (1997). Bias in test statistics when restrictions in randomization are caused by factors. *Communications in Statistics—Theory and Methods*, **26**(1), 47–63.

Ganju, J. and J. M. Lucas (1999). Detecting randomization restrictions caused by factors. *Journal of Statistical Planning and Inference*, **81**, 129–140.

Goh, T. N. (2002). The role of statistical design of experiments in Six Sigma: Perspectives of a practitioner. *Quality Engineering*, **14**(4), 659–671.

Hahn, G. J. (1977). Some things engineers should know about experimental design. *Journal of Quality Technology*, **9**(1), 13–20.

Hahn, G. J. (1984). Experimental design in the complex world. *Technometrics*, **26**(1), 19–31.

Inman, J., J. Ledolter, R. V. Lenth, and L. Niemi (1992). Two case studies involving an optical emission spectrometer. *Journal of Quality Technology*, **24**(1), 27–36.

Jiang, W., B. W. Turnbull, and L. C. Clark (1999). Semiparametric regression models for repeated events with random effects and measurement errors. *Journal of the American Statistical Association*, **94**, 111–124.

John, P. W. M. (2003). Plenary presentation at the *2003 Quality and Productivity Research Conference*, IBM T. J. Watson Research Center, Yorktown Heights, NY, May 21–23. The talk is available at http://www.research.ibm.com/stat/qprc/papers/Peter_John.pdf.)

Joiner, B. L. and C. Campbell (1976). Designing experiments when run order is important. *Technometrics*, **18**, 249–260.

Layard, M. W. J. (1973). Robust large-sample tests for homogeneity of variances. *Journal of the American Statistical Association*, **68**(341), 195–198.

Levene, H. (1960). Robust Tests for the Equality of Variance. In *Contributions to Probability and Statistics: Essays in Honor of Harold Hotelling* (I. Olkin, et al., eds.). Palo Alto, CA: Stanford University Press, pp. 278–292.

Lucas, J. M. (1994). How to achieve a robust process using response surface methodology. *Journal of Quality Technology*, **26**(3), 248–260.

Lynch, R. O. (1993). Minimum detectable effects for 2^{k-p} experimental plans. *Journal of Quality Technology*, **25**(1), 12–17.

Manly, B. F. (1992). *The Design and Analysis of Research Studies.* Cambridge, UK: Cambridge University Press.

Mathews, P. (2004). *Designing Experiments with MINITAB.* Milwaukee, WI: Quality Press.

Montgomery, D. C. (1996). Some practical guidelines for designing an industrial experiment. In *Statistical Applications in Process Control* (J. B. Keats and D. C. Montgomery, eds.). New York: Marcel Dekker.

Myers, R. H. and D. C. Montgomery (1995). *Response Surface Methodology. Process and Product Optimization using Designed Experiments*, 2nd ed. (2002). New York: Wiley.

Nelson, P. R., M. Coffin, and K. A. F. Copeland (2003). *Introductory Statistics for Engineering Experimentation*. San Diego, CA: Academic Press.

Nelson, P. R., P. S. Wludyka, and K. A. F. Copeland (2005). *The Analysis of Means: A Graphical Method for Comparing Means, Rates and Proportions*. Society for Industrial and Applied Mathematics and American Statistical Association: Philadelphia and Alexandria, VA, respectively.

Nester, M. R. (1996). An applied statistician's creed. *Applied Statistics*, **45**, 401–410.

Oehlert, G. and P. Whitcomb (2001). Sizing fixed effects for computing power in experimental designs. *Quality and Reliability Engineering International*, **17**(4), 291–306.

Ott, E. R. (1958). Analysis of means. Technical Report #1. Department of Statistics, Rutgers University.

Ott, E. R. (1967). Analysis of means—a graphical procedure. *Industrial Quality Control*, **24**(2), 101–109.

Ott, E. R. (1975). *Process Quality Control*. New York: McGraw-Hill.

Ott, E. R., E. G. Schilling, and D. V. Neubauer (2005). *Process Quality Control: Troubleshooting and Interpretation of Data*, 4th ed. New York: McGraw-Hill.

Pearson, E. S. and H. O. Hartley (1972). *Biometrika Tables for Statisticians*, Vol. 2. Cambridge, UK: Cambridge University Press.

Roach, M. (2003). *Stiff: The Curious Lives of Human Cadavers*. New York: Norton.

Ryan, T. P. (2000). *Statistical Methods for Quality Improvement*, 2nd ed. New York: Wiley.

Ryan, T. P. (2004). *Lead Recovery Data. Case Study*. Gaithersburg, MD: Statistical Engineering Division, National Institute of Standards and Technology. (http://www.itl.nist.gov/div898/casestud/casest3f.pdf)

Shadish, W. R. (2001). Quasi-experimental designs. In *International Encyclopedia of the Social and Behavioral Sciences* (N. J. Smelser and P. B. Baltes, eds.). New York: Elsevier, pp. 12655–12659.

Shadish, W. R., T. D. Cook, and D. T. Campbell (2001). *Experimental and Quasi-Experimental Designs for Generalized Causal Inference*. Boston: Houghton Mifflin Company.

Trochim, W. M. K., ed. (1986). *Advances in Quasi-Experimental Design and Analysis*, Vol. 31. Hoboken, NJ: Jossey-Bass.

Vivacqua, C. A. and A. L. S. de Pinho (2004). On the path to Six Sigma through DOE. In *Annual Quality Transactions*, American Society for Quality, Milwaukee, WI. (available to ASQ members at http://www.asq.org/members/news/aqc/58_2004/20116.pdf)

Van Matre, J. G. and N. Diamond (1996–1997). Team work and design of experiments. *Quality Engineering*, **9**(2), 343–348. (This article is also available as Report No. 144, Center for Quality and Productivity Improvement, University of Wisconsin-Madison and can be downloaded at http://www.engr.wisc.edu/centers/cqpi/reports/pdfs/r144.pdf.)

Wheeler, R. E. (1974). Portable power. *Technometrics*, **16**, 193–201.

Youden, W. J. (1972). Randomization and experimentation. *Technometrics*, **14**, 13–22.

EXERCISES

1.1 To see the importance of maintaining processes in a state of statistical control when experiments are performed, as stressed in Section 1.7, consider the following scenario. An industrial experiment is conducted with temperature set at two levels, 300 and 400° F. Assume that no attempt at process control is made during the experiment and consequently there is a 3-sigma increase in the average conductivity from the original value of 18.6, due to some cause other than the change in temperature, which occurs right when the temperature is changed. Assume that σ is known to be 4.5 and that 20 observations were made at each of the two temperature levels.

(a) Using the assumed value of σ, what is the expected value of the appropriate test statistic for this scenario and the corresponding expected p-value?
(b) Answer these same two questions if the process went out of control, by the same amount, when the first 10 observations had been made at the second temperature. What argument do these two sets of numbers support?

1.2 Assume that Analysis of Variance calculations have been performed for a problem where there is a single factor with two levels. This would produce results equivalent to an independent sample t-test provided that the alternative hypothesis for the t-test is (choose one): (a) greater than, (b) not equal to, (c) less than, (d) none of the above.

1.3 It has been stated that "the best time to design an experiment is after it has just been run." Explain what this means. Is this a problem if we view experimentation as being sequential?

1.4 An experiment with a single factor and two levels, 1 and 2, was used and the results were as follows:

1		6.1	8.2	7.3	8.4	8.0	7.6	8.7	9.3	6.8	7.5
2		6.3	8.0	7.7	8.1	8.6	7.2	8.4	9.7	6.8	7.2

Would you use the methodology that was used in Section 1.4.4 in analyzing these data? Why, or why not? If yes, perform the analysis and state your conclusion.

1.5 Give an example of an experimental situation in your field in which repeated readings are made instead of replications. Then indicate how the replications would be performed and state whether or not this would have any nonstatistical disadvantages, such as a considerable increase in the cost of running the experiment.

1.6 Explain the difference between replication and repeated readings.

1.7 Explain why the usual practice of randomly assigning treatments to experimental units is objectionable.

1.8 Assume that an experimenter wants to use six levels for an experiment involving only one factor, but since the runs will be somewhat expensive, he can't afford more than 24 observations for the experiment. Explain what problems, if any, this poses.

1.9 Identify, if possible, a scenario in your field of application for which the failure to maintain one or more processes in a state of statistical control could seriously undermine experimentation of a specific type.

1.10 Critique the following statement regarding the data in Exercise 1.4: "Since the data vary considerably within each level but there is much less variation between corresponding values for the two levels, if the data are time-ordered, then there must be at least one statistical process that is out of control."

1.11 Assume a single factor with two levels. Use either appropriate software or the Java applet in Section 1.4.4 to determine the power of detecting a difference of the two means of at least 2.5 when $\sigma_1 = \sigma_2 = \sigma = 4, \sigma = .05$, and there are 10 observations made at each level of the factor. Will the power be greater or less if there are more observations and σ is smaller, or can that be determined? Explain.

1.12 Explain why an experimenter would consider using factors with more than two levels.

1.13 Assume that the time order of the observations from an experiment is unknown, and it is suspected that an important process may have gone out of control during experimentation with a single factor but the run sequence was lost. Could the data still be analyzed? What does this imply about the need to keep processes in a state of statistical control and to note the run sequence?

1.14 Critique the following statement: "I believe that the expected change in my response variable is approximately 2 units when a particular factor is varied from one level to another level. Thus, I know what should happen so there is no point in performing any experimentation."

1.15 Assume that an experiment with four levels of a single factor was run, with randomization, and the value of the response variable was almost strictly increasing during the runs in the experiment. What would you suspect and what would be your recommendation?

1.16 If you have a solid statistical background, read the Jiang, Turnbull, and Clark (1999) paper that was mentioned in Section 1.7. Also read case study #2 in Inman et al. (1992), which you will likely find easier to read than the other paper. Write up the use of process control methods in each paper. If you are conversant in the latter, do you believe that the methods that they used to try to ensure control were adequate? Explain.

1.17 Use available and appropriate software to determine the probability of detecting a difference in the means for two levels of a factor of 1.5σ when $n_1 = n_2 = 20$. Would the probability increase or decrease if the two sample sizes were increased? Explain.

1.18 A discussion of experimental design considerations for tree improvement is given in *Experimental Design and Tree Improvement*, 2nd ed., by E. R. Williams, A. C. Matheson, and C. E. Harwood. The second chapter of the book is available online at http://www.publish.csiro.au/samples/Experimental DesignSample.pdf. Read the chapter and comment on the nonstatistical design considerations that are involved in seeking tree improvement through experimental design.

1.19 Is sequential experimentation feasible in your field of study? If not, explain why not. If it is possible, can a sequence of experiments be performed in a short or at least reasonable period of time? Explain.

1.20 If possible, give an example of an experiment from your field where complete randomization and the design of a one-shot experiment could be risky.

1.21 Buckner, Chin, and Henri (1997, references) performed a gauge study using a design with three factors, but we will analyze only one of the three: the operator effect. The data for the sheet resistance for the 7 kÅ tungsten film under test is shown below, along with the corresponding operator, which is labeled here simply as 1, 2, and 3.

Sheet resistance (mΩ)	84.86	84.92	84.81	84.80	84.86	84.93	84.80	84.94
Operator	1	2	2	3	3	2	2	1
Sheet resistance (mΩ)	84.91	84.86	84.78	84.86	84.96	84.89	84.90	84.90
Operator	1	2	3	2	2	2	3	1

Notice that the design relative to this factor is unbalanced. Does that create a problem with the analysis of the data? Why, or why not? If the data can be analyzed despite this imbalance, do so and determine whether or not there was a significant operator effect.

1.22 The following data appear on the Internet, preceded by the statement "In a comparison of the finger-tapping speed of males and females the following data was [*sic*] collected."

Males	43	56	32	45	36	48			
Females	41	63	72	53	68	49	51	59	60

This is all the information that was given relative to the experiment, if it can be called such, and the data.

(a) Based on the information given, could this be called an experiment. If so, what is the factor? If not, explain why it isn't an experiment.
(b) Can the data be analyzed, based on what is given? Why, or why not? If the data can be analyzed, do so and draw a conclusion.

1.23 The example given in Ryan (2000), which was mentioned in Section 1.7, showed what can happen to the conclusions of an experiment when a process goes out of control and affects the mean of the experimental level of a factor. Would the experiment similarly be undermined if the variance of either or both of the levels was increased as a result of an out-of-control condition? Explain.

CHAPTER 2

Completely Randomized Design

In this chapter we consider the use of completely randomized designs, both with and without restrictions on randomization and with and without the use of Analysis of Means (ANOM).

2.1 COMPLETELY RANDOMIZED DESIGN

As stated in Chapter 1, complete randomization is not always possible, and when it isn't possible, the ramifications of restricted randomization must be understood (see Section 4.19). We will initially assume, however, that complete randomization *is* possible, and later in the chapter will relax that assumption and discuss the consequences of restricted randomization.

When we have a single factor, complete randomization means that (1) the levels of the factor are assigned to the experimental units in a random fashion, and (2) the order in which the experiment is carried out after the assignment has been made is also random. The latter is important when one is conducting a physical experiment, in particular, but may not even be applicable for many other types of experiments. For example, for the teacher experiment described in Section 1.4, there would be no separate runs within each level of the factor "method of instruction," as all students would be subject to a particular method of instruction at the same time. If, however, the factor was "temperature," it might be highly impractical, if not impossible, to randomly change temperatures during an experiment. If, however, three levels of temperature were used with several values of the response variable recorded at each temperature level and the temperatures in the experiment changed successively from the lowest level to the highest level, an apparent temperature effect could be confounded with the effect of one of more extraneous variables (i.e., lurking factors) that could be influencing the value of the response variable over time.

If the temperatures were randomized, a plot of the response variable over time should not show almost strictly increasing values of the response variable. If such a

Modern Experimental Design   By Thomas P. Ryan
Copyright © 2007 John Wiley & Sons, Inc.

plot did occur, this would almost certainly mean that one or more extraneous factors were affecting the results. But if the temperatures were not randomized, we couldn't tell from such a plot whether the trend was due primarily to the temperature effect (with extraneous factors perhaps having a small effect), or was the trend due almost exclusively to the effect of extraneous factors.

Of course if there were a positive temperature effect, we might observe level shifts, with points randomly scattered about the midline for each level, but we would not expect to observe the response values strictly increasing. Such a plot would likely suggest that the response values are getting a boost from the effect of at least one extraneous factor.

2.1.1 Model

The model for a completely randomized design with a single factor can be written as

$$Y_{ij} = \mu + A_i + \epsilon_{ij} \qquad i = 1, 2, \ldots, k \quad j = 1, 2, \ldots, n_i \qquad (2.1)$$

with μ the overall mean (which would be estimated by the mean of all the observations), Y_{ij} is the jth observation for the ith level of the single factor, A_i is the effect of the ith level of the factor, with n_i observations for each level, and ϵ_{ij} is the corresponding error term, which is assumed to have a normal distribution with a mean of zero and a variance that is the same for each level. The errors are also assumed to be independent. Stated compactly, the assumption is, for each i, $\epsilon_{ij} \sim \text{NID}(0, \sigma_\epsilon^2)$. (Hereinafter, σ_ϵ^2 will generally be written simply as σ^2.) As discussed in Section 1.6.2.2, the levels of the factor may be either selected at random from a range of levels that is of interest, which would make the factor a *random factor*, or specific levels of interest might be used, which would make the factor a *fixed factor*. For a single factor the analysis is the same regardless of how the factor is classified, although the inference that is drawn from the data is different.

That is, for the model given by Eq. (2.1), if the factor is fixed, the null hypothesis that is tested is H_0: $A_i = 0$ for $i = 1, 2, \ldots, k$, which is the same as H_0: $\mu_1 = \mu_2 = \cdots \mu_k = \mu$. This becomes clear if we recognize that $\mu_i = \mu + A_i$. Whether H_0 is true or not, the side condition $\sum_{i=1}^{k} A_i = 0$ is imposed. In words, this means that one or more levels of the factor will affect the response value over and above the overall mean, μ, whereas one or more levels will cause the response value to be less than μ. The need for the side condition should be apparent if we sum both sides of Eq. (2.1). We would logically estimate μ by $\sum_{i=1}^{k} \sum_{j=1}^{n_i} Y_{ij} / \sum_{i=1}^{k} n_i$ and this estimator will occur from Eq. (2.1) only if the side condition is imposed in addition to the mean of ϵ_{ij} being zero so that the sum of ϵ_{ij} is zero.

If the factor were random, then the appropriate test would be H_0: $\sigma_A^2 = 0$. That is, in the fixed effects case the interest is solely on the effect of the levels of the factor used in the experiment, whereas in the random effects case the null hypothesis states that there is no effect of *any* level of A within the range of interest. The form of these hypothesis tests also applies more generally when there is more than one factor.

Assumptions should, of course, always be checked, and the normality and constant variance assumptions should therefore be checked. Methods for doing so are illustrated in subsequent chapters. From a practical standpoint, if the means differ considerably, the variances may differ more than slightly as the variance is generally related to the magnitude of the numbers. Thus, if we suspect that the means may differ considerably, it is especially important to test the equal variances assumption.

As far as parameter estimation is concerned, it would seem appropriate for A_i to be estimated by $\overline{y}_i - \overline{\overline{y}}$, with the latter denoting the overall average, and this is how A_i is estimated. The sum of the $\widehat{A}_i$, the estimates of the A_i, is zero when the n_i are equal, as the reader is asked to show in Exercise 2.1. (When the n_i differ, $\sum_{i=1}^{k} n_i(\overline{y}_i - \overline{\overline{y}}) = 0$.) The variance of the error term, σ_ϵ^2, is estimated analogous to the way it is estimated in an independent sample t-test: by pooling the variances within each level.

2.1.2 Example: One Factor, Two Levels

Assume that an experiment is performed involving 120 patients. The objective of the experiment is to test a blood pressure medication against a purported placebo, but the placebo is actually garlic, suitably disguised. Sixty patients are randomly assigned to the medication, with the other 60 patients assigned to the placebo. The study is double blinded so that the investigators do not know the patient–medication/placebo assignment, and of course the patients don't know this either. The correct assignment is known only by the person who numbered the bottles, with this information later used to properly guide the computer analysis.

Assume further that deviation from diastolic blood pressure at the start of the experiment is used for the analysis, with three measurements taken for each of three consecutive days starting 60 days after the experiment began, with the average of those nine measurements used as a single number for each of the 120 patients. Thus there are 120 numbers. We revisit this problem briefly in Section 11.1.1 in which repeated measures designs are presented. With "P" denoting the placebo and "M" denoting the medication, the results are given below.

```
M  -2  -7   -4  -8   -7  -4  -4  -1  -2  -10  -3    0   1  -10  -4  -4   -7  -3  -7
P  -2  -5   -8  -4   -2  -2   1  -8   1   -2  -3   -8  -1   -7  -1   0   -8  -7  -7
M  -9  -2   -1   1   -3  -7  -9  -6  -4   -8   1   -1   2   -2  -1  -8   -8  -1  -9
P  -1  -5   -7  -2   -8  -2  -4  -4  -6   -5  -6   -3   0   -5  -8  -4   -8  -1  -4
M  -4  -6  -10   0  -10  -3  -6  -1  -5   -7   1   -4   2   -5  -7  -9  -10  -6  -8
P   1   1   -5   1    0  -2   1   0  -7   -4  -2   -5  -8   -7   1  -7   -5  -6  -8
M  -1  -9   -5
P  -7  -8   -3
```

2.1.2.1 Assumptions

As stated previously, the assumptions must be kept in mind when the data are analyzed. One assumption that is rarely addressed is that the observations within each level of a factor must be independent. This is because the variance of an average, such as the average deviation for the medication over the 30 patients, is assumed to be σ^2/n, but this will be true only if the observations that comprise the average are independent. As Czitrom (2003) stated, "Perhaps the single most important issue related to the application of statistics in the semiconductor industry is the frequent *lack of independence* of observations . . . . The lack of independence affects the application of such basic statistical tools as t-tests, confidence intervals, analysis of variance, and control charts." By no means is this problem confined to the semiconductor industry, so independence *between observations within a group* must be checked so that the application of methods such as those given in this book will not be undermined.

We would not expect that assumption to be violated for this example because the experimental units for each level of the factor are different people. However, this does not preclude the possible effect of a lurking variable, so the data for the medication and the placebo should each be plotted over time.

2.1.2.1.1 *Checking the Assumptions*

Our analysis proceeds as follows. Time sequence plots for each set of 60 measurements do not exhibit any unusual patterns, and an autocorrelation plot of each set reveals no significant autocorrelations. So there appears to be independence and stability within each set of measurements. The normal probability plots for each set exhibit clear evidence of nonnormality. Neither these plots nor a histogram of each set provides strong evidence of skewness, however, so with moderately large sample sizes we can proceed by realizing that the sample means will be approximately normally distributed, and by realizing that a t-test will be robust to small-to-moderate departures from normality of the sample means. Levene's test for equal variances has a p-value of .27; since this is much larger than .05 or .01, we can proceed to perform a pooled t-test. If the p-value had been, say, .02, the generalized F-test for the unequal variances case given by Weerahandi (1994) would have to be used.

The results of the test are given below.

```
Two-Sample T-Test and CI: M, P
Two-sample T for M vs P

      N    Mean   StDev   SE Mean
M    60   -4.57   3.53     0.46
P    60   -3.92   3.08     0.40

Difference = mu M - mu P
Estimate for difference: -0.650
95% CI for difference: (-1.848, 0.548)
T-Test of difference = 0 (vs not =): T-Value = -1.07
  P-Value = 0.285. DF = 118
Both use Pooled StDev = 3.31
```

The results show that there is not sufficient evidence to reject the null hypothesis of an equal mean difference from the starting blood pressures for the medication group and the placebo group. Can we then conclude that the medication is ineffective? Since garlic is known to lower blood pressure, such a conclusion cannot be drawn, and in fact a one-sample t-test for the medication yields a p-value of less than .001, thus providing strong evidence that the medicine was effective. Thus, a completely erroneous conclusion could be drawn from the two-sample test, if the placebo is not a true placebo.

```
One-way ANOVA: M, P

Analysis of Variance
Source      DF       SS      MS          F        P
Factor       1     12.7    12.7       1.15    0.285
Error      118   1295.3    11.0
Total      119   1308.0
                                  Individual 95% CIs For Mean
                                  Based on Pooled StDev
Level       N     Mean    StDev   -------+---------+--------+---
M          60   -4.567    3.529   (----------  *  ----------)
P          60   -3.917    3.082             (--------  *  --------)
                                  -------+---------+--------+------
Pooled  StDev =    3.313           -4.90     -4.20     -3.50
```

Equivalently, the two-sample data can be analyzed using Analysis of Variance (ANOVA), with the output given above. In fact, we may state, loosely speaking, that the ANOVA table contains the square of the information/data from the two-sample t-test, with the same null hypothesis tested. For example, the F-statistic value of 1.15 is, without rounding, equal to $(1.07)^2$. Similarly, the square of the "pooled standard deviation" of 3.31 from the output for the two-sample t-test is, within rounding, equal to 11.0, the MS_{error} value from the ANOVA table, which estimates σ^2.

The manner in which the numerical values for the components of the ANOVA table are computed is explained in detail in Section 2.1.3.3, in addition to a detailed discussion of what ANOVA provides.

2.1.3 Examples: One Factor, More Than Two Levels

When there is a single factor with more than two levels, the experimenter does not have the option of using a t-test, since a t-test is applicable only when there are at most two levels. This is due to the fact that with a t-test either the equality of two means is being tested or a specified value of a single mean is tested. Analysis of Variance can be used, however, when there are any number of levels.

It is of course important to understand what ANOVA provides. To some extent, the term "Analysis of Variance" is a misnomer because what is analyzed is variation— variation due to different sources. To illustrate, consider the following hypothetical data.

1	2	3
8.01	8.03	8.04
8.00	8.02	8.02
8.02	8.01	8.02
8.01	8.02	8.03
8.01	8.02	8.04
8.02	8.03	8.02
8.00	8.01	8.04

The averages for the three levels are 8.01, 8.02, and 8.03, respectively. Although these averages differ only slightly, when we test the hypothesis that the three population means (μ_1, μ_2 and μ_3) are equal (using a methodology to be given shortly), we easily reject that hypothesis because the p-value for the test is .002, with the computer output as follows. (The computations that underlie the analysis are explained in Section 2.1.3.3; statistical significance results because of the very small variability within each level.)

```
      Analysis of Variance
Source     DF       SS         MS          F         P
Factor      2    0.0014000  0.0007000    9.00     0.002
Error      18    0.0014000  0.0000778
Total      20    0.0028000
                                   Individual 95% CIs For Mean
                                   Based on Pooled StDev
Level      N      Mean       StDev  -------+---------+---------+---------+---------
1          7    8.01000    0.00816   (----------*----------)
2          7    8.02000    0.00816            (---------*----------)
3          7    8.03000    0.01000                     (--------*-------)
                                   -------+---------+---------+---------+-------
Pooled StDev = 0.00882                    8.010      8.020      8.030
```

2.1.3.1 Multiple Comparisons

In general, there is a need to determine which of the three population means differ since we are rejecting the null hypothesis of equality of the means. If we observed actual data like this, we would suspect that the difference in the means would not likely be of any practical significance. To finish the example, however, since the confidence intervals for μ_1 and μ_3 do not overlap, as the computer output shows, we would logically conclude that these two means differ, and this is why the hypothesis of equality of the three means was rejected.

Looking to see if confidence intervals for means overlap is essentially an ad hoc approach. Nevertheless, such an approach is not necessarily a bad idea because it is both simple and intuitive. Another approach that is similarly intuitive is the *sliding reference distribution* approach given by Box, Hunter, and Hunter (1978, p.191). The general idea is to construct a t-distribution with the appropriate scale factor of $\frac{s}{\sqrt{n}}$ for equal sample sizes, and an approximate scale factor of $\frac{s}{\sqrt{\bar{n}}}$ with $\bar{n}$ denoting the average sample size if the sample sizes do not differ greatly. A dotplot of the means is constructed and the idea is to see if the t-distribution can be positioned in such a way as to cover as many of the means as possible. The means that cannot be covered by the (sliding) t-distribution are said to be different from the other means. Of course it would be cumbersome to have a cutout that would be used to physically slide along a dotplot, but of course this could be handled rather easily with a Java applet, although this has apparently not been done.

For the present example, $\frac{s}{\sqrt{n}} = \frac{0.00882}{\sqrt{7}} = 0.00333$. If we center the sliding t-distribution at the mean of 8.02, then $8.02 \pm t\frac{s}{\sqrt{n}}$ are at almost exactly 8.01 and

8.03 when $t = 3$. The latter is an extreme value for 18 degrees of freedom for the error term, as in this example, with $P(t_{18} > 3) = .004$. Since the latter is a small value, we would conclude that all three means differ using this approach.

There are various *multiple comparison procedures*, as they are called, from which an experimenter can select. Some of these procedures are for use when the comparisons to be made are selected before the data are collected, and others are for use after the experimenter looks at the data. Some are conservative, some are not conservative; one method (Dunnett's procedure) is used when testing against a control. It is interesting to note that Box et al. (1978) did not present any of these methods, however, instead opting for their sliding reference distribution approach. (The same is true of Box, Hunter, and Hunter, 2005.)

A conservative multiple comparison procedure is one that is based on Bonferroni's inequality, which was named after an Italian mathematician Carlo Bonferroni (1892–1960). The inequality stated that for events $B_1, B_2, \ldots, B_q$ (which in this case will be confidence intervals),

$$P\left(\bigcap B_i\right) \geq 1 - \sum_{i=1}^{q}[1 - P(B_i)]$$

or

$$P\left(\bigcap B_i\right) \geq 1 - \sum_{i=1}^{q} P(\overline{B_i}) \tag{2.2}$$

with $P(\overline{B_i}) = [1 - P(B_i)]$ denoting the probability that event B_i does not occur.

Applying this result to experimental design, let B_i denote the event that a confidence interval for a treatment effect or a linear combination of treatment effects contains the treatment effect or linear combination of treatment effects, so that $P(B_i)$ is the probability of this occurrence, and $P(\overline{B_i})$ is the probability that the confidence interval does not include the unknown treatment effect.

It follows from the inequality given in expression (2.2) that if the objective were to have the entire set of confidence intervals cover the respective treatment effects with probability of at least $1 - \alpha$, each interval could be a $1 - \alpha/q$ confidence interval, so that the probability that each interval does not cover the treatment effect (i.e., $P(\overline{B_i})$) would be α/q. Then from expression (2.2), $P(\bigcap B_i) \geq 1 - q(\alpha/q)$ so that $P(\bigcap B_i) \geq 1 - \alpha$.

Each confidence interval would be of the general form

$$\sum_{i=1}^{k} c_i \widehat{A}_i \pm t_{n-v, \alpha/2q} \sqrt{\widehat{\text{Var}}\left(\sum_{i=1}^{k} c_i \widehat{A}_i\right)}$$

with A_i denoting the effect of the ith treatment, as in Eq. (2.1), $\widehat{A}_i$ denoting the estimator of that effect, and v denoting the degrees of freedom for the error term. The value of k is the number of treatment effects involved in the confidence interval. For two treatment effects, a logical comparison would be $A_1 - A_2$, for which the constants would be $c_1 = 1$ and $c_2 = -1$. In general, $\sum_{i=1}^{k} c_i = 0$ for each comparison, which must be planned before collecting the data. To do otherwise would be to bias the results.

Since $\text{Var}(\widehat{A}_i) = \sigma^2/n_i$ and $\widehat{\sigma}^2 = \text{MS}_{\text{error}}$, as stated in Section 2.1.2.1.1, and $\widehat{A}_i = \bar{y}_i$, we may write the confidence interval as

$$\sum_{i=1}^{k} c_i \bar{y}_i \pm t_{n-v,\alpha/2q} \sqrt{\text{MS}_{\text{error}} \sum_{i=1}^{k} c_i^2/n_i}$$

The probability that each of the q comparisons contains $\sum_{i=1}^{k} c_i A_i$ for $i = 1, 2, \ldots, q$ is at least $1 - \alpha$ and quite likely much greater than $1 - \alpha$.

One of the best known multiple comparison procedures is due to Scheffé (1953), which is for every possible contrast $\sum_{i=1}^{k} c_i \mu_i$, with $\mu_1, \mu_2, \ldots, \mu_k$ denoting the k treatment means and the c_i being arbitrary constants but with the restriction that $\sum_{i=1}^{k} c_i = 0$. The Scheffé procedure, which can be used when decisions about which comparisons to make are made after examining the data, gives a set of simultaneous $100(1 - \alpha)\%$ confidence intervals with the objective being to see if any of the intervals do not cover zero. If so, the corresponding contrast (such as $\mu_1 - \mu_2$) is significant and in this example the conclusion would be that $\mu_1 \neq \mu_2$. The confidence intervals are of form

$$\sum_{i=1}^{k} c_i \bar{y}_i \pm \sqrt{(v - 1)F_{v-1,n-v,\alpha}} \sqrt{\text{MS}_{\text{error}} \sum_{i=1}^{k} c_i^2/n_i}$$

where v is defined as it was for the Bonferroni intervals. Notice that the form for the Scheffé intervals differs from the form for the Bonferroni intervals only in terms of the first component after the $\pm$ sign. There are various other multiple comparison procedures, including those due to Tukey (1953), Duncan (1955, 1975), Dunnett (1955), Fisher (1935), Kramer (1956), and Hsu (1984).

These methods are discussed in detail in Hsu (1996) and Hochberg and Tamhane (1987), and most are also discussed in detail in Dean and Voss (1999). It is worth noting that some of these papers have been extensively cited, with Duncan (1955) being the third most cited paper in the list of the 25 most cited papers given by Ryan and Woodall (2005), whereas Dunnett (1955) is 14th, Dunnett (1964) is 21st, and Kramer (1956) is 22nd.

In general, multiple comparison procedures are fraught with problems and controversies. When used, a procedure should be carefully selected and used appropriately. A good, relatively recent online treatise on multiple comparison procedures is Dallal (2001), which is available at http://www.tufts.edu/~gdallal/mc.htm.

At the other extreme, consider the following data.

1	2	3
7.06	4.91	4.87
3.13	9.04	8.01
4.12	5.95	6.76
5.59	3.86	7.98
5.10	6.24	7.38

Here the averages are 5.0, 6.0, and 7.0, respectively, but the hypothesis of equal means is not rejected, as the p-value is .184. The computer output is as follows.

```
Analysis of Variance
Source    DF      SS      MS      F       P
Factor    2     10.00    5.00    1.96   0.184
Error     12    30.66    2.56
Total     14    40.66
                                Individual 95% CIs For Mean
                                Based on Pooled StDev
Level    N     Mean    StDev   -------+---------+--------+--------
1        5    5.000    1.489   (----------*----------)
2        5    6.000    1.941            (---------*---------)
3        5    7.000    1.296                    (--------*-------)
                                -------+---------+--------+-------
Pooled StDev =   1.598           4.5       6.0      7.5
```

Notice that the confidence intervals overlap, and notice that the standard deviations of the means in this example are larger, relative to the difference in the means, than are the standard deviations in the previous example. This helps explain why the conclusions differ. Using the applet that was used in Section 1.4.4, which is found at http://www.stat.uiowa.edu/~rlenth/Power/index.html, the probability of detecting a difference of $\Delta = 2.005$ (which is essentially the difference between the first and third means above) is only .4454 when $\sigma = 1.6$ and $\alpha = .05$. The less than 50–50 chance of detecting the stated difference is due largely to the fact that σ is large relative to the magnitude of the numbers. (Another method of comparing means is ANOM, which is described in Section 2.2.)

Thus, we reject the null hypothesis when the means differ by only 0.01, but we fail to reject the null hypothesis when the means differ by 1.0. The reason for this is that in the second example there is considerable within-level variability, which drowns out the between-level variability. The reverse occurs in the first example as the within-level variability is so small that it does not offset the between-level variability, although the latter is obviously small.

2.1.3.2   *Unbalanced and Missing Data*
Although the examples given in this chapter have the same number of observations for each level of the factor, this is not a requirement as the number of observations

per level could differ, as is implied by model (2.1). Unbalanced data can be easily handled when a completely randomized design is used, although we would naturally prefer that the mean for each level of the factor be estimated from the same number of observations. Unbalanced data do present a problem when there is more than one factor, however, and imputation methods are discussed briefly in Section 4.12.2. If such methods are not used, then methods for analyzing unbalanced data must be employed, and the reader is referred to Searle (1987) for the proposed methods of analysis. Since unbalanced data present no problem with a completely randomized design, it follows that missing data that cause unbalanced data also do not present a problem unless the missing data are numerous and/or are not missing at random.

2.1.3.3 Computations

If we were to devise a measure of the variability between the levels of a factor, one obvious choice is to use some function of the difference between the average value for each level and the overall average. We can't sum those differences, however, because the sum will always be zero, as was stated in Section 2.1.1. We can, however, sum the squares of the differences, and that sum is multiplied by the number of observations for each level, if that number is constant across all levels. (The reason for the multiplier will soon become apparent.)

For a measure of variability within the levels, the obvious choice is to compute, for each level, the square of each observation from the level average and sum the squares over the observations in each level, and then sum over the levels.

An obvious choice as a measure of the total variability would be the sum of the squares of the observations from the overall average. Adopting the notation that was used in Section 2.1.1, we have the following *Analysis of Variance Identity* when the number of observations per level (i.e., the n_i) is constant and equal to n, as in the two examples in Sections 2.1.3 and 2.1.3.1.

$$\sum_{i=1}^{n}\sum_{j=1}^{k}(y_{ij} - \overline{\overline{y}})^2 \equiv n\sum_{j=1}^{k}(\overline{y}_i - \overline{\overline{y}})^2 + \sum_{i=1}^{n}\sum_{j=1}^{k}(y_{ij} - \overline{y}_i)^2 \qquad (2.3)$$

(Only a slight modification to this expression is needed if the n_i differ.) This equivalence is very easy to derive and can be accomplished by subtracting and adding $\overline{y}_i$ within the parentheses on the left side, squaring, and then simplifying, as the reader is asked to show in Exercise 2.2.

The first term on the right side of Eq. (2.3) gives what was labeled "Factor" in the computer output, and the second term gives what was labeled "Error." It could be shown with a small amount of algebra that this term is the extension to k levels of what the numerator of s_p^2 would be in Section 1.4.4 if that numerator were written in terms of the appropriate summation expressions rather than in terms of the two variances. If this is accepted, which the reader is asked to show in Exercise 2.17, it could then be shown (also Exercise 2.17) that MS_{error} is simply the average of the variances for each level when there is an equal number of observations in each level.

Notice that *if* hand computation were performed, the error sum of squares would be obtained by subtracting the factor sum of squares from the total sum of squares that is given on the left side of the equation. Obviously only two terms in the identity would have to be computed, with the other obtained by addition or subtraction. (The total sum of squares is computed the same way for every model and experimental design; only the form of the right side is model/design dependent.)

The *degrees of freedom* (df) can be similarly partitioned. Since $\sum_{i=1}^{n}\sum_{j=1}^{k}(y_{ij} - \overline{\overline{y}}) = 0$, the sum does not contain kn independent pieces of information. Rather, only $kn - 1$ components are independent since the sum is zero. Accordingly, only $kn - 1$ components on the left side of Eq. (2.3) are independent and "free to vary." Thus, that sum has $kn - 1$ df. Similarly, $\sum_{j=1}^{k}(\overline{y}_i - \overline{\overline{y}}) = 0$, so only $k - 1$ components of the sum are free to vary; thus, the first term on the right side of the equation has $k - 1$ df. Since degrees of freedom are additive, it follows that the second sum on the right side of the equation must have $kn - k$ df.

We may summarize the degrees of freedom breakdown as follows, assuming an equal number of observations per treatment.

```
Analysis of Variance

Source          df
Treatments      k − 1
Error           k(n − 1)
Total           kn − 1
```

If the n_i are not all the same, then the total degrees of freedom is $\sum_{i=1}^{k} n_i^{-1}$ and the error degrees of freedom is $\sum_{i=1}^{k} n_i - k$.

2.1.4 Example Showing the Effect of Unequal Variances

Weerahandi (2004) gave an example that showed how the F-test for the equality of treatment means can produce a result that differs from the result obtained using a heteroscedastic ANOVA approach when the population variances are apparently unequal.

An engineer at a construction company was interested in testing the comparative strength of four brands of reinforcing bars. Although Weerahandi (2004) didn't give the unit of measurement (and the data are presumably hypothetical), the data are as follows.

```
BRAND A  21.4  13.5  21.1  13.3  18.9  19.2  18.3
BRAND B  27.3  22.3  16.9  11.3  26.3  19.8  16.2  25.4
BRAND C  18.7  19.1  16.4  15.9  18.7  20.1  17.8
BRAND D  19.9  19.3  18.7  20.3  22.8  20.8  20.9  23.6  21.2
```

The means for BRANDS A–D are 17.96, 20.68, 18.10, and 20.83, respectively, whereas the variances are 3.07, 5.28, 1.39, and 1.48, respectively.

It is useful to begin the analysis by looking at boxplots for the four brands, which are given in Figure 2.1. (Reese (2005) stated, "Make it a rule: never do ANOVA without

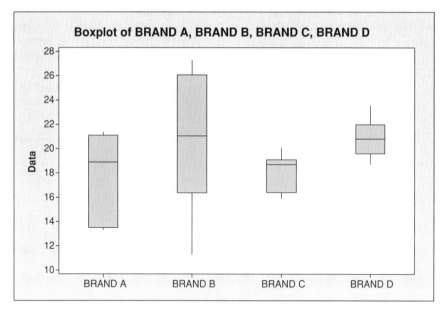

Figure 2.1   Boxplots of brand data.

a boxplot." Although such a strong statement will not be made here, especially since boxplots are of very limited value when there is more than one factor, boxplots for designs with single factors are very useful.)

The heteroscedasticity is apparent by comparing the last two brands with the first two brands.

The means are not directly comparable because the variability is so much less for the last two brands than it is for the first two brands. If this disparity is ignored and an F-test performed, $F = 1.61$ is obtained, which has a p-value of .211. Thus, the conclusion is that the means do not differ. Certainly Figure 2.1 shows that the medians do not differ greatly. (We may note that the Kruskal–Wallis nonparametric test is not applicable here, since it is based on the assumption that the populations have the same continuous distribution except for possibly different medians.) Weerahandi (2004, p. 51) obtained a p-value of .021 when applying the generalized F-test given by Weerahandi (1994).

2.2   ANALYSIS OF MEANS

Although ANOM has been in use for decades and has been part of the MINITAB software for many years, and is also included in SAS/QC 9.0 and 9.1 from SAS Software, it is still not well known to people who use designed experiments. In fact, many people would undoubtedly confuse ANOM with ANOVA, since the latter also involves an analysis involving means. It is apt to have more appeal to engineers and other industrial personnel than does ANOVA, however, since ANOM is inherently a graphical procedure and is in terms of the original unit(s) of measurement, whereas ANOVA is in the square of the original unit(s), as was stated previously in Section 1.9.

Analysis of Means is not a full substitute for Analysis of Variance, however, as ANOM can be used only for fixed factors, whereas ANOVA can be used for fixed or random factors, or for a combination of the two.

The reader will recall that with ANOVA the experimenter concludes either that all of the means are equal, or that at least one of the means differs from the others. One procedure need not be used to the exclusion of the other, however. As Ott (1967) indicates, ANOM can be used either alone or as a supplement to ANOVA.

2.2.1 ANOM for a Completely Randomized Design

It was stated previously that with ANOM one compares $\overline{x}_i$ against the average of the $\overline{x}_i$, which will be denoted by $\overline{\overline{x}}$, analogous to the notation used for an $\overline{X}$ chart. The original ANOM methodology given by Ott (1958, 1967) was based upon the multiple significance test for a group of means given by Halperin, Greenhouse, Cornfield, and Zalokar (1955), which was based upon the studentized maximum absolute deviate. That approach provided an upper bound for the unknown critical value, but will not be discussed here since it is no longer used. The interested reader is referred to Schilling (1973) for more details, including the theoretical development. The current approach is based upon the exact critical value, h, and is described in L. S. Nelson (1983).

If we were testing for the significance of a single deviation, $\overline{x}_1 - \overline{\overline{x}}$, it would stand to reason that we would want to look at some test statistic of the form

$$\frac{\overline{x}_i - \overline{\overline{x}} - E(\overline{x}_i - \overline{\overline{x}})}{s_{\overline{x}_i - \overline{\overline{x}}}} \tag{2.4}$$

where E stands for expected value. If $\mu_i = (\mu_1 + \mu_2 + \cdots + \mu_k)/k$, then $E(\overline{x}_i - \overline{\overline{x}}) = 0$, and since the former is what would be tested, we take $E(\overline{x}_i - \overline{\overline{x}})$ to be zero. (We should note that some authors have indicated that the null hypothesis that is tested with ANOM is $H_0: \mu_1 = \mu_2 = \cdots = \mu_k$. Certainly if each μ_i is equal to the average of all the means, then it follows that the μ_i must all be equal, since they are equal to the same quantity. But stating the null hypothesis in this alternative way obscures the testing that is done.)

It can be observed that Eq. (2.4) becomes a t-test when $k = 2$ since $\overline{x}_i - \overline{\overline{x}}$ is then $\overline{x}_1 - (\overline{x}_1 + \overline{x}_2)/2 = (\overline{x}_1 - \overline{x}_2)/2$ for $i = 1$ (and $(\overline{x}_2 - \overline{x}_1)/2$ for $i = 2$) so that

$$t = \frac{(\overline{x}_1 - \overline{x}_2)/2 - 0}{s_{(\overline{x}_1 - \overline{x}_2)/2}} = \frac{\overline{x}_1 - \overline{x}_2}{s_{\overline{x}_1 - \overline{x}_2}}$$

since the 2s cancel.

The two deviations $\bar{x}_1 - \bar{\bar{x}}$ and $\bar{x}_2 - \bar{\bar{x}}$ are thus equal, so we conclude that the two means differ if

$$t = \frac{|\bar{x}_1 - \bar{x}_2|}{s_{\bar{x}_1 - \bar{x}_2}} > t_\alpha$$

for a selected value of α.

When $k > 2$, the t-distribution cannot be used, however, so another procedure is needed. It can be shown that, assuming equal sample sizes, the deviations $\bar{x}_i - \bar{\bar{x}}$ are equally correlated with correlation coefficient $\rho = -1/(k-1)$. If we let $T_i = (\bar{x}_i - \bar{\bar{x}})/s_{\bar{x}_i - \bar{\bar{x}}}$, the joint distribution of $T_1, T_2, \ldots, T_k$ is an equicorrelated multivariate noncentral-t distribution, assuming that the sample averages are independent and normally distributed with a common variance (see P. R. Nelson 1982, p. 701).

Exact critical values for $k > 2$ were first generated by P. R. Nelson (1982), with a few tabular values subsequently corrected, and the corrected tables published in L. S. Nelson (1983). More complete and more accurate critical values given to two decimal places were given by P. R. Nelson (1993). These values differ by one in the second decimal place from some of the critical values given in L. S. Nelson (1983).

The general idea is to plot the averages against *decision lines* obtained from

$$\bar{\bar{x}} \pm h_{\alpha,k,\nu} \, s \sqrt{(k-1)/(kn)} \tag{2.5}$$

where n is the number of observations from which each average is computed, ν is the degrees of freedom associated with s, the estimate of σ, k is the number of averages, and $h_{\alpha,k,\nu}$ is obtained from the tables in P. R. Nelson (1993) for a selected value of α, with those tabular values given in this book in Table D. It is demonstrated in the appendix to this chapter that $s\sqrt{(k-1)/(kn)}$ is the estimate of $\sigma_{\bar{x}_i - \bar{\bar{x}}}$.

Analysis of Means can be used when the sample sizes are unequal; the expression for the decision lines is just slightly different. Specifically, the radicand is $\frac{N-n_i}{Nn_i}$, as is shown in the chapter Appendix. This causes the decision lines to be somewhat aesthetically unappealing, since the distance between the lines varies as the n_i vary. An example of an ANOM graph with varying decision lines for unequal sample sizes is given by Nelson, Coffin, and Copeland (2003, p. 262).

The value of α is the probability of (wrongly) rejecting the hypothesis that is being tested when, in fact, the hypothesis is true. (Here we are testing that each mean is equal to the average of all the k means, as indicated previously.)

2.2.1.1 *Example*

We will use the example in Section 2.1.3.1 for illustration, assuming the factor to be fixed. If we were using hand computation, the first step would be to compute the overall average, $\bar{\bar{x}}$, and then compute the decision lines from Eq. (2.5) for a selected value of α and plot the averages for the factor levels. Of course we know from Section 2.1.3.1 that the averages are 5, 6, and 7, respectively, and of course the overall average is 6. An ANOVA would generally be performed to obtain the value of s, and that output showed the value to be 1.598. The value of $h_{\alpha,k,\nu}$ is the value of

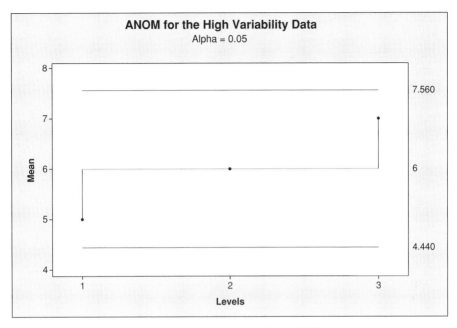

Figure 2.2 ANOM display for the high variability data.

$h_{.05,3,12}$, which is 2.67. Thus, the decision lines, as they are called, are obtained from $6 \pm 2.67(1.598)\sqrt{\frac{2}{3(5)}} = (4.44, 7.56)$, which are the numbers in the display (Fig. 2.2).

It can be observed that the means are well inside the $\alpha = .05$ decision lines, with .05 being the experiment-wise error rate (i.e., for the three tests that each of the three means is equal to the average of all of the means), so the conclusion is that no population mean differs from the average of the population means. Ott (1975) showed multiple sets of decision lines on ANOM displays, such as for .01, .05, and .10, but that option is not available with MINITAB, which is being used to produce these displays.

The ANOM display for the low variability data is shown in Figure 2.3. The conclusion is that the first and third population means differ from the averages of all the means since the first and third averages plot outside the .05 decision lines. Of course this result is very intuitive as the analysis using ANOVA showed a significant result and the two "extreme" averages of 8.01 and 8.03 are equidistant from the overall average of 8.02.

2.2.2 ANOM with Unequal Variances

There is also an ANOM procedure, due to Nelson and Dudewicz (2002), that can be used when there is evidence of unequal variances. This might be viewed as being analogous to the t-test that does not assume equal variances, which can be used when the pooled t-test cannot be used because of evidence of (highly) unequal variances.

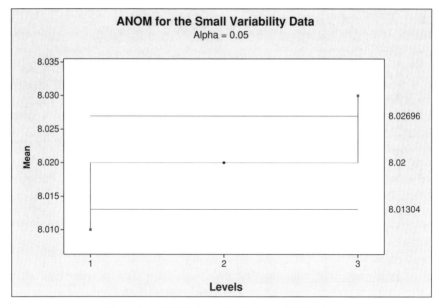

Figure 2.3 ANOM display for the low variability data.

Nelson and Dudewicz (2002) dubbed their procedure HANOM (heteroscedastic analysis of means). Before this procedure is used, the variances should be tested, and in the one-factor case it is quite possible that there will be enough observations per factor level for such a test to have reasonable power for detecting unequal variances.

One possible test would be Levene's or the modification of it, both of which were mentioned in Section 1.4.3, or if one wanted to stick strictly with ANOM procedures, the analysis of means for variances (ANOMV), due to Wludyka and Nelson (1997), might be used.

If the results from one of these methods provided evidence that the variances are more than slightly unequal, HANOM could be used. It should be noted, however, that unlike the ANOM procedure, HANOM is not a one-stage procedure. Instead, an initial sample is taken from the relevant populations for the purpose of computing the decision lines. A second sample is then taken and the sample means are computed. The latter are then compared against the decision lines. Thus, HANOM cannot be viewed as an alternative to heteroscedastic ANOVA unless the experimenter is willing to collect additional data.

If obtaining more data is not practical or feasible, heteroscedastic ANOVA (see, e.g., Bishop and Dudewicz, 1978) could be used, as could the generalized F-test due to Weerahandi (1994). (See also Weerahandi (2004), pp. 48–51). A second option would be to use Kruskal-Wallis ANOVA, a nonparametric method that assumes neither normality nor equal variances. It would be preferable to use heteroscedastic ANOVA if approximate normality seems to exist, however, since that would be a more powerful procedure than the Kruskal-Wallis procedure under (approximate) normality.

2.2.2.1 Applications

The fact that an initial sample is required reduces the practical value of HANOM. Although experimentation should be sequential, the number of factors investigated in the second stage is generally less than the number of factors investigated in the first stage. It is unlikely that experimenters will very often collect preliminary data just so that HANOM can be used. Although Nelson and Dudewicz (2002) do give an example, there is no evidence that real data were used.

2.2.3 Nonparametric ANOM

Analysis of Means is, like ANOVA, robust to small-to-moderate departures from normality. When the response values are strongly nonnormal, one possibility would be to try to transform the data to approximate normality. If that is unsuccessful, a nonparametric ANOM approach might be used. Bakir (1989) developed such a procedure based on ranks for use with a completely randomized design. The populations were assumed to be the same except for having possibly different means. The procedure is discussed in detail and illustrated by Nelson, Wludyka, and Copeland (2005, Section 9.3).

Another nonparametric ANOM approach is a permutation test. This can be done using either symmetric decision lines or asymmetric decision lines. For the former, N random permutations of the data are made and $D_{\max}^{(q)} = \max_i |\overline{Y}_{i(q)} - \overline{Y}_{\text{all}}|$ is computed for the qth permutation with $q = 1, \ldots, N$, with $\overline{Y}_{i(q)}$ denoting the ith treatment mean for the qth permutation and $\overline{Y}_{\text{all}}$ denoting the average of all the observations. This forms the randomization reference distribution for $\max_i |\overline{Y}_i - \overline{Y}_{\text{all}}|$. The treatment means, the $\overline{Y}_i$, are then plotted against decision lines given by $\overline{Y}_{\text{all}} \pm k_\alpha$, with k_α denoting the upper α quantile of the distribution of $D_{\max}^{(q)}$.

2.2.4 ANOM for Attributes Data

Although this book is primarily concerned with measurement data, it should be noted in passing that ANOM can be used advantageously with attributes data (see Ryan, 2000; Nelson et al., 2003; or Chapter 2 of Nelson et al., 2005). When this is done, it may be necessary to transform the attribute random variable to approximate normality if ANOM is to be used and if the normal approximation to the appropriate distribution is inadequate. It is well known that the rules of thumb for the adequacy of the normal approximation to the Poisson and binomial distributions, respectively, that are given in introductory statistics books fail in control chart applications, but their applicability in ANOM has apparently not been investigated, at least in the literature. The problem in control chart applications is that extreme tails are involved, which is not the case in ANOM. However, if we show .01 decision lines for proportions data, as recommended by Tomlinson and Lavigna (1983), we may be far enough out in the tails that there will often be problems relative to those decision lines, even if the binomial were the appropriate model and not even considering such possible problems as extrabinomial variation. Similar problems may exist for count data. This is something that should be researched.

2.3   SOFTWARE FOR EXPERIMENTAL DESIGN

Software must be used in analyzing data from designed experiments. The well-known software that can be used for design construction and analysis do differ somewhat, and Reece (2003) provides a very detailed and extensive comparison, although such comparisons become at least slightly outdated rather quickly, since companies introduce new releases of their software quite frequently. Nevertheless, the comparison given by Reece (2003) is worth reading, especially since it includes some software that are not widely known.

The software packages that are used to produce graphs and numerical output in this book are MINITAB, JMP, and Design-Expert, with the last two each receiving the highest possible rating by Reece (2003). Neither has ANOM capability, however, and SAS Software does not have the capability to directly (i.e., without programming) produce the ANOM displays that are used in Chapter 4. The ANOM capability in MINITAB is also somewhat limited, although an ANOM macro available at the MINITAB, Inc. Web site extends the capability that is provided by the ANOM command. The capabilities of another software package, D. o. E. Fusion Pro, are discussed in certain chapters, including Chapters 4 and 5, but output from the software is not used in the book.

2.4   MISSING VALUES

Unlike many of the designs that are presented in subsequent chapters, a missing value or two does not generally create a serious problem when a completely randomized design is used because data from an experiment that used such a design can be analyzed with unequal n_i, using either ANOVA or ANOM. Although missing values might be estimated (and methods for doing so have been proposed), an analysis using estimated values will be only approximate and thus not entirely satisfactory. If possible, the experimental run(s) that resulted in the missing value(s) might simply be repeated.

2.5   SUMMARY

A completely randomized design is a frequently used design that is attractive because of its simplicity. The numerical analysis is performed the same way regardless of whether the single factor is fixed or random, and the number of observations per factor level need not be the same. The user must check for possible heteroscedasticity, however, as this could undermine the results. Nonnormality could also be a problem, but only if it is moderate to severe.

The data from an experiment with a completely randomized design may be analyzed using either ANOVA or ANOM. The assumptions are the same for each and both can be used with unequal sample sizes.

APPENDIX

It was stated in Section 2.2.1 that $s\sqrt{(k-1)/(kn)}$ is the estimated standard deviation of $\overline{X}_i - \overline{\overline{X}}$, with the assumption that each $\overline{X}_i$ is computed from n observations. This can be demonstrated as follows:

$$\mathrm{Var}(\overline{X} - \overline{\overline{X}}) = \mathrm{Var}\left(\overline{X}_i - \frac{\overline{X}_1 + \cdots + \overline{X}_i + \cdots + \overline{X}_k}{k}\right)$$

$$= \mathrm{Var}(\overline{X}_i) - 2\,\mathrm{Cov}\left(\overline{X}_i, \frac{\overline{X}_i}{k}\right) + \mathrm{Var}(\overline{\overline{X}})$$

$$= \frac{\sigma^2}{n} - \frac{2}{k}\left(\frac{\sigma^2}{n}\right) + \frac{\sigma^2}{kn}$$

$$= \frac{\sigma^2(k-1)}{kn}$$

The result then follows after the square root of the last expression is taken, and s is substituted for σ.

Now assume that the $\overline{X}_i$ are computed from n_i observations, with the n_i not all equal. Then the corresponding derivation is

$$\mathrm{Var}(\overline{X} - \overline{\overline{X}}) = \mathrm{Var}\left(\overline{X}_i - \frac{n_1\overline{X}_1 + \cdots + n_i\overline{X}_i + \cdots n_k\overline{X}_k}{N}\right)$$

$$= \mathrm{Var}(\overline{X}_i) - 2\,\mathrm{Cov}\left(\overline{X}_i, \frac{n_i\overline{X}_i}{N}\right) + \mathrm{Var}(\overline{\overline{X}})$$

$$= \frac{\sigma^2}{n_i} - \frac{2n_i}{N}\left(\frac{\sigma^2}{n_i}\right) + \frac{\sigma^2}{N}$$

$$= \frac{\sigma^2}{n_i} - \frac{\sigma^2}{N}$$

$$= \sigma^2\left(\frac{N - n_i}{Nn_i}\right)$$

REFERENCES

Bakir, S. T. (1989). Analysis of means using ranks. *Communications in Statistics—Simulation and Computation*, **18**(2), 757–776.

Bishop, T. A. and E. J. Dudewicz (1978). Exact analysis of variance with unequal variances: Test procedures and tables. *Technometrics*, **20**, 419–430.

Box, G. E. P., J. S. Hunter, and W. G. Hunter (2005). *Statistics for Experimenters: Design, Innovation, and Discovery.* Hoboken, NJ: Wiley.

Box, G. E. P., W. G. Hunter, and J. S. Hunter (1978). *Statistics for Experimenters.* New York: Wiley.

Czitrom, V. (2003). Statistics in the semiconductor industry. In *Handbook of Statistics*, Vol. 22, pp. 459–498 (R. Khattree and C. R. Rao, eds.). Amsterdam: Elsevier Science B. V.

Dallal, G. E. (2001). Multiple comparison procedures. Online article available at http://www.tufts.edu/~gdallal/mc.htm.

Dean, A. and D. Voss (1999). *Design and Analysis of Experiments.* New York: Springer-Verlag.

Duncan, D. B. (1955). Multiple range and multiple F tests. *Biometrics*, **11**, 1–42.

Duncan, D. B. (1975). t-Tests and intervals suggested by the data. *Biometrics*, **31**, 739–759.

Dunnett, C. W. (1955). A multiple comparison procedure for comparing several treatments with a control. *Journal of the American Statistical Association*, **50**, 1096–1121.

Dunnett, C. W. (1964). New tables for multiple comparisons with a control. *Biometrics*, **20**, 482–491.

Fisher, R. A. (1935). *The Design of Experiments.* London: Oliver & Boyd.

Halperin, M., S. W. Greenhouse, J. Cornfield, and J. Zalokar (1955). Tables of percentage points for the studentized maximum absolute deviate in normal samples. *Journal of the American Statistical Association*, **50**, 185–195 (March).

Hochberg, Y. and A. C. Tamhane (1987). *Multiple Comparison Procedures.* New York: Wiley.

Hsu, J. C. (1984). Ranking and selection and multiple comparisons with the best. In *Design of Experiments: Ranking and Selection (Essays in Honor of Robert E. Bechhofer)*, pp. 22–33 (T. J. Santner and A. C. Tamhane, eds.). New York: Marcel Dekker.

Hsu, J. C. (1996). *Multiple Comparisons: Theory and Methods.* New York: Chapman & Hall.

Kramer, C. Y. (1956). Extension of multiple range tests to group means with unequal sample sizes. *Biometrics*, **12**, 307–310.

Nelson, L. S. (1983). Exact critical values for use with the analysis of means. *Journal of Quality Technology*, **15**(1), 40–44.

Nelson, P. R. (1982). Exact critical points for the analysis of means. *Communications in Statistics—Part A, Theory and Methods*, **11**(6), 699–709.

Nelson, P. R. (1993). Additional uses for the analysis of means and extended tables of critical values. *Technometrics*, **35**(1), 61–71.

Nelson, P. R., M. Coffin, and K. A. F. Copeland (2003). *Introductory Statistics for Engineering Experimentation.* San Diego, CA: Academic Press.

Nelson, P. R. and E. J. Dudewicz (2002). Exact analysis of means with unequal variances. *Technometrics*, **44**(2), 152–160.

Nelson, P. R., P. S. Wludyka, and K. A. F. Copeland (2005). *The Analysis of Means: A Graphical Method for Comparing Means, Rates, and Proportions.* Philadelphia: American Statistical Association and Society for Industrial and Applied Mathematics.

Ott, E. R. (1958). Analysis of means. Technical Report #1, Rutgers University.

Ott, E. R. (1967). Analysis of means—a graphical procedure. *Industrial Quality Control*, **24**(2), 101–109.

Ott, E. R. (1975). *Process Quality Control: Troubleshooting and Interpretation of Data.* New York: McGraw-Hill.

Reece, J. E. (2003). Software to support manufacturing systems. In *Handbook of Statistics*, Vol. 22, chap. 9 (R. Khattree and C. R. Rao, eds.). Amsterdam: Elsevier Science B.V.

Reese, A. (2005). Boxplots. *Significance*, **2**(3), 134–135.

Ryan, T. P. (2000). *Statistical Methods for Quality Improvement*, 2nd ed. New York: Wiley.

Ryan, T. P. and W. H. Woodall (2005). The most-cited statistical papers. *Journal of Applied Statistics*, **32**, 461–474.

Scheffé, H. (1953). A method for judging all contrasts in the analysis of variance. *Biometrika*, **40**, 87–104.

Schilling, E. G. (1973). A systematic approach to the analysis of means, Part I. Analysis of treatment effects. *Journal of Quality Technology*, **5**(3), 93–108.

Searle, S. R. (1987). *Linear Models for Unbalanced Data*. New York: Wiley.

Tomlinson, L. H. and R. J. Lavigna (1983). Silicon crystal termination—an application of ANOM for percent defective data. *Journal of Quality Technology*, **15**, 26–32.

Tukey, J. W. (1953). The problem of multiple comparisons. Originally unpublished manuscript that appears in *Collected Works of J. W. Tukey*, Vol. VII (H. Braun, ed.). New York: Chapmans & Hall, 1994.

Weerahandi, S. (1994). ANOVA under unequal error variances. *Biometrics*, **51**, 589–599.

Weerahandi, S. (2004). *Generalized Inference in Repeated Measures: Exact Methods in MANOVA and Mixed Models*. Hoboken, NJ: Wiley.

Wludyka, P. S. and P. R. Nelson (1997). An analysis of means type test for variances from normal populations. *Technometrics*, **39**, 274–285.

EXERCISES

2.1 For the model given in Eq. (2.1), show that the sum of the $\widehat{A}_i$ must be zero for any value of i, assuming n observations for each value of i.

2.2 Derive the equivalence given by Eq. (2.3) in Section 2.1.3.3, using the suggestion that was given below the equation.

2.3 Fill in the blanks in the following output for data from a completely randomized, one-factor design with three levels. Do you need to know whether the factor is fixed or random for completing the table? Why, or why not?

```
            One-way Analysis of Variance

            Analysis of Variance for Y
    Source    DF    SS      MS       F       P
    Factor    —     —       186.3    —       0.005
    Error     12    —       22.9
    Total     —     647.4
```

2.4 A completely randomized design was used and part of the ANOVA table is as follows.

Source	d.f.
Factor	3
Error	27
Total	30

(a) Explain why there could not have been an equal number of observations per factor level.

(b) Give one possible combination of the number of observations per factor level.

2.5 A study is to be undertaken to compare the coagulation times for samples of blood from 16 animals receiving four different diets, so that a set of four animals receive one diet.

(a) Does a completely randomized design seem appropriate for this experiment or would there likely be any extraneous factors that would have to be accounted for in the design?

(b) Note that each diet average would be computed from four observations. Does this seem adequate? Explain.

(c) Explain in detail how the experiment would be performed if a completely randomized design were used.

(d) What are the assumptions that must be made and explain how they would be tested when the data are analyzed?

2.6 Consider the following data for a completely randomized design with four levels of a fixed factor in a one-factor experiment:

1	2	3	4
17	16	16	19.6
18	20	19	18.6
15	17	14	23.6
19	18	18	17.6
13	14	14	22.6

Assume that you decide to analyze these data using both ANOVA and ANOM (with $\alpha = 0.05$), remembering that it is reasonable to use the two methods together. Are the assumptions that must be made for each the same, or do they differ? Do the assumptions appear to be met or is it even practical to test the assumptions with this amount of data? Explain why the two procedures produce different results. Since the results differ, which result would you go by? Explain.

2.7 In a completely randomized design with unequal group numbers, that is, $n_1 = 5$, $n_2 = 7$, and $n_3 = 6$, what is the degrees of freedom for the error term?

2.8 Construct an example for a single (fixed) factor with three levels and five observations per level for which the overall F-test shows a significant result but the averages for the three levels are 19.2, 19.3, and 19.5, respectively, and all 15 numbers are different.

2.9 The following data are available for a completely randomized design: $T_1 = 20$, $T_2 = 30$, and $T_3 = 40$, with T_i denoting the total of the observations for treatment i. In like manner, $n_1 = n_2 = n_3 = 5$. If the F-statistic for testing the equality of the three treatment means equals 4,

(a) What does SS_{total} equal?

(b) Would the null hypothesis be rejected for $\alpha = .05$?

2.10 Assuming the following data have come from a completely randomized design, compute the treatment sum of squares.

Treatment

1	2	3
4	6	1
3	5	4
5	5	4
4	4	4
4		2
		5

2.11 One of the sample datasets that comes with MINITAB is RADON.MTW. The dataset consists of 80 measurements of radiation in an experimental chamber. There were four different devices: filters, membranes, open cups, and badges and 20 devices of each type were used. The response variable is the amount of radiation that each device measured and the objective is to determine if there is any difference between the devices relative to the response variable. Can these data be analyzed using one or more of the methods given in this chapter? Why, or why not?

2.12 Consider the second example in Section 2.1.3. The conclusion was that the means do not differ, despite the fact that the sample means differ by far more than in the first example in that section. Apply the sliding t-distribution approach to that example. Do you also conclude that the population means do not differ? Explain.

2.13 Assume that a one-factor design is to be used and there are two levels of the factor.

(a) What is the smallest possible value of the F-statistic and when will that occur? What is the largest possible value?

(b) Construct an example with six observations at each level that will produce this minimum value.

2.14 Consider Exercises 2.3 and 2.9. Could an ANOM display be constructed for the data in the first exercise after the blanks have been filled in? Why, or why not? Could an ANOM display be constructed using the data summary in Exercise 2.9? Why, or why not? If either or both of the ANOM displays can be constructed from the information given, construct the display(s).

2.15 It was stressed in Section 1.7 that, ideally, processes should be in a state of statistical control when statistically designed experiments are performed. Assume that a manufacturing process was improved slightly with an eye toward improving the process. The change was made near the middle of the process and management wants to see if there has been any improvement. The product must go through one of two (supposedly identical) machines near the end of the process. An experiment is performed using both the standard process and the improved process, with product from each assigned to one of the two machines in a semirandom fashion such that each machine handles the same number of production units. Now assume that the older of the two machines malfunctions in such a way that is not obvious but does affect a key measurement characteristic.

(a) If the malfunction results in a higher-than-normal reading for the measurement characteristic, will this likely affect the conclusions that are drawn from the experiment if the reading is inflated by 20% and the variance is inflated by 5%? Explain.

(b) Would your answer be different if each machine received output from only one of the two processes? Explain.

(c) Could the semirandomization be improved so that true randomization is employed? If so, how?

2.16 (Harder problem) The raw data from a study are freqently summarized after collection with the consequence that data in terms of means and variances may be all that is available to an analyst. Assume that there is interest in comparing different types of paints in terms of drying times, with three types to be compared. Four walls in a building are painted with each type or paint, with the walls considered to be essentially the same. The average drying times in hours and the variances of the drying times are given below.

	Paint Type		
	1	2	3
Average	7.23	8.44	8.67
Variance	2.34	1.97	2.21

Perform an Analysis of Variance. What do you conclude and what would you recommend?

2.17 Assume that there is an equal number of observations, n, per level. Show that for k levels the second term on the right side of Eq. (2.3) is equal to the sum of the variances of the levels, multiplied by n. Then show that MS_{error} is equal to the average of the variances.

CHAPTER 3

Designs That Incorporate Extraneous (Blocking) Factors

Frequently, experiments are run in the presence of extraneous factors that can affect the value of the response variable. Unless the effect of these factors is accounted for, misleading results may occur. For example, if suppliers of a particular raw material are to be compared in terms of the effect on a measure of roundness, a supplier effect could be confounded with an operator effect if different process operators take part in the experiment.

If the effect of one or more such extraneous factors is anticipated, a design can be constructed that will isolate such factors and separate them from the error term so that they will not affect the outcome of the computations and testing for the factor(s) of interest.

In this chapter we assume that there is one factor of interest and one or more blocking factors, although there is no reason why two or more factors could not be used with the design that is given in the next section, for example.

3.1 RANDOMIZED BLOCK DESIGN

Assume that the experiment to compare the suppliers must be run in one day so as to minimize the disruption on the regular production. Assume further that there is a morning/early afternoon shift, a late afternoon/evening shift, and a night shift. Assume that a different operator will be on duty for each shift at a critical part of the process while the experimentation is performed. If there were three suppliers and the raw material from the first supplier were used during the first shift, the raw material from the second supplier used during the second shift, and the raw material from the third supplier used during the third shift, the supplier effect and a possible operator effect would be completely confounded.

Modern Experimental Design   By Thomas P. Ryan
Copyright © 2007 John Wiley & Sons, Inc.

The possible effect of the operator necessitates the use of a design with a *blocking factor.* If this is the only extraneous factor with which the experimenters are concerned, a *randomized complete block design* could be used. The word "complete" refers to the fact that every level of a factor (or every combination of factor levels if there is more than one factor) appears in each block, which are of equal size. We will occasionally refer to this as an RCB design for short. (Incomplete block designs are covered in Section 3.2.)

With this design, the extraneous factor is the blocking factor, and since there are three operators, there are three blocks. The general idea is to have experimental units that are more homogeneous within blocks than between blocks. Assume that there are also three suppliers. The number of levels of the factor of interest does not have to be the same as the number of blocks, however. Nine production units would thus be necessary for this experiment, and these production units would be the experimental units for the experiment. The raw material from the three suppliers would be assigned at random to the units within each block, and hence the name *randomized* block design. The design layout might appear as follows, with A, B, and C denoting the three suppliers.

Blocks

1	2	3
A	C	B
C	A	A
B	B	C

A sum of squares for blocks would be computed using the block totals, analogous to the way that the treatment (factor) sum of squares was shown to be computed in Section 2.1.3.3, as would the sum of squares for the suppliers since that is the treatment effect in this case.

Before proceeding further with this example, there are some questions that should be raised, and the assumptions for the design must also be considered. We can write the (typical) model for the design, analogous to the model for the completely randomized design in Eq. (2.1), as

$$Y_{ij} = \mu + A_i + B_j + \epsilon_{ij} \qquad i = 1, 2, \ldots, k \quad j = 1, 2, \ldots, t \qquad (3.1)$$

As in Eq. (2.1), A_i is the effect of the ith treatment and B_j is the effect of the jth block, with Y_{ij} denoting the observation on the ith treatment in the jth block and ϵ_{ij} the corresponding error term, which is assumed to have a normal distribution with a mean of zero and a constant variance of σ^2 for each i, j combination.

3.1.1 Assumption

One obvious difference between Eq. (2.1) and Eq. (3.1) is that there is no true experimental error term, since ϵ_{ij} is the error term for the ith treatment and the jth block. That is, there cannot be a true experimental error term unless the (i, j)th combination is repeated. (That *could* be done, but the resultant design would not be a conventional RCB design.) Instead, there is the assumption that there is no interaction between the treatment effect and the block effect. (Interaction is illustrated and discussed in detail in Section 4.2.) Simply stated, the assumption means that if we construct a scatterplot of blocks against treatments and appropriately connect the points to form lines that represent the response values for each block, the lines should be close to parallel. The crossing of lines, especially at sharp angles, would suggest that the no interaction assumption may not be valid. When this is the case, a formal analysis is not possible but the treatments might be ranked and compared within each block in an effort to extract some information from the data.

It is also generally assumed that the factor of interest is a fixed factor, whereas blocks are usually random (see Section 1.6.2.2 for a discussion of fixed and random factors). This means that the RCB model is a mixed model; that is, one factor is fixed and the other is random. Blocks might be fixed in some applications, however. Giesbrecht and Gumpertz (2004) have a lengthy discussion and some illustrations of blocks being random versus blocks being fixed. Somewhat similarly, even though the factor of interest is always assumed to be fixed, this does not have to be the case. For example, perhaps a large company is interested in the comparative performance of its machine operators, but doesn't want to test all of them. So a sample is selected and since the performance of each worker is suspected to depend on the age of the machine that is used, the workers are rotated (randomly assigned to) among the machines, which serve as the blocking factor.

If interaction between blocks and treatments is anticipated, either (a) the design should not be used, or (b) replication, as described above, should be used. Although it is good to be able to separate interaction from error if the former is feared and to be able to test for interaction, more than slight interaction will complicate the analysis, just as it complicates the analysis when designs presented in forthcoming chapters are used (see, e.g., Section 4.2.1). The main problem with a significant interaction is that the factor of interest (assuming a single factor) is tested against the error in the Analysis of Variance (ANOVA) table, with the error term representing both pure error and interaction, if the latter exists. A real interaction will thus inflate the error sum of squares, with the consequence that a significant factor effect may be declared not significant. As stated previously, the treatments in each block could be replicated if there were a concern about possible interaction, with the interaction then separated from error and tested against it. Again, however, this is not the way an RCB design is defined, but Exercise 3.26 contains an ANOVA table taken from the literature that has the two sources separated.

Relative to this experiment, the assumption means that, as a bare minimum, the ranking of the three suppliers and the separation between them (in terms of, say, closeness to a target roundness measure) are the same for each block (operator). That may or may not be a reasonable assumption. It would not be reasonable if one of the raw materials contained an ingredient to which one of the operators was slightly allergic, but the operator did not have such a problem with the other two raw materials, and the other two operators did not have a problem with any of the three raw materials.

If such a problem were to occur during the experiment, there would then be a true interaction, which should obviously not be used as the error term. In general, interaction terms should never be used as a substitute for the experimental error term if this can be avoided. If it is economically and practically feasible to use multiple observations for each (i, j) combination, then this should be done.

Another factor to consider that will usually be more important is the number of observations needed to detect an effect of a given size, using an appropriate approach. Since n is fixed in the usual (unreplicated) randomized block design as it is determined by the number of blocks, Eq. (1.3) could be used (if at least a rough estimate of σ is available) to see if n is large enough to detect an effect of at least a desired size. Generally n will not be large enough unless at least a moderate number of blocks is used. The number of blocks that can be used will be determined by the design scenario, and cannot be freely chosen. For example, for the experiment being discussed, the number of blocks is fixed at 3, since that is the number of operators that are involved.

Thus, the usual unreplicated randomized block design should generally be eschewed in favor of a replicated design, with the number of replicates determined appropriately. The current example can be used to support that recommendation. Even though Eq. (1.1) is flawed, we will, because of its simplicity, still use it as a starting point in our analysis of the current problem and see how it performs relative to other methods. With three blocks and three suppliers, from Eq. (1.1) we have

$$\Delta = \frac{4r\sigma}{\sqrt{n}}$$

$$= \frac{4(3)}{\sqrt{9}}\sigma$$

$$= 4\sigma$$

remembering that this is (purportedly) for a power of .90 and a significance level of .05. By comparison, if we use Russ Lenth's sample size calculator (http://www.stat.uiowa.edu/~rlenth/Power/index.html) and assume the use of Scheffé's method for comparing means, we find that the multiplier is 4.59, which does not differ greatly from the result obtained using Eq. (1.3), part of which is due to the difference in tests assumed.

This will be much too large a multiple of σ in a typical application; it would be necessary to use 12 blocks (four replicates) just to bring Δ down to 2σ, using both Eq. (1.1) and Lenth's applet. Thus, it should be apparent that there will erroneously be a failure to detect a significant difference in the means of a factor in many if not most applications of randomized block designs because of the number of blocks that are typically used.

A somewhat similar message is given by Dean and Voss (1999). They discuss a cotton-spinning experiment described by Peake (1953) that involved blocking. There were two treatment factors, "flyer" and "twist," with two levels of the first factor and four levels of the second factor. Two of the eight combinations were not observed and Dean and Voss (1999) subsequently converted this to, in essence, a one-factor design with six levels for the purpose of their analysis. Each experimental unit was the production of a full set of bobbins on a single machine with a single operator, and the experimenters decided to use an RCB design with each block representing the condition of a single machine, a single operator, and a single week. Of course, if there were, say, an operator effect and a machine effect, they would be confounded with this blocking scheme, but that would be a minor problem as the intent was to compare the levels of the factor(s) free of extraneous effects.

The experimenters wanted to be able to detect a true difference of at least 2 (breaks per 100 pounds of material).

The block size was chosen to be 6, since this is the number of observations that could be made on a single machine in a single week, and of course was also equal to the number of treatment combinations. The experimenters decided to analyze the data after the first 13 blocks had been run, which of course took 13 weeks.

Dean and Voss (1999) addressed the question of how many blocks should have been used, and we will also address this question, but will use a different approach to arrive at the answer. With the assumption of $\sigma^2 = 7$, their approach was based on Scheffé's simultaneous confidence intervals and required a small amount of trial and error to arrive at 40 blocks, so that the experiment would require 40 weeks to run. Using Lenth's sample size calculator (http://www.stat.uiowa.edu/~rlenth/Power/index.html) and also using the Scheffé method, we arrive at 75 blocks if we are satisfied with a power of .90. (Power was not specified directly in the example given by Dean and Voss (1999); we should also note that the familywise error rate is what is controlled at the selected value (e.g., .05) when Lenth's applet is used.) These differences notwithstanding, it is apparent that a very large number of blocks is required.

Dean and Voss (1999, p. 307) showed that after 13 runs, the minimum significant difference in the means that could be detected was 3.57—almost double the desired value. Since only six observations could be made in a week, the block size could not be increased. The point to be made here is that a very large number of blocks was needed to accomplish the experimenter's objective, and this number far exceeded the number of levels of the factor. Thus, a simple textbook layout of a randomized block design with the number of blocks equal to or at least approximately equal to the number of levels of the factor would have been totally inadequate.

3.1.2   Blocking an Out-of-Control Process

It may also be necessary to block because of a process being out of control, as discussed in Section 1.7. Box, Bisgaard, and Fung (1990) recommended that processes be brought into as good a state of process control as possible and then blocking be used. This would not create a randomized block design, however, because the blocking could not be performed ahead of time, because times when processes go out of control and then stabilize cannot be forecast. (If one knew well ahead of time when a process would likely go out of control, then such events could likely be prevented. That is, the factors that cause an out-of-control process would be known, so there would be no point in conducting an experiment for the purpose of identifying them.) Instead, a practical way to view this use of blocking would be to form blocks posts-experimentation and hope that there is balance within each block. It isn't easy to approximate the point at which a process has gone out of control, although some methods have been proposed for doing this.

3.1.3 Efficiency of a Randomized Block Design

The efficiency of an RCB design relative to a completely randomized design (CRD) has long been given in experimental design books. Efficiency figures in general are given as the ratio of two variances and in this case the ratio is the variance of a comparison of treatment levels without blocking divided by that variance when blocking is used. The efficiency expression is

$$\frac{(t-1)MS_{blocks} + t(k-1)MS_{error}}{(tk-1)MS_{error}}$$

If we write the fraction equivalently as

$$\frac{(t-1)MS_{blocks} + (tk-1)MS_{error} - (t-1)MS_{error}}{(tk-1)MS_{error}}$$

we can see that the ratio exceeds 1 if $MS_{blocks} > MS_{error}$. Since the F-test for testing whether there is a block effect is $F = MS_{blocks}/MS_{error}$, this means that the F-statistic will have to exceed 1.0, something that will frequently happen due to chance even when the blocking is not very effective.

There is a price that is paid if the blocking is ineffective, however, as the degrees of freedom for error is reduced when an RCB design is used. Specifically, with the CRD the error degrees of freedom is $n - k$, as shown in Section 2.1.3.3, whereas in an RCB design the error degrees of freedom is $n - k - (t - 1)$, with $n = kt$. Thus, the degrees of freedom for the RCB is smaller than for the CRD with the difference being the number of blocks minus one. When n is small, this reduction in the error degrees of freedom could cause a considerable loss of power to detect mean differences when blocking is not effective. Consequently, blocking should not be used injudiciously.

The main consideration, however, should be avoiding the wrong conclusion that could result from having a significant effect erroneously included in the error term. As with modeling in general, the emphasis should be on using a good model, and the randomized block model will often be the appropriate one.

3.1.4 Example

We consider the following experiment, described by Natrella (1963). Conversion gain, the ratio of available current-noise power to applied direct current power expressed in decibel units, is measured in six test sets, with four resistors used with each test set. The former served as blocks and the latter were the treatments. The data are in Table 3.1 with blocks and test sets numbered with consecutive integers, rather than using the identification numbers given by Natrella (1963).

TABLE 3.1   Dataset from Natrella (1963)

	Test Set					
Resistor	1	2	3	4	5	6
1	138.0	141.6	137.5	141.8	138.6	139.6
2	152.2	152.2	152.1	152.2	152.0	152.8
3	153.6	154.0	153.8	153.6	153.2	153.6
4	141.4	141.5	142.6	142.2	141.1	141.9

The first thing we can observe is that the blocking has been beneficial, since the conversion gain readings within each block are similar, but are dissimilar between blocks. We should plot blocks against treatments to check the interaction assumption, and the plot is given in Figure 3.1.

The configuration of points for the first resistor is curious and should perhaps lead to an investigation of the data for that resistor since this pattern does not exist for any of the other resistors. The overall configuration, however, is certainly acceptable as there is no strong indication of interaction.

The ANOVA table is given below. Computationally, the $SS_{resistor}$ (i.e., SS_{blocks}) is computed in the same general way as $SS_{test\ sets}$ is computed, that is, as $\sum_{i=1}^{4} B_i^2 - (GT)^2/24$, with B_i denoting the total of the ith block and GT denoting the

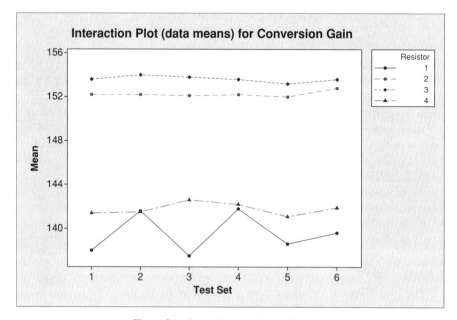

Figure 3.1   Interaction plot for Natrella data.

total of all the observations. The other computations are the same as those for a CRD.

```
        Analysis of Variance for Conversion Gain

    Source      DF      SS       MS       F       P
    Resistor     3   927.66   309.22   344.40   0.000
    Test Set     5     5.60     1.12     1.25   0.336
    Error       15    13.47     0.90
    Total       23   946.73

      S= 0.947555    R-Sq = 98.58%    R-Sq (adj) = 97.82%
```

We can see that the blocking was clearly effective, as evidenced by the large *F*-value for blocks (resistors) and by the fact that almost all of the variability in the conversion gain values is explained by blocks. The treatment factor (test set) is not close to being significant, but is closer than it would have been if blocking had not been used.

Curiously, Natrella (1963) stated, "We are interested in possible differences among treatments (test sets) and blocks (resistors)." Normally the only interest in blocks is to see if the blocking was effective, with there being only one factor of interest, the treatment effect, when an RCB design is used.

This raises the question of how the experiment should have been conducted if there had been two factors of interest. Designs for two or more factors are discussed in Chapter 4, but we will remark here that the randomization would have been different.

As stated previously, blocks are generally random in an RCB design, although fixed blocks are also used. In this application it may be appropriate to regard the blocks (resistors) as random, although that isn't clear. The statistical analysis is the same for both cases; only the inference is different, with the inference extending to a population of resistors if resistor is a random factor and to only the resistors used in the study if resistor is a fixed factor.

3.1.4.1 Critique

This is an example of an experiment for which replication was surely necessary, because the hypothesis of equality of six population means was tested with only four observations used to compute each sample average. Although there are enough degrees of freedom for estimating the error variance, the power might not be very good for detecting a mean difference that would be of practical significance. Whether or not this experiment should have been replicated by using either more blocks or multiple observations within blocks depends on the smallest difference in the treatment means that the experimenters wanted to detect, and there is no mention of this.

Surely the failure to recognize the need for adequate replication and the failure to determine how much replication is needed in a given experiment are among the major misuses of experimental designs.

If the hypothesis of equal treatment effects had been rejected, there would have been a need to determine which treatment levels differ. As with a CRD, the user can choose between a multiple comparison procedure, the sliding reference distribution approach mentioned in Section 2.1.3.1, or Analysis of Means (ANOM).

3.1.5   ANOM

As stated by Nelson (1993), ANOM can be used with any complete design. The only additional assumptions beyond those for ANOVA are that the effect estimates must be equicorrelated with correlation $-1/(k-1)$ for k means to be compared and the factor(s) must be fixed. Clearly the equicorrelation assumption is met for an RCB design since the effect estimators for the nonblocking factor have the same equicorrelation structure as in a CRD.

Examples of the ANOM used in conjunction with an RCB design are scarce in the literature; one example is given by Nelson, Coffin, and Copeland (2003, p. 330), and another example is given by Nelson, Wludyka, and Copeland (2005, p. 95). The use of ANOM with an RCB design presents no real complications or new considerations, however, as the means are plotted as in a CRD, with decision lines obtained from Eq. (2.2). Of course, blocks are considered to be random with an RCB design, so ANOM could not be applied to the blocks unless the blocks are fixed. Block analysis is generally not of any great interest, however. We would simply like to see evidence that the blocking has been beneficial after the analysis using blocks has been performed.

There are complications, however, in trying to use software to construct ANOM displays for this type of design. For example, MINITAB cannot be used to construct an ANOM display for an RCB design because it treats the data as having come from a two-factor design with interaction and there is no option for suppressing the two-factor interaction (which of course is assumed to not exist with the RCB design) and using that interaction as the error term.

Therefore, a MINITAB macro would have to be written (essentially starting from scratch) and programming would also have to be performed if SAS were used to construct an ANOM display.

Although not a suitable substitute for a graph, the test set means can be simply compared against the decision lines using Eq. (2.4), although that is hardly necessary here since the ANOVA showed the test set factor to be well short of significance. Nevertheless, we will illustrate the computations:

$$\bar{\bar{x}} \pm h_{\alpha,k,\nu}\, s\sqrt{(k-1)/(kn)} = 146.8 \pm h_{.05,6,15}\sqrt{0.90}\sqrt{5/(6*4)}$$

$$= 146.8 \pm 2.97(0.433)$$

$$= (145.514, 148.086)$$

The test set means are 146.30, 147.33, 146.50, 147.45, 146.23, and 146.98, all of which are well within the decision limits, as expected.

3.2 INCOMPLETE BLOCK DESIGNS

There will usually be physical constraints that will make it impossible to use all of the treatments in each block. For example, if days are blocks and different treatments are run through an industrial furnace in a manufacturing experiment, the furnace may not be capable of handling all the different treatments in one day. Although we usually associate physical and industrial experiments with the use of incomplete block designs, the designs are actually used in a wide variety of applications, including education (see, e.g., van der Linden, Veldkamp, and Carlson, 2004). Incomplete block designs are also often used in forestry, specifically in mixedwood and silviculture systems studies, for which it can be difficult to find blocks that are large enough to accommodate all the treatments. Incomplete block designs are also used in diallel cross experiments (Singh and Hinkelmann, 1999) because homogeneous experimental units cannot be achieved when there is at least a moderately large number of crosses. (The diallel cross is a cross-classified mating design in which a specified number of inbred lines that serve as male parents are crossed with the same lines that serve as female parents. Partial diallel cross experiments are covered in Section 12.9 of John, 1971.)

When incomplete block designs are used, they should ideally be balanced. For example, if there are six treatments but only four could be used in each block and there are six blocks, with the blocks of equal size, we would want each treatment to appear the same number of times in the six blocks combined, and pairs of treatments to appear in blocks the same number of times. If these requirements are met, the design is a *balanced incomplete block* (BIB) *design*.

3.2.1 Balanced Incomplete Block Designs

The model for a BIB design is essentially the same as the model for an RCB design. The model must differ slightly since not all block–treatment combinations are used in the design. Therefore although the model is

$$Y_{ij} = \mu + A_i + B_j + \epsilon_{ij} \qquad i = 1, 2, \ldots, k \quad j = 1, 2, \ldots, t \qquad (3.2)$$

which is the same as Eq. (3.1) with the components of the equation thus defined the same way, the model does not apply to all (i, j) combinations, since only a subset of them are used in the experiment. Therefore, the model applies only for the (i, j) combinations that were used.

As with an RCB design, there is the assumption that there is no treatment–block interaction. (We will later see the difficulty in trying to test this assumption, however.)

To illustrate the construction of a BIB design, assume that there are four treatments to be run in blocks but only two treatments can be used in each block. How many blocks are needed for the design to be a BIB design? Since $\binom{4}{2} = 6$, the answer is 6 or a multiple of 6 if more than six blocks could be used. With six blocks, each pair of treatments would occur in a block once (i.e., the design is *balanced* regarding pairs of treatments), as is shown below with A, B, C, and D denoting the treatments.

Blocks

1	2	3	4	5	6
A	C	A	A	B	B
B	D	C	D	C	D

This is easy to see but the construction problem is more difficult when there are more than two treatments per block, as then the "pairs" cannot be visualized quite so easily.

Furthermore, it is not always possible to construct a BIB design for a given number of treatments and a given block size. For example, consider Table 3.1 and assume that a block size of 6 is not possible, with 5 being the largest possible block size. Obviously 5 won't work, however, because we won't have six treatments occurring an equal number of times with 20 observations, since it is not a multiple of 6. We would need a minimum of six blocks, not four, for this balance requirement to be met, but the requirement that pairs of treatments occur the same number of times over the blocks would not be met. It would be necessary to drop down to a block size of 4 and use 15 blocks in order for both balance requirements to be met, with each treatment occurring 10 times and every pair of treatments occurring 6 times. This is undoubtedly not intuitively apparent, however, nor should it be.

As with the other designs that are presented in this chapter and with experimental designs in general, we should be mindful of the number of observations that each treatment level mean is computed from, as well as the number of degrees of freedom for the error term. These considerations will generally mandate the use of replicates of the designs in this chapter. We should also be mindful that BIB designs cannot always be constructed for a given number of treatments, replicates of treatments, and block size, as is illustrated in Section 3.2.2.

3.2.1.1   Analysis
The analysis of data from BIB designs is slightly more involved than the analysis of data from RCB designs. As a simple example, consider the BIB design given in the preceding section, with response values as indicated.

Blocks

1	2	3	4	5	6
A(8)	C(7)	A(10)	A(6)	B(10)	B(5)
B(4)	D(6)	C(9)	D(6)	C(5)	D(7)

One thing that should be immediately apparent is that we cannot easily separate a block effect from a treatment effect since all treatments do not appear in each block. For example, the total for the third block is more than 50% higher than the total for the first block. Is this because extraneous factors affected the two blocks differently, or does it mean that treatment C has a more pronounced effect on the response than does treatment B? This is, and should be, disturbing because designs in general are not blocked unless the block totals would be expected to differ more than slightly.

It seems as though the treatment totals should be adjusted for differences in the block totals, and the block totals should be adjusted for differences in the treatment totals, which would lead to circular reasoning. As with an RCB design, our interest in the block effect is solely to tell us whether or not blocks should have been used in the experiment; our primary concern is the treatment effect.

The adjustment is made as follows. Let $W_i = 2T_i - B_{(i)}$, with "2" representing the block size, T_i the total for the ith treatment, and $B_{(i)}$ the total of all blocks that contain the ith treatment. (The reason for the "2" is that we are summing only half of the observations in the blocks that contain the ith treatment, whereas $B_{(i)}$ is the total sum of the observations.) For example, $W_A = 2(24) - 43 = 5$. Note that this number is the sum of the differences between each A and the other treatment that is in each block with A. The other adjusted totals are -1, -5, and 1, for B, C, and D, respectively. Note that the sum of the adjusted treatment totals is zero.

This can be explained as follows. Note that $\sum T_i = GT$, the sum of all the observations, so $\sum 2T_i = 2GT$. The $\sum B_{(i)}$ also equals $2GT$ because the block size is 2 and the design is balanced, so that every block total is used twice. The adjusted treatment sum of squares is then obtained as $\sum W_i^2 / kt\lambda$, with k and t as previously defined and λ denoting the number of times each pair of treatments occurs together in blocks, which in this case is 1. Thus, $SS_{\text{treatments(adjusted)}} = 52/8 = 6.5$. Notice that the correction factor $(GT)^2/12$ is not used because a correction is being made in the computation of the $SS_{\text{treatments(adjusted)}}$.

The other, unadjusted, sums of squares are computed in the usual way, producing the ANOVA table given below.

```
General Linear Model: Y versus Treatments, Blocks

Factor          Type    Levels          Values
Treatments      fixed     4        1, 2, 3, 4
Blocks          fixed     6        1, 2, 3, 4, 5, 6

Analysis of Variance for Y, using Adjusted SS for Tests

Source                  DF     SS       MS      F       P
Treatments (adjusted)    3    6.500   2.167   0.38    0.775
Blocks (unadjusted)      5   19.417   3.883   0.685   0.669
Error                    3   17.000   5.667
Total                   11   42.917
```

The analysis shows that neither blocks nor treatments are significant. This is the usual form of the analysis, although it is potentially misleading regarding the block effect since blocks have not been adjusted. The blocks component must be adjusted to obtain the proper test for blocks. This can be accomplished very easily simply by reversing the order in which the model is specified. For example, in MINITAB the treatment sum of squares was adjusted because the treatment term was the first term specified in the model. Since the first term entered is the one that is adjusted, blocks simply have to be entered first. Doing so produces the following output.

```
Analysis of Variance for Y, using Adjusted SS for Tests

Source                      DF    SS       MS       F       P

Blocks (adjusted)            5    20.333   4.067    0.72    0.653
Treatments (unadjusted)      3     5.583   n.a      n.a     n.a
Error                        3    17.000   17.000   5.667
Total                       11    42.917
```

It can be observed that blocks and treatments are adjusted by the same amount, and in fact it can be shown that

$$SS_{blocks(adjusted)} - SS_{blocks(unadjusted)} = SS_{treatments(adjusted)} - SS_{treatments(unadjusted)}$$

This relationship would be an aid in hand computation, although that of course is not recommended other than to become acquainted with what underlies the numbers obtained in the ANOVA table.

3.2.1.2   *Recovery of Interblock Information*

Blocks are usually considered to be random, although this will not always be the case since in many experiments the blocks that are formed are the only ones that can be formed. When this is the case, the analysis given in the preceding section, which has been termed the *intrablock analysis*, is appropriate. When blocks are random, however, the experimenter has the option of performing an *interblock analysis*. A complete explanation of the latter would be involved and lengthy, and so will not be given here. Perhaps the simplest and most lucid explanation of an interblock analysis is given by Montgomery (2005), which includes a comparison of the treatment effect estimates for both the interblock and intrablock analysis for a particular example. (See also Cochran and Cox (1957, p. 382) and Johnson and Leone (1977).

3.2.1.3   *ANOM*

Since a BIB design is, as the name indicates, an incomplete design, the correlation structure of the estimators of the treatment means must be determined to see if ANOM can be applied to data obtained from the use of such designs. Nelson (1993) showed that the set of estimators of the treatment means does have the requisite correlation structure. Those estimators do not have the form of the estimators of the treatment means in a CRD or RCB design, however, so Eq. (2.5) cannot be used to obtain the decision lines.

As the reader might surmise, this is due to the fact that the treatment totals are adjusted, as was illustrated in Section 3.2.1.1. Nelson (1993) gave the decision lines for a BIB design as

$$0 \pm h_{\alpha,I,\nu} s \sqrt{\frac{I-1}{IJ(b-1)}} \qquad (3.3)$$

with I = the number of treatments, J = the number of blocks, b = block size, and $s = \sqrt{\text{MS}_{\text{error}}}$. Although Nelson (1993) didn't write the expression in this form, when written in this manner we see the close similarity to Eq. (2.5), as the only difference (other than the different symbols) is the $b - 1$ in the denominator of the fraction.

The form for the decision lines given in Eq. (3.3) results from the fact that the form of the estimators of the A_i is given by

$$\widehat{A_i} = \frac{(I - 1)}{J(b - 1)} \left(w_i - \frac{T_i}{b} \right)$$

where I, J, and b are as previously defined, w_i is the ith treatment total, and T_i is the sum of the block totals in which the ith treatment appears.

As with the RCB design, however, there is no software that will directly produce a BIB ANOM display. Therefore, any such display would have to essentially be produced manually, using Eq. (3.3) to compute the decision limits and computing the average for each treatment (i.e., level of the factor) and displaying the averages on the graph analogous to Figures 2.2 and 2.3.

The use of ANOM for BIB designs is discussed in more detail, with examples, in Nelson et al. (2005, Section 6.3).

3.2.2 Partially Balanced Incomplete Block Designs

Partially balanced incomplete block (PBIB) designs are an obvious, less restrictive, alternative to BIB designs. They are also almost a necessary alternative when the latter cannot be constructed for a given combination of treatments and blocks. (Another, less known alternative is to seek an A- or D-optimal design, using methods such as those given in Reck and Morgan, 2005.) For example, assume that there are six treatments that are to be repeated four times in blocks of size 4, which would obviously require six blocks. To be balanced, each pair of treatments occurs in a block the same number of times, λ, with the latter defined as $r(k - 1)/(t - 1)$, with r denoting the number of repeats of each treatment, k the block size, and t the number of treatments. Of course, λ must be an integer but here $\lambda = 4(3)/5$, which is not an integer. Clearly, either r or $k - 1$ must be a multiple of $t - 1$ for this example. Although a block size of 5 might be feasible, it might not be possible to use a block size of 10. Similarly, $r = 5$ would not be possible with a block size of 4 for any number of blocks.

A PBIB design has at least two values of λ, such as λ_1 and λ_2, with some pairs of treatments occurring λ_1 times and the other pairs occurring λ_2 times.

Although these designs are less restrictive than BIB designs, they are also less efficient. Specifically, for a BIB design the variance of the difference of two treatment effect estimates will be the same for all pairs of treatments. For a PBIB design with two values of λ, there will be two variances and the average of the two variances is higher than the variance for the BIB design, so the latter is a more efficient design.

PBIB designs are covered in considerable detail in Hinkelmann and Kempthorne (2005), and Chapter 12 of John (1971) is devoted to the subject. The reader is referred to these sources for detailed information on the construction of these designs. Because

BIB designs and PBIB designs cannot be constructed with most statistical software, the use of catalogs of these designs is highly desirable. One such catalog was given by Raghavarao (1971) and an extensive listing of PBIB designs with two associate classes was given by Bose, Clatworthy, and Shrikhande (1954). See also Sinha (1989). Tables of BIB designs were given by Cochran and Cox (1957) and more recently by Colbourn and Dinitz (1996). Nineteen BIB designs were given in Natrella (1963), but unfortunately that has long been out of print.

A recent and most extensive list of BIB designs is given in Hinkelmann and Kempthorne (2005), which lists all known BIB designs and also lists many PBIB designs.

3.2.2.1 Lattice Design

Lattice designs were introduced by Yates (1936) for use in large-scale agricultural experiments and were originally known as quasi-factorial designs. A lattice design, which is a special type of incomplete block design, is related to a Latin square design (covered in Section 3.3) in the sense that two orthogonal Latin square designs (defined in Section 3.4) constitute a simple lattice design. There are different types of lattice designs, but a balanced lattice draws its name from the fact that the treatment numbers can be written at the intersections of lines that form a square lattice. Using results from Kempthorne and Federer (1948), John (1971, p. 262) showed that a lattice design is more efficient than an RCB design that uses the same experimental material. The extent of the efficiency depends, however, on the values of certain variance components, which of course are unknown.

There are various types of lattice designs, including balanced lattices, partially balanced lattices, rectangular lattices, and cubic lattices. A balanced lattice is restrictive in that the number of treatments must be an exact square.

An actual "home improvement" example of a lattice design is given later in Section 3.3.5. Lattice designs are discussed in detail in Chapter 18 of Hinkelmann and Kempthorne (2005), Chapter 10 of Cochran and Cox (1957), and in Federer (1955).

3.2.3 Nonparametric Analysis for Incomplete Block Designs

The analysis of data from most experimental designs is problematic because there generally won't be enough data available to test the assumptions. Consequently, as with statistical procedures in general, it is desirable to perform a nonparametric analysis whenever possible and compare the results with the parametric analysis. Skillings and Mack (1981) gave a nonparametric analysis procedure for data from general incomplete block designs, which was discussed in detail by, for example, Giesbrecht and Gumpertz (2004), and the reader is referred to these sources for information on the approach.

3.2.4 Other Incomplete Block Designs

A class of incomplete block designs known as α-designs was introduced by Patterson and Williams (1976) and found some favor with experimenters. These designs can be constructed with the Gendex software (see, e.g., http://www.designcomputing.net/gendex).

3.3 LATIN SQUARE DESIGN

Often there will be a need to protect against the possible effect of more than one extraneous factor. For example, in an agricultural experiment there may be soil variation both north/south and east/west relative to a plot of land. There may also be multiple extraneous factors in industrial applications. As discussed in Section 3.1.2, blocking will often be necessary because of processes being out of control. If it is necessary to block on two process characteristics, then a Latin square design could be used. In general, such a design is used when there is a single factor and there are two suspected extraneous factors, the effect of which must be isolated and separated from the error term.

A Latin square, introduced by the famous mathematician Euler in 1783, is as the name implies, a square, an example of which is the following.

```
A   B   C
B   C   A
C   A   B
```

The rows would represent one extraneous factor and the columns the other one, with this 3×3 Latin square thus having three levels of each of the suspected extraneous factors and three levels of the factor of interest, A, B, and C. Notice that each letter occurs once in each column and once in each row, as the design would be unbalanced if this were not the case.

There are 12 ways to construct a 3×3 Latin square such that this requirement is met, as the reader is asked to show in Exercise 3.22. There is only one standard Latin square of this size, however, with a standard Latin square being one that has the letters in the first row and first column in alphabetical order. This should be apparent upon careful inspection of the Latin square given above.

Many of the designs given in this book have been invented in recent years, and virtually all of the designs have been invented within the past 100 years. The Latin square design is an exception, however. As explained by Freeman (2005), the first use of the design (under a different name) may have occurred in 1788, as a French agriculturalist named Cretté de Palluel reported on an experiment that involved the comparison of different types of food for sheep. Sixteen sheep, four from each of four breeds, were fed such that each sheep from each breed received one of four different types of food. The sheep were then weighed at four different times during the winter. Thus, the blocking variables were breed and time. We generally don't think of time as a blocking variable, however, unless the times are close together such as morning and afternoon or consecutive days. So, saying that this is a Latin square design is stretching things a bit, as one could contend that this is a design with a single blocking variable and multiple response variables corresponding to the four measurement times during the winter.

Since the designation of the levels of the factor (i.e., the letters) is arbitrary, it has been claimed (e.g., Giesbrecht and Gumpertz, 2004, p. 120) that it is necessary to put Latin squares in standard form to see if they really differ, which is why Latin squares are generally written in standard form, at least before any permuting of rows or columns occurs, which is recommended to ensure randomization, as this is the only type of randomization that can be performed. For example, the design

$$
\begin{array}{ccc}
A & C & B \\
B & A & C \\
C & B & A
\end{array}
$$

becomes the same as the previous design if the second and third columns are switched.

It shouldn't make any difference which of these Latin square designs is used if the assumption of no interaction between rows and columns is met. (This assumption is discussed in detail later in the next section.) That assumption won't be met exactly in many applications, however, and then it will make a difference which design is used. Furthermore, the usual recommendation is to take a selected Latin square design and randomly permute rows and columns, then use the design that results from the rearrangements. The method of O'Carroll (1963) might also be used.

Since the assumption of no interaction probably won't be met exactly, it seems highly questionable to say that Latin squares are equivalent if the rows or columns of one design can be permuted so as to create the other design.

The number of Latin squares that are clearly different when in standard form grows rapidly with the order of the square if we think of the order increasing. For example, there are four such squares of order 4, 56 of order 5, and so on.

One obvious restriction of Latin squares, which could be a hindrance, is that the number of levels of all three factors must be the same in order to have a square.

Although selecting a Latin square at random might seem to be a reasonable strategy, this isn't necessarily true. Copeland and Nelson (2000) showed that for Latin squares whose order is 2^k, there is *one* standard square that corresponds to a 2^{3k-k} fractional factorial design, with "2" denoting the levels of each factor, $3k$ denoting the number of factors, and the second k denoting the degree of fractionization. That is, a 2^{3k-k} design is a $1/2^k$ fraction of a 2^{3k} design. (Fractional factorial designs are covered in Chapter 5, as is the relationship between these designs and certain Latin square designs.) From this it follows that not all Latin squares of a given order have equivalent properties.

3.3.1 Assumptions

Another restriction is the assumption that there is no interaction between rows and columns. How often this assumption is violated in practice is conjectural since the true state of nature is never known, but Box, Hunter, and Hunter (2005, p. 160) state

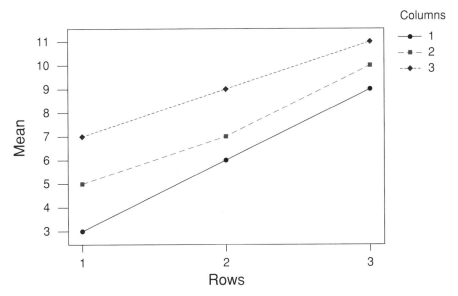

Figure 3.2   A desirable relationship between rows and columns for a Latin square design.

that a Latin square design "has frequently been used inappropriately to study process factors that can interact." Similarly, Nelson et al. (2005, p. 134) stated "... using a Latin square design if the assumption of no interactions were reasonable. Unfortunately, in most physical science and engineering experiments that is not the case, and Latin square designs should be used with great caution since the presence of interactions can invalidate the results."

The assumption means, for example, that if a plant experiment were run on different machines and using different operators, and using these as the row and column blocking factors in an experiment to compare processing methods, that there is no interaction between operators and machines such that, say, one operator seems to work much better on one particular machine than on the other machines. Similarly, for an agricultural experiment the assumption is violated if a noticeable difference in soil fertility occurs in the upper right corner of the design area, or if there is some other irregularity such that a scatterplot of soil fertility, using rows and columns, deviates greatly from parallel lines when the points are connected, as in the graph given in Figure 3.2, for which we might regard soil fertility as the "Response" in that graph.

Another assumption is that there is no interaction between treatments and either rows or columns. This means that if we replaced Rows or Columns in Figure 3.2 with Treatments, we would also have lines that are essentially parallel. (Practically speaking, parallel lines, which would correspond to zero interaction, would not be likely to occur, just as it is unlikely that a parameter estimate could be equal to the corresponding parameter value.) Jaech (1969) illustrates the consequences of interaction in a Latin square design, which is seen by using expected mean squares. (The latter are discussed in detail in Appendix D to Chapter 4.) The effect of interaction between rows and columns is that the error term is of course inflated, with the consequence that a significant treatment effect may not be detected. See Wilk and Kempthorne (1957) for a detailed discussion of the effect of interactions in a Latin square design.

3.3.2 Model

The model for an $n \times n$ Latin square design is similar to that of a randomized block design, the only difference being that there is an additional blocking variable, so there is an additional term in the model, which is

$$Y_{ijk} = \mu + A_i + R_j + C_k + \epsilon_{ijk} \quad i = 1, 2, \ldots, t \quad j = 1, 2, \ldots, t \quad k = 1, 2, \ldots, t$$

$$(3.4)$$

with R_j denoting the effect of the jth row, C_k representing the effect of the kth column, and as for a randomized block design, A_i denoting the effect of the ith treatment. The three subscripts on Y represent the fact that each observation is classified by treatment, row, and column.

3.3.3 Example

A classic example, one that has been used by many writers, is the design of an experiment to compare brands of tires in a driving test, with four cars, four drivers, and four tires of each of four brands available for the experiment. How should the experiment be designed? Obviously all four tires of one brand should not go on each car as then the differences between the brands could not be separated from the car differences. Similarly, each brand should not be solely assigned to wheel position as then brand differences would be confounded with differences due to wheel position.

For the moment we will assume that there is no interest in trying to separate the car effect from the driver effect, then we will address this issue in Section 3.4 in the context of Graeco–Latin squares. By not trying to separate the two (possible) effects, we will actually be measuring the sum of their effects, similar to what happens when fractional factorial designs are used, which are presented in Chapter 5.

With two blocking variables we can use a Latin square design, or perhaps use sets of Latin squares, as is discussed in Section 3.3.5

Assume for the sake of illustration that a single Latin square is to be used, with the following being the one that is randomly chosen (Table 3.2). The numbers in parentheses denote a coded measure of tire wear.

TABLE 3.2   A 4 × 4 Latin Square

Wheel Position	Cars			
	1	2	3	4
LF	A(13)	B(6)	C(12)	D(19)
RF	B(9)	A(9)	D(11)	C(19)
LR	C(9)	D(11)	A(9)	B(8)
RR	D(7)	C(9)	B(3)	A(12)

The analysis of the data is as follows.

```
          Analysis of Variance for Tread Wear

Source              DF        SS        MS       F       P
Brands               3     85.250   28.417    7.41   0.019
Cars                 3     92.250   30.750    8.02   0.016
Wheel Position       3     61.250   20.417    5.33   0.040
Error                6     23.000    3.833
Total               15    261.750


S = 1.95789      R-Sq = 91.21%   R-Sq(adj) = 78.03%


Means for Tread Wear

Brands    Mean    SE Mean
1        10.750   0.9789
2         6.500   0.9789
3        12.250   0.9789
4        12.000   0.9789
```

Each of the first three sums of squares in the ANOVA table is computed analogous to the way that the treatment sum of squares is computed for a CRD. That is, for this example SS$_{\text{brands}}$ is computed as $\sum_{i=1}^{4} B_i^2/4 - (\sum_{i=1}^{4} B_i)^2/n$, with B_i denoting the total of the observations for brand i, again invoking the general rule that whenever a number is squared in a computing formula, the squared number is divided by the number of observations that add to the number that is squared. As with a CRD, it can be shown that this expression is equivalent to $4\sum_{i=1}^{4}(\bar{B}_i - \bar{B}_{\text{all}})^2$, with $\bar{B}_i$ denoting the average of the observations for the ith brand, and $\bar{B}_{\text{all}}$ denoting the average of those averages. Since each average is computed from the same number of observations, this is the same as the average of all of the observations.

The sums of squares for Cars and Wheel Position are computed in the same general way, and as stated in Section 2.1.3.3, the computation of the total sum of squares is not design dependent.

The degrees of freedom is $n - 1$ for each of the three factors, and $(n - 1)(n - 2)$ for the error term, with "total" of course having $n^2 - 1$ degrees of freedom.

The analysis shows that 91 percent of the variability in tread wear is accounted for by the model with brands and the two blocking variables. The relatively small p-value for brands shows that there is apparently a difference between brands. Note, however, that the mean for the second brand is far less than the means for the other brands, and there is also a moderate difference between the first and third brand. Yet, the p-value of .019 is not extremely small, which is due to the fact that each average is based on only four observations. (The standard error for each mean of 0.9789 that is given in the output is $\sqrt{MS_{error}/4} = \sqrt{3.833/4} = \sqrt{0.958}$.)

The analysis also shows that it was important to use both blocking variables as each is significant at a .05 level of significance (i.e., each p-value is less than .05). Obviously the wheel position is "fixed," as there are four wheel positions on a car, but we would logically regard the cars as a random factor. For a Latin square design the manner of analysis does not depend on whether a blocking factor is fixed or random, although this does matter when various other types of designs are used.

Multiple comparison tests can be used on data from a Latin square design—and the user has the same general options as with the designs given previously—but in this case that is clearly unnecessary as the last part of the output shows that the average tread wear for brand "B" is much less than the average tread wear for the other three brands, which differ very little, so it is obvious why there was a significant result from the F-test.

If, for example, an experimenter wanted to use ANOM, Eq. (2.4) would again be appropriate; it is simply a matter of identifying and specifying the number of observations from which each average is computed, in addition to using the appropriate constant as determined by the number of means that are being compared, the degrees of freedom for the error term, and the significance level. Specifically, for $\alpha = .05$

$$\bar{\bar{x}} \pm h_{\alpha,k,v}\, s\sqrt{(k-1)/(kn)} = 10.375 \pm 3.31\sqrt{3.833}\sqrt{3/(4)(4)}$$

$$= 10.375 \pm 2.81$$

$$= (7.565, 13.185)$$

The average for the second brand is below 7.565, so we conclude that the mean for that brand is different from the average of all four of the means and the second brand is thus preferable.

Assume that the information on blocking was lost after the experiment was conducted so that only the tread wear figures for each brand were available. Preferably the experiment should be repeated, but assume that a data analyst not knowing that the data were from a Latin square design analyzes the data as having come from a CRD. We can see what the conclusion would have been by adding the sums of squares and degrees of freedom for the two blocking factors to the error degrees of freedom and error sum of squares. It is obvious that doing so would produce a value for the F-statistic of approximately 2, which would result in a p-value well in excess of .05. Thus, a wrong conclusion would have been drawn.

In addition to industrial applications, Latin square designs have been used extensively in medical applications, as indicated by Armitage and Berry (1994). Therefore, despite the somewhat gloomy picture painted at the end of Section 3.3.1 regarding the likelihood of Latin square assumptions being met, the design has been used successfully in various applications.

3.3.4 Efficiency of a Latin Square Design

Whenever blocking is used with any design, the blocking factor(s) should have a significant effect. Otherwise, the degrees of freedom for the error term is unnecessarily reduced, resulting in a more variable estimator of σ_ϵ^2 than should be the case and reducing the capability of the design to detect differences that are deemed significant. If neither blocking factor in a Latin square design is significant, then the Latin square design is a very inefficient design.

   No general conclusion can be drawn about the efficiency of a Latin square design relative to any other design because the efficiency is obviously application- and data dependent. Cochran and Cox (1957, p. 127) gave an expression that would be used for estimating what the error mean square would have been if the blocking variable represented by the rows had not been used, so that the design would have been an RCB design. A similar expression could be derived if the factor represented by the columns had not been used.

3.3.5 Using Multiple Latin Squares

One potentially serious problem when a single Latin square is used is that the number of observations made at each factor level is equal to the number of levels. We should expect this to almost always be an inadequate number to detect a difference between levels of a factor that would be deemed significant, especially when there is only a small number of levels.

   This also presents a problem relative to the normality assumption, as that assumption cannot be practically tested when each level mean is computed from only three or four observations.

   To illustrate the power problem for the tread wear example given in the preceding section, assume that the experimenter wants to be able to detect a difference in mean tread wear of at least 2 units. With $\hat{\sigma} = 1.96$ in that example, Lenth's calculator (see Section 1.4.4) shows that the probability of detecting a difference of that size is only .23, with the familywise error rate for a set of t-tests controlled at .05.

   This problem can be remedied by using multiple Latin squares. Consider the example in Section 3.3.3, with each average that was displayed in the output computed from only four observations since it was a 4×4 Latin square. If two other Latin squares of that size had been randomly selected and combined with the first one, with different rows and columns, the averages would have been computed from 12 observations, which would be much better. Furthermore, the error degrees of freedom is increased to 18, the number of degrees of freedom for the single Latin square times the number of squares.

   This increases the power to .68; using four Latin squares of this size produces a power of .81 and five Latin squares gives a power of .88. An experimenter should use some computing device to obtain these numbers and determine how many Latin squares to use after looking at the power numbers.

The way that degrees of freedom are allocated to the components of the ANOVA table depends upon whether or not there is any relationship between the squares when multiple squares are used.

Example 3.1

To illustrate this, Jaech (1969) gave a nuclear production reactor example involving process tubes, with 10 tubes used and 20 positions on each tube used for test purposes. The 200 tube–position combinations were to be utilized in eight 5×5 Latin squares. The first five tubes were used in square #1 and the second five tubes were used in square #6, but they utilized the same five tube positions, with five different tube positions used for each of squares 2 and 7, 3 and 8, and so on. With this in mind, Jaech (1969) gave the degrees of freedom breakdown for this experiment as follows.

Source	df
Squares	7
Treatments	4
Positions	16
Tubes	8
Error	164
Total	199

The df for squares and for treatments should be obvious since there were eight squares and five treatments, respectively, but the other df may not be obvious. The df for positions is obtained as four degrees of freedom for each of positions 1–5, 6–10, 11–15, and 16–20. The df for tubes consists of four df for each of tubes 1–5 and 6–10. These df should be intuitively apparent when we remember that each Latin square is 5×5. The total df must of course be $200 - 1 = 199$, regardless of how the squares are constructed.

If the squares were unrelated, with randomization performed for each square, there would be no degrees of freedom for positions and tubes. Instead, there would be, as we would expect, 32 df for rows and 32 df for columns, this resulting from 4 df for rows and columns from each of the eight squares. This results in 124 df for the error term—much less than the 164 df under the assumption that the squares are related.

A much smaller df for error will result if the treatment × square interaction is isolated. Although interactions are assumed negligible when a single Latin square is used, the treatment × square interaction could be isolated. For this example it would have $(8 - 1)(5 - 1) = 28$ df, so the error would then have 96 df. In general, it is best not to assume that any particular interaction is negligible unless there is a strong reason for doing so. Therefore, it would be desirable to plot the treatment averages against the squares, roughly analogous to Figure 3.1 for an RCB design. That is, for this example there would be $5 \times 8 = 40$ plotted points and the objective would be to see if the lines that connect averages for each treatment over the squares cross, and if so, how severe is the crossing, as extreme crossing would suggest an interaction that is sufficiently large that it should not be combined with the error term.

The interaction could be tested with a hypothesis test, using the appropriate numbers in the ANOVA table. In general, we would not want to see a significant treatment × square interaction as this would mean that the values for the factor of interest differ significantly across the squares, something that should theoretically not happen. If this interaction is significant, that should trigger an investigation. If it is not significant, it might be pooled with the error term, as some authors suggest, but the error term will generally have enough degrees of freedom when multiple Latin squares are used that such pooling shouldn't be necessary.

There are various other possibilities for relationships between squares, and these various cases are discussed in considerable detail by Giesbrecht and Gumpertz (2004, pp. 127–134) and the reader is referred to their discussion for additional details.

Although "Squares" is one component of the ANOVA, we would hope this would not be significant. If it is significant, there might be a lurking variable involved, such as some variable that is affecting the response values over time for all three factors. It is interesting to note that the use of multiple Latin squares, although not generally stressed in textbooks on experimental design, was in use as far back as the 1930s. For example, in the famous spindle study of Tippett (1936), two identical Latin square designs were used.

More recently, two mathematics professors at Lafayette College, Gary Gordon and Liz McMahon, came up with a novel application of two orthogonal 4 × 4 Latin squares. As they describe at http://ww2.lafayette.edu/~math/Gary/Doors.pdf, when they moved into their new house in 1986, each of their two garage doors was white. Deciding to do something colorful to cheer up the neighborhood, they decided to paint their garage doors—in a novel way. Each door had four sections and each section had four raised rectangular panels. Thus, each door had 16 panels, arranged in a 4 × 4 grid. They selected four colors—purple, light blue, dark blue, and teal—and painted each door in such a way that the doors were orthogonal Latin squares. From the discussion in Section 3.2.2.1 we can also say that their configuration was a lattice design—a somewhat interesting coincidence since the word "lattice" is often used in describing an aspect of a house and means "a window, door, or gate having a lattice," with the latter being "a framework or structure of crossed wood or metal strips." The "rest of the story" can be gleaned at the URL given at the beginning of this discussion.

3.3.6 ANOM

Just as data from an RCB design can be analyzed with ANOM by using decision lines given by Eq. (2.4), so too can data from a Latin square design. As with these other designs, $s = \sqrt{MS_{error}}$. Unfortunately, as with data from an RCB design, there is at present no software that can be used to directly generate an ANOM display with data from a Latin square design.

3.4   GRAECO–LATIN SQUARE DESIGN

A variation of a Latin square design that allows for three extraneous factors to be blocked and isolated is called a *Graeco–Latin square design.* Considering the tire experiment in Section 3.3.3, an obvious question to ask is whether or not the experiment should have been designed so that the effect of cars could be separated from the effect of drivers. The answer is "no" because such separation is not of interest. The blocking factor actually measured the sum of those two effects and that was good enough; there would be a problem if an extraneous factor that was thought to be possibly important was not incorporated into the design.

The design derives its name from the fact that Greek letters are typically used to represent the third blocking factor. For example, the first Latin square design given in Section 3.3 can be converted to a Graeco–Latin square design by superimposing a "Graeco square" on top of the Latin square in such a way that each pair of letters occurs only once. The following design is one example.

$$
\begin{array}{ccc}
A\alpha & B\gamma & C\beta \\
B\beta & C\alpha & A\gamma \\
C\gamma & A\beta & B\alpha
\end{array}
$$

A more formal way to state how a Graeco–Latin square design is constructed is to say that it is constructed using mutually orthogonal Latin squares of the indicated size, with mutually orthogonal Latin squares being those that produce a design such that each pair of Latin and Greek letters occurs only once, as in the design above. Thus, in order for a Graeco–Latin square design of a given size to exist, a pair of mutually orthogonal Latin squares of that size must exist. No such pair of 6×6 Latin squares exists, so it is not possible to construct a Graeco–Latin square design of that size.

Although it isn't discussed to any extent in the literature, it is possible to view and use a $k \times k$ Graeco–Latin square design as a k^2 design with two blocking factors, with of course $k > 3$. In general, especially with hyper-Graeco–Latin squares, a treatment factor can be substituted for a blocking factor, with the result that the design is a factorial or fractional factorial design with blocking. Factorial and fractional factorial designs are discussed in Chapters 4 and 5, respectively.

3.4.1   Model

The model for the Graeco–Latin square design is just a slight variant of the model for a Latin square design, with the model for the former given by

$$Y_{ijkl} = \mu + A_i + R_j + C_k + G_l + \epsilon_{ijkl} \qquad i = 1, 2, \ldots, t \qquad (3.5)$$

$$j = 1, 2, \ldots, t \quad k = 1, 2, \ldots, t \quad l = 1, 2, \ldots, t$$

with G_l denoting the effect of (the blocking factor level corresponding to) the *l*th Greek letter and the other components of Eq. (3.5) defined as in Eq. (3.4).

TABLE 3.3 **Breakdown of Degrees of Freedom for** $t \times t$
Graeco–Latin Square Design

Source	Degrees of Freedom
Row	$t - 1$
Column	$t - 1$
Other Blocking Factor (Greek letters)	$t - 1$
Treatments	$t - 1$
Error (Residual)	$(t - 3)(t - 1)$
Total	$t^2 - 1$

3.4.2 Degrees of Freedom Limitations on the Design Construction

It is easy to construct a 3×3 Graeco–Latin square, and such designs are given in books on design (e.g., Wu and Hamada (2000, p. 74), Kempthorne (1973, p. 187), and Box et al. (2005, p. 161)), with Euler (1782) being the first person to construct the design. Data from experiments in which such designs are used cannot be analyzed using ANOVA—the standard and essentially the only method of a formal analysis in the absence of a prior estimate of σ. This is because there are no degrees of freedom for error, something that must be kept in mind when these types of designs are used. (Of course we might notice that the average response at one or more levels is clearly superior to the average response at other levels, but some type of formal or semiformal testing would clearly be preferable to eyeing the numbers.)

We can see the problem if we look at the degrees of freedom for a $t \times t$ Graeco–Latin square design (Table 3.3).

We see that we "run out" of degrees of freedom by the time we get to the error term when $t = 3$, as all eight degrees of freedom are used by the three blocking factors plus the factor of interest. Thus, unless the design is replicated, the smallest Graeco–Latin square design that permits analysis using ANOVA is a 4×4 design.

Such a design has very little utility when only one square is used, just as is the case with a single Latin square.

(Readers who are returning to this chapter after having read the next chapter might wonder why we can't use methods illustrated in that chapter when there are no degrees of freedom for error. Such methods were developed for unreplicated factorials and it would not be possible, or at least practical, to try to develop a similar method for Latin-square-type designs as the successful use of methods discussed in the next chapter depends upon many effects not being significant.)

The analysis of a Graeco–Latin square is the same as that of a Latin square except that the ANOVA table contains one more blocking factor, represented by the Greek letters, with the sum of squares for it computed in the same general way as the sum of squares for the single factor of interest using the Latin letters.

It is possible to use more than three blocking variables and construct a *hyper-Graeco–Latin* square but such designs will not be discussed here, other than to point out that the degrees of freedom for the error term must also be considered for this design, which does impose a restriction on the size of the design. For example, unless the design is replicated, a 5×5 design will be necessary if there are four blocking factors, a 6×6 design if there are five blocking factors, and so on. It is easy to show that the number of blocking factors must be one less than the order of the design. Specifically, with s denoting the number of blocking factors, the number of degrees of freedom for the blocking factors will be $s(t - 1)$ and the degrees of freedom for the factor of interest is $t - 1$. Clearly we must have $s(t - 1) + t - 1 < t^2 - 1$. Thus, $(t - 1)(s + 1) < t^2 - 1$ will be satisfied only if $s < t$. Thus, the order of the design must be one greater than the number of blocking factors—a severe restriction if there are only a few levels of the factor of interest that are to be considered.

Another potential problem—and a serious one—is that although it may seem desirable to block on *all* expected extraneous sources of variation, the more blocking factors that are used, the more likely the assumption of no interactions is to be violated, and the violation(s) may occur in such a way as to seriously undermine the analysis.

3.4.3   Sets of Graeco–Latin Square Designs

In addition to the degrees of freedom problems with Graeco–Latin square designs noted in the preceding section, such designs have the same problem as a Latin square design regarding the power to detect differences in factor level effects because of the small number of observations from which each factor level average is computed unless a large design is used. Therefore the motivation and need to use sets of Graeco–Latin squares is essentially the same as for a Latin square design, and we would expect that the gains that are achieved in power by using multiple Graeco–Latin squares would parallel the gains realized by using multiple Latin squares discussed in Section 3.3.5.

For example, using the same scenario as was used in discussing single and multiple Latin squares, the power is only .176 when a single Graeco–Latin square design is used, but increases to .489 when two squares are used, to .680 when three squares are used, to .807 when four squares are used, and to .888 when five squares are used. Notice that these are essentially the same values that were obtained for the Latin square design as the number of squares was allowed to increase, as would be expected.

3.4.4   Application

An application of a Graeco–Latin square in the semiconductor industry was described by Michelson and Kimmet (1999). The purpose of the experiment was to evaluate the relationship between the size and density of defects introduced in a wafer fab at specific masking levels and the resultant failures of the individual die at circuit probe for electrical overstress (EOS) related tests.

Defects were intentionally created by placing a large number of known defects on the die during specific levels of manufacturing. The known defects could then be detected at circuit probe EOS testing and linked to the specific process levels.

As Michelson and Kimmet (1999) stated, however, "...the Graeco–Latin square was used as a setup tool rather than as an analysis tool." Specifically, the rows and columns of the 4×4 Graeco–Latin square were simply used to form a grid of 16 cells. Sixteen die were used within each of the cells. The response variables were the yield of each cell (the count of good die divided by 16), and the count of EOS failures in each cell. The factors within the square were the size of the defects (1, 2, 4, and 8 μm) and the density of the defects (1, 3, 9, and 27 defects/cm^2). Obviously these levels are powers of 2 and 3, respectively, which is presumably how the levels were determined. Thus, unlike a traditional Graeco–Latin square design, here the two factors inside the grid were the ones of interest, rather than having one factor serve as a blocking factor. Thus, this was really a 4^2 factorial with two blocking factors, neither of which was significant.

Since the response variable was percent yield, the variable is thus not normally distributed, so ANOVA is not applicable unless the variable is transformed. The variable was transformed with a logit transformation, however, so the response that was analyzed was actually log [yield/(1 − yield)]. A model was fit that had three terms: size, density, and the size × density interaction. (Note that technically an interaction term cannot be fit (and tested) from an ANOVA standpoint since the interaction term requires nine degrees of freedom and interactions cannot be fit anyway with one observation per cell. A regression approach can be used, however, just as such an approach can be applied to data from a nonorthogonal design, and regression was indeed used in analyzing these data rather than an ANOVA approach.)

Additional Examples

An example for which the design configuration was actually a Graeco–Latin square was described by Cochran and Cox (1957, p. 132), for an experiment described in the literature (Dunlop, 1933), although it was stated that the third blocking factor may have been unnecessary. Box et al. (2005, p. 161) gave an example of a Graeco–Latin square design that is at least realistic, if not real. Overall, however, it is probably safe to say that the use of Graeco–Latin squares has been quite limited.

As stated previously, the more blocking variables that are used, the more likely it is that there will be interactions among them or with the treatment factor. An example of this is the second experiment described by Sheesley (1985). There was a single factor of interest and three factors that were viewed as blocking factors. The data were analyzed as having come from a $2^3 \times 3$ design with four replications. It was a good thing that it was not run as a Graeco–Latin square design because there were some moderately large interactions among the blocking factors. This was not a problem because of the way that the data were analyzed, however, and Sheesley (1985) stated, "The significance of the blocking/process variables and interactions among them is not of concern since the purpose of the test was to simply account for their existence in order to evaluate the effect of lead types."

This is undoubtedly a common occurrence and militates against the use of Latin square type designs with more than two blocking factors.

3.4.5   ANOM

As with a Latin square design, data from a Graeco–Latin square design can be analyzed with ANOM by using the decision lines given by Eq. (2.4), but there is apparently no software that will directly construct the display. From a practical standpoint, there would be a problem even if such software did exist since it is virtually impossible to check for normality of populations with one or a few observations from each population. Specifically, assume that a single 3×3 Graeco–Latin square design is used so that the mean of each level of the factor is computed from three observations. There is no way to assess the normality assumption that underlies ANOM with only three observations. Of course this problem must also be addressed when ANOVA is used, which is one reason why multiple Graeco–Latin squares should generally be used rather than a single Graeco–Latin square.

3.5   YOUDEN SQUARES

In addition to the small number of degrees of freedom for the error term, another major shortcoming for nonagricultural use is the fact that the number of treatments and levels of the two blocking factors must be the same. (Obviously this is not a problem for agricultural experiments in which soil fertility in the two directions forms the blocking factors and the number of plots of land could be easily selected to equal the square of the number of treatments.)

This restriction will be unrealistic in many industrial experiments, however. Consider, for example, the tread wear experiment of Section 3.3.3. Obviously the number of wheel positions is fixed, but perhaps there are only three tire brands under consideration instead of four. Or perhaps only three cars are available for the experiment instead of four.

A prominent industrial statistician, W. Jack Youden (1900–1971), once was confronted with a situation where a Latin square design would seem to be appropriate because of the need to use two blocking factors, but there was a physical limitation that prevented the use of a square. This led him to use selected rows of a Latin square design. Such a design might be called an incomplete Latin square, but was termed a Youden square by R. A. Fisher. This is discussed in Youden (1937), which is apparently the first presentation of a Youden design. Most of the published Youden squares are due to Youden.

The term "Youden square" is of course a misnomer because if we use part of a Latin square, we no longer have a layout that is square. Therefore, we will henceforth refer to the type of design as the "Youden design."

An obvious question is "Can the rows of a Latin square design be selected at random to produce a Youden design, or can only certain rows of a given Latin square design be used?" The rows cannot be selected at random in arriving at a desired number of rows as there are requirements relating to balance that must be met. This is because a Youden design is essentially a BIB design. That is, the columns become incomplete blocks since not all of the rows are used. In order to have balance, each pair of treatments must occur the same number of times, and of course no treatment occurs more than once in a column, with not all treatments occurring in any column. That is, the columns are balanced incomplete blocks, but a BIB design doesn't exist for all sizes of rectangular configurations.

To use a simple example as an illustration, if we start with a 3×3 Latin square and delete the last row, we have six observations and the design is as follows.

$$\begin{array}{ccc} A & B & C \\ B & C & A \end{array}$$

The columns will form incomplete blocks with two treatments per block. In order for the balance requirement to be met, the number of pairs of treatments, which is obviously $\binom{3}{2} = 3$, must occur the same number of times per block over the set of blocks. Notice that each of the pairs (AB, AC, and BC) occurs once, so the balance requirement is met.

Similarly, if we start with the 4×4 Latin square in Table 3.2 and delete the last row, we will also obtain a Youden design with each pair of treatments occurring twice. (Notice that this balance requirement is also met if we delete any one row from that design.)

From these simple examples it might *seem* as though we could delete any row from a Latin square design and obtain a Youden design and that is true. The balance property can be easily established when the Latin square is a 4×4 design, and the reader is asked to establish the result in general in Exercise 3.5. For the 4×4 design, since one row is deleted, each block contains three treatments, and since the deleted row contains different treatments, any pair of blocks in the Youden design will have two treatments in common. Hence every pair of treatments will occur in two blocks, thus satisfying the balance property.

This won't necessarily be true for larger designs, however. For example, Dean and Voss (1999) point out that there is no BIB design for eight treatments in eight blocks of size 3. Therefore, it follows that we cannot delete *any* group of five rows from an 8×8 Latin square and obtain a Youden design. Thus, since it isn't always possible to arbitrarily delete multiple rows from a Latin square and produce a Youden design, catalogs of Youden designs are useful, and these are discussed in Section 3.5.2.

3.5.1 Model

Since a Youden design is a BIB design, the model for a Youden design is essentially the same as the model for a BIB design. We will write it slightly differently, however, to conform to the notation for a Latin square design since it is also an incomplete Latin square. The model is

$$Y_{ijk} = \mu + A_i + R_j + C_k + \epsilon_{ijk}$$
$$i = 1, 2, \ldots, t \quad j = 1, 2, \ldots, m \quad k = 1, 2, \ldots, t \quad (3.6)$$

with $m < t$, and Eq. (3.6) is otherwise the same as Eq. (3.4).

3.5.2 Lists of Youden Designs

Since most Youden designs cannot be simply constructed, it will generally be necessary to refer to lists/catalogs of such designs. Unfortunately such lists are not readily
available. Natrella (1963) listed 29 such plans and gave the actual design for 9 of
them, but unfortunately that book has long been out of print, as noted previously, and
its Internet successor, the *NIST/SEMATECH e-Handbook of Statistical Methods* at
http://www.itl.nist.gov/div898/handbook, does not cover Youden designs.

Cochran and Cox (1957), which is in print, gives more of these designs than did
Natrella (1963), however, and lists the actual design layout for almost all of these 29
plans. At present this is the best source for Youden designs.

3.5.3 Using Replicated Youden Designs

Just as there will generally be a need to use multiple Latin squares, there will also
generally be a need to use a replicated Youden design. Unlike the case with Latin
squares, however, there will generally not be multiple Youden designs from which to
select. Therefore, it will usually be necessary to replicate a given Youden design. The
Java applets used previously do not have the capability to perform power calculations
for a Youden design. Since a Youden design is really a BIB design, as stated previously,
one can use the same approach as is used for determining a suitable sample size for
that design. This is discussed by Dean and Voss (1999, p. 407), who illustrate how to
solve for the number of replicates.

3.5.4 Analysis

Not surprisingly, the analysis of data from a Youden design is the same as the analysis
for a BIB design, which was illustrated in Section 3.2.1.1. Therefore, the reader is
referred to that section for details. The analysis using ANOM is of course also the
same, and the application of ANOM to Youden squares is described in Section 6.4 of
Nelson et al. (2005).

3.6 MISSING VALUES

Missing values create a major problem when a design in this chapter is used. In
particular, a missing value in a Latin square causes the "square" to be lost. In general,
missing values cause a major problem whenever a design is constructed for more than
a single factor. Of course, a value can also become "missing" when there is a botched
experimental run and the recorded value is nonsensical.

Assume that an observation is missing from an RCB design. One approach is to
estimate the missing value by minimizing the error sum of squares, using a standard
calculus approach. The problem with this method is that it biases the results because
the value that is substituted is the value that most closely fits the model that is used.
The actual value will almost certainly be different and will not adhere as closely to
the fitted model as would the imputed value.

An unsatisfactory approach would be to delete the observations on the same treatment that is missing in one block from the other blocks. This would result in fewer treatments being compared and would defeat the purpose of the experiment.

A better approach is to analyze the unbalanced data that results when there are missing observations. An RCB design becomes "incomplete" when there is a single missing observation and also becomes unbalanced because the block sizes will be different. An "adjusted analysis" can be performed, similar to what is done with a BIB design, while recognizing that the power of the design is lessened by the missing observation.

For example, assume that we have the following example, with A, B, C, and D denoting the four treatments and the numbers in parentheses denoting the response values, with "*" denoting a missing value.

Blocks

1	2	3
C(10)	E(16)	A(19)
D(12)	C(18)	E(18)
E(14)	D(20)	D(11)
B(16)	A(21)	C(10)
A(13)	B(24)	B(*)

A "standard" analysis of these data as a two-way classification of the data would be met with an error message. Instead, the data must be analyzed as unbalanced data, which can be done in MINITAB, for example, by using the general linear model capability. The output for this example is shown below, indicating that the blocking was needed but there is no significant treatment effect. Note that none of the data are discarded as the total df is 13.

```
General Linear Model: Y versus Blocks, Treatments

Factor       Type      Levels      Values
Blocks       random    3           1,2,3
Treatments   fixed     5           1, 2, 3, 4, 5

Analysis of Variance for Response, using Adjusted SS for Tests

Source       DF     Seq SS     Adj SS     Adj MS     F      P
Blocks       2      125.914    118.267    59.133     8.67   0.013
Treatments   4      74.067     74.067     18.517     2.72   0.118
Error        7      47.733     47.733     6.819
Total        13     247.714

    S = 2.61133    R-Sq = 80.73%    R-Sq (adj) = 64.21%
```

As a second example, consider the example in Section 3.1.4 and assume that the last observation (i.e., in the lower right corner) is missing. The analysis is given below. (As indicated, blocks are assumed to be random but that classification has no effect on the analysis since the error term is really the interaction and the random effect is tested against the error in a two-factor mixed model, with the fixed effect tested against the interaction. But since interaction and error are inseparable, both effects are tested against error.)

```
General Linear Model: Conversion Gain versus Test Set, Resistor

     Factor     Type     Levels    Values
     Test Set   fixed       4      1,2,3,4
     Resistor   random      6      1,2,3,4,5,6

Analysis of Variance for Conversion Gain, using Adjusted SS for Tests

     Source    DF    Seq SS    Adj SS    Adj MS      F       P
     Test Set   3    902.67    893.33    297.78    309.68   0.000
     Resistor   5      5.59      5.59      1.12      1.16   0.375
     Error     14     13.46     13.46      0.96
     Total     22    921.72

     S = 0.980585      R-Sq = 98.54%        R-Sq(adj) = 97.70%
```

Notice that the loss of the observation has only a very slight effect on the numerical results and the conclusions are unaffected.

Similarly, the analysis of data from a Latin square design with a single missing observation is also straightforward. The computer output below is for the example in Section 3.3.3, with the observation in the lower right corner of the square assumed to be missing.

```
General Linear Model: Tread Wear versus Brands, Cars, Wheel Position

Factor           Type      Levels    Values
Brands           fixed        4      1,2,3,4
Cars             random       4      1,2,3,4
Wheel Position   fixed        4      1,2,3,4

Analysis of Variance for Tread Wear, using Adjusted SS for Tests

Source           DF    Seq SS    Adj SS    Adj MS     F      P
Brands            3    84.517    85.389    28.463    6.23   0.038
Cars              3   104.861    65.722    21.907    4.80   0.062
Wheel Position    3    46.722    46.722    15.574    3.41   0.110
Error             5    22.833    22.833     4.567
Total            14   258.933

S = 2.13698   R-Sq = 91.18%        R-Sq(adj) = 75.31%
```

With only 16 original observations, we might expect that removing one observation could make a noticeable difference, and we see that it does happen here. In particular, there are major changes in the p-values with the p-value for Brands exactly doubled and a greater percentage p-value increase occurring for Cars and Wheel Position. Here it would have been better to simply repeat the experiment, if possible and practical, than to try to analyze the data with the missing value. In general, the percentage of missing observations should be kept in mind before trying to draw conclusions from the remaining observations.

The analysis by ANOVA would proceed similarly for other designs discussed in this chapter, such as Graeco–Latin squares and Youden designs, when there is missing data.

Missing data do create major problems when ANOM is used, however, because even one missing data point destroys the equicorrelation structure that is necessary for ANOM to be used. Subramani (1992) proposed a step-by-step ANOM approach for use with missing data and illustrated the technique by applying it to data obtained from the use of a randomized block design.

3.7 SOFTWARE

Experimental design software is generally not oriented toward Latin square designs and its variants. For example, Latin square designs are not explicitly incorporated into Design-Expert and the user has to do some work to construct one, as is shown in a help file, which is discussed later in this section. Similarly, a Latin square design cannot be constructed using MINITAB, although a help file does explain how the model should be specified in MINITAB so as to allow data from such a design to be analyzed. The design also cannot be constructed directly using JMP, but can be constructed indirectly as a D-optimal design with one k-level factor and two blocks. (Optimal designs are covered in Section 13.7.) A Latin square design can be constructed using SAS, although SAS code must be written to accomplish this. (SAS code for randomly selecting a 5×5 Latin square is given by Giesbrecht and Gumpertz, 2004, p. 121.)

As stated, a Latin square design also cannot be constructed directly using Design-Expert, but can be constructed with some work. That is, as with the other software, it is not possible to select the design from a menu of designs. Rather, it is necessary to build the design using the general factorial design capability and specifying three categorical factors—for rows, columns, and treatments. The steps that must be followed in creating a 5×5 Latin square design and analyzing the resultant data are given at http://www.statease.com/rocket.html. In essence, a 5^3 design is initially constructed, with the last 100 rows of the design then being deleted. It might thus look as though a 5^{3-1} design is being constructed, but that is not the case as the Latin square is actually entered manually. The row and column factors that are specified simply serve to provide the template for the design construction. Since the design (i.e., the letters for the treatments, A–E) is entered manually, it would be a bit of a stretch to state that the software is constructing the design. Furthermore, a Latin square is obviously not randomly selected when the user is instructed to enter a particular Latin square. Although the steps are easy to follow, most software users would undoubtedly prefer to be able to construct the design directly and in fewer steps, and without having to actually enter the design themselves.

There is clearly a need for software developers to devote some attention to Latin square designs and their variants. It is relatively easy to go to a catalog of designs and randomly select a Latin square design of a particular size, but multiple Latin squares will generally be needed, which requires a mechanism for selecting them and software to analyze the data.

The situation is similar, although not as dire, regarding randomized block designs. The latter is essentially a two-factor design with no interaction and with randomization within each level of the blocking factor. Since the design layout is essentially that of a replicated one-factor design, the design can be easily constructed in Design-Expert, although as with a Latin square design, there is no menu that lists "randomized block design" as one of the menu items. Similarly there is no such menu item when JMP is used, but the design can easily be constructed by specifying one of the two factors to be a blocking factor.

The situation is different with incomplete block designs as such designs cannot be constructed indirectly. Specifically, a BIB design will be balanced only if it is constructed so as to be balanced, and similarly for PBIB designs. Balanced incomplete block and partially balanced incomplete block designs cannot be directly constructed with JMP (but can be constructed indirectly), MINITAB, or Design-Expert. Balanced incomplete block designs can be constructed using PROC OPTEX in SAS/QC, however, with appropriate code, including a BLOCKS statement. As stated in http://support.sas.com/techsup/faq/stat_key/k_z.html, "No procedure creates these specifically, but SAS/QC PROC OPTEX may find such designs if they are optimal according to the criterion used."

There *is*, however, statistical software that can be used to construct the designs given in this chapter. For example, Statgraphics can be used to construct RCB designs, BIB designs, Latin squares, Graeco–Latin squares, and hyper-Graeco–Latin squares, and Gendex, which is a DOE toolkit, will create certain types of incomplete block designs, including α-designs.

Although construction of incomplete block designs with most software packages is obviously a problem, there are various statistical packages that will analyze data from such designs. The extent of the use of incomplete block designs today seems questionable, however, at least in certain fields. For example, John (2003) stated "... incomplete block designs—a mathematical subject of theoretical use in agriculture and, so far as I can tell, of little interest to engineers." Incomplete block designs are generally not taught to engineers in short courses, so this may be one of the primary reasons for the fact that such designs are not used to any extent in engineering experimentation.

3.8 SUMMARY

It is important to isolate and extract extraneous factors from the error term, so that the latter is not improperly inflated, with the consequence that the factor(s) of interest may not be judged significant, as was discussed and illustrated in Section 3.3.3.

There are potential problems involved in the use of the designs given in this chapter, however, as no interactions are permitted for these designs. This assumption should always be checked graphically, as was illustrated in Figure 3.1. Statistical tests are also available for testing specific types of interactions (see Hinkelmann and Kempthorne, 1994).

The assumption of normality also presents a problem, especially when a single Latin square or Graeco–Latin square design is used, as it is not practical to test for normality with only a few observations. This is another reason why multiple Latin squares, Graeco–Latin squares, and hyper-Graeco–Latin squares should be used.

It is also important to bear in mind degrees of freedom restrictions and considerations, which will generally lead to the use of multiple Latin squares, Graeco–Latin squares, or hyper-Graeco–Latin squares. For additional reading on Latin squares and their variations, the reader is referred to Chapter 6 of Giesbrecht and Gumpertz (2004); for further reading on incomplete block designs the reader is referred to Chapter 11 of Dean and Voss (1999) and Chapter 8 of Giesbrecht and Gumpertz (2004). See also Montgomery (2005). Also worth noting is the Internet project at www.designtheory.org, which is intended to eventually be a source and catalog of a large number of designs, and the Web site includes software that is available as freeware. Much of what is there and will be there in the future will be of interest only to researchers in design theory, however, such as the 26×26 Latin square design that can be found there.

Finally, the general untestable assumption of normality underlies an ANOVA approach for each of these designs. Consequently, it would be desirable to use a nonparametric approach and compare the results, but other than the nonparametric approach given by Skillings and Mack (1981) for incomplete block designs that was mentioned in Section 3.2.3, nonparametric approaches for designs with blocking factors seem to be almost nonexistent.

REFERENCES

Armitage, P. and G. Berry (1994). *Statistical Methods in Medical Research*, 3rd ed. Oxford, UK: Blackwell.

Bose, R. C., W. H. Clatworthy, and S. S. Shrikhande (1954). Tables of partially balanced designs with two associate classes. Technical Bulletin #107, North Carolina Agricultural Experiment Station.

Box, G. E. P., S. Bisgaard, and C. Fung (1990). *Designing Industrial Experiments*. Madison, WI: BBBF Books.

Box, G. E. P., J. S. Hunter, and W. G. Hunter (2005). *Statistics for Experimenters: Design, Innovation and Discovery*. Hoboken, NJ: Wiley.

Cochran, W. G. and G. M. Cox (1957). *Experimental Designs*, 2nd ed. New York: Wiley.

Colbourn, C. J. and J. H. Dinitz, eds. (1996). *The CRC Handbook of Combinatorial Designs*. Boca Raton, FL: CRC Press.

Copeland, K. A. F. and P. R. Nelson (2000). Latin squares and two-level fractional factorial designs. *Journal of Quality Technology*, **32**(4), 432–439.

Dean, A. and D. Voss (1999). *Design and Analysis of Experiments*. New York: Springer-Verlag.

Dunlop, G. (1933). Methods of experimentation in animal nutrition. *Journal of Agricultural Science*, **23**, 580–614.

Euler, L. (1782). Reserches Sur Une Novelle Espece des Quarres Magiques. Verh. Zeeuwsch. Genooot. *Weten Vliss*, **9**, 85–239.

Federer, W. T. (1955). *Experimental Design: Theory and Application*. New York: MacMillan.

Freeman, G. (2005). Latin squares and su doku. *Significance*, **2**(3), 119–122.

Giesbrecht, F. G. and M. L. Gumpertz (2004). *Planning, Construction, and Statistical Analysis of Comparative Experiments*. Hoboken, NJ: Wiley.

Hinkelmann, K. and O. Kempthorne (1994). *Design and Analysis of Experiments. Vol. 1: Introduction to Experimental Design*. New York: Wiley.

Hinkelmann, K. and O. Kempthorne (2005). *Design and Analysis of Experiments. Vol. 2: Advanced Experimental Design*. Hoboken, NJ: Wiley.

Jaech, J. L. (1969). The Latin square. *Journal of Quality Technology*, **1**(4), 242–255.

John, P. W. M. (1971). *Statistical Design and Analysis of Experiments*. New York: MacMillan (reprinted in 1998 by the Society for Industrial and Applied Mathematics).

John, P. W. M. (2003). Plenary presentation at the *2003 Quality and Productivity Research Conference*, IBM T. J. Watson Research Center, Yorktown Heights, NY, May 21–23, 2003. (The talk is available at http://www.research.ibm.com/stat/qprc/papers/Peter_John.pdf.)

Johnson, N. L. and F. C. Leone (1977). The recovery of interblock information in balanced incomplete block designs. *Journal of Quality Technology*, **9**(4), 182–187.

Kempthorne, O. (1973). *Design and Analysis of Experiments*. New York: Robert E. Krieger Publishing Co. (copyright held by John Wiley & Sons, Inc.)

Kempthorne, O. and W. T. Federer (1948). The general theory of prime power lattice designs. *Biometrics*, **4** (Part I), 54–79; (Part II), 109–121.

Mead, R., R. N. Curnow, and A. M. Hasted (1993). *Statistical Methods in Agriculture and Experimental Biology*. London: Chapman & Hall.

Michelson, D. K. and S. Kimmet (1999). An application of Graeco–Latin square designs in the semiconductor industry. In *Proceedings of the American Statistical Association*, American Statistical Association, Alexandria, VA.

Montgomery, D. C. (2005). *Design and Analysis of Experiments*, 6th ed. Hoboken, NJ: Wiley.

Natrella, M. (1963). *Experimental Statistics*, National Bureau of Standards Handbook 91. Washington, DC: United States Department of Commerce.

Nelson, P. R. (1993). Additional uses for the analysis of means and extended tables of critical values. *Technometrics*, **35**(1), 61–71.

Nelson, P. R., M. Coffin, and K. A. F. Copeland (2003). *Introductory Statistics for Engineering Experimentation*. San Diego, CA: Academic Press.

Nelson, P. R., P. S. Wludyka, and K. A. F. Copeland (2005). *The Analysis of Means: A Graphical Method for Comparing Means, Rates, and Proportions*. Philadelphia: Society for Industrial and Applied Mathematics (co-published with the American Statistical Association, Alexandria, VA).

Nyachoti, C. M., J. D. House, B. A. Slominski, and I. R. Seddon (2005). Energy and nutrient digestibilities in wheat dried distillers' grains with solubles fed to growing pigs. *Journal of the Science of Food and Agriculture*, **85**(15), 2581–2586.

O'Carroll, F. (1963). A method of generating randomized Latin squares. *Biometrics*, **19**, 652–653.

Patterson, H. D. and E. R. Williams (1976). A new class of resolvable incomplete block designs. *Biometrika*, **63**, 83–92.

Peake, R. (1953). Planning an experiment in a cotton spinning mill. *Applied Statistics*, **2**, 184–192.

Raghavarao, D. (1971). *Constructions and Combinatorial Problems in Design of Experiments.* New York: Wiley. (published in paperback in 1988 by Dover Publications).

Reck, B. and J. P. Morgan (2005). Optimal design in irregular BIBD settings. *Journal of Statistical Planning and Inference*, **129**, 59–84.

Sarrazin, P., A. F. Mustafa, P. Y. Chouinard, and S. A. Sotocinal (2004). Performance of dairy cows fed roasted sunflower seed. *Journal of the Science of Food and Agriculture*, **84**(10), 1179–1185.

Sheesley, J. H. (1985). Use of factorial designs in the development of lighting products. In *Experiments in Industry: Design, Analysis, and Interpretation of Results*, pp. 47–57 (R. D. Snee, L. B. Hare, and J. R. Trout, eds.). Milwaukee, WI: Quality Press.

Singh, M. and K. Hinkelmann (1999). Analysis of partial diallel crosses in incomplete blocks. *Biometrical Journal*, **40**(2), 165–181.

Sinha, K. (1989). A method of constructing PBIB designs. *Journal of the Indian Society of Agricultural Statistics*, **41**, 313–315.

Skillings, J. H. and G. A. Mack (1981). On the use of a Friedman-type statistic in balanced and unbalanced block designs. *Technometrics*, **23**(2), 171–177.

Subramani, J. (1992). Analysis of means for experimental designs with missing observations. *Communications in Statistics—Theory and Methods*, **21**(7), 2045–2057.

Tippett, L. H. C. (1936). Applications of statistical methods to the control of quality in industrial production. *Transactions of the Manchester Statistical Society*, Session 1935–1936, 1–32.

van der Linden, W. J., B. P. Veldkamp, and J. E. Carlson (2004). Optimizing balanced incomplete block designs for educational assessment. *Applied Psychological Measurement*, **28**(5), 317–331.

Wilk, M. B. and O. Kempthorne (1957). Non-additivities in a Latin square design. *Journal of the American Statistical Association*, **52**, 218–236.

Wu, C. F. J. and M. Hamada (2000). *Experiments: Planning, Analysis, and Parameter Design Optimization.* New York: Wiley.

Yates, F. (1936). A new method of arranging variety trials involving a large number of varieties. *Journal of Agricultural Science*, **26**, 424–455.

Youden, W. J. (1937). Use of incomplete block replications in estimating tobacco-mosaic virus. *Contributions from Boyce Thompson Institute*, **9**(1), 41–48 (reprinted in the January, 1972 issue of the *Journal of Quality Technology*).

EXERCISES

3.1 Assume that a randomized complete block (RCB) design was used and either (a) the block totals were approximately the same but the treatment totals differed considerably, or (b) the treatment totals were approximately the same but the block totals differed considerably. Does either (a) or (b) suggest that this type of design should have been used? Explain.

3.2 Assume that you are given a 4 × 4 hyper-Graeco–Latin square design to analyze with four blocking factors. What would you tell the experimenter who gave you the data?

3.3 Explain how an RCB design for five factors and four blocks should be constructed.

3.4 Assume that an experimenter deletes two rows of a 3 × 3 Latin square and claims to have constructed a Youden design. Do you agree that this is a Youden design or is there another name for the design? Explain.

3.5 Prove that the properties of a Youden design will be met whenever the design is formed by deleting one row of a Latin square.

3.6 Consider the following experiment, a modification of one found on the Internet. An engineer is studying the effect of five illumination levels on the occurrence of defects in an assembly operation. Because time may be a factor in the experiment, he decided to run the experiment in five blocks, with each block corresponding to a day of the week. The department in which the experiment is conducted has (only) four workstations and these stations represent a potential source of variability. The engineer decided to run an experiment with the layout given below, with the rows representing days, the columns representing workstations, and five treatments denoted by the letters A through D. The data, coded for simplicity, are shown below.

```
                        Work Station
          Day   1       2       3       4

           1    A(6)    B(1)    C(2)    D(3)
           2    B(2)    C(2)    D(3)    E(7)
           3    C(4)    D(5)    E(4)    A(3)
           4    D(7)    E(6)    A(4)    B(2)
           5    E(5)    A(2)    B(2)    C(5)
```

(a) What type of design was used?
(b) Perform the appropriate analysis, but first state what assumptions must be made and check those assumptions.
(c) What is your conclusion regarding the five illumination levels?

3.7 The following problem can be found on the Internet: "The following experiment was designed to find out to what extent a particular type of fabric gave homogeneous results over its surface in a standard wear test. In a single run the test machine could accommodate four samples of fabric, at positions 1, 2,

3, and 4. On a large sheet of the fabric, four areas A, B, C, and D were marked out at random at different places over the surface. From each area four samples were taken, and the sixteen samples thus obtained were compared in the machine with the following results, given in milligrams of wear." The design layout was a Latin square with the test runs denoting the columns and the rows denoting the positions. What is the factor of interest and what are the blocking factors?

3.8 (a) Construct a 5 × 5 Latin square design.

(b) Write out the model when this design is used and state the assumptions that must be made, then state how you would check those assumptions.

(c) Describe an application in your field of study in which this design might be successfully employed.

3.9 Assume that an experiment is to be conducted to examine the effect of four different gasoline additives A, B, C, and D on reduction of nitrogen oxides. Four cars are available for the experiment, as are four drivers. Assume that differences are expected between the cars and between the drivers, so these are to serve as blocking factors.

(a) How would you design the experiment if 16 runs are to be made, assuming an identical driving course (terrain)?

(b) Now assume that because of time considerations, it isn't practical to use the same driving course for each run, as that would require various combinations of cars and drivers to be constantly "in waiting," so this would not be an efficient use of resources. To alleviate this problem, four driving courses (routes) are laid out. Now explain how the experiment should be designed and performed.

(c) Considering the number of observations, would you suggest that the design that you named in part (b) be replicated? If so, how would the replication be performed?

3.10 Assume that a 3 × 3 Latin square design is to be used and a particular square is randomly selected. Before the experiment is conducted, however, the experimenter decides that one of the blocking factors is actually a factor of interest. With this is mind, how should the design be described and should the analysis of the data be affected by this decision? Explain.

3.11 Nelson (1993, references) gave an example to illustrate the application of data from an experiment with a Graeco–Latin square design. A manufacturer of disk drives was interested in studying the effect of four substrates (aluminum, nickel-plated, and two types of glass) on the amplitude of the signal that is received. There were four machines, four operators, and four days of production that were to be involved, with machines, operators, and days to serve as blocking variables. The 4 × 4 Graeco–Latin square design is given below, with columns corresponding to operators, rows representing machines, Greek letters representing days, and Latin letters representing the substrates. The numbers in parentheses are the coded response values.

Aα (8)	Cγ (11)	Dδ (2)	Bβ (8)
Cδ (7)	Aβ (5)	Bα (2)	Dγ (4)
Dβ (3)	Bδ (9)	Aγ (7)	Cα (9)
Bγ (4)	Dα (5)	Cβ (9)	Aδ (3)

(a) State the assumptions that must be made if ANOM is to be applied to these data. Do these assumptions appear to be met? Explain.

(b) If the assumptions appear to be met, analyze the data with ANOM by comparing the substrate means against the decision lines (without a display).

3.12 Assume that one or more 4 × 4 Latin squares are to be used and σ is known to be approximately 1. If the power is to be .95 of detecting a difference between two means of the factor that is at least 2 units using the Scheffé approach, how many Latin squares should be used?

3.13 Construct an example of a 3 × 3 Latin square with data for which the sum of squares for the factor of interest is zero.

3.14 The abstract of an article in the journal *Arthritis and Rheumatism* (**50**(2), 458–468, February 2004) contains the following sentence: "Forty-two physical signs and techniques were evaluated using a 6 × 6 Latin square design." Is there anything wrong with that statement? Read the article and determine if the analysis of the data and inference drawn therefrom are correctly performed. If so, explain the analysis relative to the quote given above. If the analysis and/or inference are incorrectly performed, perform the correct analysis, if possible. If a proper analysis is not possible, explain why not.

3.15 Assume that a 5 × 5 Latin square is used and one of the graphical checks on the assumptions shows a moderate columns × treatments interaction. Explain how this could affect the determination of whether there is a treatment effect.

3.16 Assume that a Latin square design was used and part of the ANOVA table is as follows.

ANOVA

Source	DF	SS
Rows	3	122.4
Columns	3	26.9
Treatments	3	167.8
Error	6	61.2

Do the numbers suggest that a Latin square design should have been used? Explain. If not, what design could have perhaps been used instead?

3.17 In his article "The Latin square" in the October 1969 issue of the *Journal of Quality Technology*, J. L. Jaech gave the following data, in parentheses, for a Latin square design example, but with data that were not generated using the model for a Latin square.

C(19)	A(9)	B(−4)	D(9)
B(13)	D(9)	A(6)	C(16)
A(18)	C(15)	D(3)	B(8)
D(10)	B(11)	C(11)	A(12)

What model assumption(s) for a Latin square design appears to be violated, if any? Explain.

3.18 If a randomized block design would be useful in your field of engineering or science, give an example. If such a design would not be useful, explain why.

3.19 Assume that there is a need to detect fairly small differences between treatments, and the BIB design given at the start of Section 3.2.1.1 is to be used. What would you suggest to the experimenters?

3.20 **(a)** Complete the following ANOVA table for a Latin square design, assuming that all assumptions are met.

```
Analysis of Variance for Yield

Source        DF      SS      MS       F

Rows           4      12
Columns              16
Treatments
Error                36
Total               112
```

(b) Use $\alpha = .05$ and test the hypothesis that the treatment means are equal.
(c) Assume for the moment that the hypothesis is not rejected. What would you recommend to the experimenter(s)?

3.21 Determine the treatment sum of squares for the following Latin square design.

A(3.7)	B(3.9)	C(6.4)	D(5.1)
B(4.0)	A(3.5)	D(5.1)	C(4.8)
C(6.1)	D(3.8)	A(3.3)	B(2.9)
D(4.1)	C(5.5)	B(4.7)	A(3.1)

Is there evidence that any of the assumptions were violated? Could the assumptions be formally tested for this number of observations and design layout? Explain.

3.22 Show that there are 12 possible 3×3 Latin squares.

3.23 Could a PBIB design be constructed for five treatments to be repeated four times with a block size of 4? Why, or why not? If not, indicate a change in the number of repeats and/or block size that would allow a PBIB design to be constructed.

3.24 Assume that the blocks that are used for one of the blocking factors in a Latin square design are selected at random from a population of blocks. Does that create a problem with the analysis of the data? Explain.

3.25 Consider the following example given by Natrella (1963, pp. 13–14). An experiment was performed to determine the effects of four different geometrical shapes of a certain film-type composition resistor on the current-noise of the resistors. A BIB design was used because only three resistors could be mounted on one plate. In the design layout below it can be observed that each pair of treatments occurs twice in a block, so $\lambda = 2$. The observations are logarithms of the noise measurements.

Shapes (Treatments)

Plates (Blocks)	A	B	C	D
1	1.11		0.95	0.82
2	1.70	1.22		0.97
3	1.66	1.11	1.52	
4		1.22	1.54	1.18

Analyze the data by performing the intrablock analysis (using either hand computation or software) and draw appropriate conclusions.

3.26 The following ANOVA table appears in "Induced resistance in agricultural crops: Effects of jasmonic acid on herbivory and yield in tomato plants" by J. S. Thayer (*Environmental Entomology*, 28(1), 30–37). The response variable was the level of the enzyme polyphenol oxidase 8 d.

Source	MS	F	df	p
Treatment	69.68	14.51	1	0.001
Block	149.42	31.11	1	<0.001
Treatment X Block	5.68	1.18	1	0.29
Error	4.80		26	

(a) Is there anything odd or incorrect about this table, relative to an RCB design? If so, what is it? In particular, explain why this table could not be right if these are the only sources of variation and these sources are not a subset of a larger set of sources (which would be misleading).

(b) How many treatments and how many blocks were used, and what was the block size?

(c) The raw data were not given by the author, which precludes a full analysis of the data. Does the table above suggest that it was appropriate to use the RCB design? Explain.

3.27 Mead, Curnow, and Hasted (1993, references) gave an example of an RCB design for a study that involved three drugs and a placebo, with the experimental units being mice—four from each of five litters for a total of 20 mice. The objective was to determine if the drugs affected lymphocyte production. The data are as follows, with the letters A, B, C, and D representing the three drugs plus the placebo, respectively.

		Litter				
		1	2	3	4	5
Mouse within	1	6.7 (B)	5.4 (D)	6.2 (C)	5.1(B)	6.2 (C)
Litter	2	7.1 (C)	6.1 (A)	5.9 (B)	5.2 (D)	5.8 (B)
	3	6.7 (D)	5.8 (C)	6.9 (A)	5.0 (C)	5.3 (D)
	4	7.1 (A)	5.1 (B)	5.7 (D)	5.6 (A)	6.4 (A)

Perform an appropriate analysis, treating the drugs as fixed and the litters as random. Could both Treatment $\times$ Block and Error be shown in the ANOVA table, as in the preceding problem? Why, or why not?

3.28 Construct an example of an RCB design with five blocks and four levels of the factor such that the F-statistic for testing the factor effect is zero. Then construct an example (possibly different) such that both the F-statistic for testing the factor effect and the F-statistic for testing the block effect are zero.

3.29 Analyze the data given in Exercise 3.25 using ANOM.

3.30 Nyachoti, House, Slominski, and Seddon (2005, references) used a replicated 3×3 Latin square to determine the digestibilities of nutrients that comprise different diets. There were thus three diets and "...the pigs were randomly divided into two groups of three pigs each and within groups assigned to the experimental groups in a 3×3 Latin square design to give six observations per diet." Would you have designed this experiment differently if considerably more than six pigs were available for the experiment? If so, how would you have proceeded? If not, explain why not. In particular, do you believe that having multiple observations per pig could cause a problem, even though "period" is one of the blocking variables? Why, or why not?

3.31 Sarrazin, Mustafa, Chouinard, and Sotocinal (2004, references) studied the effects of dietary treatments on yield and composition of milk by using nine Holstein cows in three 3×3 Latin squares. These were not three replications, however, as different effects were studied in each Latin square, although two Latin square replications were used in one of the studies. In one of the other studies, "three multiparous lactating Holstein cows ... were used in a 3×3 Latin square design ...." This meant that there were repeated measures, with the different sampling times constituting the repeated measures. Read the article, if possible. Would you have used something other than Latin square designs since the overall study was multifaceted? Explain. Since repeated measures were involved, this problem is revisited in Exercise 8.7, in Chapter 8, which covers repeated measures designs.

CHAPTER 4

Full Factorial Designs with Two Levels

In this chapter we extend the discussion of designed experiments to experiments involving more than one factor, but restrict consideration to designs with two levels. The most frequently used design with multiple factors is a *factorial design.*

4.1 THE NATURE OF FACTORIAL DESIGNS

A factorial design has no direct connections with factorials per se (such as 4!). All possible combinations of factor levels are used (ideally) in a factorial design, so one could argue that "combination designs" would be a better term. The term "crossed design" is also used because each level of each factor is "crossed" with the levels of all the other factors.

To illustrate, the simplest example of a factorial design would be a 2^2—a design with two factors, each at two levels. This design would hardly ever be used in practice, but it has considerable illustrative value due to its simplicity. For the rest of this chapter, we will assume, unless stated otherwise, that each factor is *fixed*, that is, the levels used in the experiment are the only ones of interest. (Recall the discussion in Section 2.1.1 regarding fixed and random factors.)

In general, factorial designs are represented by the form s^k, with s denoting the number of levels and k denoting the number of factors. When $s = 2$, the two levels are usually denoted as "high" and "low" if the factor is quantitative, such as temperature. Of course if the factor is qualitative, with the two "levels" being two operators involved in a production process, such a designation would have no meaning.

For a 2^2 design there are obviously four combinations of the high and low levels of the two factors and these combinations are often designated by using the presence and absence of lowercase letters to indicate the high level and low level, respectively, of the corresponding factor, which is designated by the appropriate capital letter. For example, a would mean that factor A is at the high level and factor B is at the low level. This takes care of combinations a, b, and ab, but if the combination with each

TABLE 4.1   Design Layout for the 2^2 Design

Treatment Combination	A	B	AB
(1)	-1	-1	1
a	1	-1	-1
b	-1	1	-1
ab	1	1	1

factor at its low level was similarly designated, we would just have a blank space. Since that obviously wouldn't work, we use (1) to indicate each factor at its low level and this is also used when there are more than two factors.

Using a 2^2 design and estimating all the effects that can be estimated (A, B, and AB, the interaction term to be explained later) implies a tentative Analysis of Variance (ANOVA) model of the form

$$Y_{ij} = \mu + A_i + B_j + (AB)_{ij} \quad i = 1, 2 \;\; j = 1, 2 \tag{4.1}$$

with Y_{ij} denoting the response when the ith level of A is used in combination with the jth level of B, and "1" and "2" denoting the low level and the high level, respectively, of each factor. Note that there is no error term in this model, as there was, for example, in Eq. (2.1). This is because there is an exact fit when this design is used and the design is not replicated. This is illustrated later in the section. There is an error term when the design is replicated, so then the model is

$$Y_{ijk} = \mu + A_i + B_j + (AB)_{ij} + \epsilon_{ijk} \quad i = 1, 2 \;\; j = 1, 2 \;\; k = 1, 2, \ldots, c$$

for c replicates.

For the moment, we can view $(AB)_{ij}$ as a product term, analogous to a product term in a regression model, because this is how the "levels" of the term are formed. Specifically, the design layout, including the values of the product AB, which of course are forced rather than specifically selected, is as given in Table 4.1.

The high level of each factor is denoted by $+1$ and the low level by -1. The levels could have been denoted by 0 and 1, or by 1 and 2, and these designations are used in certain literature articles, but the $(+1, -1)$ designation has some advantages that will become evident shortly, and it is also a natural designation. Regarding the latter, recall that the numerator of a two-sample t-test, as discussed in Section 1.4.3 is $\bar{y}_1 - \bar{y}_2$, with the coefficients of the two sample averages thus being $+1$ and -1. In general, whenever we compare data from two populations or compare the effects of two levels of a factor, the obvious choice for the coefficients is $+1$ and -1.

Notice that the dot product of any pair of these columns is zero. (For example, the dot product of A and B is $(-1)(-1) + (1)(-1) + (-1)(1) + (1)(1) = 0$.) This implies that the design is *orthogonal*, which means that the effects of each factor and the product of the two factors can be estimated independently of each other. The A and B effects are referred to as *main effects*, and the product term will, in the next section, be explained as an *interaction effect*. Note that two levels result when the multiplication is carried out to produce the product term.

The treatment combinations in Table 4.1 are listed in *Yates order*. If there had been three factors, for example, the rest of the sequence would have been c, ac, bc, abc. That is, whenever a new letter is listed, it is followed by all possible combinations of it with the previously listed letters. So if there had been a fourth factor, the sequence would have continued with d, ad, bd, abd, cd, acd, bcd, $abcd$, with the entire list having $2^4 = 16$ treatment combinations.

The numbers in the columns in Table 4.1 indicate how the effect of each factor would be estimated. Even if we didn't have these numbers to look at, a logical way to estimate, say, factor A would be to take the average of the observations that result when A is at one of the two levels and subtract from this number the average of the two responses when A is at the other level.

Example 4.1

To illustrate, let's assume that the observations of some physical characteristic (tensile strength, perhaps) that result when the experiment is run (with A and B perhaps denoting temperature and pressure, respectively) are 70, 62, 59, and 71, corresponding to the order of the treatment combinations listed in the table.

Assume that the temperature levels are 350 and 370°F. Then we would code these levels as $(370 - 360)/10 = 1$ and $(350 - 360)/10 = -1$, and similarly for the other factor. It is important that analyses be performed using coded values when interaction effects are being estimated, because misleading results can be obtained when analyses are performed using raw values, and these results will not be in agreement with the results obtained using the coded values. This can be illustrated as follows.

As long as we have a single factor, the analysis, in terms of the conclusions that are reached, is invariant to whatever coding, if any, is used. This is also true when there is more than one factor, provided that the model does not contain any interaction terms. To see this, assume that we have two factors, X_1 and X_2, and an unreplicated 2^2 design. In coded form, X_1, X_2, and X_1X_2 are of course pairwise orthogonal. That is, the dot product of the following columns

A	B	AB
-1	-1	1
1	-1	-1
-1	1	-1
1	1	1

is zero, as stated previously. Assume that the two levels of factor A are 10 and 15 and the two levels of factor B are 20 and 30. The raw-form representation of the two factors and their interaction is then

A	B	AB
10	20	200
20	20	400
10	30	300
20	30	600

Notice that the last column of numbers cannot be transformed to the corresponding set of coded-form numbers with any method of coding (i.e., any linear transformation). Furthermore, if the analysis were performed in raw form there would be a major multicollinearity problem as the correlation between A and AB is .845 and the correlation between B and AB is .507. If we let the four response values be 6.0, 5.8, 9.7, and 6.3, respectively, it can be shown that the ordinary least squares (OLS) coefficient of A is positive when the raw form is used and negative when the coded form is used.

We can see that the A effect is negative because $6.0 + 9.7$ exceeds $5.8 + 6.3$, using the signs in the A column, which is used to obtain the A effect estimate, and similarly for the other effect estimates. Specifically, the effect estimate is $(5.8 + 6.3)/2 - (6.0 + 9.7)/2 = -1.8$. Therefore, since the effect estimate is negative, the regression coefficient should also be negative. The sign is (correctly) negative when the coded form is used but not when the raw form is used. Thus we have a wrong sign problem, and unlike the "wrong sign" discussed in the regression literature (see, e.g., Ryan, 1997, p. 131), there really *is* a wrong sign problem!

Returning to the temperature data, since A is a quantitative factor, the estimate of the A effect would logically be computed as the average response value at the higher of the two temperatures minus the average response at the lower temperature, and notice that this is what happens when we multiply the column of numbers for A by the corresponding response values.

Of course "higher" is denoted by "1" and "lower" by "-1" in terms of the coded levels. The B effect would be computed in the same manner as the A effect. Note that two levels result when A and B are multiplied together to produce the interaction AB. Accordingly, it seems reasonable to compute the interaction effect as the average response at the "1 level" minus the average response at the "-1 level," and this is how the interaction effect is computed if we multiply the AB column by the corresponding response values.

Thus, the estimate of the A effect is $(62 + 71)/2 - (70 + 59)/2 = 2$. Similarly, the estimate of the B effect is $(59 + 71)/2 - (70 + 62)/2 = -1$, and the estimate of the AB effect is $(70 + 71)/2 - (62 + 59)/2 = 10$.

If we write the fitted model in the form of a regression model, with b_0, b_1, b_2, and b_{12} denoting the OLS regression coefficients, we then have

$$\widehat{Y} = b_0 + b_1 A + b_2 B + b_{12} AB \tag{4.2}$$

with $b_0 = \widehat{\mu} = \overline{Y} = 65.5$ and b_1, b_2, and b_{12} are, by definition, half the effect estimates for the terms for which they serve as coefficients. (See chapter Appendix B for details.) That is, the A effect estimate was calculated to be 2, so $b_1 = 1$. The effect estimates of B and AB are -1 and 10, respectively, so $b_2 = -0.5$ and $b_{12} = 5$.

Notice that the fitted values are equal to the observed values. For example, the first fitted value is $\hat{Y} = 65.5 + (1)(-1) + (-0.5)(-1) + (5)(1) = 70$, and it can be similarly shown that the other fitted values are the same as the observed values.

Of course this does not generally happen when a regression model is fit to a set of data, but in such applications of regression the number of observations is invariably greater than the number of parameters that are estimated. Here the number of coefficients is equal to the number of observations, so there is an exact fit.

Although it might seem like utopia to be able to exactly reproduce the observed values with a fitted model, this is actually a problem because it stands to reason that there would not be an exact fit if there were more observations. For example, if there were two observations per treatment combination, there would not be an exact fit unless the two observations at each treatment combination were the same, which would be highly unlikely and would suggest that there was a problem with the experiment.

An unreplicated factorial design, be it a 2^2 design or a design with more factors, is a *saturated design*. That is, the number of effects to be estimated is equal to the number of observations minus one. That this will be the case for all two-level full factorial designs regardless of the number of factors can be seen as follows. The number of observations for a 2^k design will of course be 2^k since this defines the number of observations. Now think of the "2" as representing the two possibilities, presence or absence, of each factor in an effect to be estimated. If all factors are "absent," then there is nothing to estimate, so the total number of effects is $2^k - 1 = n - 1$.

Hypothesis tests, as were performed in Section 2.1.2, for example, by using information from an ANOVA table, are not available for unreplicated designs because the variance of the error term cannot be estimated because there isn't enough data. Instead, a normal probability plot approach is typically used, which is illustrated and discussed in Sections 4.10 and 4.11 and is also used in subsequent chapters.

Example 4.2

Recall the example with the runner, which was given in Section 1.3. Assume that the runner conducts an experiment over a period of several months with the two levels of the training factor being "heavy" and "moderate" and the two levels of nutritional supplementation being "moderate" and "increased." The runner's time in minutes is recorded over a prescribed course for each of the four combinations. Those times are 57, 52, 53, and 58 for the combinations (1), a, b, and ab, respectively, with factor A being the training factor and factor B the nutritional factor. It is easy to see that the estimate of the A effect is 0, and the estimate of the B effect is 1.0. The estimate of the AB interaction effect is 5.0, which complicates the analysis since it is much bigger than the A effect and B effect. Clearly there is an A effect at low $B(52 - 57)$ and an A effect at high $B(58 - 53)$—the effects simply add to zero. This necessitates the use of conditional effects, which are discussed in Section 4.2.1.

Although the 2^2 design is the ideal design to use for illustrating basic concepts in factorial designs, we can use Eq. (1.3), appropriately modified for a two-tailed test, to show that it really has no practical value. For an unreplicated 2^2 design, we obtain $\Delta \doteq \frac{((1.96+1.28)2\sigma)}{\sqrt{n}} = \frac{6.48\sigma}{\sqrt{n}} = 3.24\sigma$ as the smallest effect of A, B, or AB that could be detected, if σ were known, with a probability of .90 and a significance level of $\alpha = .05$. This is far too large a multiple of σ. The multiplier would have to be much smaller than this for a design to have practical value. We apply this formula to larger factorial designs in later sections.

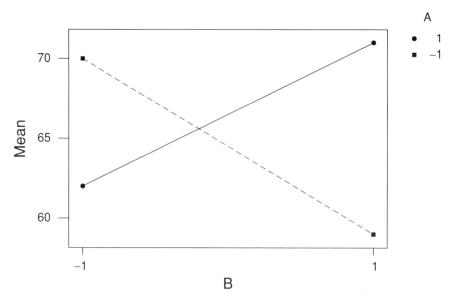

Figure 4.1 Interaction plot for 2^2 example.

So while recognizing that the design has no practical value, we will continue to use it for another section or two to illustrate basic concepts.

4.2 THE DELETERIOUS EFFECTS OF INTERACTIONS

In the previous section, *AB* was viewed as a product term, and it could be seen from Table 4.1 that it is indeed a product term. In this section we will view it, equivalently, as an *interaction term*, with the term "interaction" unrelated to the interaction of physical variables as in a chemical interaction. Interaction simply means that the effect of a factor depends upon the level(s) of the other factor(s). This will be illustrated later in the section.

The presence of interaction, particularly extreme interaction, can easily result in completely erroneous conclusions being drawn if an experimenter is not careful. This can be seen in the example in the preceding section, as the interaction effect estimate was much larger than either of the main effect estimates. Daniel (1976, p. 21) stresses that data from a designed experiment should not be reported in terms of main effects and interactions if an interaction is more than one-third of a main effect.

In the example in Section 4.1, the main effect estimates are not even $1/3$ of the interaction effect estimate! This is a serious interaction problem, which can also be seen graphically. The four points are plotted in Figure 4.1.

As was indicated by the numerical calculations, this plot illustrates severe interaction, with the interaction effect estimate being nonzero if the lines are not parallel. Here the plot deviates sharply from parallelism, almost forming an "X." If an "X" had been formed, the main effect estimates would have been zero and an analysis using main effects would have been totally misleading. In particular, note that there is a very large A effect at the high value of B, as represented by the vertical distance between the two lines at $B = 1$. Similarly, there is a very large B effect at the low level of A, as reflected by the slope of the dotted line.

4.2.1 Conditional Effects

These effects have been called *simple* effects in the literature, as in Glasnapp and Sauls (1976), Bohrer, Chow, Faith, Joshi, and Wu (1981), Woodward and Bonett (1991), Winer, Brown, and Michels (1991), Bonett and Woodward (1993), Kirk (1995), Schabenberger, Gregoire, and Kong (2000), Kuehl (2000), and Hinklemann and Kempthorne (2005, p. 260), and Schabenberger et al. (2000) referred to collections of simple effects as "slices." A more appropriate term, however, probably is *conditional effects* because effect estimates are computed by conditioning on a particular level of one of the factors, which for a two-level factor means that each conditional effect is computed using half of the data, with the computation formed as for a regular (i.e., unconditional) effect estimate, except that only half of the data are being used. The term "conditional effect" has also been used by Kao, Notz, and Dean (1997) and Wu and Hamada (2000). See also Taguchi (1987, p. 279), in which it is termed the "trans-factor" technique.

It is worth noting that the data are also split when CART (Classification and Regression Trees) methodology is used to analyze data from designed experiments, but there the objectives are different, as one objective might be to determine the combination of factor levels that maximizes the response. Another objective is to uncover interactions that might not be easily detectable when small designs are used. The use of CART in analyzing data from factorial designs is illustrated by Wisnowski, Runger, and Montgomery (1999–2000).

Consider the following example, which illustrates the need for examining conditional effects.

Example 4.3

As a very simple example, consider an unreplicated 2^2 design and the following data.

A	B	Response
-1	-1	14
1	-1	14
-1	1	13
1	1	22

The interaction effect estimate is $1/2(14 + 22 - 14 - 13) = 4.5$.

Splitting on B, the conditional effects of A are 0 at low B and 9.0 at high B.

Splitting on A, the conditional effects of B are -1.0 at low A and 8.0 at high A.

The conditional effect estimates are just linear combinations of two numbers for this very small example, but the point to be made here is that none of the conditional effects well-represent the corresponding main effects, which are $A = 4.5$ and $B = 3.5$, and this is due to the large interaction. This could be a major problem if, several months after an analysis was performed, it became desirable or necessary to fix one of the factors at a particular level, with another factor being hard to maintain at a desired level due to physical considerations, so that the factor level is apt to vary within a particular range. If a conditional effects analysis had not been performed, an engineer might have a hard time determining the expected effect on the value of the response variable as the hard-to-maintain factor is varied within its range. This problem might be avoided if all significant interactions are included in the model and the model is used to try to determine the effect of a particular scenario on the response. Interactions can have a disruptive effect on analyses when they are relatively large but not large enough to be declared significant and included in a model.

In this example the interaction effect was positive, which means that the conditional effect at the high level of each factor is greater than the corresponding conditional effect at the low level of each factor. Conversely, if the interaction effect estimate is negative, the conditional effect at the high level of each factor must be less than the conditional effect at the low level of each factor. This result follows from the derivation in chapter Appendix A, since the interaction effect is added to the main effect to obtain the conditional effect at the high level of the other factor and subtracted from the main effect to obtain the conditional effect at the low level of the other factor.

To illustrate the conditional effect at the high level of each factor being less than the conditional effect at the low level, we can simply exchange the last two response values in the example. We then have the following.

The interaction effect estimate is $1/2(14 + 13 - 14 - 22) = -4.5$.

Splitting on B, the conditional effects of A are 0 at low B and -9.0 at high B.

Splitting on A, the conditional effects of B are 8.0 at low A and -1.0 at high A.

Again, none of the conditional effects are representative of the corresponding main effects because of the size of the interaction relative to the main effects, which here are -4.5 for A and B is still 3.5.

It appears as though simple/conditional effects and the need for their use in analyzing data from designed experiments has been largely ignored in the applied literature, however. Exceptions include Toews, Lockyer, Dobson, Simpson, Brownell, Brenneis, MacPherson, and Cohen (1997), Stehman and Meredith (1995), and Peeler (1995).

Whatever descriptive label is used, the concept should be applied only in the case of fixed factors, since for random factors variance components are estimated rather than effects.

These conditional effects are easy to see and compute for a 2^2 design, but it becomes much more involved when there are more than two factors. This is illustrated in Sections 4.6 and 5.1.4, for example.

IMPORTANT POINT

Conditional effects should be used routinely because at least some interaction effect estimates are likely to at least moderately differ from zero in practically any application, and even moderate departures from zero will cause the corresponding conditional effects to differ considerably. As with any average, we would like to have conditional effects computed from a reasonable number of observations. If not, and this could happen with small designs, the variance of the conditional effects estimates could be high and possibly be misleading.

The failure to recognize this has resulted in some bad advice in the literature. For example, Emanuel and Palanisamy (2000) discuss sequential experimentation but state "... two-factor interactions containing at least one insignificant main effect are negligible." The same message can be found in Chipman (1996), who referred to the "strong heredity assumption," which holds that an interaction is likely to be significant only if both main effects are significant (see Wu and Hamada, 2000, p. 365 for additional discussion). Many researchers adopt an opposing position, including Bingham (2001) who argues against the strong heredity assumption. The use of the strong heredity assumption overlooks the fact that large two-factor interactions can *cause* insignificant main effects, with both the interactions and the conditional effects being real. Similarly, Box, Meyer, and Steinberg (1996) and Box and Tyssedal (2001) have contended that the "weak heredity assumption," which requires that an interaction cannot be present in a model unless at least one of the factors in the interaction is present, is not valid in every experimental situation.

One obvious question to ask is the following: Should these interaction plots be constructed by having the factor that comes second in alphabetical order plotted on the horizontal axis, or does this matter? It does matter because we tend to associate the magnitude of an interaction with the extent to which the lines deviate from parallelism, and visually we see stronger evidence of extreme interaction when the lines cross than when they don't cross. Whether or not they cross, however, will often depend on which factor is plotted on the horizontal axis, as the reader is asked to show for the data given in Exercise 4.3. Because of this, it is a good idea to construct interaction plots both ways. Of course this is easy to do with a small number of factors—such as two, as we have here—but becomes more of a chore for a moderate-to-large number of factors. (For the data in Figure 4.1, the lines will cross regardless of which factor is plotted on the horizontal axis.)

Fortunately, there is an efficient way to do this with software. For example, the FFINT command in MINITAB can be used to construct a matrix of interaction plots for two-level designs. To reverse the axis on each plot in the scatterplot, the order in

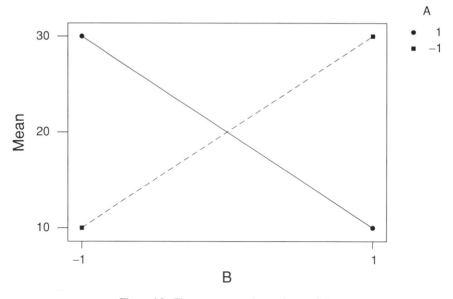

Figure 4.2 The most extreme interaction possible.

which the factors are listed in the command are simply reversed. For example, FFINT
C3 C2 C1 in command mode will give the reversal of the interaction plots produced
by FFINT C1 C2 C3.

As a slightly more extreme example, using the "X" configuration mentioned previously, consider Figure 4.2.

It is clear that the value of the response variable varies by 20 units when either A
or B is set at one of the two levels and the other factor is varied between its two levels.
Yet, when the data are analyzed we would find that the main effect estimates for each
of the two factors are exactly zero. This is because the conditional effects differ only
in sign and thus add to zero. When this occurs, the main effect estimate will be zero,
as the reader is asked to show in Exercise 4.1.

This result shows that a "blind" analysis, such as unthinkingly relying on computer
output, would lead an experimenter to conclude that there is neither an A effect nor a
B effect, although each clearly has an effect on the response variable when viewed in
the proper perspective. It is important that interaction plots such as those in Figures
4.1 and 4.2 be used in addition to other graphical displays in analyzing data from
multifactor designs.

As in the graphical analysis of the data in Figure 4.1, we might ask if the configuration of points and the strength of the signal of the interaction depend on which
factor is plotted on the horizontal axis. In order to have a perfect "X," there can be
only two distinct values of the response variable. The "X" means that both main effect
estimates are zero, and the only way this can happen when the other factor is plotted

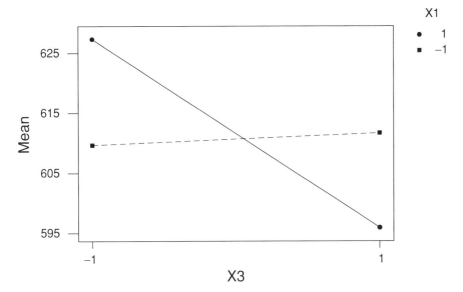

Figure 4.3 Interaction plot of data from the NIST ceramics experiment.

on the horizontal axis is when the other graph is the same "X," which must occur since there are only two distinct values of the response variable. The practical significance of this is that if we have anything approaching such a configuration of points, then we shouldn't bother to construct the other graph.

In general, however, it is a good idea to construct both graphs, especially if conditional effects are being used.

Although the configuration in Figure 4.2 is an extreme example, and would seem not likely to occur exactly in practice, we should not be surprised to find significant interactions and nonsignificant main effects. We may, however, observe interaction plots that approach the configuration in Figure 4.2, and Figure 4.3 is an interaction plot from an actual experiment. Furthermore, Box (1999–2000) described an actual experiment such that "all the variables were clearly active ... but our main effects were all essentially zero." So a scenario that is essentially the same as Figure 4.2 *has* occurred in practice and has perhaps occurred in a higher fraction of experiments than we might suspect.

The data graphed in Figure 4.3 are given in the case study in Section 1.4.2.10.1 of the *NIST/SEMATECH e-Handbook of Statistical Methods* (http://www.itl.nist.gov/div898/handbook/eda/section4/eda42al.htm). The data used in that case study were from a larger set of data collected by Said Jahanmir of the NIST Ceramics Division in 1996 for a NIST/industry ceramics consortium. Three factors were used in the case study, from the 12 factors in the study of Jahanmir.

Interaction Plot - Data Means for Response

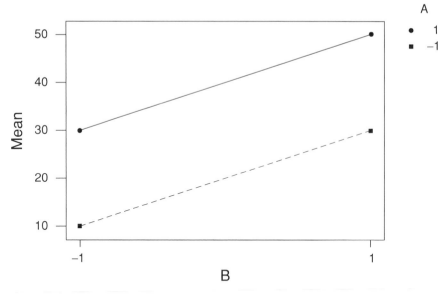

Figure 4.4   Interaction plot illustrating no interaction.

Two batches were used and the overall objective was to optimize the strength of a ceramic material. The factors denoted as X1 and X3 were table speed and wheel grit, respectively. The former was not identified in the case study as an important factor in the analysis of the data for batch 2, and we can see from Figure 4.3 that the effect estimate must be close to zero, relative to the magnitude of the numbers. (It is actually 0.949.)

If the dotted line in Figure 4.3 had been horizontal and the solid line had crossed that line halfway between -1 and 1 for X3, the estimate of the main effect for X1 would have been zero, as the two sides of the graph would have been mirror images. What occurred is very close to this, which illustrates that an extreme interaction plot configuration can occur in a case study of data from an experiment.

Since the X1X3 interaction effect estimate was -16.711, the conditional effects for X1 are $0.949 \pm (-16.711) = -15.762$ and 17.660. The conclusion from the case study was that the main effect of X1 was essentially zero, but X1 clearly has an effect, as is indicated by the magnitude of the conditional effects, and the different signs should also be noted.

Another type of extreme interaction plot is given in Figure 4.4. This illustrates no interaction and we would like to see something close to this as then the main effect estimates can be interpreted unambiguously.

To this point in our discussion of the 2^2 design the main effect estimates have been defined in terms of the average response value at one level of a factor minus the average response value at the other level of the factor, and the way that the interaction effect estimate is obtained was illustrated in Table 4.1.

Example 4.4

Sztendur and Diamond (2002) gave an example in which the coefficients of conditional effects were estimated using a regression approach applied to data from an experiment given in Hsieh and Goodwin (1986).

We will work up to what they did by building on the discussion of conditional effects in this chapter and by using results in the chapter Appendix A.

There is a way of expressing conditional effects in terms of the main effect and interaction effect that is derived in chapter Appendix A and which was used earlier in this chapter. Applying that result, let A^+ denote the effect of factor A at the high level of factor B, while A^- denotes the effect of factor A at the low level of factor B, with A denoting the main effect and AB denoting the interaction effect. Then $A^+ = A + AB$ and $A^- = A - AB$. In chapter Appendix B, the derivation is given for the general result: regression coefficient $= \frac{1}{2}$ (effect estimate).

This suggests that the "regression variables" to be used for the conditional effects of A should be $\frac{(A+AB)}{2}$ and $\frac{(A-AB)}{2}$. If we apply this to the 2^2 example in Example 4.1 and list the response values, we obtain the following if we elect to use the three degrees of freedom for estimating the conditional effects of A and the B effect.

A^+	A^-	B	Y
0	-1	-1	70
0	1	-1	62
-1	0	1	59
1	0	1	71

If we fit a regression model to this, we obtain $\widehat{Y} = 65.5 + 6A^+ - 4A^- - 0.5B$, which tells us indirectly that the conditional effects of A are 12 and -8, respectively, and the B effect is -1. Of course we can see these results just by looking at the columns above.

Sztendur and Diamond (2002) used a fitted model that contained a conditional effect estimate (of the effect of factor B at the high level of factor G). No additional computing is necessary if B and BG have already been fit in a model since the conditional effect coefficient is simply the sum of the coefficients of B and BG because the conditional effect of B at high G is simply $B + BG$. Since this was done in the context of determining a path of steepest ascent in a response surface approach, we will return to this example in Section 10.10.

4.2.1.1 Sample Sizes for Conditional Effects Estimation

The size of an effect that can be detected (with high probability) was discussed in Section 1.4.4. Some consideration should be given to the number of observations that each conditional effect estimate will be computed from when 2^k designs are used.

Of course this number will be 2^{k-1} when an unreplicated design is used and $r2^{k-1}$ when r replicates are used. For an unreplicated design, 2^{k-1} will of course be small when k is small and may not be large enough to obtain conditional effect estimates with an acceptably small variance. This is something that should be kept in mind if interactions are anticipated or at least thought to be quite possible, so that a replicated design might be chosen if extra experimental runs are not expensive or otherwise impractical.

4.2.2 Can We "Transform Away" Interactions?

Since interactions will usually necessitate a conditional effects analysis and thus complicate the analysis, we might ask whether we should try to transform the data in an attempt to remove an interaction. Sometimes it is possible to transform the data so as to essentially remove an interaction by making the interaction on the transformed scale quite small. However, if the interaction is large on the original scale, it will not be small on the transformed scale.

To illustrate, assume that a two-factor interaction is as shown in Figure 4.1. Common transformations such as square root and logarithmic transformations will preserve the ranking of the numbers. To illustrate, assume that the numbers represented by Figure 4.1 are 70, 62, 58, and 72. The only way the lines in the interaction profile for the transformed numbers would not cross would be if the ranking of the first two numbers is the same as the ranking of the last two numbers when the data in each pair are ranked separately. That is, the ranking of the untransformed numbers is 1, 2, 2, 1. If we use a logarithmic transformation or a square root transformation, for example, the rankings will be maintained and the lines will still cross. If we use a reciprocal transformation, however, we "transform" the rankings so that they become 2, 1, 1, 2. The lines will still cross since we have "flipped" the rankings and the ranking of the pairs is still different.

Similarly, if the interaction is extreme, but not so extreme that the lines cross, transforming the data will not remove the interaction. This is not to suggest that one shouldn't consider removing interactions through transformations, as the analysis of no-interaction models is much simpler than the analysis of models with interactions. Not only is a conditional effects analysis obviated, but hypothesis testing is greatly simplified since interaction terms are the denominators in F-tests for experiments with random factors as well as experiments with both fixed and random factors.

4.3 EFFECT ESTIMATES

In this section we take a more formal look at effect estimates, complementing what was done in Section 4.1. There it was stated that the effect of a two-level factor would logically be estimated as the average response at the high level of the factor minus the average response at the low level.

Another (mathematically equivalent) way of obtaining a main effect estimate is as follows. Consider Figure 4.4. It would be logical to estimate the A effect by taking the average of the A effect at the high level of B and the A effect at the low level of B. Since the lines are parallel, we have the same number for each, namely, 20, which is the vertical distance between the lines. (The reader is asked to show in Exercise 4.2 that this is mathematically equivalent to what was used in Section 4.1.)

It is comforting that the two numbers are the same because if they differ greatly, they will each differ from the average of the numbers that is used to represent the effect.

If we similarly view the estimate of the B effect in the same way, we see that the estimate of the B effect is the average of the "slopes" of the lines, and that each number is 20. (This must be the case if the numbers that we average to obtain the A effect estimate are the same.)

An obvious way to measure the interaction effect is to use the *difference* between the two vertical distances in Figure 4.4. The reader may wish to verify that this is equivalent to using the coefficients in Table 4.1. Thus, the set of numbers in the AB column has a physical interpretation.

Whereas the scenario in Figure 4.4 is ideal from the standpoint of assessing the effect of the factors A and B, the scenarios depicted in Figure 4.2 and 4.3 present major problems. In particular, it is imperative that an analysis using conditional effects be performed for the data shown in Figure 4.2 since both main effect estimates are zero. Similarly, conditional effects must also be used when data plot as in Figure 4.3. The situation illustrated in Figure 4.1 is almost as bad as the conditional effects differ in sign, which will always be the case when the lines cross.

So only for the Figure 4.4 scenario does it make sense to do a traditional analysis. Although we of course use software to perform these analyses, it is useful to see how the effect estimates are computed, and the computing formulas are given below, with "2" denoting the high level of the factor and "1" denoting the low level.

$$A = \tfrac{1}{2}(A_2 B_1 - A_1 B_1 + A_2 B_2 - A_1 B_2)$$
$$B = \tfrac{1}{2}(A_1 B_2 - A_1 B_1 + A_2 B_2 - A_2 B_1)$$
$$AB = \tfrac{1}{2}[A_2 B_2 - A_2 B_1 + (A_1 B_2 - A_1 B_1)]$$
$$= \tfrac{1}{2}(A_2 B_2 - A_2 B_1 - A_1 B_2 + A_1 B_1)$$

4.4 WHY NOT ONE-FACTOR-AT-A-TIME DESIGNS?

As we consider statistically designed experiments and consider more than one factor, it seems logical to pose the following question: "Why not study each factor separately rather than simultaneously?" Indeed, this is frequently done. We can use Figure 4.1 to show why this won't work, and indeed it won't work in general, although one-factor-at-a-time designs can sometimes be useful. (See Section 13.1 for a detailed discussion of such designs.)

Assume that a company's engineering department is asked to investigate how to maximize process yield, where it is generally accepted that temperature and pressure have a profound effect on yield. Let factor A denote temperature and factor B denote pressure. Three of the engineers are given this assignment, and their initials are BW, JC, and LM, respectively. Each engineer conducts his own experiment. Assume that BW and JC each investigate only one factor at a time, whereas LM decides to look at both factors simultaneously. Assume further that Figure 4.1 depicts what can be expected to result when both factors are studied together.

If engineer BW had used the low temperature and varied the pressure from low to high, he would conclude that the best way to increase the yield is to increase the pressure, whereas he would have reached the opposite conclusion if he had used the high temperature. Similarly, if engineer JC had set the pressure at the high level, he would have concluded that the best way to increase yield is to reduce the temperature, whereas he would have reached the opposite conclusion if he had used the low pressure level.

Engineer LM, on the other hand, would be in the proper position to conclude that interpreting a traditional main effects analysis would not be possible because of the interaction effect of the two factors.

This type of feedback is not available when factors are studied separately rather than together. These "one-at-a-time" plans have unfortunately been used extensively in industry. They are considered to have very little value, in general, although Daniel (1973, 1976, p. 25) discusses their value when examining three factors. Qu and Wu (2005) also discuss conditions under which one-at-a-time plans might be used. In particular, the designs can have value when there are hard-to-change factors. These designs are discussed extensively in Section 4.19.

4.5   ANOVA TABLE FOR UNREPLICATED TWO-FACTOR DESIGN?

The data from a designed experiment is typically presented, whenever possible, in an ANOVA table, as was done starting in Chapter 2. In this section we show why this won't work for an unreplicated 2^2 design, and indeed won't work for any unreplicated two-level factorial design.

Let's try to construct the ANOVA table for the data in Figure 4.2. We could compute the sum of squares corresponding to each effect analogous to the way that sums of squares were computed in Chapter 1. Since we already have the effect estimates, however, it would be easier to obtain the sums of squares by squaring the effect estimates. In general, for any two-level factorial, $SS_{\text{effect}} = r(2^{k-2})(\text{effect estimate})^2$, with r denoting the number of replicates and k denoting the number of factors. Thus, for an unreplicated 2^2 design this reduces to $SS_{\text{effect}} = (\text{effect estimate})^2$. The sums of squares for the Figure 4.2 data are thus 0, 0, and 400 for A, B, and AB, respectively.

Since we now have these sums of squares, we might attempt to construct an ANOVA table. Remembering from Section 2.1.3.3 that the degrees of freedom (df) for "Total" is always the total number of observations minus one, and that the df for a factor is always the number of factor levels minus one, we thus have df(Total) = 3, df(A) = 1, and df(B) = 1. The df for any interaction effect is always obtained as the product of the separate df of each factor that comprises the interaction. Thus, in this case we have df(AB) = (1)(1) = 1.

TABLE 4.2 ANOVA for the Data in Figure 4.2

Source of Variation	df	SS	MS	F
A	1	0	0	
B	1	0	0	
AB (residual)	1	400	400	
Total	3	400		

If we add the df for A, B, and AB, we recognize immediately that we have a problem. Specifically, there is no df left for estimating σ^2. Thus, unless we have an estimate of σ^2 from a previous experiment (remember that experimentation should generally be thought of as being sequential) we have a case in which the interaction is said to be "confounded" (i.e., confused or entangled) with the "residual," where the latter might be used in estimating σ^2. We can summarize what we know to this point in the ANOVA table given in Table 4.2.

Notice that the F-values are not filled in. It is "clear" that there is no A effect and B effect since the sum of squares for each is zero. (Remember, however, that we recently saw that each does have an effect on the response variable; their effect is simply masked by the interaction effect.)

It was stated in the section on one-factor ANOVA that the analysis is not influenced by whether the factor is fixed or random. This is not true when there is more than one factor, however. In general, when both factors are fixed, the main effects and the interaction (if separable from the residual) are tested against the residual. When both factors are random, the main effects are tested against the interaction effect, which, in turn, is tested against the residual. When one factor is fixed and the other random, the fixed factor is tested against the interaction, the random factor is tested against the residual, and the interaction is tested against the residual. (By "tested against" we mean that the mean square for what follows these words is used in producing the F-statistic.)

In this example the interaction is not separable from the residual because the experiment has not been "replicated"; that is, the entire experiment has not been repeated so as to produce more than one observation ($r > 1$) per treatment combination. This should be distinguished from *multiple readings* obtained within a *single* experiment, which does *not* constitute a replicated experiment (i.e., the entire experiment is not being replicated). This may seem like a subtle difference, but it is an important distinction that is discussed further in Section 4.7 (see also Box, Hunter, and Hunter, 1978, p. 319).

If a prior estimate of σ^2 is available, possibly from a previous replicated experiment with perhaps slightly different factor levels, that estimate could be used in testing for significance of the main effects. If a prior estimate is not available, we might still be able to obtain an estimate of σ^2.

Tukey (1949) proposed a test for detecting interaction of a specific functional form for an unreplicated factorial. The general idea is to decompose the residual into an interaction component and an experimental error component, and perform an F-test on the interaction. If the test is not significant, then σ^2 might be estimated using the residual. It should be recognized, however, that this test will detect only an interaction that can be expressed as a product of main effects times a constant.

It should be noted that there is a difference, conceptually, between "experimental error" and "residual," and the latter cannot be used, in general, as a substitute for the former. Experimental error should be regarded as the variability that results for a given combination of factor levels in a replicated experiment, and is composed of variability due to factors not included in the experiment, sampling variability, and perhaps variability due to measurement error. A residual (as in residual sum of squares) may consist of various interaction terms that are thought to be not significant, in addition to experimental error. To estimate σ^2 using MS_{AB} in Table 4.2 would be to act as if no model fits the data because the only term that has a nonzero sum of squares is the AB term. But we have a perfect fit using that term and there is also much information in the conditional effects for A and B.

It is interesting to note that Tukey's test would not detect this interaction, since it can detect only interactions that are a constant times the product of main effects, as stated previously. In this case the test would indicate that the interaction is zero because the main effects are zero. We should remember that the test is not a general test for detecting the presence of interaction, nor can there be such a test for an unreplicated experiment.

This idea of experimental error versus residual is a very important one, and it is indicated in Section 4.6 how we can go wrong by using an interaction to estimate σ^2 for a particular set of actual data.

Can the analysis begun in Table 4.2 be completed? The analysis was actually completed *before* the (attempted) construction of the ANOVA table, as the data are not appropriate for analysis by an ANOVA table. We have seen that there is indeed a temperature effect and a pressure effect, and the interaction profile in Figure 4.1 clearly shows the strong interaction.

In the absence of a prior estimate of σ^2, the only course of action that would allow completion of the ANOVA table would be to use the interaction as the residual, and to test the two main effects against it. This, of course, would be sheer folly for these data as it would lead to the conclusion that nothing is significant, whereas in actuality all three effects are of significance.

This example was given for the purpose of illustrating how a routine analysis of data could easily lead to wrong conclusions. This message can also be found in other sources such as Box et al. (1978, p. 329) and Daniel (1976). The reader is referred to Daniel (1976, p. 20) for additional reading on the interpretation of data from a design of this type when a significant interaction effect exists.

It might appear that one solution to the problem of not being able to separate an interaction from a residual is simply to replicate the experiment. Although this is generally desirable, it is not always practical. One possible impediment is, of course, money, the data may be so expensive to collect as to preclude replication.

There are, however, some methods for assessing the possible significance of main effects and interactions in unreplicated experiments. One of these methods is due to Daniel (1959) and consists of plotting effect estimates on normal probability paper. This is illustrated in a later example.

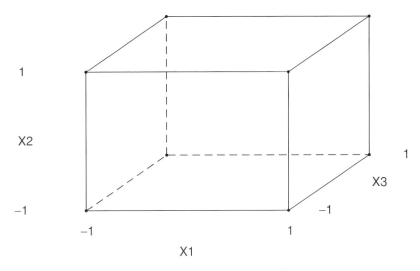

Figure 4.5   Configuration of points in a 2^3 design.

4.6   THE 2^3 DESIGN

The 2^3 design has been used in many applications and George E. P. Box, in lamenting the relative lack of use of experimental designs by engineers, has stated on occasion that just getting them to use a 2^3 design would be a big step in the right direction. Quite frankly, this would not be a big enough step, however, because 8-point designs do not have good power for detecting differences between means.

The 2^3 design is just an extension of the 2^2 design to one additional factor, which would be denoted as factor C. There are seven estimable effects, although the ABC interaction generally will not be significant. If it is significant, that could be caused by one or more data points. This is illustrated in Section 4.10. There are various ways in which we can think of the ABC interaction. Specifically, $ABC = AB(C) = AC(B) = BC(A)$. The expression $AB(C)$ means the interaction between the AB interaction and factor C (and similarly for the other expressions). This interaction will be significant if the two AB interaction profiles (as in Figure 4.1, for example) differ considerably over the two levels of C.

The points in the design form a cube, as is shown in Figure 4.5.

Recall the discussion in Section 4.1 regarding the magnitude of effects that can be detected with a 2^2 design. Does an unreplicated 2^3 design have the same shortcoming? Only a slight problem, because the smallest difference that can be detected is 2.29σ (using the expression for Δ given in Example 4.2), with as before, a probability of .90 of detecting the difference, a significance level of $\alpha = .05$, and σ assumed to be known. We can relate this to a z-value or a t-value from a simple hypothesis test in an introductory course for testing $H_0 : \mu_1 = \mu_2$. We would generally reject the hypothesis when the test statistic, z or t, has a value much greater than 2, and we would certainly reject the hypothesis for a value greater than 3. Relative to these

**TABLE 4.3 Number of Replicates
Needed to Detect a Minimum Detectable
Effect of 2.0σ for 2^k Designs**

k	Number of Replications
3	2
4	1
5	1
6	1

numbers, we would prefer to see a multiplier less than 2.29, but this is not a serious shortcoming of the 2^3 design.

A 2^3 design would have to be replicated in order for the multiplier of σ to be clearly acceptable. For example, the multiplier drops from 2.29 to 1.62 for two replicates (i.e., two observations per treatment combination), 1.32 for three replicates, and 1.15 for four replicates. The last two, in particular, are much more reasonable values than the multiplier 2.29 for the unreplicated 2^3 design.

If a practitioner considers, for example, a multiplier of 2.0 to be acceptable, Table 4.3 could be used to determine how many replications are needed for a 2^k design with various values of k. Again, this assumes a probability of .90 of detecting the difference and a significance level of $\alpha = .05$. For extensive tables, the reader is referred to Lynch (1993).

Note, however, that the tables of Lynch (1993) assume that σ is estimated, and more specifically that it is estimated from a certain number of degrees of freedom. Hence, the numbers given there will not agree with formulas that assume σ to be known. Of course, typically σ is unknown and the likelihood of detecting effects that are real and of considerable magnitude will depend upon whether or not the design is replicated. This is because whereas the analysis is straightforward for a replicated design, the estimate of σ for an unreplicated design is affected by the magnitude and number of the "small" effects, and a poor estimate will result when the number of such effects is small. These methods are discussed and illustrated in Section 4.10.

These methods generally depend upon the assumption that high-order interactions are not significant, but the folly of using the highest-order interaction in an unreplicated 2^3 factorial design was illustrated by Daniel (1976) and discussed in Ryan (2000).

Example 4.5

In describing a cement experiment, Daniel (1976) stated that σ was known to be about 12, whereas the estimate of σ obtained using the ABC interaction is 3.54— approximately $1/3$ of the assumed value.

Other problems with those data that indicate the need for a careful analysis include the large interactions, with the effect estimates shown in Table 4.4.

There are two numbers that preclude a customary main effects analysis: the BC effect estimate of 47.5 and the AB effect estimate of 13.5. That is more than half the absolute value of the C effect estimate, which means that the conditional main effects

TABLE 4.4 Effect Estimates from Example in Daniel (1976)

Effect	Estimate
A	15.5
B	−132.5
C	−73.5
AB	13.5
AC	1.5
BC	47.5
ABC	2.5

of C will differ considerably. These are easy to compute because, as stated previously, they are simply the main effect estimate plus and minus the interaction effect estimate (see chapter Appendix A for details). That is, $C \pm BC = -73.5 \pm 47.5$, so the C conditional main effect estimates are −121 and −26. These are the C effect estimates that would be obtained by splitting the data into two halves on factor B, using first the low level of B and then the high level of B, respectively.

In like manner it can be seen that the B conditional main effects are −180 and −85, which also differ greatly. Similarly, the conditional main effects of A that are obtained by splitting the data on B, since the AB interaction is large relative to the A effect, are 2 and 29.

In this example all of the conditional effects had the same sign. This will often not be the case and they will be of opposite signs for interaction plots of the form shown in Figure 4.1 and anything that resembles such a plot. For less extreme configurations, either or both sets of the two conditional main effects could have different signs. Exercise 4.3 is an example of this, and Wu and Hamada (2000, p. 109) give another example and make an important point that when the conditional main effects differ in sign, there may be a point between the levels in the experiment for which the response is a flat function of the factor. As they point out, this could have important ramifications if robust parameter design (see Chapter 8) were used, as we would then expect the response to not vary much over, say, a manufacturing variable that could not be tightly controlled during production, but which could be controlled within a reasonable range.

Although it is apparent from the list of averages in Table 4.4 that a conditional effects analysis must be used, the same message would come from an ANOM display, with the latter showing not only the magnitude of the effect estimates but also the average response values that determine the numerical value of the effect estimate. Of course practitioners would want to know these average response values, in addition to the effect estimates.

Example 4.6

Kinzer (1985) described an experiment that utilized a 2^3 design plus two centerpoints, with the latter used to detect curvature of the response if it existed since curvature cannot be detected with only two levels. The objective of the experiment was to determine influence of curing conditions on the strength of a composite material product. The development engineers identified three factors that they felt contributed most significantly to the product strength: autoclave temperature, autoclave time, and air cure. Each point was replicated five times, but only the average of the five values was given in Kinzer (1985).

The analysis was performed in the raw units of the factors, something that should not be done if interaction terms are to be fit, as discussed in Section 4.1. This is because an interaction term will not be uncorrelated with the factors in a 2^3 design when raw form is used, but will be uncorrelated when coded form is used. This can create contradictory results between the two forms, as was illustrated in Section 4.1 and by Ryan (2000, p. 401).

For this example, Kinzer (1985) used stepwise regression to arrive at a model, which had the following terms: X1, X3, and X1 * X3. This model was selected from an initial fit that had all the linear terms, all four of the possible interaction terms, and a single term that represented the sum of the squares of the factors. There are high correlations between some of these terms, however, which can cause problems when a technique such as stepwise regression is used. Nevertheless, when the reported data were used (not the data for all replicates), the same conclusion was reached regardless of whether the analysis was performed using coded form or raw form. Specifically, X1 and X3 are significant main effects, and the model that uses these terms has an R^2 value of .934. This is considerably higher than the R^2 value of .794 given by Kinzer (1985) for his model with three terms, when all the data are used.

There are two important points here: (1) a different model is obtained when averages are used instead of all the raw data, and (2) R^2 is much lower when all the data are used. The latter is generally true and we don't want the selected model to depend upon whether or not averages were used.

4.7 BUILT-IN REPLICATION

Although the analysis of unreplicated factorials can be challenging when there is a high percentage of real effects, as will be illustrated starting in Section 4.9.1, it is worth noting that under certain circumstances it is reasonable to proceed as follows. Assume that a 2^5 design is used and one of the main effect estimates is practically zero, the conditional effects for the factor are also very close to zero, and all interaction effect estimates involving the factor are also close to zero. It would then be safe to drop the factor from the analysis. Doing so then produces a replicated 2^4 design with two replicates, which could then be analyzed using ANOVA.

The situations under which this approach can be used will be quite limited, however, and indiscriminate dropping of factors can lead one astray. For example, let's assume that an experimenter generally uses a 2^5 design and drops the factor with the smallest main effect estimate, reasoning that all five factors are not going to be significant. But a small main effect estimate could be caused by a large interaction, with the conditional main effects differing considerably from zero. If so, dropping the factor would be the wrong thing to do. In this case it would be better to rely on a normal probability plot analysis (see Section 4.9.1).

How often will we be better off dropping a factor or two and performing the analysis using ANOVA as opposed to using a probability plot analysis? Any answer to that question is of course conjectural, but let's consider some possible scenarios. Again assume that a 2^5 design is used and the main effect estimates are A (16.82), B (24.84), C (31.16), D (19.43), and E (1.16). Looking at only these numbers, it might seem safe to eliminate factor E and analyze the data as a replicated 2^4, although of course the experiment wasn't designed that way. We can't do this, however, without looking at the magnitude of all interactions involving factor E. If any of those are large, then factor E cannot be removed from the analysis because then at least one conditional effect will differ considerably from zero. If all four of the two-factor interactions involving E are close to zero, then there should be enough terms to use in constructing a meaningful pseudo-error term (PSE), since hardly any of the higher-order interaction terms could be expected to be large.

When the importance of examining conditional effects is considered and the discoveries that might be made when all factors are used in the initial analysis are also considered, it is somewhat difficult to imagine scenarios where the probability plot approach will fail and dropping a factor or two and using an ANOVA analysis would be the right thing to do.

4.8 MULTIPLE READINGS VERSUS REPLICATES

It seems safe to say that true replicated experiments rarely exist in practice, even though probably every experiment that even resembles a replicated experiment is analyzed as if it were a replicated experiment.

The conditions that must be met for an experiment to be termed a replicated experiment are rather stringent. In general, everything must be reset. For example, if factor levels in an experiment were set by turning knobs, then the knobs must be turned to some neutral position and then set at the desired level before there is an additional run within a set or runs or between replicate runs. To obtain observations on the response variable without resetting any factors is to obtain multiple readings, not replicate values, and the errors will not be independent, as pointed out by Lucas (1999). As stated in Lucas (1999, p. 29) in his 1997 Annual Quality Congress talk, Lucas asked three questions of the over 200 people who attended his talk. All of them raised their hands when he asked them if they were involved in running experiments and if they used randomization. However, when they were asked if they set each factor to a neutral level and then reset it, only four people raised their hands.

Based on the results of this informal survey, it seems safe to assume that very few experiments in which randomization is used actually have complete randomization.

Even if the proper procedures are followed, it still may not be possible to have true replicates. For example, Ryan (2004) described a lead recovery from paint experiment in which true replicates were not possible because particle sizes could not be ground to be an exact size. Since the extent of the departure from true replicates may not be known under such conditions, nor the effect of that departure on the statistical analysis, it is best to also analyze the data by obtaining an average for each replicated point and performing an analysis using the averages, as was done in that case study.

4.9   REALITY VERSUS TEXTBOOK EXAMPLES

Experiments do not always work out as designed, and sometimes practical consider-ations can result in some unusual designs. One such example is the study described by Ryan (2004). Five factors were examined, each at two levels. There were 112 experimental units (specimens) available, and 7 could be used in each run. Of course, 112 is not an integer multiple of 2^5, so the only way to produce 112 runs starting from a 2^5 design is to use something other than an equal number of replicates for each treat-ment combination. That is what was done, with half of the treatment combinations replicated three times and the other half replicated four times.

The fact that an unequal number of replicates was used creates problems, regardless of how the 32 combinations are split into two halves. The problems can be minimized if the split is performed in an optimal manner, however. This is explained when we return to this example in Section 5.12. (A detailed discussion is appropriate for Chapter 5 because splitting the data in half strongly relates to fractional factorials.)

4.9.1   Factorial Design but not "Factorial Model"

Ideally, the design that is used should facilitate the estimation of the coefficients of the model that results from analysis of the data from the experiment. When a factorial design is used, there is the tacit assumption that if all the important factors have been identified and are included in the experiment, a good model will be a linear model that includes some or all the factors used in the experiment. That isn't necessarily true, however. Bisgaard, Vivacqua, and de Pinho (2005) illustrated this by using an example from Brownlee (1953). A 2^4 design was used and a normal probability plot of the effect estimates (given in Fig. 4.6) shows that no effect is significant. (Lenth's PSE, which is shown on the plot, is explained in detail in Section 4.10.) At first thought, this might seem to suggest that the experimenters did a poor job of selecting the effects to be studied, but we need to look further.

The Pareto chart of effect estimates exhibits some real oddities (Fig. 4.7). In particular, although no effect is even close to being significant, the stairstep appearance of the chart is odd, as is the fact the four-factor interaction is much larger than two of the main effect estimates. (Note that the absolute values of the effect estimates are plotted here on the Pareto chart, whereas the actual values of the effect estimates are plotted on the normal probability plot.)

The determination of the threshold value that is denoted by the vertical line needs some explanation. The line is drawn at the value of $t \times$ PSE with the latter being Lenth's PSE that is explained in Section 4.10, as stated, and $t = t_{.025,\nu}$ with ν denoting the degrees of freedom defined by MINITAB, which was used to produce this graph, as the total number of effects divided by 3. The value of $t \times$ PSE also determines whether

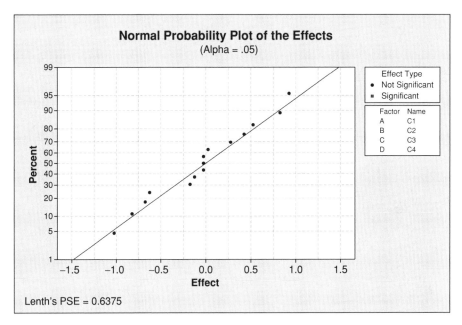

Figure 4.6 Normal probability plot of effect estimates.

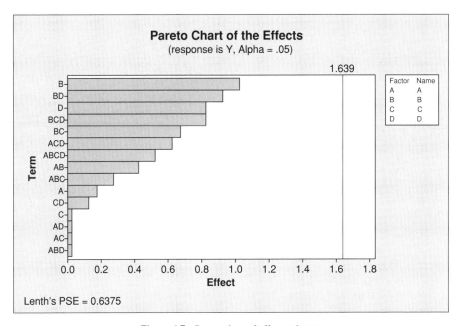

Figure 4.7 Pareto chart of effect estimates.

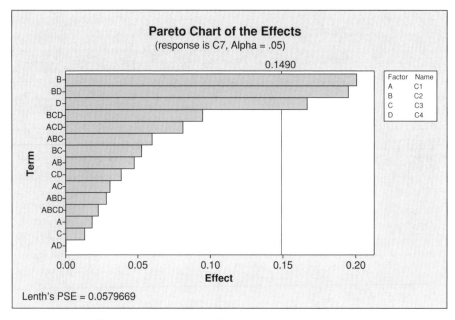

Figure 4.8   Pareto chart of effect estimates after transformation.

or not any effects in a normal probability plot are declared significant, although the threshold values are $\pm t \times$ PSE, since effect estimates can be negative and some of them are negative in this example. We can see why no effect was declared significant in Figure 4.6 as the PSE value is given as 0.6375 and all but one of the effect estimates is in the range $(-1, 1)$ and the one that is outside the range is just very slightly outside the range. Since $t_{.025,v} > 1.96$ for any value of v, it follows that $0.6375 \times t_{.025,v}$ will exceed 1.2, which clearly exceeds the absolute values of any of the plotted effect estimates. (More specifically, $(0.6375)(2.5706) = 1.639$.)

Since the ratio of the largest response value to the smallest response value exceeds 3, there is the potential for improving the fit of the model by transforming the response variable. It should be noted, however, that the response variable is generally transformed for the purpose of trying to meet the model assumptions, and not for improving the fit of the model. Nevertheless, Bisgaard et al. (2005) elected to use the reciprocal transformation, stating that in attempting to find a suitable transformation "...for most practical purposes it is easier to simply proceed by trial and error." A trial-and-error approach is reasonable only if we are willing to restrict consideration to common transformations, of which there is a small number.

The Pareto chart that results from the analysis using the transformation is shown in Figure 4.8.

We see immediately that this is much better, as there are three effects that are clearly larger than the rest of the effects and are judged as being significant. Thus, we might wish to fit a model with the B and D main effects and the BD interaction. We should not be quick to dismiss the main effect of C, however, since the AC interaction is the fourth largest effect. Nevertheless, since the interaction does not show as being significant, we will proceed with these three effects.

If we fit a regression model with the transformed response variable, Y^{-1}, we obtain $\widehat{Y}^{-1} = 0.414 + 0.100\,B - 0.0833\,D - 0.0975\,BD$. Bisgaard et al. (2005) took a different approach, however, concluding that this is a "critical mix situation," meaning that the only treatment effect occurs when factor B is at the high level and factor D is at the low level. Both factors have presence/absence levels with the high level of factor B signifying that the crude material used in the experiment was "boiled" and the low level signifying that it was "not boiling"; and the low level of factor D means that the crude material was precipitated from either of two solvents, with the high level signifying that this did not occur.

The response values at 3 of these 4 treatment combinations of the total of 16 do support this view, but the response at the fourth of these treatment combinations (0.53) is not much larger than the next two largest values, both of which are 0.45. Bisgaard et al. (2005) used an indicator variable approach and arrived at the model $\widehat{Y}^{-1} = 0.318 + 0.377\,\mathbf{I}(x_{\mathrm{B}}, x_{\mathrm{D}})$, with $\mathbf{I}(x_{\mathrm{B}}, x_{\mathrm{D}}) = 1$ if $B = 1$ and $S = -1$, and 0 otherwise. Not surprisingly, the standardized residual at the response value of 0.53 is -2.13. Although this is not a large value, it is large enough to raise questions about the critical mix assumption. Furthermore, when the first model is fit with the three terms, the fit is slightly better as the residual sum of squares is 0.1136 compared to 0.1163 for the indicator variable approach.

4.10 BAD DATA IN FACTORIAL DESIGNS

If factors are chosen judiciously and an experiment run properly, a reasonable number of effects should be significant. Something is generally wrong when we obtain extreme results such as no effects being significant or almost all effects being significant.

The latter will often be caused by bad data. As an example, consider the following data resulting from use of a 2^3 design, with the data given in Yates order: 16, 22, 18, 24, 19, 23, 20, and 78. Assume that the last observation should have been 28 but was misread as 78. One extreme bad data point such as this will result in none of the effects estimates being small, which is nothing being declared significant by Lenth's (1989) method because a reasonable pseudo-error cannot be constructed. Indeed the normal probability plot for this example is given below and we can see that (1) the plot looks very peculiar, and (2) no effects are declared significant.

This plot was generated using MINITAB; the method that MINITAB uses for determining if an effect is significant is due to Lenth (1989). The latter uses a PSE that is defined as follows. Let $s_0 = 1.5 \times \text{median}\, c_j$, with c_j denoting the estimate of the jth estimable effect. Then the PSE is computed as

$$\text{PSE} = 1.5 \times \underset{|c_j| < 2.5 s_0}{\text{median}}\, |c_j|$$

with the PSE essentially computed from a trimmed median of $|c_j|$ values, as the median is computed using only values of $|c_j|$ that are less than $2.5s_0$.

Lenth's method and alternatives were investigated by Haaland and O'Connell (1995), whose modification was not as simple as Lenth's method. They used the general form

$$\widehat{\sigma}_{\text{PSE}}(q, b) = a_{\text{PSE}}(q, b) \cdot \text{median } \{|\hat{\theta}_i| : |\hat{\theta}_i| \leq b \cdot s_0(q)\} \qquad (4.3)$$

with $a_{\text{PSE}}(q, b)$ denoting a consistency constant, b denotes a tuning constant, the $|\hat{\theta}_i|$ are the absolute values of the ordered effect estimates, and q is obtained from the initial robust estimator of scale $s_0(q)$. The latter has the general form

$$s_0(q) = a_0(q) \cdot \text{quantile } \{q; |\hat{\theta}_i| \, i = 1, \ldots, k\} \qquad (4.4)$$

with $a_0(q) = 1/\Phi_0^{-1}(q)$ and $\Phi_0^{-1}(q) = \Phi^{-1}[(q + 1)/2]$, with Φ denoting the cumulative distribution function of the standard normal distribution. Thus, for example, if $q = 0.5$, then $s_0(q) = 1.48 \cdot \text{median } (|\hat{\theta}_i|)$. Lenth's method is essentially equivalent to $\widehat{\sigma}_{\text{PSE}}(0.5, 2.5)$. As explained by Haaland and O'Connell (1995), Daniel (1959) used $q = 0.683$ in s_0, and also used this value of q in $\widehat{\sigma}_{\text{PSE}}$. He did not, however, have a trimming threshold of $b \cdot s_0(q)$ but rather trimmed large effects by inspection. We will return to the paper by Haaland and O'Connell (1995) in Section 5.3.1, since their paper was on unreplicated, 16-point fractional factorials, but the results also apply to unreplicated full factorials.

Of course the objective with any PSE and with any method for identifying significantly large effects in an unreplicated factorial (or fractional factorial, as in Chapter 5) is to have a method that works well when there are few real effects and also when there are many real effects, yet has a tolerable false alarm rate.

Consider the form of $s_0(q)$. A single bad data point (as in this example) can cause none of the effects to be small without any of the effects being particularly large, as will be seen in Figure 4.9. When this occurs, $s_0(q)$ will break down regardless of what value of q is used. Assume that $q = 0.5$ since this is a common choice. As is stated in Section 4.11, Daniel (1976, p. 75) stated that seven significant effects is about average for a 2^5 design. Notice that this is far fewer than half of the estimable effects. In general, less than half of the estimable effects should be real, so the median of the $|\hat{\theta}_i|$ should generally be estimating an effect that is zero.

Notice that when $s_0(q)$ is inflated, the value of $b \cdot s_0(q)$ in $\widehat{\sigma}_{\text{PSE}}(q, b)$ will also be inflated. The consequence of this is that certain large effects will not be trimmed in determining median $(|\widehat{\beta}_i|)$, with the consequence that certain large effects will not be trimmed in computing $\widehat{\sigma}_{\text{PSE}}(q, b)$. The consequence of this would be that $\widehat{\sigma}_{\text{PSE}}(q, b)$ is inflated, which would reduce the number of effects that are identified as real. If the inflation of $\widehat{\sigma}_{\text{PSE}}(q, b)$ is severe, then multiple real effects may not be identified as such.

For the present example, with 28 being recorded as 78, we can regard the effects computed using 28 as the truth and compare the results with those obtained using 78. Using the value 28, Lenth's PSE = 1.5 and A is identified as a real effect. As

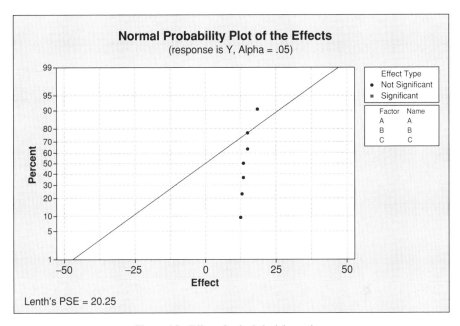

Figure 4.9 Effect of a single bad data point.

is shown in Figure 4.9, PSE = 20.25 when 78 is used and no effect is identified as significant. This shows how bad data can result in an erroneous conclusion when a normal probability plot of effects estimates is used.

There is an important point that should be made regarding this plot. Before hypothesis test statistics became a part of probability plots of effects, the recommended use of such plots was to construct a line through the majority of the points. In a typical experiment most effects will *not* be significant, so points that lie more than slightly off the line would represent significant effects. Notice that most of the points do practically lie on a line, but the line in the plot is not drawn through the center of the points. When the axes are labeled in this manner, the line must go through the point (0, 50 percent), but notice that there is no way to draw a line through this point and have it lie even close to the majority of the points.

We might ask why none of the effects are identified as being significant since most of the points lie well off the line. As indicated previously, Lenth's method, and indeed any method for obtaining a pseudo error from an unreplicated factorial, implicitly uses small effects in obtaining the pseudo error, but here practically all the effects are of virtually the same size, as can be seen from Figure 4.9. This causes none of the effects to be identified as significant, despite the fact that one of the effect estimates is larger than two of the observations, and the average effect estimate is more than half of the average of the seven good observations. Thus, the effects are actually large relative to the magnitude of the observations but the error causes none of the effects to be identified as significant.

In this example the bad data point was obvious just from inspection of the data, but bad data points won't always be so obvious. In general, whenever, the results differ from what might be expected, it is important that the data be checked carefully. If no bad data can be detected, it would be wise to perform a follow-up experiment and compare the results with the original experiment.

A follow-up experiment may be necessary even when bad data are detected and discarded, as discarding a bad data point will cause the orthogonality of an orthogonal design to be lost. For example, if the bad data point in this example is deleted, the correlations between the columns of the design matrix are all $-.167$. These are not large correlations, but the definition of an effect estimate becomes somewhat shaky as there will be an unequal number of observations at the high and low levels of each factor, and there will also not be a direct relationship between the effect estimates and the regression coefficients, as there is when all the data are available. Furthermore, the ABC interaction could not be estimated because of the loss of one degree of freedom.

Bad data that are detected and discarded present the same general type of problem as is caused by data that are simply missing (one or more runs that were not performed because of equipment that broke, etc.). Missing data and incomplete data that result from bad data being detected and discarded are discussed in the following sections.

Box (1990–1991b) illustrated a simple method, due to Daniel (1976), for finding bad data in factorial designs. The method consists of constructing a normal probability plot of effect estimates, which would normally be done anyway for an unreplicated factorial design. For illustration, Box used a 2^4 design. When we think about the configuration of $+1$ and -1 components in the table for a 2^4 design, analogous to Table 4.1, it should be apparent that each row of the table has eight plus signs and eight minus signs. Therefore, if a number like 10.3 is erroneously recorded as 103, eight of the effect estimates will be greatly overestimated and the other eight will be greatly underestimated, perhaps resulting in two distinct lines of plotted points on a normal probability plot.

Assume that the data, in Yates order, are supposed to be 4.8, 5.9, 10.3, 6.4, 11.6, 12.6, 9.8, 12.4, 6.2, 8.4, 4.5, 7.8, 7.0, 8.1, 10.5, and 10.3 but the 6.2 is erroneously recorded as 62. Of course this number is so much bigger than the other numbers that we would know something is wrong just by looking at the numbers. With such a large error, this should cause two distinct lines on the normal probability plot, which is what we see in Figure 4.10.

Of course the signal from the plot would not be as strong if the bad data value was considerably smaller, as the reader is asked to show in Exercise 4.34.

An interesting case study that illustrates the steps that led to the detection of bad data in a designed experiment is described by Mark Anderson and Pat Whitcomb of Stat-Ease, Inc., in their article "How to use graphs to diagnose and deal with bad experimental data," which can be viewed at http://www.statease.com/pubs/baddata.pdf. The authors make an important point that an outlier is an outlier relative to the fitted model, which might not be an appropriate model. They illustrate this with the first of two datasets from Box (1990) that they use for illustration. Specifically, they show that two outliers disappear when a log transformation of the response variable is used.

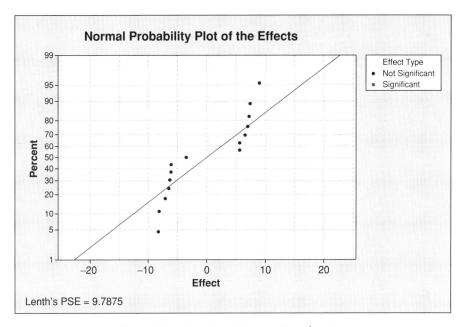

Figure 4.10   Effect of a bad data point in a 2^4 design.

Example 4.7

Yin and Jillie (1987) described the application of a 2^4 design in developing a nitride etch process. Although the authors apparently did not identify any bad data, suspicions arise when we examine the data. In particular, the absolute values of the effect estimates given in Table 4.5 exhibit some peculiarities.

What is most disturbing about this list is the presence of a two-factor interaction as the second largest effect, and quite surprisingly, the four-factor interaction being the fifth largest effect. Indeed seven of the nine largest effects are interaction effects. This is quite surprising and greatly complicates the analysis. Since the *ABCD* interaction effect is an order of magnitude larger than the seven smallest effects, we should not be surprised if it is declared significant in a normal probability plot assessment. The plot is given in Figure 4.11.

Seven of the 15 effects are judged significant by Lenth's method that is used in this output produced by MINITAB, with the default value of $\alpha = .10$ used. (The *ABCD* effect is also significant when $\alpha = .05$ is used.) It is disturbing that five of the seven significant effects are interaction effects. One or more extreme observations could cause such an anomaly, so it is desirable to search for extreme observations. Accordingly, a dotplot of the response values is given in Figure 4.12.

Obviously there are some extreme points, in addition to small groups of points that are clearly separated from other groups. The four largest values, which seem to be outliers, all occurred at the four treatment combinations of the high level of factor *D* and the low level of factor *A*. Since the estimate of the *AD* effect is obtained as

**TABLE 4.5   Rank Order of Effect
Estimates by Largest Absolute Value**

Effect	Estimate
D	303.1
AD	153.6
A	101.6
BC	43.9
ABCD	40.1
BCD	25.4
AC	24.9
ABC	15.6
AB	7.9
C	7.4
ACD	5.6
ABD	4.1
CD	2.1
B	1.6
BD	0.6

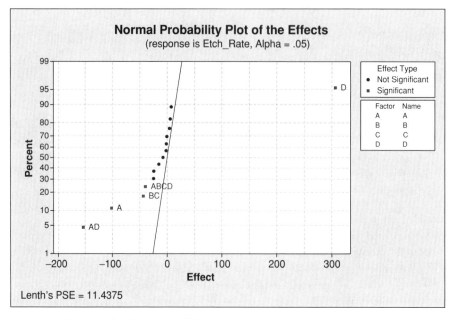

Figure 4.11   Normal probability plot of effects for Yin and Jillie (1987) data.

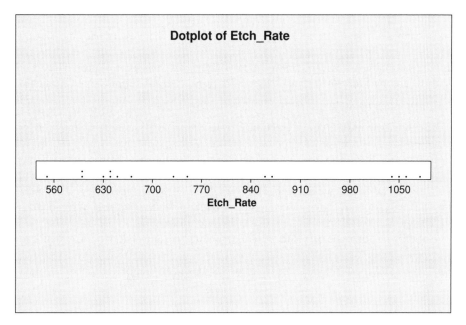

Figure 4.12 Dotplot of etch rate data.

the average of the eight observations for which, using the coded levels, A times D is positive minus the average of the other eight observations for which A times D is negative, having four large observations occur at the treatment combinations for which A times D is negative is likely to cause a large negative value, as occurs here.

Clearly the authenticity of these four values should be verified. It may simply be that the specific levels of these two factors "interact" as, say, in a chemical experiment to produce extreme values. Or perhaps there are some major measurement issues for this combination of levels.

If these four values are valid, then a conditional effects analysis *must* be performed, especially since the AD interaction is considerably larger than the A effect. The conditional main effects of A are -52 and 255.2. Clearly the unconditional effect estimate of 101.6 does not well represent either of these two numbers. It is especially troublesome that they differ in sign. A conditional main effect being negative and not close to zero while the unconditional main effect is a large positive number could result in an erroneous conclusion being drawn, although this sign difference could be of value in robust parameter design, which is covered in Chapter 8.

The significance of the $ABCD$ interaction cannot be explained by the four large values because two of them occur at treatment combinations for which the product of the four factor levels is positive and two at treatment combinations for which the product is negative. This makes the significance of the interaction rather perplexing. The significance of that interaction is due in large part to the fact that the response value that is by far the smallest occurs at a treatment combination for which the product is positive, as does the second smallest value.

If all the observations were valid data points—and that seems highly questionable—it would be necessary to do a full conditional effects analysis, starting with the three-factor interactions, which are affected by the large four-factor interaction. In particular, the BCD effect was judged significant in the normal probability plot analysis and the two conditional effects differ in sign because of the large four-factor interaction.

In essence, large high-order interactions create a domino effect because they render meaningless the lower-order interaction effect estimates and main effect estimates. Unfortunately, the analysis of conditional effects is complicated by the fact that conditional effects are correlated across different effects, and conditional effects and unconditional effects have different variances. Nevertheless, if these were valid data, such an analysis would have to be attempted, but will not be attempted here since the significant four-factor interaction casts aspersions on the data.

Not only can there be bad data with experimental designs, but of course there can also be errors in the listing of a design layout, as is illustrated in Section 13.4.1.2, along with a method for detecting the error and how it should be corrected.

4.10.1   ANOM Display

ANOM displays for a single factor were used in Chapter 2. When there are multiple factors, and especially when there are four or more factors, some thought must be given to how the display should be constructed. We want to see the magnitude of interaction effects relative to the corresponding main effects, and of course we want to determine which effects are significant. There are some obvious options. One possibility would be to display the effect estimates in decreasing order of magnitude (as in Table 4.5) without regard to the order of the effects regarding main effects and interactions. Another possibility would be to use Yates order (see Section 4.1). A third possibility would be to list all main effects in decreasing order of magnitude, followed by all two-factor interactions ordered in the same way.

When there are only two factors, the appropriate order is obvious: Show the main effects followed by the single interaction, as is shown in Figure 4.13.

The vertical lines for factors A and B are self-explanatory as each line connects the average responses at the low and high levels of the factor. The line representing the interaction requires some explanation, however. To facilitate this explanation, we will let "1" represent the low level and "2" the high level. Then Lbar = $(1/2)(\overline{A_1B_1} + \overline{A_2B_2})$ and Ubar = $(1/2)(\overline{A_1B_2} + \overline{A_2B_1})$, with the bar over the treatment combination designating the average response value for that treatment combination. Obtaining the interaction effect estimate as Lbar $-$ Ubar is in accordance with the expression for the AB effect given in Section 4.3. The Lbar/Ubar notation is essentially the same notation that is used in Ryan (2000) and was originally used by Ellis Ott.

The display shows that although no effect is significant at the $\alpha = .05$ level, the interaction is large relative to the main effects, as the interaction dwarfs the B effect estimate and is almost the same size as the A effect estimate. The fact that the A effect is not far from being significant means that the conditional effects of A will

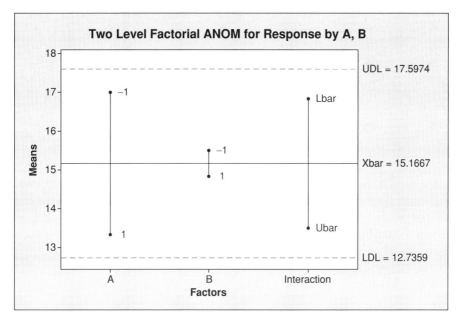

Figure 4.13 ANOM display for two factors.

differ considerably and thus be "of significance." It can be shown that the conditional effects are -7.00 and -0.33, compared to the main effect of -3.667. Thus, one of the conditional effect estimates is almost double the main effect estimate, whereas the ratio of the standard deviations of the effect estimates is 1.4:1. Thus, the conditional effect estimate of -7.00 looks "more significant" than the main effect estimate. Of course, the conditional effects of B also differ considerably, whereas the main effect estimate is obviously quite small.

It is important to note that such displays for factorial designs can be constructed only if σ is assumed to be known or if the design is replicated. We may also note that even though this information is available from ANOVA output, it is not available in quite as convenient a form as it is with an ANOM display. In particular, software with design of experiments capability does not produce an ANOVA equivalent to an ANOM plot. Dataplot, developed primarily by Jim Filliben at the National Institute of Standards and Technology and now available as freeware, does come close, however, as it can be used to produce a matrix plot that includes both the main effects scatterplots and the two-factor interaction scatterplots (see the matrix scatterplot at http://www.itl.nist.gov/div898/handbook/pri/section5/pri594.htm). MINITAB can be used to produce a matrix scatterplot for main effects with the FFMAIN command, whereas the FFINT command will produce the matrix scatterplot for two-factor interactions.

Although Figure 4.13 shows the general form of an ANOM display for two factors, there are occasions when a different type of display would be more appropriate. For example, if the objective is to seek the best combination of two factors to maximize or minimize the response, then plotting each combination would be useful, acting as if there was a single factor with the number of levels given by the product of the levels of the two factors. This is illustrated by Nelson, Coffin, and Copeland (2003, p. 313), with the decision limits obtained from Eq. (2.4). See also Figure 5.24 in Nelson, Wludyka, and Copeland (2005, p. 102), which clearly identifies the best factor-level combination for two 3-level factors and shows that combination to be statistically different from the other eight treatment combinations.

4.11 NORMAL PROBABILITY PLOT METHODS

In order for normal probability plot methods such as the one due to Lenth (1989) to be effective in assessing effect significance, it is essential that effect sparsity exists. That is, the number of significant effects should be less than half of the estimable effects. When this condition is met, the various methods that have been proposed (which were reviewed and compared by Berk and Picard (1991) and Hamada and Balakrishnan (1998)) will have approximately the same degree of success in identifying significant effects. When most effects are significant, or at least seem to be significant, all the methods can be expected to fail.

Should this be of concern? Assume that five factors are being investigated and a 2^5 design is being used. How many of the 31 effects that could be estimated should be significant? As stated previously, Daniel (1976, p. 75) stated that seven significant effects is about average for a 2^5 design. What I strongly suspect has never been investigated is the variability of the number of significant effects over a large number of datasets for a specified design, such as a 2^5 design, after the bad data have been identified and removed. If the experimenters are very good at identifying factors to study in an experiment, we should not be surprised that most, if not all, of them turn out to be significant (the main effects, that is). The possible breakdown of the methods due to Lenth (1989) and others will then depend on how many interactions are significant.

Loughin and Noble (1997) proposed a permutation test as an alternative to these normal probability plot methods, and compared the performance of their test with Lenth's test over 36 datasets in the literature. Of course one weakness in making comparisons over actual datasets is that "the truth" is never known, so it isn't possible to say that one method gives the correct answer whereas the other one does not. Nevertheless, the comparison is of interest. Loughin and Noble (1997) found that their test performed the same as Lenth's test for the vast majority of datasets. There were some significant differences, however. In particular, there was a huge difference for an example in Montgomery (1991, p. 314), as the permutation test identified nine significant effects in the 2^4 design whereas Loughin and Noble (1997) stated that the properly calibrated Lenth's test did not come close to identifying any effects as being significant. This is in reference to the calibration of Lenth's test given by Loughin (1998).

This result is disturbing, so we will examine the data. An interesting feature of this dataset, which may or may not be real data, is that this is a replicated design, so ANOVA can be applied and the results compared with the results obtained using Lenth's test and the permutation test. Of course ANOVA also isn't "the truth," but it does provide a useful benchmark. Loughin and Noble (1997) gave the ANOVA results for this example, which showed 10 effects as being significant at the .05 level, including the replication that was treated as a factor.

This might at first appear to be a somewhat amazing result since effect sparsity obviously does not exist for this dataset, as we assume that it does in general. This seems to be much ado about nothing, however, at least for this dataset, since it seems as though this probably isn't actual data since the four-factor interaction is significant in the ANOVA table given by Loughin and Noble (1997), with a p-value of .0001. Such a highly significant four-factor interaction will occur only very, very rarely. Furthermore, Lenth's method is for unreplicated factorials, as the title of the paper states, whereas the dataset under discussion is a replicated factorial.

An enhancement to Lenth's method, although it was not directly presented as such, was given by Tripolski, Benjamini, and Steinberg (2005). They considered large factorial experiments, for which there is potentially a large number of real effects. They proposed the use of the false discovery rate (FDR) for such experiments and showed how to combine the control of FDR with methods for obtaining a PSE, such as Lenth's method. The FDR, which was introduced by Benjamini and Hochberg (1995) with an adaptive version given by Benjamini and Hochberg (2000), is defined as the number of inert effects falsely identified as real effects divided by the total number of effects identified as real effects.

Let q denote the nominal level at which the FDR is to be controlled, k denotes the number of effects that are tested, with k_0 denoting the number of effects that are inert. Let $p_{(1)} \leq p_{(2)} \leq \cdots \leq p_{(k)}$ denote the p-values for the k effects. (Of course there are no p-values, as such, available for an unreplicated factorial; this will be addressed shortly.) Benjamini and Hochberg (1995) showed that the FDR is controlled at $q(k_0/k) \leq q$ when the l largest effects (i.e., those with the smallest p-values) are declared active with

$$l = \max \left\{ i : p_{(i)} \leq \tfrac{i}{k} q \right\} \qquad (4.5)$$

with no effects being declared active if the condition is not met for any value of i. This rule might be applied to replicated factorial designs for which p-values are available. For unreplicated factorial designs (i.e., designs for which Lenth's method would be used), a "p-value" of sorts can be computed by using the pseudo-error to compute the value of a t-statistic for each effect estimate, with the p-value being obtained from the value of that statistic. With the p-values thus obtained arranged in ascending order, all effects with a p-value smaller than the largest p-value that satisfies $p_{(i)} \leq \tfrac{i}{k} q$ would result in the corresponding effects being declared active, which of course is the same condition to be met for a replicated factorial.

Tripolski et al. (2005) recommended using Lenth's procedure in conjunction with the rule for controlling the FDR, with their recommendation based on simulations with 16-run and 32-run designs. They found that this improved the power of Lenth's procedure by more than a small amount. That is, Lenth's procedure would be used to obtain the pseudo-error, then Eq. (4.5) would be applied. That reduces the role of Lenth's procedure in the decision-making process. Although practitioners might accept the simplicity of the decision rule, they might still want to see the normal probability plot of effect estimates and perhaps a Pareto chart of the effect estimates as well. Although the decision rule might seem to be somewhat arbitrary, the effectiveness of the rule is supported by the authors' simulation results.

A method that is competitive with normal probability plot methods is a Bayes plot (Box and Meyer, 1986). The objective with this plot, which assumes normality for the real effect estimates and the effects that are just noise, is to separate those two sets of effects using a Bayesian analysis.

4.12 MISSING DATA IN FACTORIAL DESIGNS

Missing data are a fact of life. Statisticians and lawmakers have focused considerable attention on data missing from the decennial U.S. census. Missing data also occur when designed experiments are run (due to botched runs, etc.), but what to do about data missing from designed experiments has been much less debated.

Missing data can occur in a variety of ways. Sometimes bad parts can result when undesirable treatment combinations are used, thus resulting in no measurement being taken. This occurred in an experiment described by Lin and Chananda (2003), as the treatment combination consisting of the low level of each of four factors in a replicated 2^4 design resulted in defective parts. Of course this results in a slight loss of orthogonality as all the pairwise correlations are $-.071$. (A general result on correlations that are created by a single missing observation is given later in this section.)

Missing data do destroy the orthogonality of a design, assuming the design was originally orthogonal. For a design with a large number of runs, such as a 2^6, a single missing observation will not be a major problem as the design will be only slightly nonorthogonal. Furthermore, although orthogonality is a desirable property, it is not absolutely essential. Supersaturated designs (see Section 13.4.2), for example, are not orthogonal.

One obvious consequence of missing observations is that fewer effects can be estimated. For example, if a single observation is missing from a 2^3 design, one effect less can be estimated and that would logically be the ABC interaction. That generally isn't a problem, but the other consequences are more serious. The missing observation causes the correlations between the factors to all be $.167$ in absolute value, regardless of which observation is missing.

This can be easily explained, as follows, as well as generalized. The correlation between any pair of factors will have a denominator of $n(n-2)/(n-1)$, regardless of the observation that is missing. If the missing observation is in a position that has opposite signs ($+1$ and -1 or the reverse) for the pair of factors for which the correlation is computed, the numerator will be of the form $1-(-1)/(n-1) = n/(n-1)$. If the signs are the same, the numerator is $-1-(1)/(n-1) = -n/(n-1)$. Thus, the correlation is either $1/(n-2)$ or $-1/(n-2)$. If n is at least of moderate size (e.g., 32), this correlation should not be of any great concern, but could result in interpretation problems when n is small.

Another possible concern is the imbalance between the number of observations of a factor and the number of observations at the low level. As with the correlations between factors, this will be a problem for designs with a small number of runs. Specifically, there will be a 4–3 imbalance between high and low levels with an 8-run design with one missing observation, whereas a 16–15 imbalance should not be a major problem.

Probability plots of effect estimates can, strictly speaking, be constructed only when the variances of the effect estimates are equal and the effect estimates are independent. The variances of the effect estimates will often be the same when there is missing data, but that won't always be the case. For example, when data for the bc and abc treatment combinations are missing when a 2^3 design is used, the variances are not all equal. Furthermore, regardless of whether or not the variances are equal, the effect estimates will not be independent because missing observations cause correlations between the factors. If those correlations are quite small, however, one might still use a probability plot.

Draper and Stoneman (1964) gave a simple method for dealing with this problem in two-level factorial and fractional factorial designs (the latter are discussed in Chapter 5). Box (1990–1991a) discussed and illustrated this method, using a missing observation from a 2^4 design as an example. The method consists of setting the $ABCD$ interaction equal to zero and solving for the missing value. This is a reasonable approach if the highest-order interaction is at least a four-factor interaction. It would not be a good idea to use that method for, say, a 2^3 design because the ABC interaction could be significant.

4.12.1 Resulting from Bad Data

Of course missing data also result when bad data are discovered and subsequently removed before most of the data analysis is performed. Anderson and Whitcomb (1997) used an example of a 2^{5-1} design given by Box et al. (1978) to illustrate the deleterious effects of missing data. Box et al. (1978) first analyzed all (2^5) of the data points and concluded from a probability plot analysis that B, D, E, BD, and DE were significant. Using the same type of analysis, but using Lenth's method, which of course had not been invented by 1978, we additionally identify ACE as a significant effect, although it is clearly borderline.

The term is significant at the .05 level ($p = .03$) when it is added to the model, however. Justifying the term might be difficult, however, because using it in the model creates a nonhierarchical model since the model does not include the main effects of A and C, nor does it include the AC interaction. (See the discussion of hierarchical versus nonhierarchical models in Section 4.18, however.) This is not necessarily justification for discarding this interaction term, although there is obviously no strong statistical support for including it.

In terms of the analysis, missing data have the same effect as bad data that are subsequently discarded. It is important that not very many observations be missing, however, as severe imbalance could result and effect estimates could be misleading. For example, if the last two observations in Yates order are missing when an unreplicated 2^3 design is used, the correlation between the estimates of the B and C main effects is $-.50$. This disturbs the relationship between the effect estimates and the regression coefficients, as the 2:1 ratio for effect estimate : regression coefficient does not exist when the orthogonality property has been disturbed. Furthermore, with these two missing values the B and C main effect estimates would have to be computed with four observations at one level and two observations at the other level. Clearly such imbalance is highly undesirable.

4.12.2 Proposed Solutions

Various solutions have been proposed for the missing data problem. If a single observation is missing, one proposal is to solve for the missing value to minimize the residual sum of squares (Hicks and Turner, 1999), which of course requires a model assumption that might be difficult to make until the data have been analyzed. This approach is also suggested by Montgomery (1997, p. 189) for a randomized complete block design, in addition to recommending that the data simply be analyzed as unbalanced data, as in Searle (1987). The suggestion of Giesbrecht and Gumpertz (2004) is the same, except stated differently, as they suggest, for a randomized complete block design, imputing a value that leaves the error sum of squares unchanged. Nelson (2003) suggested a trial-and-error iterative approach that is based on using either a single interaction or pooled interactions to estimate the error variance with an unreplicated factorial. Of course this could be risky unless four-factor interactions and higher are used for this purpose.

Although these may seem to be reasonable solutions to the problem, phony data are being used with the first three suggested approaches, which of course is objectionable unless the phony data well represent the missing data.

Of course, the simplest approach is to not impute the missing value and to proceed with the data that are present, just as Lin and Chananda (2003) did.

4.13 INACCURATE LEVELS IN FACTORIAL DESIGNS

It isn't always possible to maintain intended levels when a design is carried out. We will consider a few possible scenarios. Assume that temperature was to be set at 350 and 400°F but the actual settings were 360 and 390°F. This may affect the outcome of the experiment, but it does not affect the analysis, regardless of whether the actual settings are known or not. If the actual settings were not known, the analysis would be performed in the usual way and conclusions would be drawn, based on the assumption that the actual temperature settings were the nominal settings. If the range of the actual settings is too small to show if the factor has an effect, the conclusion may differ from the conclusion that would have been reached if the nominal settings had been used. Furthermore, if the regression equation were converted to raw form, the equation would be assumed to be valid for 350 and 400°F, whereas the equation does not strictly apply for those values.

If the mistakes were discovered during the course of the experiment but not soon enough to correct the problem, the coded-form analysis would be unaffected, except for the fact that the inference would apply to 360 and 390°F rather than to 350 and 400°F. Somewhat similarly, the raw-form regression equation would be applicable only for 360 and 390°F.

A bigger problem occurs when the settings vary uncontrollably during the course of the experiment. Orthogonality of the design will be lost (almost certainly), and the conclusions could be erroneous, depending upon the degree of departure of the actual settings from the nominal settings. The method of analysis will certainly have to change, as will the interpretation. For example, assume that a 2^4 design is being used but it is learned during the experiment that the level of the fourth factor cannot be held steady. Consequently, the actual number of levels of that factor might turn out to be, say, 8 or even 16, which would preclude analyzing the data as having come from the use of a factorial design. Instead, regression or a generalized linear model approach would have to be used, and the term "effect estimate" would not have any meaning for the fourth factor. If the departure from the nominal levels is not great, however, the analysis will probably agree with the factorial design analysis that would have been performed if the levels of the fourth factor had been held steady.

Donev (2004) considered the properties of experimental designs where the factor levels cannot be set precisely and found that the properties of the design with the inaccurate levels could be better or worse than the properties of the design without the inaccuracies and gave recommendations for selecting a design region so as to minimize the risk of losing observations.

4.14 CHECKING FOR STATISTICAL CONTROL

It was stated in Section 1.7 that processes should be in a state of statistical control when experiments are run. The extent to which this is important depends upon the likelihood of factors extraneous to the experiment (e.g., temperature) going outside of acceptable boundaries and affecting the values of the response variable.

As stated in Section 1.7, one way to check for this is to make runs at the current operating conditions at the beginning, in the middle, and at the end of an experiment. An obvious question is "Can the data from these runs be utilized in some manner, and if so, how?" Specifically, since there are no degrees of freedom for estimating the error variance when an unreplicated factorial is used, can this problem be resolved by using these data points to estimate the error variance? The answer is "yes," provided that the experimenter is willing to make the somewhat strong (and perhaps untenable) assumption that the variance at each design point is the same as the variance at the standard conditions, and provided that the replicates at the standard conditions are true replicated points.

As an aside, it should be noted that failure to maintain a state of statistical control is often what motivates a designed experiment, with the experiment designed to identify the factors that are causing the problem. An example of this is given in Hare (1988), where the objective was to conduct an experiment to identify the factors that were causing unacceptable variation.

4.15 BLOCKING 2^k DESIGNS

It may not be possible to make all the 2^k runs under the same conditions. When faced
with this reality, a blocked 2^k factorial can be constructed. Because of the analogy
with fractional factorials, this topic is covered in more detail in Section 5.6, but we
will also make some comments about it here.

Let's consider a simple scenario in which a 2^5 design is to be used, but only 16
runs can be made per day. Consequently, the 32 runs would have to be split up in
some way, and the most obvious option would be to use two blocks with 16 units
per block. Doing so would create a block component in an ANOVA table, just as is
the case when a randomized block design is used. With two blocks, there would be
$2 - 1 = 1$ degree of freedom for the block effect. This means that one of the factorial
effects could not be estimated since a 2^5 design is a saturated design with all of the
31 degrees of freedom available for estimating the 31 effects.

The obvious choice for the interaction to relinquish, which will be confounded (i.e.,
confused with the block effect), is the five-factor interaction $ABCDE$, since there is
virtually no chance that a five-factor interaction will be significant, and interpreting
it would be quite difficult even if it were significant.

The usual assumption when blocking is used is that there is no interaction between
the blocking factor(s) and the factors under study. Assuming something to be true
doesn't make it true, however, so it is a good idea to check this assumption. One
way to do so is to simply treat blocks as a regular factor and see if any interactions
involving blocks are significant.

Example 4.8

An example of a 2^3 factorial that was blocked out of necessity was given by Chapman
and Roof (1999–2000). Three factors were studied and the primary objective was
to find a method to prevent or reduce potential hypoglycemia rather than treat its
effects after it occurs. The authors sought to accomplish this by trying to minimize
the variation in glucose levels that results from exercise.

What made the study quite unusual is that the second author, a diabetic, was the
experimental unit—the *only* experimental unit! A single experimental unit, although
quite unusual in an experimental setting, is what often happens when student experimental design projects are performed (as in Hunter, 1977), as the experimental unit is
often the student, or the student's dog, and so on. The concern, of course, when this
is done is that there can be carryover effect (see Section 11.4). If there is no carryover
effect, there is still the problem of the inference not being extendible beyond the person who was the experimental unit. Both of these issues are discussed later relative
to this experiment.

The response variable was blood glucose level and the factors were (1) volume
of juice intake before exercise (4 ounces and 8 ounces), (2) amount of exercise on
a Nordic Track cross-country skier (10 min and 20 min), and (3) delay between the
time of juice intake and the beginning of the exercise period (none and 20 min). The

runs were made in the morning and in the evening, and since blood glucose levels were suspected of varying between morning and evening under constant conditions, it was decided to use time of day as the blocking factor. (*Note*: Blood glucose levels are considered to be higher in the morning than during the rest of the day; see http://www.diabetic-lifestyle.com/articles/feb01_healt_1.htm.)

The data from the experiment are given below, with the runs listed in standard order rather than the randomized order that was used (but not given in the article).

A JUICE (oz)	B EXERCISE (min)	C DELAY (min)	Block TIME OF DAY	PRE	POST	AVG
4	10	0	pm	78	65	71.5
8	10	0	am	101	105	103
4	20	0	am	96	71	83.5
8	20	0	pm	107	145	126
4	10	20	am	128	123	125.5
8	10	20	pm	112	147	129.5
4	20	20	pm	111	79	95
8	20	20	am	83	103	93

There are some important questions that were not addressed by Chapman and Roof (1999–2000). In particular, no information was provided regarding the length of time covered by the experiment. Perhaps all the runs were made in one day, but such information was not provided. The authors used the average blood glucose level for each treatment combination, with the "PRE" reading being before exercise and the "POST" reading being after exercise. Thus, the reading *before* exercise has a weight of 0.5 in determining the (average) response value that is used *after* exercise.

It would have been very helpful to know whether or not all of the runs were performed in one day and to know the sequence of the runs because carryover effect might be a real problem.

Curiously, although the authors stated that the experiment was blocked, it was not analyzed as such when using average glucose level as the response variable. We can observe, after writing down the treatment combinations that are in each block, that the *ABC* interaction is confounded with blocks. In their analysis, Chapman and Roof (1999–2000) created an error term with two df using Blocks(*ABC*) and the *AB* interaction. Looking at the data, the latter seems to be a reasonable assumption but when we use blocking we would like to see evidence that the blocking was necessary. The authors did not use a normal probability plot. This could be done although such plots applied to small designs may not give reliable results. Indeed, the plot constructed using MINITAB shows no significant effects, whereas the ANOVA table given by the authors shows that all of the estimable effects except Exercise (factor *B*) is significant at the .05 level. As stated in Section 4.11, normal probability plot methods fail when there is no effect sparsity. Here the authors' ANOVA table shows that four of the seven estimable effects are significant at $\alpha = .05$. That is *not* effect sparsity!

Their regression model was constructed using uncoded units, so the results in terms of p-values for effects do not agree with the p-values in their ANOVA table. Doing the regression analysis on the uncoded units induces correlations between main effects and interaction effects and essentially renders meaningless the regression analysis p-values. All analyses should be performed with coded units when interaction terms are used, as was illustrated in Example 4.1 (see also Ryan, 2000, Section 13.11).

Since hypoglycemia can result from exercise, a major goal was to minimize the variation in blood glucose levels due to exercise. Accordingly, the authors did the following. They computed the standard deviation (of two numbers), s, at each treatment combination, used the variance-stabilizing transformation $\ln(s)$ because of the large variation in s, and then used $\ln(s)$ as the response variable.

The ANOVA table for $\ln(s)$ showed that Exercise and Time of Day were significant at the .05 level. Again, however, there was a discrepancy between the regression analysis and the ANOVA results since the former was performed on the uncoded units. The authors concluded that the best strategy was to drink 8 ounces of juice immediately before exercise (i.e., no delay) and exercise for 10 minutes. This treatment combination does have the smallest value of $\ln(s)$.

This conclusion is highly sensitive to the assumption of no carryover effects, however, and then there is the problem of not being able to apply these results to other diabetics. The second author recognized this with the statement, "Co-author Roof suspects that the optimum conditions achieved by this experiment might not be universally applied at all times nor to all diabetics."

Blocking factorial designs is also discussed in Section 5.8.

4.16 THE ROLE OF EXPECTED MEAN SQUARES IN EXPERIMENTAL DESIGN

This section is placed near the end of the chapter because the topic is one that many experimenters would prefer to skip. The way that expected mean squares determine the manner in which hypothesis testing is performed in replicated designs is practically hidden in computer output. The user of experimental design software indicates whether each factor is fixed or random, with fixed generally being the default, and the software performs the appropriate analysis for the indicated classification of factors.

We will look at a simple example to see what effect that classification can have. Assume that we have a 2^2 design with three replications and the data are as follows.

Treatment Combination	Data
(1)	12.1, 11.8, 12.6
a	10.7, 11.1, 11.0
b	12.0, 12.0, 11.7
ab	11.9, 12.3, 12.6

We will first assume that both factors are fixed. With this assumption, we obtain the output given below.

```
         ANOVA:      Y versus   A, B

    Factor    Type     Levels      Values
      A       fixed      2         -1, 1
      B       fixed      2         -1, 1

             Analysis of Variance for Y

    Source    DF       SS        MS        F        P
      A        1     0.5633    0.5633     6.26    0.037
      B        1     0.8533    0.8533     9.48    0.015
      A*B      1     1.9200    1.9200    21.33    0.002
      Error    8     0.7200    0.0900
      Total   11     4.0567

  S = 0.3    R-Sq = 82.25%     R-Sq (adj) = 75.60%
```

Notice that all three effects are significant at the $\alpha = .05$ level.

Now we will assume that both factors are random. This results in the output given below.

```
         ANOVA:   Y versus   C1,  C2

      Factor    Type     Levels      Values
        A       random     2         -1,  1
        B       random     2         -1,  1

               Analysis of Variance for Y

    Source    DF       SS        MS        F        P
      A        1     0.5633    0.5633     0.29    0.684
      B        1     0.8533    0.8533     0.44    0.626
      A*B      1     1.9200    1.9200    21.33    0.002
      Error    8     0.7200    0.0900
      Total   11     4.0567

  S = 0.3    R-Sq =  82.25%     R-Sq(adj)  =  75.60%
```

Now the A and B effects are not even close to being significant, as the p-value for testing each effect is quite large. The data have not changed, however, and if the model coefficients had been part of this output, it would be apparent that they do not change. So the only changes are in the F-statistics and p-values.

The difference is that, assuming replication, in the fixed factor case all effects are tested against the error term, meaning that each of the three mean squares is divided by the mean square error in producing the F-statistics, whereas in the random factor case the main effects are tested against the interaction term and the interaction term is tested against the error. The reason for this is given in Appendix C at the end of the chapter. Since the interaction term in this example is large, the results thus differ greatly.

The user of software for experimental design does not have to be concerned with how expected mean squares are obtained because these are produced with software. For the interested reader, however, Lenth (2001) gave a simple method for determining expected mean squares. For each effect in a restricted model (e.g., a restriction that components of an interaction sum to zero over the levels of a fixed factor, with the other factor being random), the terms that are part of the expected mean square for a given effect are terms that contain the effect in question and that involve no interactions with other fixed factors. The coefficient for a variance component is the number of observations at each distinct level of that component. (Of course, σ^2 is part of every expected mean square.)

For example, assume that there are three random factors, A, B, and C, with each factor having two levels, and two replicates are used. The model with all possible interaction terms is fit. The expected mean square for the AB interaction is then $\sigma^2 + 2\sigma^2_{ABC} + 4\sigma^2_{AB}$. In accordance with Lenth's (2001) method, σ^2_{ABC} is one of the terms because it contains AB and the coefficient is 2 because the design is replicated. The coefficient of the σ^2_{AB} term is 4 because there are two levels of C at each AB combination and there are also two observations because the design is replicated. Thus, $2 \times 2 = 4$.

For the unrestricted model, the expected mean square contains all terms that contain the effect and at least one other random factor.

We may note that although expected mean squares can be produced with MINITAB, this must be done with the ANOVA command, not the FFACT command.

4.17   HYPOTHESIS TESTS WITH ONLY RANDOM FACTORS IN 2^k DESIGNS? AVOID THEM!

It may be disturbing that the classification of factors as fixed or random can have such a dramatic effect on the results. We will show why hypothesis tests with random factors in 2^k designs should be avoided. The reason for this is that an F-test should never be performed when the denominator degrees of freedom is 1, as will be the case whenever any effect, main effect or interaction, is tested against an interaction effect for two-level designs, as in the example above with both factors random.

Since $F_{1,1,.05} = 161.44$, an effect estimate would have to be extremely large in order to have significance declared at the .05 level. This can be seen as follows. Since the mean squares are the same as the sum of squares when the degrees of freedom is 1, this means that the numerator sum of squares divided by the denominator sum of squares must exceed 161.44.

We can express this relationship in terms of effect estimates. This can be done by using the fact that for a replicated (or unreplicated) 2^k design, the relationship is

$$\text{Effect estimate} = \sqrt{\frac{\text{SS}_{\text{effect}}}{r(2^{k-2})}}$$

Thus, for an unreplicated 2^2 design, the relationship is effect estimate $= \sqrt{SS_{\text{effect}}}$. (This last result is the general form of the result in Exercise 4.17 that the reader is asked to derive.) Thus, the ratio of the effect estimates would have to be at least 12.71:1 in order for significance to be declared.

Consider the modified data given below, with the numbers in the first two rows the same as previously given but the numbers in the last two rows increased so that the average of those numbers is almost twice the average of the numbers in the first two rows.

```
Treatment Combination         Data
        (1)            12.1, 11.8, 12.6
         a             10.7, 11.1, 11.0
         b             20.0, 20.0, 19.5
        ab             19.8, 20.5, 21.0
```

Since the difference of these two averages estimates the B effect, there is obviously such an effect, but the analysis using ANOVA results in a p-value of .068, so the ANOVA result doesn't make any sense, and this is because of the low power to detect significant differences because of the use of an F-test with one degree of freedom in the denominator of the F-statistic.

It is worth noting that there is no "Daniel rule" violation with the modified data since the ratio of the effect estimates is almost 10:1. Thus, it is strictly a problem with the denominator degrees of freedom; a problem that exists regardless of the number of replications since the latter does not affect the degrees of freedom for the effects, which will be 1 for each effect, for every two-level design.

4.18 HIERARCHICAL VERSUS NONHIERARCHICAL MODELS

A *hierarchical model* is one in which all factors appear as main effects in a model when those factors appear in higher-order terms. For example, a model with a three-factor interaction would contain two-factor interaction terms in the factors that comprise the three-factor interaction, and a model that contains the AB interaction would also have to contain a main effect term in A and one in B. A model that does not meet this requirement is a nonhierarchical model.

It was stated in Section 4.12.1 that a nonhierarchical model could be difficult to justify. However, Montgomery, Myers, Carter, and Vining (2005) take a different position on the matter, stating that "Most industrial experiments involve systems driven by physical or chemical mechanisms. These underlying mechanisms, while unknown, are unlikely to be hierarchical." Furthermore, they point out that it is not uncommon to encounter two-factor interactions without both main effects being significant when one factor is quantitative and the other factor is qualitative, and especially when both factors are qualitative. At the other extreme, there are undoubtedly many users who will be influenced by software packages that give warning messages about nonhierarchical models and are also influenced by articles such as Franks (1998).

Recall the discussion in Section 4.2.1 that a signal to use a nonhierarchical model is also a signal to do a conditional effects analysis. In other words, we should not simply settle for a model that does not contain a main effect but does contain an interaction term in that factor, but rather should do a conditional effects analysis.

4.19   HARD-TO-CHANGE FACTORS

Assume that a factor can be varied, with great difficulty, in an experimental setup (such as a pilot plant), although it cannot be freely varied during normal operating conditions. Assume further that it is imperative that the number of changes of the hard-to-change factor be minimized, each factor has two levels, and the design is to have factorial structure.

Again assume that we have two factors. We can minimize the number of level changes of one factor simply by keeping the level constant in pairs of consecutive runs. That is, either the high level is used on consecutive runs and then the low level on the next two runs, or the reverse. This means that we have *restricted randomization*, however, as there are six possible run orders without any restrictions, but with the restriction there are only two possible run orders (high, high, low, low, or the reverse). Thus, there is restricted randomization in regard to the run orders. Restricted randomization increases the likelihood that extraneous factors (i.e., not included in the design) could affect the conclusions that are drawn from the analysis. Furthermore, this will also cause bias in the statistics that are used to assess significance, as shown by Ganju and Lucas (1997). Although restricted randomization and the problems caused by it might seem to have been seriously considered only since the late 1990s, the issue was discussed in the literature much earlier, dating at least from Youden (1964, 1972). (See also Bailey (1985, 1987), White and Welch (1981), Monod and Bailey (1993), and Bowman (2000)).

The need to address the issue of hard-to-change factors dates at least from Joiner and Campbell (1976). Design plans in the presence of such factors have been given more recently by Webb and Lucas (2004). See also Ju and Lucas (2002), who considered designs when there is one easy-to-change factor or one hard-to-change factor.

Although hard-to-change factors have not been discussed extensively in the journal literature or in textbooks, it seems safe to assume that such factors occur very frequently in practice. Some examples of hard-to-change factors described in the literature are as follows. Czitrom (2003) described an experiment originally given in Czitrom, Mohammadi, Flemming, and Dyas (1998) for which one of the three factors used in the experiment was a heat plug in a furnace. Time-consuming hardware changes were necessary whenever the plug was removed and subsequently put back in. Consequently, the first four experimental runs were made with the plug in and the next four were made with the plug out, which created a split-plot structure relative to that factor. (Split-plot and related designs are discussed in Chapter 9, and Example 9.1 illustrates the effect of ignoring the randomization restriction relative to the results using the proper analysis.)

Other examples of hard-to-change (or impossible to change) factors in experiments described in the literature include an experiment described by Inman, Ledolter, Lenth, and Niemi (1992), as it was not possible to change temperature between individual runs, so temperature was used as a blocking variable. This is an example of an impossible-to-change factor creating an "unnatural" blocking factor, because if there were a time or day effect it would be confounded with the temperature effect. This required that care be exercised to ensure that laboratory conditions did not vary from day to day. The authors stated, "In the analysis, we ignored the restriction on randomization with respect to temperature." Undoubtedly this was also done for countless unpublished experiments during that era, but in light of recent research, analysts should now assess the possible consequences in terms of bias in ignoring hard-to-change or impossible-to-change factors.

Eibl, Kess, and Pukelsheim (1992) described a sequence of experiments with the response variable being paint coat thickness. All factors could be easily changed except paint viscosity. Consequently, the levels of paint viscosity were kept constant as long as possible in one of the three experiments. The authors recognized that this probably caused the experimental error to be underestimated, but they decided not to adjust for the lack of randomization when they noticed that the error standard deviations were practically the same in each experiment.

As a final example of a hard-to-change factor, Prat and Tort (1989) described an experiment conducted in a pet food manufacturing company that had four factors and one of them, compression zone in die, was hard to change. This experiment is described in considerable detail in Section 5.9.1.2.

Simpson, Kowalski, and Landman (2004) found from their consultation with engineers in the aerospace and defense industries that there is a strong need for something less than complete randomization. They show, using MINITAB, how to analyze data from a design that has four factors with two being hard to change. Similarly, there is a good tutorial on using MINITAB with one hard-to-change factor at www.minitab.com/support/docs/OneHardtoChangeFactor.pdf.

If a cost can be assigned to the changing of levels of hard-to-change factors, a design with minimum cost might be sought, although such a design might of course be viewed as undesirable in other ways. Taihrt (1971) and Taihrt and Weeks (1970) gave run orders for two-level factorial designs that require a change in only one factor between successive runs. More recently, Joseph (2000) discussed the construction of minimum cost designs for hard-to-change factors, the objective being to determine the experimental sequence that minimizes the cost of adjusting the factor levels during experimentation. Of course this requires that the cost of adjusting each hard-to-change factor is obtainable, or at least that factors can be ranked in terms of difficulty of adjustment. Joseph (2000) gave an example for which it was difficult to obtain the adjustment costs from speaking to the operators, so it was necessary to settle for a rank order of the factors in terms of the difficulty of adjustment. Goos and Vandebroek (2004) showed that factorial designs with hard-to-change factors run in a split-plot arrangement can be superior, in terms of the design efficiencies that they use, to factorial designs run without a split-plot arrangement.

Using an experimental run sequence determined from cost considerations is of course a form of restricted randomization and thus makes the use of hypothesis tests using ANOVA rather shaky, although they still might be used, recognizing their limitations.

4.19.1 Software for Designs with Hard-to-Change Factors

Since the consideration of hard-to-change factors is of somewhat recent origin, the way in which the designs are generated and analyzed with popular statistical software can best be described as a workaround with some software. For example, although the designs can be constructed and analyzed with MINITAB, doing so is rather involved (e.g., see http://www.minitab.com/support/docs/OneHardtoChangeFactor.pdf for information on how to proceed with one hard-to-change factor).

4.20 FACTORS NOT RESET

In addition to experiments involving hard-to-change factors, there are many experiments for which at least one factor is not reset when the same level is to be used for the next run. Resetting a factor that is to have the same level for the next run (e.g., the same temperature) would seem to be a waste of time and money. Therefore, it is not surprising that a sizable fraction of industrial experiments have at least one factor that is not reset and in many experiments none of the factors are reset (Lucas, 1999).

There would seem to be no harm in not resetting a factor if the same level is to be used in the next run. For example, if temperature is one of the factors, why lower it to some level only to raise it back to the level at which it was set? This would seem to be rather impractical, but the argument in favor of resetting is that not resetting the temperature could induce a correlation between consecutive values of the response variable, which would invalidate the outcome of statistical tests. The question that must be addressed, however, is whether or not the bias and prediction variance inflation (see Webb, Lucas, and Borkowski, 2004) that results from an incorrect analysis is more than offset by the cost of resetting factors. See Webb et al. (2004) for a detailed discussion of these issues.

4.21 DETECTING DISPERSION EFFECTS

The detection of dispersion effects can be at least as important as the detection of location effects. Designs for detecting both are discussed in Chapter 8; in this chapter we simply wish to emphasize the importance of identifying dispersion effects and also to point out some problems in attempting to do so. In particular, Schoen (2004) explains how the detection of dispersion effects can be hampered by unidentified location effects; McGrath and Lin (2001) showed that the presence of dispersion effects may make it difficult to detect other dispersion effects and to estimate those effects. Bisgaard and Fuller (1995–1996) showed how to identify factors that affect dispersion by using a 2^4 design in conjunction with the logarithm of the sample variance as the response variable. Fuller and Bisgaard (1996) compared methods for identifying dispersion effects using two-level factorial designs.

4.22 SOFTWARE

Software for factorial designs is quite plentiful and all of the leading statistical software packages have this capability. Design-Expert is unusual, however, in that commentary on analyses is provided, as long as the "Annotated ANOVA" option is selected, and there is also a "Tips" button that can be used to obtain general advice on two-level factorial designs. For example, when the cement dataset from Daniel (1976) is analyzed with the software, the following direction and advice is shown after the saturated model is fit.

> Proceed to Diagnostic Plots (the next icon in progression). Be sure to look at the:
>
> 1) Normal probability plot of the studentized residuals to check for normality of residuals.
> 2) Studentized residuals versus predicted values to check for constant error.
> 3) Externally Studentized Residuals to look for outliers, i.e., influential values.
> 4) Box-Cox plot for power transformations.
>
> If all the model statistics and diagnostic plots are OK, finish up with the Model Graphs icon.

This has the flavor of expert systems software for design of experiments, which, although developed and used internally by certain companies for many years, has not been a part of the best known, general purpose statistical software.

Another software package that is exclusively for design of experiments, D. o. E Fusion Pro, also has some features of expert systems software. The software is not well known but received a rating of "Excellent +" in the comparison study of Reece (2003). One feature of this software that might raise some eyebrows, however, is that one of the four modules is Data Mining/Analysis. We generally don't think of the analysis of data from a designed experiment as data mining since the latter term is usually reserved for the analysis of large datasets.

D. o. E. Fusion Pro users can follow one of two paths, the "Design Navigation Wizard" or the "Design Menu Wizard," with the latter used when a user-specified design is to be constructed, and the former selected when the user does not know the most appropriate design to use. One interesting feature is that the number of design points to be replicated can be selected, but the points cannot be specified unless centerpoint replication is specified or all the points are to be replicated.

The comparison study of Reece (2003) was mentioned in Section 2.3. It is also mentioned here since the software packages that were included in the study were rated on factorial and fractional factorial designs, with the latter covered in Chapter 5.

4.23 SUMMARY

Although they have been used extensively, full factorial designs should be used only under certain conditions. When there is a small number of factors, such as 3, the design should be replicated so as to have a reasonable chance of detecting something less than large effects. Designs with at least 16 runs are more useful (see Box, 1992). When there is at least a moderate number of factors, a full factorial design may be too expensive to run, depending on the field of application. (An exception is computer experiments, which are generally inexpensive to conduct.)

This is not to suggest that one-factor-at-a-time designs should be used instead of factorial designs, as the former will generally be inferior to the latter. There are conditions under which well-constructed, one-factor-at-a-time designs can be useful, but such conditions do not occur very often. An extended discussion of these designs is deferred to Section 13.1.

Unless interaction effects are quite small, a conditional effects analysis should be performed. Unfortunately, such an analysis could become complicated when there is at least a moderate number of factors and more than a few moderate-to-large interactions. A step-by-step guideline for handling such scenarios has not been developed, however.

It is important to properly identify random factors and fixed factors as the analysis of factorial designs is determined by the declaration of fixed and random factors, as was seen in Section 4.16.

The importance of having processes in statistical control, or at least approximate statistical control, cannot be overemphasized, as spurious results can be obtained when processes are badly out of control.

Analysis of Means (ANOM) is a statistical tool that should undoubtedly be used more often. It has the advantage over ANOVA of the plotted points being in the original units, and there is also the advantage of the procedure being inherently graphical.

APPENDIX A

Derivation of Conditional Main Effects

We wish to show that $\delta_C = \delta \pm \delta_{int}$, with δ denoting a main effect, δ_{int} a two-factor interaction that contains the factor represented by δ plus another factor, and δ_C the conditional effects that result from splitting the data on the other factor in the two-factor interaction represented by δ_{int}.

Let δ^+ denote the conditional main effect of the factor of interest at the high level of the other factor (i.e., the factor that the data are being split on), and similarly define δ^- for the low level of the other factor. Using the result $\delta_{int} = (\delta^+ - \delta^-)/2$, which follows from the definition of a two-factor interaction (see, e.g., Wu and Hamada (2000, p. 106, Eq. 3.10) and/or consider how the interaction effect would be computed from an interaction plot), and $\delta = (\delta^+ + \delta^-)/2$, which follows from the definition of an effect with a plus sign being attached to the high level of the factor and a minus sign to the low level, we obtain $\delta_C = \delta^+$ and $\delta_C = \delta^-$ when these substitutions are made in the postulated expression. This establishes the result since δ^+ and δ^- designate the two conditional effects.

The ramification of this result is that the data should be split on the factor whose interaction with the factor of interest is the largest, as this will give the greatest difference in the conditional effects for that factor and thus give the maximum insight.

APPENDIX B

Relationship between Effect Estimates and Regression Coefficients

It was stated in Section 4.1 that the OLS regression coefficients are half the effect estimates. This can be explained as follows. Define a matrix $\mathbf{X}$ such that the first column is a column of 1s (which provides for estimation of b_0, the constant) and the other columns contain the coded values (i.e., $+1$ and -1) of the effects to be estimated. Since all but the first column sum to zero and pairwise dot products of all columns are zero, $\mathbf{X}'\mathbf{X}$ is a diagonal matrix with n as each element on the main diagonal. Thus, $(\mathbf{X}'\mathbf{X})^{-1}$ is a diagonal matrix with each diagonal element $1/n$. The regression coefficients are obtained from $(\mathbf{X}'\mathbf{X})^{-1}\mathbf{X}'\mathbf{Y}$ with $\mathbf{Y}$ a vector that contains the response values. The vector obtained from $\mathbf{X}'\mathbf{Y}$ then has the sum of the response values as the first element and every other element of the form $\sum_{j=1}^{n}(X_{ij}Y_{j})/n$ for each effect X_i that is being estimated. Since each X_{ij} is either $+1$ and -1, each sum is thus of the form $(\sum y_{(+)} - \sum y_{(-)})/n$, with $y_{(+)}$ denoting a value of Y for which X_{ij} is positive, and similarly $y_{(-)}$ denotes a value of Y for which X_{ij} is negative. Note that the divisor is n instead of $n/2$, with the latter being the number of terms in each of these last two sums. Thus, this simplifies to $\frac{1}{2}(\bar{y}_{(+)} - \bar{y}_{(-)}) = \frac{1}{2}$ (effect estimate).

APPENDIX C

Precision of the Effect Estimates

Since the regression coefficients are obtained from $(\mathbf{X}'\mathbf{X})^{-1}\mathbf{X}'\mathbf{Y}$, it follows that the variance of each one is $\mathrm{Var}[(\mathbf{X}'\mathbf{X})^{-1}\mathbf{X}'\mathbf{Y}] = (\mathbf{X}'\mathbf{X})^{-1}\mathbf{X}'[\mathrm{Var}(\mathbf{Y})]\mathbf{X}(\mathbf{X}'\mathbf{X})^{-1} = (\mathbf{X}'\mathbf{X})^{-1}\mathbf{X}'(\sigma^2\mathbf{I})\mathbf{X}(\mathbf{X}'\mathbf{X})^{-1} = \sigma^2(\mathbf{X}'\mathbf{X})^{-1}$. As noted in the previous section, $(\mathbf{X}'\mathbf{X})^{-1}$ is a diagonal matrix with each diagonal element $1/n$. Thus, each regression coefficient is estimated with a variance of σ^2/n, so that the standard deviation is $\sigma/\sqrt{n}$. Since an effect estimate is twice the corresponding regression coefficient, it follows that the standard deviation of an effect estimate is $2\sigma/\sqrt{n}$. (This result could also be obtained without using matrices, by writing each effect estimate as the appropriate function of the $y's$, converting that to the appropriate function of the error terms and then computing the variance of that expression (see, e.g., Bisgaard and de Pinho, 2004).

   Of course σ must be estimated, which would lead to the standard error of an effect estimate expression of $2\hat{\sigma}/\sqrt{n}$. Thus, each effect is estimated with the same precision with a 2^k design and of course the precision increases with larger n, which might result either from replication, or from increasing the value of k.

APPENDIX D

Expected Mean Squares for the Replicated 2^2 Design

The determination of expected mean squares in general is a laborious task and there is no need for an experimenter to do so as these can be produced with software, including MINITAB. Design-Expert, however, does not give expected mean squares. Nevertheless, experimenters who wish to "see for themselves" how the expected mean squares are obtained could proceed in one of two ways: either derive the results directly (not recommended routinely although we will do so here for illustration), or use general rules for obtaining the results, such as those given by Dean and Voss (1999, p. 616) or Lenth (2001).

(The derivations sketched out and given in the rest of this section require a knowledge of expected value, variance, and covariance. Readers without this knowledge can obtain it from any introductory book on mathematical statistics or statistical theory, such as Casella and Berger, 2001.)

When derived from scratch rather than by using general rules, expected mean squares are first obtained by computing the corresponding expected sum of squares, since the mean square and sum of squares of course differs by only a constant. We consider the model

$$Y_{ijk} = \mu + A_i + B_j + (AB)_{ij} + \epsilon_{ijk} \quad i = 1, 2 \ \ j = 1, 2 \ \ k = 1, 2, \ldots, r \quad \text{(D.1)}$$

corresponding to a 2^2 design with r replicates, with, as before, "1" denoting the low level of each factor and "2" denoting the high level, with both factors assumed to be random.

We will use the result stated in Exercise 4.17, $SS_A = r(\overline{A}_2 - \overline{A}_1)^2$, for a design with r replicates, which can be easily derived, as the reader is asked to do in that exercise. Here $\overline{A}_2$ is the average of the Y-values at the high level of factor A.

For an unreplicated 2^2 design, this would be the average of two numbers, whereas for the replicated design it is the average of $2r$ numbers.

Assume $r = 2$, for the sake of illustration. Then there will be four observations at the high level of A and four observations at the low level of A. From Eq. (D.1), the former are $Y_{211}, Y_{212}, Y_{221}$, and Y_{222}, and the latter are $Y_{111}, Y_{112}, Y_{121}$, and Y_{122}. The $E(SS_A)$ is then $E[2(\overline{A}_2 - \overline{A}_1)]^2 = (1/8)E(Y_{211} + Y_{212} + Y_{221} + Y_{222} - Y_{111} - Y_{112} - Y_{121} - Y_{122})^2$. By definition, $E(W^2) = \text{Var}(W) + [E(W)]^2$, for any random variable W. Since $E(Y_{ijk}) = \mu$ for all i, j, k, which follows from the assumption that the effect terms in Eq. (D.1) are assumed to have a mean of zero, $E(Y_{211} + Y_{212} + Y_{221} + Y_{222} - Y_{111} - Y_{112} - Y_{121} - Y_{122}) = 0$. Thus, the desired expected value is given by $(1/8)[\text{Var}(Y_{211} + Y_{212} + Y_{221} + Y_{222} - Y_{111} - Y_{112} - Y_{121} - Y_{122})]$.

Obtaining the variance expression without the use of rules or dot notation (which can be confusing) is both tedious and unorthodox, but that is the only way to literally "see" how the final result is obtained. The variance of each Y_{ijk} is the same, namely, $\sigma^2 + \sigma_A^2 + \sigma_B^2 + \sigma_{AB}^2$, which follows directly from Eq. (D.1) and the independence of the effects listed in the model. (Note that the interaction term is random because both factors are random, thus there is a variance component for the interaction term.) There are 28 covariance terms since there are eight Y-values and $\binom{8}{2} = 28$, and eight of these are zero. The 20 nonzero covariance terms are determined by what is common to each pair of Y-values. For example, $\text{Cov}(Y_{211}, Y_{221}) = \sigma_A^2$ because A is at the high level in each of the two expressions so that A_2 is common to each Y-expression. Similarly, $\text{Cov}(Y_{211}, Y_{212}) = \sigma_A^2 + \sigma_B^2 + \sigma_{AB}^2$ since both A and B are at the same level in each Y-value, and hence AB has the same combination of values.

Combining the 20 nonzero covariance expressions with the variance part, which is $8(\sigma^2 + \sigma_A^2 + \sigma_B^2 + \sigma_{AB}^2)$ and then multiplying by the constant $(1/8)$, given two paragraphs earlier, produces the final result, which is $\sigma^2 + 4\sigma_A^2 + 2\sigma_{AB}^2$, which the reader is asked to verify in Exercise 4.22.

The other mean squares could be similarly obtained, and it can be shown that $E(\text{MS}_{AB}) = \sigma^2 + 2\sigma_{AB}^2$. The A effect would thus be tested against the AB effect because the two mean squares differ only by $4\sigma_A^2$, and $\sigma_A^2 = 0$ is the hypothesis that would be tested. Since $E(\text{MS}_{\text{Error}}) = \sigma^2$, the AB effect would thus be tested against the error term.

APPENDIX E

Expected Mean Squares, in General

The expected mean square for an effect depends upon the classification of factors as fixed or random. This applies not only to any factor involved in an effect whose expected mean square is being computed, such as the A effect in Appendix D, but it also applies to the other factors in the design and the effects in the model.

To see this, recall from the preceding section that when we start from scratch, we see that expected mean squares depend upon $E(Y)$, which in turn depends partly on the model and partly on the classification of factors. When we have a replicated 2^2 design, with both factors being random, we have four variance components: $\sigma^2, \sigma_A^2, \sigma_B^2$, and σ_{AB}^2. If we had an unreplicated 2^2 design, then we could only have three variance components because we could not separate σ^2 from σ_{AB}^2.

If both factors are fixed, then we cannot have a variance component associated with those factors because the levels were not chosen at random from a population of levels. To illustrate the difference, let's consider a simple example. Assume that there are two students who perform the following experiment. One student randomly generates two integers in the interval 1–10 and for each random integer so generated, generates a value for Y using a simple linear regression model with known parameter values *and without an error term*. The other student picks two integers in that interval, but not at random, and uses the same regression model to generate the Y-values. If the second student were to do this again, using the same two integers, then the same pair of Y-values would be generated. Thus, there would be no variance component; there would simply be the difference between two Y-values, and that difference is caused by the different results when each of the two integers is used in the model. If the first student randomly generates another pair of integers and solves for the corresponding Y-values through the model, the new pair of Y-values will almost certainly be different from the first pair since the second pair of integers will almost certainly be different.

Hence, the first student would generate a *variance* of Y through repeated experimentation, whereas the second student would be forever generating the same two values, with the *effect* of the two integers represented by the difference of the Y-values.

Now think about doing this *with* an error term used in the simple linear regression model, with the errors randomly generated. The second student will now be generating different pairs of Y-values, due only to the random errors in the regression model, whereas the first student will be generating different Y-values that are due to two variance components: the variance due to the variability in the factor levels and the second due to the variance of the error term.

Thus, there will be an "effect" associated with a fixed factor, but there will not be a variance component. Consider again the replicated 2^2 design, but this time we will assume that both factors are fixed. We can proceed with the same general approach that was used in the random effects case, for which $E(Y_{211} + Y_{212} + Y_{221} + Y_{222} - Y_{111} - Y_{112} - Y_{121} - Y_{122}) = 0$ and the variance required tedious but straightforward calculations. We have the reverse in the fixed effects case in that the variance is a simple expression but the expected value is nonzero and requires some work.

Since the only random component in Eq. (D.1) in the fixed effects case is the error term, all of the covariances in the linear combination of Y-values are zero, so the variance of the linear combination is simply $8\sigma^2$. Unlike the random effects case, the expected value of the linear combination is not zero, and this is because the fixed effects translate into constants and there are different constants involved. For example, $E(Y_{211}) = \mu + A_2 + B_1 + (AB)_{21}$, whereas $E(Y_{221}) = \mu + A_2 + B_2 + (AB)_{22}$. Only when the first two subscripts are the same are the expected values the same. When we examine the linear combination for estimating the A effect, $Y_{211} + Y_{212} + Y_{221} + Y_{222} - Y_{111} - Y_{112} - Y_{121} - Y_{122}$, we note that the levels of the other factor add to zero when the appropriate sign of each term is used. All of the $(AB)_{ij}$ are different, but when the side condition $\sum_{i=1}^{2}(AB)_{ij} = \sum_{j=1}^{2}(AB)_{ij} = 0$ is imposed, we see that the interaction term also drops out of the computation, as of course does μ since the coefficients add to zero. So all that is left from the expected value expression is $(4A_1 - 4A_2)^2$. The correct answer for two replicates is $E(\text{SS}_A) = \sigma^2 + 4\sum_{i=1}^{2} A_i^2$, so it would seem as though we are about to obtain an incorrect answer because, for one thing, $(4A_1 - 4A_2)^2$ will have a cross-product term. We recall from Section 2.1.1, however, that we also have the side condition $\sum_{i=1}^{2} A_i = 0$. From this it follows that $A_1 = -A_2$. We may thus write our result as $(1/8)(8\sigma^2 + 64A_2^2) = \sigma^2 + 8A^2$, which can be seen to be equivalent to the conventional way of writing the result.

In the mixed effects case with one random factor and one fixed factor, two models have been proposed: the restricted model and the unrestricted model. The difference between the two models is that the interaction term for the two-factor design is treated differently in each case. Assume that factor A is random and factor B is fixed. If an interaction term includes at least one random factor, then the interaction term, $(AB)_{ij}$, is considered to be random. More specifically, $(AB)_{ij} \sim N(0, \sigma_{AB}^2)$. This is the *unrestricted* model.

The other approach is to impose the restriction that the sum of the interaction effects over the fixed factor is zero, thus treating part of the interaction as being fixed. For this example that means the restriction $\sum_{j=1}^{2}(AB)_{ij} = 0$. If this approach is adopted, then $E(\text{MS}_B)$ will not contain a term in σ_{AB}^2, so that B would be tested for significance using the statistic MS_B/MS_E instead of $\text{MS}_B/\text{MS}_{AB}$. Thus, the approach that is adopted will affect the results in a way similar to the effect of classifying a factor as fixed or random.

Opinions vary on the approach that should be adopted. Dean and Voss (1999, p. 629) prefer the unrestricted model; Montgomery (1997, p. 480) states, "Most statisticians tend to prefer the restricted model, and it is the most widely encountered in the literature." Kuehl (2000, p. 243) points out that the restricted model cannot be used with unbalanced data and concludes that the choice between the two models depends on the scenario (see also Cobb, 1998, Section 13.3).

There is also a difference in the approaches adopted by the major statistical software. For example, MINITAB fits the unrestricted model by default but will fit the restricted model if specified by the user. Both JMP and SAS use only the unrestricted approach and Design-Expert similarly does not give a user the choice between a restricted and unrestricted model analysis.

All things considered, including software, the unrestricted model seems preferable since regarding part of an interaction as fixed is somewhat impractical.

It is important to recognize that all of these expected value computations are based on the assumption that the fitted model is the correct model, which is almost certainly not going to be true. Recall that $E(Y_{ijk})$ underlies the expected mean squares computations, and this expected value will not, in general, be μ when the wrong model is used. Some model must be assumed if a parametric approach is used, however, and we would hope that the postulated model is a good representation of the true, unknown model.

REFERENCES

Anderson, M. and P. Whitcomb (1997). How to analyze two-level factorials with missing data. Slide presentation. Stat-Ease, Inc.

Bailey, R. A. (1985). Restricted randomization versus blocking. *International Statistical Review*, **53**, 171–182.

Bailey, R. A. (1987). Restricted randomization: A practical example. *Journal of the American Statistical Association*, **82**, 712–719.

Benjamini, Y. and Y. Hochberg (1995). Controlling the false discovery rate: A practical and powerful approach to multiple testing. *Journal of the Royal Statistical Society, Series B*, **57**, 289–300.

Benjamini, Y. and Y. Hochberg (2000). On the adaptive control of the false discovery rate in multiple testing with independent statistics. *Journal of Educational and Behavioral Statistics*, **25**, 60–83.

Berk, K. N. and R. R. Picard (1991). Significance tests for saturated orthogonal arrays. *Journal of Quality Technology*, **23**, 79–89.

Bingham, D. (2001). Discussion of "Factor screening and response surface exploration" by Cheng and Wu. *Statistica Sinica*, **11**, 580–583.

Bisgaard, S. and A. L. S. de Pinho (2004). Quality Quandaries: The error structure of split-plot experiments. *Quality Engineering*, **16**(4), 671–675.

Bisgaard, S. and H. T. Fuller (1995–1996). Reducing variation with two-level factorial experiments. *Quality Engineering*, **8**(2), 373–377. (This article is available as Report No. 127,

Center for Quality and Productivity Improvement, University of Wisconsin-Madison and can be downloaded at http://www.engr.wisc.edu/centers/cqpi/reports/pdfs/r127.pdf.)

Bisgaard, S., C. A. Vivacqua, and A. L. S. de Pinho (2005). Quality Quandaries: Not all models are polynomials. *Quality Engineering*, **17**(1), 181–186.

Bohrer, R., W. Chow, R. Faith, V. M. Voshi, and C.-F. Wu (1981). Multiple decision rules for factorial simple effects: Bonferroni wins again! *Journal of the American Statistical Association*, **76**, 119–124.

Bonett, D. G. and A. J. Woodward (1993). Analysis of simple main effects in fractional factorial experimental designs of Resolution V. *Communications in Statistics A*, **22**, 1585–1593.

Bowman, D. T. (2000). TFPlan: Software for restricted randomization in field plot design. *Agronomy Journal*, **92**, 1276–1278.

Box, G. (1990). George's Column: Do interactions matter? *Quality Engineering*, **2**(3), 365–369.

Box, G. (1990–1991a). A simple way to deal with missing values in designed experiments. *Quality Engineering*, **3**(2), 249–254. (This is also Report No. 57, Center for Quality and Productivity Improvement, University of Wisconsin-Madison and can be downloaded at http://www.engr.wisc.edu/centers/cqpi/reports/pdfs/r057.pdf.)

Box, G. (1990–1991b). Finding bad values in factorial designs. *Quality Engineering*, **3**(3), 405–410.

Box, G. (1992). What can you find out from sixteen experimental runs? *Quality Engineering*, **5**(11), 167–178. (This article is also Report No. 78, Center for Quality and Productivity Improvement, University of Wisconsin-Madison and can be downloaded from http://www.engr.wisc.edu/centers/cqpi/reports/pdfs/r078.pdf.)

Box, G. (1999–2000). The invention of the composite design. *Quality Engineering*, **12**(1), 119–122.

Box, G. E. P. and N. R. Draper (1969). *Evolutionary Operation*. New York: Wiley.

Box, G. E. P., W. G. Hunter, and J. S. Hunter (1978). *Statistics for Experimenters*. New York: Wiley.

Box, G. E. P. and R. D. Meyer (1986). An analysis of unreplicated fractional factorials. *Technometrics*, **28**(1), 11–18.

Box, G. E. P., R. D. Meyer, and D. Steinberg (1996). Follow-up designs to resolve confounding in multifactor experiments. *Technometrics*, **38**, 303–332.

Box, G. E. P. and J. Tyssedal (2001). Sixteen run designs of high projectivity for screening. *Communications in Statistics—Simulation and Computation*, **30**(2), 217–228.

Brownlee, K. A. (1953). *Industrial Experimentation*. New York: Chemical Publishing.

Casella, G. and R. L. Berger (2001). *Statistical Inference*. Belmont, CA: Brooks/Cole.

Chapman, R. E. and V. Roof (1999–2000). Designed experiment to stabilize blood glucose levels. *Quality Engineering*, **12**(1), 83–87.

Chipman, H. (1996). Bayesian variable selection with related predictors. *Canadian Journal of Statistics*, **24**, 17–36.

Cobb, G. W. (1998). *Introduction to Design and Analysis of Experiments*. New York: Springer-Verlag.

Czitrom, V. (2003). Guidelines for selecting factors and factor levels for an industrial designed experiment. In *Handbook of Statistics*, Vol. 22, Chap. 1 (R. Khattree and C. R. Rao, eds.). Amsterdam: Elsevier Science B. V.

Czitrom, V., P. Mohammadi, M. Flemming, and B. Dyas (1998). Robust design experiment to reduce variance components. *Quality Engineering*, **10**(4), 645–655.

Daniel, C. (1959). Use of half-normal plots in interpreting factorial two-level experiments. *Technometrics*, **1**, 311–341.

Daniel, C. (1973). One-at-a-time plans. *Journal of the American Statistical Association*, **68**, 353–360.

Daniel, C. (1976). *Applications of Statistics to Industrial Experimentation.* New York: Wiley.

Dean, A. and D. Voss (1999). *Design and Analysis of Experiments.* New York: Springer-Verlag.

Donev, A. N. (2004). Design of experiments in the presence of errors in factor levels. *Journal of Statistical Planning and Inference*, **126**, 569–585.

Draper, N. R. and D. M. Stoneman (1964). Estimating missing values in unreplicated two-level factorial and fractional factorial designs. *Biometrics*, **20**(3), 443–458.

Eibl, S., U. Kess, and F. Pukelsheim (1992). Achieving a target value for a manufacturing process: A case study. *Journal of Quality Technology*, **24**(1), 22–26.

Emanuel, J. T. and M. Palanisamy (2000). Sequential experimentation using two-level fractional factorials. *Quality Engineering*, **12**(3), 335–346.

Franks, J. (1998). The importance of hierarchy in design of experiments. *Quality in Manufacturing*, March–April issue. Available at: http://www.manufacturingcenter.com/qm/archives/0398/398doe.htm.

Fuller, H. T. and S. Bisgaard (1996). A comparison of dispersion effect identification methods for unreplicated two-level factorials. Report No. 132, Center for Quality and Productivity Improvement, University of Wisconsin-Madison. This can be downloaded at http://www.engr.wisc.edu/centers/cqpi/reports/pdfs/r132.pdf.)

Ganju, J. and J. M. Lucas (1997). Bias in test statistics when restrictions in randomization are caused by factors. *Communications in Statistics—Theory and Methods*, **26**(1), 47–63.

Giesbrecht, F. G. and M. L. Gumpertz (2004). *Planning, Construction, and Statistical Analysis of Comparative Experiments.* Hoboken, NJ: Wiley.

Glasnapp, D. R. and J. Sauls (1976). Comparative magnitude of simple effects as an interpretive index in factorial ANOVA interactions. *Journal of Experimental Education*, **45**(2), 42–46.

Goos, P. and M. Vandebroek (2004). Outperforming completely randomized designs. *Journal of Quality Technology*, **36**(1), 12–26.

Haaland, P. D. and M. A. O'Connell (1995). Inference for contrast-saturated fractional factorials. *Technometrics*, **37**, 82–93.

Hamada, M. and N. Balakrishnan (1998). Analyzing unreplicated factorial experiments: A review with some new proposals. *Statistica Sinica*, **8**, 1–41.

Hare, L. B. (1988). In the soup: A case study to identify contributors to filling variability. *Journal of Quality Technology*, **20**(1), 36–43.

Hicks, C. R. and K. V. Turner, Jr. (1999). *Fundamental Concepts in the Design of Experiments*, 5th ed. Oxford, UK: Oxford University Press.

Hinklemann, K. and O. Kempthorne (2005). *Design and Analysis of Experiments. Vol. 2: Advanced Experimental Design.* Hoboken, NJ: Wiley.

Hsieh, P. I. and D. E. Goodwin (1986). Sheet molded compound process improvements. In *Fourth Symposium on Taguchi Methods*, pp. 13–21. American Supplier Institute, Dearborn, MI.

Hunter, W. G. (1977). Some ideas about teaching design of experiments with 2^5 examples of experiments conducted by students. *The American Statistician*, **31**(1), 12–17.

Inman, J., J. Ledolter, R.V. Lenth, and L. Niemi (1992). Two case studies involving an optical emission spectrometer. *Journal of Quality Technology*, **24**(1), 27–36.

Joiner, B. L. and C. Campbell (1976). Designing experiments when run order is important. *Technometrics*, **18**, 249–260.

Joseph, V. R. (2000). Experimental sequence: A decision strategy. *Quality Engineering*, **12**, 387–393.

Ju, H. L. and J. M. Lucas (2002). L^k factorial experiments with hard-to-change and easy-to-change factors. *Journal of Quality Technology*, **34**(4), 411–421.

Kao, L.-J., W. I. Notz, and A. M. Dean (1997). Efficient block designs for estimating main effects contrasts. *Journal of the Indian Society of Agricultural Statistics*, Special Golden Jubilee Issue, **49**, 249–258.

Kinzer, G. R. (1985). Application of two-cubed factorial designs to process studies. In *Experiments in Industry: Design, Analysis, and Interpretation of Results* (R. D. Snee, L. B. Hare, and J. R. Trout, eds.). Milwaukee, WI: Quality Press.

Kirk, R. E. (1995). *Experimental Design: Procedures for the Behavioral Sciences*, 3rd ed. Belmont, CA: Duxbury Press.

Kuehl, R. O. (2000). *Design of Experiments: Statistical Principles of Research Design and Analysis*, 2nd ed. Pacific Grove, CA: Brooks/Cole.

Lenth, R. V. (1989). Quick and easy analysis of unreplicated factorials. *Technometrics*, **31**, 469–473.

Lenth, R. V. (2001). Review of book by Weber and Skillings. *The American Statistician*, **55**(4), 370.

Lin, T. and B. Chananda (2003). Quality improvement of an injection-molded product using design of experiments: A case study. *Quality Engineering*, **16**(1), 99–104.

Loughin, T. M. (1998). Calibration of the Lenth test for unreplicated factorial designs. *Journal of Quality Technology*, **30**, 171–175.

Loughin, T. M. and W. Noble (1997). A permutation test for effects in an unreplicated factorial design. *Technometrics*, **39**, 180–190. (A SAS program for performing this test can be found at http://www-personal.ksu.edu/~loughin/permtest.sas.)

Lucas, J. M. (1999). Comparing randomization and random run order in experimental design. In *AQC Annual Quality Congress Transactions*, pp. 29–35. American Society for Quality, Milwaukee, WI.

Lynch, R. O. (1993). Minimum detectable effects for 2^{k-p} experimental plans. *Journal of Quality Technology*, **25**(1), 12–17.

McGrath, R. N. and D. K. J. Lin (2001). Testing multiple dispersion effects in unreplicated fractional factorial designs. *Technometrics*, **43**, 403–414.

Monod, H. and R. A. Bailey (1993). Valid restricted randomization for unbalanced designs. *Journal of the Royal Statistical Society, Series B*, **55**, 237–251.

Montgomery, D. C. (1997). *Design and Analysis of Experiments*, 4th ed. New York: Wiley. (6th ed. in 2004).

Montgomery, D. C., R. H. Myers, W. H. Carter, Jr., and G. G. Vining (2005). The hierarchy principle in designed industrial experimentation. *Quality and Reliability Engineering International*, **21**, 197–201.

Natrella, M. (1963). *Experimental Statistics*, National Bureau of Standards Handbook 91. Washington, DC: United States Department of Commerce.

Nelson, L. (2003). Designed experiments with missing or discordant values. *Journal of Quality Technology*, **35**(2), 227–228.

Nelson, P. R., M. Coffin, and K. A. F. Copeland (2003). *Introductory Statistics for Engineering Experimentation*. San Diego, CA: Academic Press.

Nelson, P. R., P. S. Wludyka, and K. A. F. Copeland (2005). *The Analysis of Means: A Graphical Method for Comparing Means, Rates, and Proportions*. Philadelphia: American Statistical Association and Society for Industrial and Applied Mathematics.

Peeler, D. F. (1995). Shuttlebox performance in BALB/CBYJ, C57BL/6BYJ, and CXB recombinant inbred mice—environmental and genetic-determinants and constraints. *Psychobiology*, **23**(2), 161–170.

Prat, A. and X. Tort (1989). Case study: Experimental design in a pet food manufacturing company. Report No. 37, Center for Quality and Productivity Improvement, University of Wisconsin-Madison. (This can be downloaded at www.engr.wisc.edu/centers/cqpi.)

Qu, X. and C. F. J. Wu (2005). One-factor-at-a-time designs of Resolution V. *Journal of Statistical Planning and Inference*, **131**, 407–416.

Reece, J. E. (2003). Software to support manufacturing systems. In *Handbook of Statistics*, Vol. 22, Chap. 9 (R. Khattree and C. R. Rao, eds.). Amsterdam: Elsevier Science B. V.

Ryan, T. P. (1997). *Modern Regression Methods*. New York: Wiley.

Ryan, T. P. (2000). *Statistical Methods for Quality Improvement*, 2nd ed. New York: Wiley.

Ryan, T. P. (2004). *Case Study of Lead Recovery Data*. Gaithersburg, MD: National Institute of Standards and Technology, Statistical Engineering Division. (see http://www.itl.nist.gov/div898/casestud/casest3f.pdf)

Schabenberger, O., T. G. Gregoire, and F. Kong (2000). Collections of simple effects and their relationship to main effects and interactions in factorials. *The American Statistician*, **54**, 210–214.

Schoen, E. D. (2004). Dispersion-effects detection after screening for location effects in unreplicated two-level experiments. *Journal of Statistical Planning and Inference*, **126**(1), 289–304.

Searle, S. R. (1987). *Linear Models for Unbalanced Data*. New York: Wiley.

Simpson, J. R., S. M. Kowalski, and D. Landman (2004). Experimentation with randomization restrictions: Targeting practical implementation. *Quality and Reliability Engineering International*, **20**, 481–495.

Stehman, S. V. and M. P. Meredith (1995). Practical analysis of factorial experiments in forestry. *Canadian Journal of Forest Research—Revue Canadienne de Recherche Forestiere*, **25**(3), 446–461.

Sztendur, E. M. and N. T. Diamond (2002). Extensions to confidence region calculations for the path of steepest ascent. *Journal of Quality Technology*, **34**(3), 289–296.

Taguchi, G. (1987). *System of Experimental Design*, Vol. 1. White Plains, NY: UNIPUB.

Taihrt, K. J. (1971). Randomization for 2^{n-p} factorials in sequential experimentation. *Journal of Quality Technology*, **3**, 120–128.

Taihrt, K. J. and D. L. Weeks (1970). A method of constrained randomization for 2^n factorials. *Technometrics*, **12**, 471–483.

Toews, J. A., J. M. Lockyer, D. J. G. Dobson, E. Simpson, A. K. W. Brownell, F. Brenneis, K. M. MacPherson, and G. S. Cohen (1997). Analysis of stress levels, among medical students, residents, and graduate students at four Canadian schools of medicine. *Academic Medicine*, **72**(11), 997–1002.

Tripolski, M., Y. Benjamini, and D. M. Steinberg (2005). The false discovery rate for multiple testing in large factorial experiments. In *Proceedings of the Fifth Annual European Network of Industrial and Applied Statistics*, Newcastle upon Tyne, UK. (submitted for publication)

Tukey, J. W. (1949). One degree of freedom for non-additivity. Biometrics, **5**(3), 232–242.

Webb, D. F. and J. M. Lucas (2004). Blocking strategies for factorial experiments with hard-to-change factors. In *Proceedings of the Joint Statistical Meetings*, pp. 2181–2188. American Statistical Association, Alexandria, VA.

Webb, D. F., J. M. Lucas and J. J. Borkowski (2004). Factorial experiments when factor levels are not necessarily reset. *Journal of Quality Technology*, **36**(1), 1–11.

White, L. V. and W. J. Welch (1981). A method for constructing valid restricted randomization schemes using the theory of D-optimal design of experiments. *Journal of the Royal Statistical Society, Series B*, **43**, 167–172.

Winer, B. J., D. R. Brown, and K. M. Michels (1991). *Statistical Principles in Experimental Design*, 3rd ed. New York: McGraw-Hill.

Wisnowski, J. W., G. C. Runger, and D. C. Montgomery (1999–2000). Analyzing data from designed experiments: A regression tree approach. *Quality Engineering*, **12**(2), 185–197.

Woodward, A. J. and D. G. Bonett (1991). Simple main effects in factorial designs. *Journal of Applied Statistics*, **18**, 255–264.

Wu, C. F. J. and M. Hamada (2000). *Experiments: Planning, Analysis, and Parameter Design Optimization*. New York: Wiley.

Yin, G. Z. and D. W. Jillie (1987). Orthogonal design for process optimization and its application in plasma etching. *Solid State Technology*, **30**, May, 127–132.

Youden, W. J. (1964). Inadmissible random assignments. *Technometrics*, **6**, 103–104.

Youden, W. J. (1972). Randomization and experimentation. *Technometrics*, **14**, 13–22.

EXERCISES

4.1 Show that when the conditional effect estimates for one of the factors in a 2^2 design differ only in sign, the corresponding main effect estimate must be zero.

4.2 Show that the approach for obtaining a main effect estimate given in Section 4.3 is equivalent to the method used in Section 4.1.

4.3 Consider the following data, with the data ordered in accordance with Table 4.1: 70, 62, 65, and 63. Show that the lines cross in the interaction for one of the two ways to construct the plot but not for the other way. What does this suggest about how interaction plots might be used?

4.4 An experiment is run using a 2^3 design. The eight response values are as follows.

	C_{low}		C_{high}	
	B_{low}	B_{high}	B_{low}	B_{high}
A_{low}	21	23	27	29
A_{high}	24	19	26	32

(a) Construct the BC interaction plot.
(b) Based solely on this plot, would you recommend that the B and C main effects be reported? Why, or why not?

4.5 Consider a 2^2 design with three observations per treatment combination. If the AB interaction is very close to zero, what will be the relationship between the conditional effects of factor B?

4.6 Critique the following statement: "The coefficients obtained from use of a 2^3 design can sometimes be unreliable because of multicollinearity."

4.7 Bill Hunter believed that statisticians should "do statistics." Similarly, students and others who want to learn how to design experiments should "practice" designing experiments and carrying them out. A manufacturing plant is not the place to learn, however, as costly mistakes could be made. Instead, practice should be gained in scenarios where mistakes will do no harm. Many examples of innocuous 2^3 experiments performed by students were given in a paper by Hunter (1977, *The American Statistician*, **31**(1), pp. 12–17). Conduct an actual experiment using a 2^3 design and do a thorough analysis of the data. Ideally, this should be in a subject area that interests you; so consider one of the 32 examples in the paper by Hunter, only if nothing else comes to mind. If data points are both inexpensive and easy to obtain, then conduct a replicated experiment.

4.8 Critique the following statement: "I'm not worried about the effect of large interactions in data that results from my use of factorial designs because if I had such interactions, I would simply include them in the model along with the factors that comprise large interactions."

4.9 Assume that Analysis of Means is to be applied to data from a 2^3 design with four replicates. The estimate of σ using s thus has 24 degrees of freedom. If any main effect or interaction effect is to be significant using $\alpha = .01$, the *difference* between the two means for any of the effects must exceed _____ _____ when $s = 4$.

4.10 Analyze the following data from a 2^2 design with three replicates, assuming that A and B are fixed factors.

		A	
		Low	High
	Low	18 14 17	8 9 12
B			
	High	13 18 16	11 13 14

4.11 Assume that ANOM is used for data from a 2^4 design. How would you suggest that the effects be arranged (ordered) in a single ANOM display so as to permit an easy visual comparison of interactions relative to main effects?

4.12 Consider the following interaction plot.

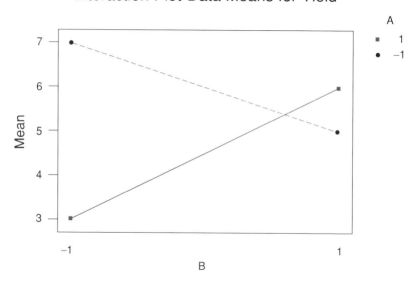

Interaction Plot-Data Means for Yield

(a) Assuming that this is for an unreplicated 2^2 design, what is the estimate of the B effect? How would you report the effect of factor B to management?

(b) If this plot were for an unreplicated 2^4 design, each point in the plot would actually be the average of how many observations?

4.13 Natrella (1963, p. 12–14, references) gave part of the data from a larger experiment that was designed to evaluate the effect of laundering on certain fire-retardant treatments for fabrics. The 16 observations given are from a 2^4 factorial design with the factors being A – fabric, B – treatment, C – laundering condition, and D – direction of the test. The response variable is inches burned, measured on a standard size sample after a flame test. The data are as follows: (1) (4.2), a(3.1), b(4.5), ab(2.9), c(3.9), ac(2.8), bc(4.6), abc(3.2), d(4.0), ad(3.0), bd(5.0), cd(4.0), abd(2.5), acd(2.5), bcd

(5.0), $abcd$(2.3). Analyze the data, paying particular attention to the possible need to compute conditional effects.

4.14 Assume that the levels of a factor are 250 and 375 (degrees Fahrenheit). What is the coding transformation that would be used to convert the levels of the factor to $+1$ and -1?

4.15 Explain the difference between a 2^3 design and a 3^2 design.

4.16 Assume that an experiment with a 2^3 design is used and only the main effects are significant, and (only) those terms are to be used in the model. Indicate the relationship between the model coefficients with the coded data and those for the raw data, using whatever symbolism you prefer.

4.17 Show that for a 2^2 design with r replicates, SS_A can be written as $r(\overline{A}_{\text{high}} - \overline{A}_{\text{low}})^2$, with of course "high" and "low" denoting the high and low levels, respectively.

4.18 Under what conditions, if any, would it be safe to use high-order interactions in estimating σ^2?

4.19 Consider the interaction plot in Exercise 4.12. Construct the plot with factor A plotted on the horizontal axis. (All of the plotted points in the graph in Exercise 4.12 are integers, which may not be obvious from the graph.)

4.20 Considering the nature of the design that was used to produce the data given in Exercise 4.4, what would be necessary before an ANOM display of the effect estimates could be constructed? Assuming that condition is met, if all of the effect estimates are shown in an ANOM display, what will be the length of the line segment that connects the two plotted points for the ABC effect?

4.21 Assume that a data analyst has constructed an ANOM display for data from an experiment in which a 2^3 design was used, with two of the factors fixed and the other factor random. Does this present a problem? If so, explain what the problem is in regard to testing hypotheses.

4.22 Verify $E(\text{MS}_A)$ that was given in Appendix C.

4.23 Designed experiments have been performed by very young students for class projects. Eric Wasiloff and Curtis Hargitt, a pair of ninth grade students, used a 2^3 design to determine factors that affect AA battery life. The results of their experiment were reported in their article "Using DOE to determine AA battery life" (*Quality Progress*, March, 1999, pp. 67–71). The second page of their article contains the all-too-familiar-words "... the typical one-factor-at-a-time method." The response variable that they used was time to discharge in minutes and the three factors were battery type (Durall high-cost alkaline batteries and Panasonic low-cost dry cell batteries), connector design type (gold-plated and standard), and battery temperature (ambient and cold). The data are given below.

Battery type	Connector design type	Temperature	Time to Discharge
High	Gold-plated	Ambient	493
High	Gold-plated	Cold	490
High	Standard	Ambient	489
High	Standard	Cold	612
Low	Gold-plated	Ambient	94
Low	Gold-plated	Cold	75
Low	Standard	Ambient	93
Low	Standard	Cold	72

The students claimed that the "battery type" factor was significant (which is obvious from looking at the data, and concluded that the other two factors are not significant). They obtained F-statistics of 170, 0.7, and 0.4 (rounded off) for these three effects, which led to their conclusion.

(a) Since this is an unreplicated factorial, how were those F-statistics obtained? What advice would you give the students/authors?

(b) Do you feel that a conditional effects analysis of these data is necessary? Explain.

4.24 Although a 2^2 design is seldom used in practice, it was used in an experiment described by D. F. Aloko and K. R. Onifade in their paper, "The stability of the adsorption of some anions on chemical manganese dioxide," *Chemical Engineering and Technology*, **26**(12), 2003, pp. 1281–1283. There were three response variables, so three regression models were developed, with the two main effects and interaction effect used in each model. The design was unreplicated, so there were four values for each response variable. The value of R^2, which was termed the "regression coefficient," was given for each model and the values were .54, .67, and .84. Something is wrong here. What is it? If possible, read the paper and try to determine the cause of the problem.

4.25 In the paper cited in the preceding problem, the authors gave the model with both main effects, interaction term and intercept term, which they referred to as Eq. (3) and stated, "The 2^2 arrangement, when applied to only two variables, permits uncorrelated, low variance estimates of the four coefficients indicated in Eq. (3)." Apart from the fact that the word "estimates" should have been "estimators," is there anything else that is erroneous or suspect in the quote?

In particular, what is the relationship between the expression for the variances of the estimators when a 2^2 design is used relative to the variances of the appropriate estimators when a 2^3 design is used?

4.26 Various books and journal articles have listed the two levels of a factor as (0, 1) or (1, 2) rather than $(-1, 1)$. How would you code a temperature factor with levels of 450 and 470 so as to produce (a) the first pair and (b) the second pair. Does the use of (0, 1) or (1, 2) present any special problems in the analysis of data?

4.27 Usher and Srinivasan (*Quality Engineering*, **13**(2), 2000–2001) describe a 2^4 experiment for which there were six observations per treatment combination, but the data were not analyzed as having come from a replicated experiment. Specifically, the authors stated, "Note that these six observations do not represent true replicates in the usual sense, because they were not manufactured on unique runs of the process." Consequently, the average of the six values at each treatment combination was used as the response value, and the data were thus analyzed as if it had come from an unreplicated experiment. Explain how analyzing the data as if the data had come from a replicated experiment could produce erroneous results.

4.28 Somewhat similar to the scenario in Exercise 4.27, Ryan (2004, references) analyzed data from a case study using both averages and individual observations and compared the results, with both approaches necessary because the replicates weren't quite true replicates. Read Ryan (2004), which is available on the Web, and comment.

4.29 Assume that you have data from a 2^5 design with two replicates. What is the first step that you will take in analyzing the data? Would your answer be the same if the data in hand were from an unreplicated 2^2 design? Explain.

4.30 (Requires MINITAB). The sample data file YIELDPLT.MTW that comes with MINITAB contains data from a 2^3 design with two response variables: yield of a chemical reaction and cost. The factors are reaction time, reaction temperature, and type of catalyst. There were enough resources for 16 runs, but only 8 could be made in a day. Therefore, two replicates of the 2^3 design were used, with replicates being days. Analyze the data, using yield as the response variable. (You will notice that a two-factor interaction is significant. Does this necessitate the use of conditional effects? Why, or why not?)

4.31 Photolithography is the process of transferring a pattern from a photomask onto the surface of a silicon wafer or substrate. In photolithography it is important to achieve a certain line width of the etched grid. One of the key chemicals of this photolithographic process is polyvinyl alcohol (PVA). In addition, the other components of the emulsion, which we will label as A, B, and C, are thought to be important factors in controlling line width. In order to study the effects of these three components on line width, a 2^3 factorial was chosen as the design. The levels of each factor used in the experiment were as follows.

Factor	Low	High
A	0%	6%
B	8%	16%
C	0%	3%

The line widths for the eight runs, written in the usual (Yates) order were as follows: 6.6, 7.0, 9.7, 9.4, 7.2, 7.6, 10.0, and 9.8. Of course the order of the runs was randomized, and assume that the factors were reset after each run. Determine the effects that are significant.

4.32 Assume that you have data from a 2^4 design with three replicates. Do you need to assume homogeneity of variance if you use ANOM or does that apply only to ANOVA? Explain.

4.33 Referring to Exercise 4.32, will the decision limits for assessing each effect if ANOM is used be computed using a t-value? If a t-value is appropriate, what will be the degrees of freedom? If a t-value is inappropriate, how should the decision limits be computed?

4.34 Consider the data for the 2^4 design in Section 4.10. It was shown in that section that a badly misrecorded value can result in two distinct lines on a normal probability plot, thus providing a signal that something is wrong. Does this same phenomenon occur when the number is recorded as say, 26.2, or 36.2 or 46.2 instead of 62.0? What does this suggest about using normal probability plots to detect bad data?

4.35 Consider Example 4.5, in which the ABC interaction was small. Does this mean that the two AB interaction graphs for each level of C will be similar? Construct the displays and comment.

CHAPTER 5

Fractional Factorial Designs with Two Levels

The numerical value of 2^k increases rapidly as k increases, so the number of design points for a 2^k design could be quite large when there are more than just a few factors of interest. As Steinberg and Hunter (1984) indicated, there might be as many as 50 or 100 potentially important factors in some applications.

A 2^k full factorial allows an experimenter to estimate all of the $2^k - 1$ effects, but when k is greater than 3 or 4, most of these effects will be high-order interaction effects that would not be significant, as the higher the order of an interaction effect, the less likely it is to be significant. Therefore, using a full factorial would be wasteful when the value of k is not small. A *fractional factorial* has, as the name suggests, a fraction of the number of design points of the corresponding full factorial, with the objective being to not waste design points that allow the estimation of effects that will almost certainly not be real.

We shall focus attention on two-level fractional factorial designs as this type of fractional factorial has been used extensively in practice. Fractional factorials were first presented by Finney (1945) and popularized in the landmark papers of Box and Hunter (1961a,b). We shall adopt the notation used by the latter.

A two-level fractional factorial can be written in the general form 2^{k-p}, where as indicated previously, k denotes the number of factors, and the fraction of the full 2^k factorial that is to be run is $1/2^p$. The *Resolution* of a fractional factorial design, first discussed in detail by Box, Hunter, and Hunter (1978), who assigned resolution numbers to categories of effects that could be estimated, indicates the effects that can be estimated with the design. Specifically resolution III indicates that only main effects can be estimated in the absence of interaction effects; IV indicates that only main effects are estimable even if two-factor interactions exist (provided that higher-order interactions do not exist); and V means that both main effects and two-factor interactions are estimable, provided that higher-order interactions do not exist. We

Modern Experimental Design   By Thomas P. Ryan
Copyright © 2007 John Wiley & Sons, Inc.

will use notation such as 2_{IV}^{6-2} in representing fractional factorial designs, which here is a $1/4$ fraction of a 2^6 design that is of resolution IV. In the foregoing we will sometimes indicate the resolution of a design in this manner.

The term *defining relation* is explained and illustrated in Section 5.1. Briefly, it is composed of one or more interactions, whose inclusion in the defining relation determines effects that are confounded (i.e., confused). With this in mind, we have the following.

IMPORTANT POINT

The Resolution of a fractional factorial design is generally defined as the length of the shortest word in the defining relation, with the number of words in the defining relation determined by the degree of fractionization. This definition will suffice in almost all cases, although exceptions can be found when blocking is involved (see Giesbrecht and Gumpertz, 2004, pp. 381–382).

The last part of the Important Point will be illustrated by the examples in Sections 5.1 and 5.2. (*Note*: Fractional factorial designs are "orthogonal arrays," meaning that the columns for the effects to be estimated are orthogonal. There are also orthogonal arrays that are not fractional factorials. It is common for the "strength" of an orthogonal array to be given. This is simply one less than the resolution of the design.)

It is very important to realize that the properties of a fractional factorial design (or any design in general) for all the factors and effects that are estimable with a given design is not what is important, as generally less than half of the effects that are being examined will be significant. Therefore, the focus must be on the properties of the design for the effects that are significant. Thus, the *projective properties* of designs are important, as a design is projected onto the number of dimensions corresponding to the number of significant factors. For example, the properties of a design for investigating six factors are not particularly important when only three factors turn out to be significant. The projective properties of 2^{k-p} designs are discussed in Section 5.11.

The term "regular design" is used extensively in the design literature. A full factorial design is a regular design because all the effects can be estimated independently of each of the other effects. The 2^{k-p} designs are also regular designs because although not all effects are estimable, the effects that are not estimable are completely confounded with other effects. A design for which at least some effects are neither independently estimable nor completely confounded is called a *nonregular design*. The classification of designs in this way is of interest because we would naturally like to know whether there is at least partial information on all effects that are of interest.

5.1 2^{k-1} DESIGNS

Thus, a 2^{3-1} design would be a $1/2$ fraction of a 2^3 design. Sixteen point designs (so that $k - p = 4$) are the ones that have been used most often in industry, although designs with a much higher number of points are sometimes needed, especially when there is a large number of factors. An example with 47 factors is discussed in Section 5.12.

For simplicity, however, we shall first illustrate a 2^{3-1} design, which, although of very limited usefulness, does have value for illustrative purposes. We should first recognize that with only four design points (here we are assuming that the design is not replicated), we will have only three df, so we can estimate only three effects. Which three effects do we choose to estimate? Although in rare instances a two-factor interaction will be of more interest to an experimenter than a main effect, we would generally choose to estimate main effects over interactions, if we had to select one over the other. Thus, the logical choice would be to estimate the three main effects: A, B, and C. Before we can undertake that task, however, we must determine what four design points to use. We cannot just randomly select four treatment combinations from the eight that are available. For example, we obviously could not estimate the main effect of A if we happened to select four treatment combinations in which A was at the high level and none in which A was at the low level, and similarly for B and C. Thus, we would clearly want to have two treatment combinations in which A is at the high level and two in which A is at the low level, and the same for B and C.

With a little trial and error we could obtain four treatment combinations that satisfy this property without too much difficulty, but it would obviously be preferable to use some systematic approach. Whenever a $1/2$ fraction is used, we have to select one effect to "confound" with the difference between the two fractions, that is, that particular effect would be estimated by the difference of the averages of the treatment combinations in each fraction (which of course is the way that we would logically estimate the difference between the two fractions). If we have to "give up" the estimate of one effect in this way (which is obviously what we are doing since we will run only one of the two fractions), it would be logical to select the highest-order interaction to relinquish. For a 2^3 design that is obviously the ABC interaction.

One simple way to construct the two $1/2$ fractions (from which one could be randomly selected, although this is normally not done in practice) would be to assign those treatment combinations with an even number of letters in common with ABC to one fraction, and those with an odd number of letters in common with ABC to the other fraction. This has the desired effect of creating the two fractions with ABC confounded between the two fractions, which can be explained as follows. As in the case of a full factorial design, the sign of an interaction for a particular treatment combination is a product of the signs of the factors represented by the treatment combination. The "even" fraction must have all minus signs because the number of letters in common with ABC must be either 0 or 2. In each case there will be an odd number of minus signs (e.g., the treatment combination ab has two letters in common with ABC, and $(+1)(+1)(-1) = -1$, so the "even" fraction will be the "minus fraction").

The two fractions are given in Figure 5.1.

There is a *defining relation* associated with each fraction, with $I = -ABC$ for the minus fraction and $I = ABC$ for the plus fraction. The first defining relation, which corresponds to the second fraction, states that the estimate of the ABC interaction is

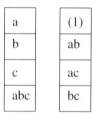

Figure 5.1 Two 1/2 fractions of 2^3.

confounded with minus the estimate of the overall mean. That is, we obtain a minus for each treatment combination (e.g., ab is $(+)(+)(-) = -$), so the observations for this fraction would estimate the negative of the ABC effect. Of course we would estimate the mean using all plus signs, and "I" refers to the mean effect. Thus, the estimate of the mean effect will be the negative of the estimate of the ABC effect.

One way to gain insight into the 2^{3-1} design is to write out the manner in which each effect is estimated using the full 2^3 design and then look at the treatment combinations in Figure 5.1 that comprise each fraction. Although Yates' algorithm, which can be used for obtaining effect estimates, no longer has much practical value since effect estimates are routinely obtained using software, we can use it effectively here. The treatment combinations are written in a specified order, and for a 2^3 design the order is given in Table 5.1, using the data from Daniel (1976) that was used previously in Section 4.5 in illustrating conditional effects and the potential hazards of using high-order interactions in estimating σ^2.

Here we will use it to illustrate half fractions, and since we have data from the full factorial, we will be able to compare the results from the two fractions. For

TABLE 5.1   Yates' Algorithm Calculations Using the Data in Daniel (1976)

Treatment Combination	Response	(1)	(2)	(3)	Column (3) Representation
(1)	297	597	834	1374	(Sum of all responses)
a	300	237	540	62	$a + ab + ac + abc$ $-b - c - bc - (1)$
b	106	355	28	-530	$b + ab + bc + abc$ $-a - c - ac - (1)$
ab	131	185	34	54	$c + ab + abc + (1)$ $-a - b - ac - bc$
c	177	3	-360	-294	$c + ac + bc + abc$ $-a - b - ab - (1)$
ac	178	25	-170	6	$b + ac + abc + (1)$ $-a - c - ab - bc$
bc	76	1	22	190	$a + bc + abc + (1)$ $-b - c - ac - ab$
abc	109	33	32	10	$a + b + c + abc - ac$ $-ab - bc - (1)$

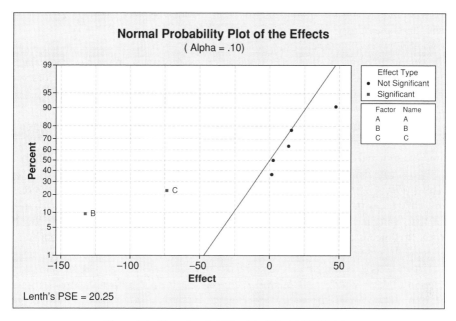

Figure 5.2 Normal probability plot of Daniel (1976) data.

convenience, the normal probability plot analysis of that data, using Lenth's (1989) method for determining significant effects, is repeated (Fig. 5.2), since it will be referred to in subsequent sections. (Haaland and O'Connell (1995) concluded that Lenth's (1989) method is a reasonable approach.)

The plot shows only the B and C effects to be significant. (It is worth noting that even though the BC interaction effect doesn't show as being significant, it is more than half the C effect and more than $1/3$ the B effect, thus rendering those estimates almost meaningless.)

Using adjacent treatment combinations in Table 5.1, there are four pairs, and additions and subtractions are performed in pairs, with the first number in each pair subtracted from the second number in the pair. This should be clear from inspection of the column labeled (1). The number of times that this set of operations is performed is equal to the number of factors, with the last column in Table 5.1 indicating how each number in column (3) is obtained.

Those numbers are used in obtaining the estimates of the various effects. For example, the AB interaction effect is estimated as $54/4 = 13.5$, with the number 54 being used because that is in the row for the ab treatment combination and 4 is used as the divisor because there are four plus signs and four minus signs.

It can be observed that the 10, which is the last number in column (3) and corresponds to the abc treatment combination, is obtained by adding the response values of the treatment combinations in the first fraction in Figure 5.1 and subtracting the response values of the treatment combinations in the second fraction. This shows that the ABC interaction is confounded with the difference between the two fractions, as was stated previously.

Before proceeding any further, we should think about what we are giving up in addition to an estimate of the ABC interaction when we run one of the fractions in Figure 5.1. We assume that we are relinquishing information on the three two-factor interactions by using one of the fractions in Figure 5.1 (since we have only three df), but we also assume that we will be able to estimate the three main effects.

Let's verify this by looking at some of the other parenthetical expressions in Table 5.1. Assume that we have randomly selected the first fraction in Figure 5.1 to run. The sum and difference of the various treatment combinations beside the number -294 indicate that we would estimate the main effect of C by $(abc + c - a - b)/2 = (109 + 177 - 300 - 106)/2 = -60$, which does not differ greatly from -73.5 that was obtained using all eight treatment combinations. Further, we can see from Table 5.1 that, using the first fraction, we would also estimate AB by $(abc + c - a - b)/2$. What this means is that the estimate of the C effect is *confounded* (i.e., confused) with the AB effect.

(*Note*: This is complete confounding. In subsequent chapters, and especially in Section 13.4.1, we will speak of *partial aliasing*. As the name suggests, this is a state of affairs between orthogonality and complete confounding. With the former, the dot product of two columns that represent effects to be estimated is zero; with the latter the dot product is n because the signs are the same. With partial aliasing the dot product is something between 0 and n. If the sum is $n/3$, then the correlation (using the term somewhat loosely) between the two columns is $1/3$; if the sum is $-n/3$, then the correlation is $-1/3$. It should be noted, however, that the term partial *confounding* is used in conjunction with designs that are run in blocks to mean that an effect is confounded with some blocks but not with others. That is different from what is discussed in this section, so that terminology is not used here.)

Similarly, it could be shown that A is confounded with BC and B is confounded with AC, as can be verified from Table 5.1. Fortunately, there are easier ways to determine which effects are confounded. One way is to write out how each effect is estimated using plus and minus signs, and then identify those effects that have the same configuration of signs. The other method is much easier and is simply a matter of multiplying each effect by the effect that was confounded with the difference of the two fractions (ABC in this example), and removing any letter whose exponent is a 2. This applies to *any* two-level fractional factorial. This is modular arithmetic and it is modulo 2 because the factors have two levels. To see why a letter whose exponent is 2 is removed, notice that if we square a column of numbers that are all $+1$ and -1, the result is a column of numbers that are all $+1$, and any such column multiplied by any effect will simply give the column corresponding to that effect.

Thus, $A(ABC) = A^2BC = BC$, $B(ABC) = AB^2C = AC$, and $C(ABC) = ABC^2 = AB$. The effects that are confounded with each other are said to be *aliases* of each other, and the set of such aliases is said to be the *alias structure*.

Another way to view the alias structure is to use Yates' algorithm after filling in zeros for the treatment combinations that are in the fraction that is not used. Factors that are aliased will then have the same totals in column (3), as the reader is asked to demonstrate for these data in Exercise 5.3.

This is not a recommended approach for determining the alias structure, however, as effect estimates (and, hence, numbers in column (3)) can be the same without the effects being confounded. It is also far more time-consuming than the multiplication approach just illustrated. It is simply another way of viewing the alias structure. Of course software will readily provide the alias structure for a fractional factorial, and obviously that is the preferred approach for obtaining the alias structure.

We saw from the analysis of the full factorial that the AB interaction was not significant, and we can also see from Table 5.1 that the estimate of C plus the estimate of AB equals -60 that we just obtained using the four treatment combinations in the first fraction. Thus, we are actually estimating $C + AB$ rather than just C, and the extent to which our estimate of C is contaminated depends upon the size of the AB interaction. Here there is no serious problem because the AB interaction is not large.

What if we had randomly selected the other fraction? A little arithmetic would reveal that our estimate of C would be -87 and that we would really be estimating $C-AB$.

Although moderately large, the BC interaction was declared not significant by the normal probability plot analysis of Figure 5.2, and we now know that BC is aliased with A. Therefore, we can see to what extent the estimate of the A effect is contaminated by the presence of a moderate BC interaction. Again assuming that we had used the first fraction, our estimate of the A effect would be obtained from $(abc + a - b - c)/2 = (109 + 300 - 106 - 177)/2 = 63$, which differs dramatically from our estimate of 15.5 obtained from the full factorial. As the reader might suspect, we are actually estimating $A + BC$ (and $A - BC$ with the other fraction), so that, in this example, we would erroneously conclude that there is a strong A effect when in fact there was not. (We remember, however, that our detective work did reveal that A was somewhat influential when B was at its high level.)

The upshot of all of this is that when we run a fractional factorial we do take a risk, and the severity of the risk depends upon the order of the interactions that are lost. First-order interactions (i.e., involving two factors) and the second-order interaction were lost in the 2^{3-1} example; so that design should be considered only if there is a strong prior belief that none of the interactions will be significant. Even then, the advantage of the fractional factorial would be minimal, as four design points would be run instead of eight—not much of a saving.

The picture changes considerably, however, when there are more than three factors. What about a 2^{4-1}? The alias structure would certainly be more palatable in that the fractions could be constructed in such a way that the main effects would be aliased with second-order interactions, but, unfortunately, the first-order interactions are aliased in pairs. The $ABCD$ interaction would be confounded with the difference between the two fractions, and we would then have $A = BCD, B = ACD, \ldots, AB = CD, AC = BD$, and so on. Although it is a small design, 8-point designs have been used in practice—an application in which a 2^{4-1} design was used was discussed by Hill and Wiles (1975).

From a practical standpoint, however, we should ask whether or not such small designs should be used. When Design-Expert is used to construct a 2^{4-1} design, the user receives a *warning message* which states that effects must be at least 2 standard deviations in size in order to be detected. (Actually the discussion of a 2^3 design in

TABLE 5.2   The Two 2^{3-1} Fractions

First Fraction				Second Fraction			
Treatment Combination	A	B	C	Treatment Combination	A	B	C
a	+	−	−	(1)	−	−	−
b	−	+	−	ab	+	+	−
c	−	−	+	ac	+	−	+
abc	+	+	+	bc	−	+	+

Section 4.6 applies here since the number of design points is the same. Recall that it was stated there that 2.29σ is the smallest effect that can be detected, assuming that σ is known.) Lynch (1993) gave tables for minimum detectable effects for 2^{k-p} designs, with most of the entries for a small number of runs based on the assumption that there were at least three degrees of freedom for estimating σ. Those tables show that for $\alpha = .05$ and a power of .90 for detecting significant effects, the minimum detectable effect (in standard deviation units) for a 2^{4-1} design is 3.5. A 2^{5-1} design fares somewhat better as the minimum detectable effect is 2.5, but this will also be too large for many practitioners. (Of course these results cannot be compared directly with results stated under the assumption that σ is known. The result of 2.29σ given earlier in this paragraph is certainly more in line with the warning message given by Design-Expert.)

One way to view the 2^{3-1} design—and this applies in a general way to all 2^{k-p} designs—is that each fraction contains a full factorial, namely a 2^2 design in any pair of the factors. This can be seen by writing out the treatment combinations as shown in Table 5.2.

Notice that all four combinations of plus and minus signs occur with each pair of factors in each fraction. This result extends generally in that there is a full factorial in $k - p$ factors in each fraction whenever a 2^{k-p} design is constructed. The importance of this is that if, by chance, only $k - p$ factors seem to be important, there is a full factorial in those factors that can be used for analysis. If the number of apparently significant factors, w, is less than $k - p$, then there is a replicated 2^w design within the 2^{k-p} design. This is illustrated in Section 5.11.

5.1.1   Which Fraction?

Software for DOE generally provides the user with the capability to specify a particular fraction. For example, in MINITAB the FRACTION subcommand for the FFDESIGN command can be used to specify the fraction that is to be used. The following commands would be used to select the second fraction in Figure 5.1.

```
MTB>  FFDESIGN 3 4 ;
SUBC>  FRAC 2 .
```

Without the subcommand, MINITAB will simply list the first fraction, as the fraction that contains all plus signs for the confounded interaction(s) is the default fraction in MINITAB, and is termed the principal fraction, in general. JMP automatically produces the same fraction but does give the user certain options, as discussed in Section 5.2.1. Design-Expert allows any desired fraction to be constructed when the Make Generators Editable option is invoked. The user can then specify the generator(s). For the 2^{3-1} design being discussed in this section, the user would enter $C = -AB$ to produce the other fraction (i.e., not the principal fraction).

The obvious question to address at this point is "Does it make any difference which fraction we use?" We can address this question relative to (a) having no prior information, and (b) having certain prior information. If we have no prior information, we would have no reason to select one fraction so as to not "miss anything," whereas prior information might lead us to select a particular fraction. For example, Daniel (1976, p. 41) stated that in his experience interactions are usually "one-cell interactions." That is if an additive model is fit to the data, only one cell will usually have a large residual that stands out. If desired, a significance test could be used to determine a "large residual." This is discussed by Daniel (1976, p. 40) and elsewhere.

If we anticipated this and had no other a priori information, we would want to select the fraction that included that corner point. For example, let's assume an experiment was run and the AC interaction was found to be significant, and that this was due primarily to a large value for the abc treatment combination. If this outcome had been anticipated, then the fraction that contains treatment combination abc could have been (nonrandomly) selected. If, however, this was not anticipated and the other fraction was used, then there is a strong possibility that the conclusion regarding the AC interaction would be incorrect. Therefore, prior information should be utilized in design construction if it is available and reliable.

If, however, no prior information is available and in fact no unusual values would have been observed if the full factorial had been run, then one or more conclusions from using one fraction that differ from the set of conclusions obtained using the other fraction would expectedly be due only to random variation.

This obviously raises the question of how much random variation might be expected between the results obtained using the two fractions. This question is addressed in Section 5.1.4.

5.1.2 Effect Estimates and Regression Coefficients

The manner in which effect estimates and the regression coefficients for the selected model are obtained for a full factorial was explained in Section 4.1 and in Appendix B of Chapter 4. It works essentially the same way for a fractional factorial, except that we cannot estimate all interaction effects up to one order less than the number of factors. Specifically, we know that we cannot estimate the ABC interaction in a 2^{3-1} design since the interaction is confounded with the difference between the two fractions.

5.1.3 Alias Structure

Let's return to the experiment described by Daniel (1976) and look at the numbers in a somewhat different way, still thinking about what would have happened if a 2^{3-1} design had been used.

In particular, let's look for extreme observations and see what their effect will be. With only eight observations, as given in Table 5.1, it is easy to see that there are two numbers that are much larger than any of the other numbers. We would be especially concerned if these numbers were in the same fraction, but that does not happen as treatment combination a is in the principal fraction and (1) is in the other fraction.

Thus if a 2^{3-1} design had been used instead of the 2^3 design we would have had one large number with which to be concerned. What is common to those two treatment combinations is that the other two factors are each at their low levels. The comparison of, say, b versus (1) and c versus (1) suggests that the main effects B and C may be significant (as they are), as each effect estimate is the average of four differences and one of the differences is very large relative to the magnitude of the numbers. Since hardly any designed experiments are run flawlessly, these two large observations should have been checked out. There is no evidence from Daniel (1976), however, that the experiment resulted in any bad data.

Therefore, we will assume that all of the data points are valid. The estimates of the effects from each fraction, some of which were given in Section 5.1, are as follows.

Fraction

I = ABC		I = −ABC	
Effect	Estimate	Effect	Estimate
A + BC	63	A + (−BC)	−32
B + AC	−131	B + (−AC)	−134
C + AB	−60	C + (−AB)	−87

Notice that the A effect differs greatly between the two fractions, which results from the fact that the two large values are in different fractions, as mentioned previously, with one at the high level of A and the other at the low level. This means that when the A effect estimate is computed, the large value at the high level of A has a plus sign attached to it in the computation, whereas the other large value has a minus sign attached to it, in the other fraction. Although a 2^{3-1} design is not a particularly useful design, as stated previously, this does illustrate how the estimates of effects from the two fractions can differ greatly on data from an actual experiment.

Furthermore, since the experiment was actually run as a 2^3, we know the separate main effect and interaction effect estimates. These estimates can be obtained from the results given above, as we might expect. For example, since $B + AC = -131$ and $B + (-AC) = -134$, if we add the two equations together we obtain $2B = -165$, so the estimate of the B effect from the two fractions combined is $-265/2 = -132.5$. Similarly, if we subtract $C + (-AB) = -87$ from $C + AB = -60$, we obtain $2AB = 27$, so the estimate of the AB interaction effect is $27/2 = 13.5$.

As is explained in Section 5.4, if we ran one of these fractions and then later decided to run the other fraction because we became concerned about possible interaction effects, we would be using a *foldover design*. That is, we would be "folding over" the first fraction (i.e., changing all of the signs in the first fraction) to obtain the second fraction, with the foldover design being the full 2^3 design.

In this case, folding over one of the fractions is the appropriate thing to do because there is a large BC interaction effect (the estimate being 47.5), which causes the estimates from the two fractions to differ greatly since the estimate of the A effect in one fraction is really the estimate of $A + BC$, whereas it is $A - BC$ in the other fraction. Obviously the estimates would not differ by very much if the BC effect were small.

As a practical matter, if a fractional factorial were run and the sign of an effect was unanticipated, with the magnitude of the effect not being small, more runs, such as in a foldover design, should be made. (Note that this problem caused by a large interaction effect is different from the problem of a large interaction effect necessitating an analysis using conditional effects.)

5.1.4   What if I Had Used the Other Fraction?

Cochran and Cox (1958) gave an example of a field experiment performed in 1936 at the famous Rothamsted Experimental Station that had four factors, each at two levels. The experimental area was divided into two blocks of land, with each block containing eight plots because the experimenters suspected that the blocks of land might be dissimilar. So the experiment was run with two blocks and the $ABCD$ effect was confounded with the difference between the blocks.

Since it can be shown that there is no evidence of a block effect (see Section 4.15 for the analysis of factorial designs run in blocks), we will proceed to analyze the data as two half fractions of a 2^4 design. We would look at only the main effect estimates since the two-factor interactions are confounded in pairs, with the main effects confounded with three-factor interactions.

I = ABCD		I = −ABCD	
Effect	Estimate	Effect	Estimate
A	−6.25	A	4.75
B	−5.75	B	−10.25
C	2.25	C	−1.75
D	−4.25	D	−0.25

Although a normal probability plot analysis is somewhat shaky with only eight data points, using such an analysis with the data in the second fraction results in the B effect identified as significant, whereas no effect is identified as being significant for the first fraction. Thus, the conclusion depends on which fraction is used.

One problem we can immediately recognize is that we can't realistically perform a conditional effects analysis for a resolution IV design because we can't determine how to split the data to look at the conditional effects. This is because we perform the splits by looking at the magnitude of the two-factor interactions, but the two-factor interactions are confounded in pairs in resolution IV designs.

It is instructive to compare the results obtained using each fraction with the results obtained using all of the data and analyzing it as a 2^4 design. When this is done, the normal probability plot analysis identifies the main effect of B as the only significant effect.

Pareto Chart of the Effects

(response is Yield, Alpha = .10)

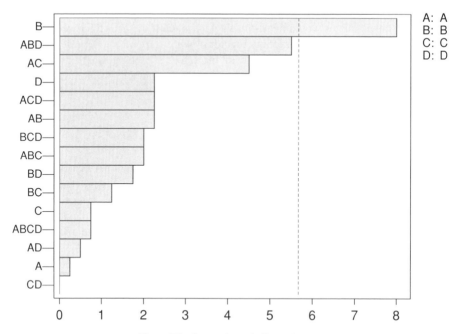

Figure 5.3   Pareto chart of effect estimates.

What we see with this analysis that of course isn't available with either fraction is that the BCD interaction is the second largest effect, as is shown in the Pareto chart of the effects that is given in Figure 5.3. (Of course this interaction is confounded with the A effect in each fraction, and we know that three-factor interactions are generally not real.)

If the 2^{4-1} design had been used, it would have probably seemed safe to have the A effect confounded with the BCD interaction, but we see from the full analysis that the BCD interaction is just barely below the line that separates effects that are judged to be significant from those that are not so judged. Similarly, if we do a normal probability plot analysis and use $\alpha = .11$ in MINITAB instead of the default value of .10, the BCD interaction shows as being significant. Therefore, it is not an effect that can be safely ignored.

It is of interest to determine the effect, if any, that the large BCD interaction has on the B effect estimate. Although normally we don't think of going directly from a three-factor interaction to conditional main effects, we can do so in two steps. The large BCD interaction means that all two-factor interactions involving those letters are unreliable. The largest of these interactions is the BC interaction, whose conditional effects, splitting on the D factor, are $2.25\pm(-5.50)$. We would then obtain the conditional B effects from the four observations that produce the larger of these two numbers (7.75), which occurs at the low level of D. Specifically, the conditional effects of B are -14 at the low level of C and 1.5 at the high level of C.

Although each of these last two numbers is computed from averages of only two observations, and would both be more meaningful and have a smaller variance if the design were replicated, the numbers do suggest that there may be a considerable B effect at the low level of C and the low level of D.

This analysis simply shows the region of the factor space where the B effect is the greatest; this information is not available with traditional analyses. Of course the B effect was significant in the full factorial analysis and in the second of the two half fractions, so this simply adds to that information.

It would be of greater interest to discover a conditional main effect of a similar magnitude, also based on four numbers, for which the main effect was not significant in the full factorial or with either of the half fractions. From inspection of Figure 5.3 we can see that this is not going to happen, although we note in particular that the small unconditional A and C effects in the full analysis are very misleading because of the large AC interaction. We can observe that the conditional main effects of A and C, with the data split, in turn, on each, will not differ much from plus and minus the AC interaction since both main effect estimates are close to zero.

Those conditional main effects would each be computed using averages based on four numbers. We could, if desired, obtain the conditional main effects based on averages computed using two numbers each. We will not, however, obtain a conditional main effect of C that is anywhere near the largest conditional main effect of B since the unconditional effect of C is close to zero.

Nevertheless, to illustrate the computations, our starting point would again be the BCD interaction and we would split on D and again focus on the low level of that factor for the same reason that we did so in looking at the conditional main effects of B. This leads to -4.0 and 7.5 as the conditional main effects of C. It should be noted that these numbers bear no relationship to the unconditional main effects since only half of the data are being used. This differs from the case when the conditional main effects are obtained from only a single split because in that case all of the data are being used.

Accordingly, we could, if desired, designate conditional main effects in accordance with the number of splits that were performed in producing the estimates, so these conditional effects for B and C might be called *two-split conditional main effects*.

Although the largest, in absolute value, two-split conditional main effect of C of 7.5 is only about half of the largest two-split conditional main effect of B, it is nevertheless large enough to tell us that we should not overlook factor C.

We will pursue the topics of the "fraction effect" and multi-split conditional main effects further in Section 5.2.

5.2 2^{k-2} DESIGNS

In this section we will use an experiment performed using a 2^5 design and analyze the data as four 2^{5-2} designs and also as two 2^{5-1} designs to further illustrate and reinforce the concepts described in Section 5.1. This is somewhat akin to seeing how certain methodology works when we know the right answers, which is generally the case only when simulated data are used. Of course we don't exactly have the "answers" when a 2^5 design is used, but this has enough experimental runs, especially when hidden replication is considered, to give an experimenter reasonable confidence in the results.

Example 5.1

Kempthorne (1973, p. 267) gave the results of an experiment performed at the Rothamsted Experimental Station to determine the effect of certain fertilizers on the yield of mangolds. A 2^5 design was used and the five factors were sulphate of ammonia, superphosphate, muriate of potash, agricultural salt, and dung. Each of these was a presence–absence factor in that one of the two levels of each factor was "none." The experiment was apparently run in four randomized blocks of size 8, and we will analogously analyze each of the blocks as if each block were a $1/4$ fraction, in addition to analyzing the data as having coming from a 2^5 design without blocking, which would be defensible if there were no block effect, as there wasn't in this example.

Kempthorne (1973) did not do a full analysis of these data, but did construct an error term with 13 df using the 13 three-, four-, and five-factor interactions that are unconfounded with blocks. Of course, this analysis was performed over 50 years ago as it was in the original 1952 edition of the book. Today it would be better not to assume that all three-factor interactions, in particular, do not exist. In fact, we will see for these data that some high-order interactions show as being significant in a normal probability plot analysis.

We will first analyze the data as an unblocked 2^5 design, ignoring the way that the design was run. (Assume that this information was initially lost.) The normal probability plot for this analysis is given in Figure 5.4.

We should be suspicious of these results simply because a four-factor interaction is identified as being significant, in addition to two three-factor interactions. Now assume that the blocking was discovered. The proper analysis is then to isolate the block effect and the three effects that are confounded with the difference between blocks: the ABD, BCE, and $ACDE$ interactions. Of course this will affect the assessment of significant effects in two ways: the BCE interaction will not be a significant effect, as it was previously, since it will be confounded with blocks, and the pseudo-error term implicit in the assessment of significant effects will be affected by the fact that the very small ABD interaction won't be part of the pseudo-error computation since it is confounded with blocks.

The "corrected" results are given in Figure 5.5.

We note that the ACE interaction is still significant, and since three-factor interactions generally are not significant, this should make us somewhat suspicious of the data. Kempthorne (1973) gave the bd treatment combination as having a yield of 864 in his Table 14.11, and that was the value used in these two analyses. However, 964 was the value for that treatment combination given in Table 14.10. Furthermore, the

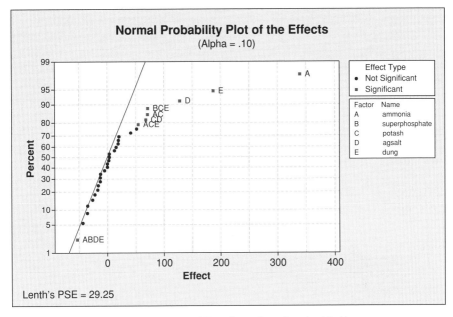

Figure 5.4 Analysis of Kempthorne data—ignoring blocking.

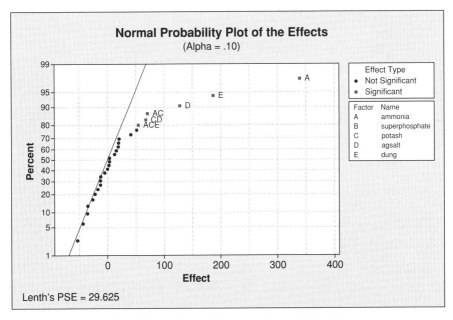

Figure 5.5 Analysis of Kempthorne data—with blocking.

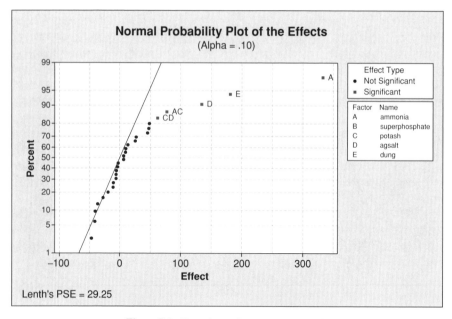

Figure 5.6 Kempthorne data—error corrected.

numerical results given in his Table 14.12 are those that result when 964 is used, so this appears to be the correct value. The normal probability plot with the corrected value is shown in Figure 5.6.

These results look more reasonable than the previous results. When the individual fractions are analyzed, the results are as follows. For the first fraction, only the A effect shows as being significant in the normal probability plot analysis, even though several other points are well off the line. No effect is identified as significant in the second fraction, despite the fact that there are some large effect estimates, and the same thing happens with the third fraction. Only the A effect is identified as significant in the fourth fraction.

Thus, there is agreement only between the first and fourth fractions and second and third fractions. All of these fractions fail to identify four effects in the full factorial analysis that were significant, and of course the second and third fractions miss all five effects. Why does this occur? Notice that there are many effect estimates in Figure 5.6 that are practically on the line. As has been emphasized previously (e.g., Section 4.10), a normal probability plot analysis works only if there is a sufficient number of small effect estimates. The normal plots for each of the four fractions are quite different from Figure 5.6, with no point being very close to the line. The alias structure is not the culprit as none of the significant effects in Figure 5.6 is aliased (with each other) in any of the fractions.

This example is important because it illustrates that (1) the results obtained from different fractions can easily disagree, and (2) the results from the fractions can differ greatly from the result that would be obtained from the full factorial as the

TABLE 5.3 Treatment Combinations of Four 1/4 Fractions of a 2^5 ($I = ABC = BDE = ACDE$)

(1)	(2)	(3)	(4)
(1)	ab	a	b
ac	acd	c	cd
de	ace	abcd	ce
acde	d	abce	bde
abd	e	bd	abc
abe	bc	be	ae
bcd	bcde	ade	ad
bce	abde	cde	abcde

necessary requirement of effect sparsity for a normal probability plot analysis to work is obviously more likely to be met when a small number of effect estimates are plotted than when there is a much larger number of effects.

5.2.1 Basic Concepts

In general, when the fraction is of the form $1/2^p$, there will be 2^{p-1} effects that must be confounded with the difference between the fractions, and their product(s) will also be confounded. Thus, for the 2^{5-2} design in the preceding example, we must select two effects to confound, and their product will also be confounded. So in that example, $ACDE$ was the product of ABC and BCE.

Since, as stated in the Important Point at the beginning of the chapter, the resolution of a two-level fractional factorial is defined as the number of letters in the shortest word in the defining relation, it might seem as though we should select $ABCDE$ and one of the four-factor interactions. This would be disastrous, however, as their product would contain only a single letter, and we would consequently lose a main effect. Similarly, a little reflection should indicate that any pair of four-factor interactions will have three factors in common, and will thus produce a two-factor interaction when multiplied together. We can, however, choose a pair of three-factor interactions in such a way that the product is a four-factor interaction. For example, ABC and BDE would produce $ACDE$ when multiplied together. (There is no advantage in selecting any particular pair, as only main effects are estimable, anyway.)

The alias structure is then obtained by multiplying each of the effects by the three defining contrasts. For example, $A = BC = ABDE = CDE$. The four 1/4 fractions could be constructed by determining the treatment combinations that have (1) an even number of letters in common with both ABC and BDE, (2) an even number in common with ABC and an odd number in common with BDE, (3) an odd number in common with ABC and an even number in common with BDE (i.e., the reverse of (2)), and (4) an odd number in common with both ABC and BDE.

The four fractions are given in Table 5.3.

The five main effects would be estimated in the usual way—the average of the four values in which the factor is at the high level minus the average of the four values in which the factor is at the low level. For example, if the first fraction were used, the main effect of A would be estimated by $(ac + abe + abd + acde)/4 - [bcd + bce + de + (1)]/4$.

The estimation of the interaction effects with this design also follows the same rule that is used in estimating interaction effects with full factorials and other fractional factorials. That is, the average of the response values for which the product of the signs of the factors that comprise the interaction is positive is computed, and from that average is subtracted the average of the response values for which the product of the signs is negative.

It should be noted that with this fractional factorial there will be two df for the residual, so the plotting of effect estimates on normal probability paper would not be absolutely essential, although generally desirable since the two df correspond to two pairs of two-factor interactions that are not being estimated. Rarely will there be sufficient prior information to strongly suggest that none of the four interactions is real.

It was stated previously that obtaining the design configuration by enumerating the treatment combinations is somewhat laborious when there are more than just a few factors. For 1/4 fractions, however, that is the most straightforward approach. Another approach is described by Box et al. (1978, p. 397), who show how to obtain a 2^{5-2} design from a 2^{7-4} design by dropping columns from the latter. Of course we prefer to use software to enumerate all of the fractions.

It is of course highly desirable for experimenters to understand the mechanics of design construction and effect estimation so that they can have the knowledge about when they should construct custom designs rather than routinely depending on the default options in statistical software to construct designs. Once that knowledge has been acquired, experimenters can confidently rely on the extensive design construction capabilities of readily available statistical software.

For example, in MINITAB the principal fraction for the design with defining relation $I = ABD = ACE = BCDE$ is constructed and stored with the commands

```
MTB> FFDESIGN 5 8;
SUBC> XMATRIX C1-C5.
```

Note that the principal fraction is the one that has all plus signs in the defining relation. Another one of the four fractions would have the defining relation $I = ABD = -ACE = -BCDE$, with of course the sign for the third interaction being the product of the signs for the first two interactions.

As stated in Section 5.1, fractional factorial designs in the 2^{k-p} series can be viewed as full factorials in $k - p$ factors, so if for some reason a different alias structure were desired, such as to avoid confounding effects that are believed a priori as being possibly significant, it is easy to construct such a "custom design" with practically any software. For example, in MINITAB, a 2^{5-2} design with the alias structure in the example in Section 5.2 could be constructed with the commands

```
MTB > FFDESIGN 3 8;
SUBC > ADD D = AB E = AC;
SUBC > XMATRIX C1 - C5.
```

It is also possible to specify the alias structure in JMP by specifying how, in this case, the two additional variables are defined by using the Change Generating Rules option when the "Screening Designs" capability is used.

5.3 DESIGNS WITH $k - p = 16$

Snee (1985) considered the 2^{5-1}, 2^{6-2}, 2^{7-3}, and 2^{8-4} designs for the study of five, six, seven, and eight factors, respectively, to be the most useful fractional factorial designs. Accordingly, it is useful to examine these designs for their power to detect real effects, using the tables in Lynch (1993). Table 3 (for $\alpha = .05$) shows that if $\beta = .90$, as in Chapter 4, the minimum detectable difference between factor level means is 2.5σ, assuming that at least three degrees of freedom are used to estimate σ, with these degrees of freedom obviously coming from high-order interactions if the design is not replicated. Of course, replicating a fractional factorial design would seem to defeat the purpose of using such a design, but Table 3 in Lynch (1993) reveals that the minimum detectable difference in factor level means is $k\sigma$ with $k = 1.2, 1.3, 1.4$, and 1.8 for five, six, seven, and eight factors, respectively. Thus, if 32 design points could be afforded, these 16-point fractional factorial designs should be replicated whenever users consider 2.5σ to be unacceptable, as will often be the case.

The obvious alternative to doing this would be to use a 2^5 design for five factors, a 2^{6-1} design for six factors, a 2^{7-2} design for seven factors, and a 2^{8-3} design for eight factors—all of which are 32-point designs.

Which approach is preferable? If there are hard-to-change factors or debarred observations (see Section 4.19 and Section 13.8, respectively), then it would obviously be better to use 16 factor–level combinations than 32. The resolution of these designs must be considered, however. The respective comparisons are 2^{5-1} versus 2^5, 2^{6-2} versus 2^{6-1}, 2^{7-3} versus 2^{7-2}, and 2^{8-4} versus 2^{8-3}. The results in terms of resolution are V versus not defined, IV versus VI, IV versus IV, and IV versus IV. Thus, except in the case of six factors, for which the loss in resolution would be deemed unacceptable if a design must be at least resolution V, using two replicates of a 16-point fractional factorial design is not a bad idea.

5.3.1 Normal Probability Plot Methods when $k - p = 16$

Haaland and O'Connell (1995) gave specific recommendations for normal probability plot methods used with fractional factorial designs with 15 estimable effects. Recalling the general form of Eq. (4.3) that was given in Section 4.10,

$$\widehat{\sigma}_{\text{PSE}}(q, b) = a_{\text{PSE}}(q, b) \cdot \text{median} \{|\widehat{\theta}_i| : |\widehat{\theta}_i| \leq b \cdot s_0(q)]\} \tag{5.1}$$

and Eq. (4.4), which was

$$s_0(q) = a_0(q) \cdot \text{quantile} \{q; |\widehat{\theta}_i| \, i = 1, \ldots, k\} \tag{5.2}$$

for simplicity they recommended selecting q such that quantile $\{q; |\widehat{\theta_i}| i = 1, \ldots, k\} = |\widehat{\theta}|_{(m)}$, with the latter denoting the mth order statistic of the absolute values of the effect estimates. If more than 6 of the 15 estimable effects are expected to be real effects (which seems unlikely to occur very often), they recommend using $m = 7$ and $b = 1.25$ in Eq. (5.1). If very few real effects are expected, the test given by Dong (1993) is recommended, with Lenth's (1989) method recommended whenever a moderate number of effects are expected. Of course, there may not be any a priori beliefs about the number of real effects. If so, then Lenth's (1989) method was recommended.

A much more extensive study was performed by Hamada and Balakrishnan (1998), who preferred the methods of Daniel (1959), Schneider, Kasperski, and Weissfeld (1993), Zahn (1975), Berk and Picard (1991), and the half-normal plot version of Loh (1992) for up to 6 real effects out of 15 estimable effects. Certainly the latter should be the usual scenario. (Their study and the entire paper are available online at http://www.stat.sinica.edu.tw/statistica/oldpdf/A8n11.pdf.)

More recently, Miller (2005) proposed a totally different type of (nongraphical) procedure, one that could be used for arbitrary n, which utilizes a selection score in arriving at the best model through implicitly comparing all possible candidate models. The author showed with simulation that the method is often superior to Lenth's method. Nevertheless, new methods such as this one are not likely to be used to any extent unless they are implemented in statistical software.

5.3.2   Other Graphical Methods

Certain novel (unpublished) graphical methods for experimental design data have been developed by Jim Filliben of the National Institute of Standards and Technology (NIST). For example, in an internal NIST publication (J. J. Filliben, K. Inn, and R. Radford, "Troubleshooting plutonium contamination in NIST radionuclide shellfish measurements via experiment design," 2004) an ordered dataplot is given in which the ordered Y-values are plotted against the treatment combinations. When there are extreme observations—and three such observations were shown in their graph—the reader can easily see if those observations appear to be associated with certain factor levels, or combinations of factor levels.

5.4   UTILITY OF SMALL FRACTIONAL FACTORIALS VIS-À-VIS NORMAL PROBABILITY PLOTS

It should be apparent from previous sections that small, unreplicated, fractional factorials, such as a 2_{IV}^{4-1} design, can present problems when a normal probability plot analysis must be employed. This is because there must be a sufficient number of small effect estimates for the normal plot approach to work, as has been stated previously, but that requirement will often not be met when there simply aren't very many effects that are estimable.

It is entirely possible that some two-factor interactions exist when a 2_{IV}^{4-1} design is used. If so, since these interactions are confounded in pairs, there might not be more than one or two small interaction pairs, which would cause the normal plot to likely fail if there were more than one significant main effect.

One way to avoid this problem is to simply not use these small designs and to be guided by results such as those given in Hamada and Balakrishnan (1998). Some of their recommendations were mentioned briefly in Section 5.3.1. They also discussed the difficulty in identifying the real effects and not selecting the null effects when designs have only eight points because of the substantial variability in the plots when all effects are null.

If experimentation is so expensive in a particular application that a small, unrepli-cated fractional factorial design must be used, thus necessitating the use of a normal probability plot approach, then selective replication might be employed, combined with checks on process control. Specifically, it is important that processes be main-tained in a state of statistical control when experimentation is performed, as was stressed in Section 1.7. Therefore, runs at the standard operating conditions might be performed at the beginning, the end, and possibly in the middle of the experimental runs. Alternatively, the first run that is performed might be replicated in the middle of the experiment and at the end of the experiment, which would require one fewer run than if the standard conditions were replicated.

If an experimenter is daring, these replicated runs might be used to estimate the error term, which would be used in hypothesis tests, obviating a normal plot approach. There is a big risk in this approach, however, in that (1) it is not very practical to estimate a variance based on only a few replicated points, and (2) it would be necessary to assume that the variance is the same at each design point—an assumption that could not be checked.

The work of Carroll and Cline (1988) and others on variance estimation in regres-sion is somewhat applicable here as they showed that at least 8–10 replications of a design point are necessary in order to obtain a good estimate of the variance at that point.

Therefore, it would be unwise to try to estimate the error variance from a few replicates of either the standard conditions or one of the design points. These replicates should be used simply to provide a rough check on process control, nothing more. Unfortunately, however, probably only a tiny fraction of experiments is run with process control checks. This is undoubtedly due in part to the fact that experimenters in general are not experts in process control and are probably not well aware of the extent to which experiments can be undermined when relevant processes go out of statistical control.

Of course the replications must be true replications and not just multiple readings, with the distinction between the two having been made in Section 1.4.2. This means that factor levels must generally be reset, even though they are to be reset to the same value. If true replicates are not obtained, then an analysis using the average of the repeated values as a single observation would be appropriate. If it is uncertain whether or not true replicates have been obtained, it would be wise to analyze the data using both approaches. This was done by Ryan (2003) in analyzing some lead recovery data for the NIST. If the results differ considerably, then it would be best to draw conclusions from the analysis obtained using the averages.

5.5   DESIGN EFFICIENCY

As discussed by Mee (2004), Daniel (1956) defined the degrees-of-freedom efficiency of designs for estimating all main effects and two-factor interactions as

$$\text{df-efficiency} = \frac{\text{Number of effects estimated}}{\text{Total degrees of freedom}}$$

For k factors there are obviously k main effects and $k(k-1)/2$ interaction effects, so the total number of effects is $0.5k(k+1)$. Since the total degrees of freedom is always $n-1$, df-efficiency can thus be written as

$$\text{df-efficiency} = \frac{0.5\,k(k+1)}{n-1}$$

Designs that are df-efficient are designs for which the fraction is close to 1.0, which would be 100 percent efficiency. A 2^{5-1} design is 100 percent efficient since numerator and denominator are both 15, this being the only commonly used fractional factorial design in the 2^{k-p} series that is 100 percent efficient. As shown by Figure 1 in Mee (2004), most 2^{k-p} designs have efficiencies of about 50 percent and most designs when k is large have efficiencies well below 50 percent. As discussed by Mee (2004), nonorthogonal designs have been proposed as an efficient alternative.

Mee (2004) gives design recommendations for $5 \leq k \leq 15$ with the 2^{5-1} design being the only 2^{k-p} design recommended.

5.6   RETRIEVING A LOST DEFINING RELATION

As pointed out by Bisgaard (1993), it is not uncommon for information on the defining relation to be lost, thus necessitating that it be determined in some manner. Various researchers have addressed this problem, with Margolin (1967) being among the first to do so. His method is also described in John (1971, p. 160).

Bisgaard (1993) gave a simple, matrix-based method for retrieving a defining relation that is more efficient than previous methods. The method utilizes the alias matrix, which was originally given by Box and Wilson (1951). Assume that we have fit a model of the form $\mathbf{Y} = \mathbf{X}_1\mathbf{b}_1 + \mathbf{e}$, but the true model is $\mathbf{Y} = \mathbf{X}_1\mathbf{b}_1 + \mathbf{X}_2\mathbf{b}_2 + \mathbf{e}$, with $\mathbf{X}_1$ denoting the matrix whose columns contain the values of the terms used in the model and $\mathbf{X}_2$ denoting the corresponding columns for terms that are in the true model but not in the fitted model. These matrices will have $+1$ and -1 for their entries.

Using standard regression and linear model expressions, $E(\widehat{\mathbf{b}}_1) = (\mathbf{X}_1^T\mathbf{X}_1)^{-1}$
$\mathbf{X}_1^T E(\mathbf{Y}) = (\mathbf{X}_1^T\mathbf{X}_1)^{-1}\mathbf{X}_1^T(\mathbf{X}_1\mathbf{b}_1 + \mathbf{X}_2\mathbf{b}_2) = \mathbf{b}_1 + (\mathbf{X}_1^T\mathbf{X}_1)^{-1}\mathbf{X}_1^T\mathbf{X}_2\mathbf{b}_2 = \mathbf{b}_1 + \mathbf{A}\mathbf{b}_2$, with
$\mathbf{A} = (\mathbf{X}_1^T\mathbf{X}_1)^{-1}\mathbf{X}_1^T\mathbf{X}_2$ denoting the alias matrix.

The defining relation can be easily determined by adapting this result, although not in an obvious way. Specifically, let $\mathbf{X}_1$ represent the k factors in the design matrix and let $\mathbf{X}_2$ represent all possible interactions formed from the factors in $\mathbf{X}_1$ that are of the order that would be used in constructing a specific design. (The last point will be explained later.)

To illustrate, consider the following 2_{III}^{5-2} design.

A	B	C	D	E
−1	−1	−1	1	1
1	−1	−1	−1	−1
−1	1	−1	−1	1
1	1	−1	1	−1
−1	−1	1	1	−1
1	−1	1	−1	1
−1	1	1	−1	−1
1	1	1	1	1

As usual the first three columns form the basis columns, as they constitute a full factorial in $5-2 = 3$ factors. We could easily use trial and error to determine how D and E were formed, but such an approach would be laborious for at least a moderate number of factors. So let $\mathbf{X}_1$ be composed of the first three columns and let $\mathbf{X}_2$ be composed of all two-factor interactions since we know that such designs are constructed using two-factor interactions to represent the fourth and fifth factors.

Straightforward matrix multiplication, using software, yields

$$\mathbf{A} = \begin{pmatrix}
0 & 0 & 0 & 0 & 0 & 1 & 0 & 0 & 1 & 0 \\
0 & 0 & 1 & 0 & 0 & 0 & 0 & 0 & 0 & 0 \\
0 & 0 & 0 & 1 & 0 & 0 & 0 & 0 & 0 & 0 \\
1 & 0 & 0 & 0 & 0 & 0 & 0 & 0 & 0 & 0 \\
0 & 1 & 0 & 0 & 0 & 0 & 0 & 0 & 0 & 0
\end{pmatrix}$$

Multiplying this matrix times the vector of 10 interactions (AB through DE)

$$\begin{pmatrix}
AB \\
AC \\
AD \\
AE \\
BC \\
BD \\
BE \\
CD \\
CE \\
DE
\end{pmatrix}$$

shows that $D = AB$ and $E = AC$, in addition to the aliasing of the first three factors, which is irrelevant as we need only these two to determine that the defining relation is $I = ABD = ACE = BCDE$. Of course in this simple example we can observe that $D = AB$ and $E = AC$ from inspection of the five columns that form the design matrix.

5.7 MINIMUM ABERRATION DESIGNS AND MINIMUM CONFOUNDED EFFECTS DESIGNS

Unless there is only a small number of factors to investigate, a fraction smaller than a $1/2$ fraction will be necessary. For example, a 2^{7-1} design requires 64 points, which may be prohibitive if experimentation is expensive, so a 2^{7-2} design or a 2^{7-3} design might have to be used. The defining relation for such designs is more involved than the defining relation for a $1/2$ fraction, as the number of words (effects) in the defining relation is given by the order of the fraction plus the number of generalized interactions of those effects. For example, the highest possible resolution for a 2^{7-2} design is IV, and one possible defining relation for a 2_{IV}^{7-2} design is $I = ABCDE = DEFG = ABCFG$, with the latter being the product (modulo 2) of the first two effects.

The resolution of a fractional factorial design is given by the length of the shortest word, so many 2_{IV}^{7-2} designs could be constructed, one of which would have the defining relation $I = ADEG = BCDEFG = ABCF$. Which design is preferable? The first design is preferable because there would be fewer main effects confounded with three-factor interactions, and more importantly, there would be fewer two-factor interactions confounded among themselves. This is because the word of the shorter length occurs fewer number of times with the first design. For a given resolution and a given number of factors, the design that has the fewest number of words of the shortest length is the *minimum aberration* design, with the term due to Fries and Hunter (1980).

Minimum aberration designs have been automatically generated by Design-Expert (*Source*: http://www.statease.com/e6ug/DE06-Features-Design.pdf (screen 2)) and the two-level fractional factorial designs generated by JMP are also minimum aberration (*Source*: JMP help file for "screening designs"). Although "minimum aberration" is not covered in the MINITAB help file material, the two-level fractional factorial designs are probably minimum aberration.

It is worth noting that minimum aberration designs do not necessarily maximize the number of unconfounded effects. In the example given above the number of two-factor interactions that are not confounded with other two-factor interactions is minimized by the minimum aberration design, so this result may seem counterintuitive. The result is well known, however, and is discussed by, for example, Wu and Hamada (2000, p. 176), who showed that the minimum aberration 2_{IV}^{9-4} design has 8 two-factor interactions that are not confounded with other two-factor interactions, whereas another 2_{IV}^{9-4} design has 15 such two-factor interactions.

Consequently, the *number of clear effects* was proposed as an additional criterion by Chen, Sun, and Wu (1993), whereas others have suggested that the "maximum number

of unconfounded effects" be the criterion that is used (see http://www.statsoftinc.
com/pdf/MinimumAberrationPaper.pdf). Of course, the maximum number of clear
effects and the maximum number of unconfounded effects would be the same thing.
Designs that are efficient in terms of the minimum aberration criterion and other
objectives are discussed by Butler (2005). Jacroux (2003) gave a two-stage procedure
that first identifies 2^{k-p} designs with the maximum number of estimable effects, and
within this class the minimum aberration design is found.

In comparing the best design according to each criterion, assuming that the designs
have the same resolution, the design with the minimum number of confounded effects
should be chosen.

For a 2^{k-p} design there aren't many different combinations of k and p for which the
application of the two criteria will result in the selection of different designs, however.
To illustrate, if $p = 1$, the same design results because there is only one word. For
$p = 2$, if $k = 5$, the design cannot exceed resolution III because there is obviously no
way to construct the columns for factors D and E from the 2^3 design for factors A, B,
and C that is the interaction of three factors because there is only one triplet, ABC.
Furthermore, if we let, say, $D = ABC$ and $E = AB$, then the generalized interaction is
C, so we would not be able to estimate the main effect of that factor and the design
would not even be of resolution III. So there is no option but to construct D and E
using two-factor interactions, so that each word will be of length 3.

For a 2^{6-2}_{IV} design, the generators for E and F cannot each be a word of length 4
because there is only one such word. Furthermore, one of the two generators cannot
be that word because the other generator would have three letters in common with
that word, so the generalized interaction would be a three-factor interaction and the
design would thus be only resolution III. Therefore, E and F must each be formed
using three-factor interactions, and when this is done, the generalized interaction will
also be a word of length 4. To see this, note that the generators for E and F will have
to differ by one letter for the generalized interaction to have four letters (e.g., $E =
ABC$ and $F = ABD$) and they will differ by exactly one letter regardless of which
pair is chosen because three of the four letters are being selected. It follows that
the defining relation will have two words of length 4 that differ by two letters, so
the generalized interaction consists of the two different letters from the first word
and the two different letters from the second word. It also follows that no two-factor
interaction is estimable because six of the four-factor interactions can be formed from
the letters of the first word, five from the second one (since one is accounted for by the
first word), and four from the third word (since one is accounted for by each of the first
two words). The sum of these three numbers is 15, which is the number of two-factor
interactions.

There is no unique 2^{6-2}_{IV} minimum aberration design, and since only the six main
effects are estimable regardless of the 2^{6-2}_{IV} design that is used, there is thus no unique
design that gives the minimum number of unconfounded effects.

We won't pursue this type of analysis any further and will conclude the discussion
by stating that the minimum aberration design and the design with the maximum
number of unconfounded effects will generally be the same, as for only a few designs
will they differ. (Wu and Hamada (2000) identified six combinations of k and p for

which the two criteria lead to different designs: 2_{IV}^{9-4}, 2_{IV}^{13-7}, 2_{IV}^{14-8}, 2_{IV}^{15-9}, 2_{IV}^{16-10}, and 2_{IV}^{17-11}. The difference for the last design is the greatest as the design that maximizes the number of unconfounded effects allows 31 of the 136 two-factor interactions to be estimable, none of which can be estimated with the minimum aberration design.)

Another proposed criterion, which has been recommended when many two-factor interactions are expected, is to minimize the maximum length of two-factor interaction alias chains (Mee and Block, 2006).

5.8 BLOCKING FACTORIAL DESIGNS

The example in Section 5.2 is an example of a blocked factorial, although that example was used to illustrate variable results between fractions of a single set of data, and also to illustrate how and why the results from the fractions will generally not agree with the result from the full factorial.

In this section we revisit blocking factorial designs, which was also covered in Section 4.15, this time considering the topic relative to fractional factorials and serving as a lead-in to the blocking of fractional factorial designs. Factorial designs are generally blocked because of practical considerations, such as not being able to make all the runs on one day and, in general, not being able to make all the runs under the same conditions. As with a randomized block design (Section 3.1), the blocking should be effective, meaning that we want the conditions to be essentially the same within each block if they are different blocks.

Blocking an unreplicated factorial means that we will have $b - 1$ fewer interactions that are estimable, with b representing the number of blocks, since interactions will be confounded with differences between blocks. As in the selection of the defining relation of a fractional factorial, we don't want to relinquish the estimation of interaction effects that may be significant.

In that respect, we should be guided by the same principles that are used in constructing fractions. For example if we run a 2^4 design in two blocks, we would confound the *ABCD* interaction with the difference between blocks, just as we select that interaction to confound between the difference between the two fractions so as to maximize the resolution of the fractional factorial. The main difference of course is that when we use a fractional factorial we use only one of the fractions, whereas when we block a factorial we use all the fractions. A blocked factorial is "kinder" on a normal probability plot analysis, however, because considerably more effects are estimated with a blocked factorial than with a fractional factorial, so with a blocked factorial a better pseudo-error estimate is likely to result.

Consider what happens when a 2^{k-2} design is used. Two extra columns are created and three effects are confounded with the estimate of the mean. With a 2^{k-2} design we can regard the four fractions as blocks with three effects confounded with the differences between the blocks. Similarly, for a 2^{k-3} design three additional columns would be created and seven interaction effects would be lost because there are four generalized interactions between the effects corresponding to the three columns: $\binom{3}{2} + \binom{3}{3}$. There are of course eight available fractions with a 2^{k-3} design, and we can regard these as representing eight blocks, with seven interaction effects confounded with the differences between the blocks. And so on.

The bottom line relative to this analogy is that we can construct the blocks analogous to the way that we construct fractional factorials—the difference is simply that we use all the "fractions." Whether or not we hope to see a block effect is likely to depend upon the trouble we go through in constructing the blocks. If the blocks are days, with the blocks formed because the runs must be apportioned to different days, the blocking is natural and does not require any special effort. On the other hand, if the blocking results from careful consideration of differences in raw materials, the blocking may require considerable effort and we would then prefer to see a block effect, just as is the case when a randomized complete block (RCB) design is used. As with an RCB design, there is an assumption of no blocks × treatment interactions. That is, if raw material problems cause the yield for the second block to be less than the yield for the other blocks, it should uniformly affect all treatment combinations in the second block. If that is not the case, then the usual statistical analysis that is performed will be flawed.

Wu and Hamada (2000) gave a table of blocking schemes for 2^k factorials in 2^q blocks in their Appendix 3A.

5.8.1 Blocking Fractional Factorial Designs

The need to block a fractional factorial also frequently arises. For example, with a 2^{5-1} design there are 16 observations that would have to be run under the same conditions and that may not be possible.

Bisgaard (1994) described an application in which a team of engineers that was working on developing a new motor wanted to estimate the effect on fuel efficiency of a new carburetor design, new manifold design, and so on—eight design changes (factors) were involved, with the levels being "new" versus "present." Sixteen runs were to be made, so this was to be a 2^{8-4} design. There were 10 parallel test stands available at the time of the experiment, and parallel testing would of course reduce the time required to conduct the experiment.

As stated by Bisgaard (1994), an additional factor is required to block a fractional factorial, and in this case the additional (ninth) factor was obvious: the test sites, with two such sites used. Since the 16 runs are made in two blocks of 8 runs each, the question arises as to what effects are lost of the effects that can be estimated with the 2^{8-4} design? Bisgaard (1994) gave tables that can be used for constructing such designs. For the 2_{IV}^{8-4} design run in two blocks of size 8, Table 1 lists $B = ABCD$ as the generator for the blocks, with B being the blocking factor. Since the design is only of resolution IV, the $ABCD$ interaction is not estimable anyway; this simply indicates the assignment of the treatment combinations to the two blocks. The obvious question is, "Why use this particular generator for the blocks?" As with generators for fractional factorials without blocking, there generally won't be a single optimal choice of a generator. We should note that the design resolution drops from IV to III when the blocking is performed, even though the blocking factor is confounded with a four-factor interaction. There is a simple explanation for this. The generators for factors E, F, G, and H are all three-factor interactions, but there are only $\binom{4}{3} = 4$ of them, so all of them are used. If a two-factor interaction were used, the resolution would, by definition, be III, so nothing would be gained in terms of resolution as the same general level of aliasing would exist as when $\mathsf{B} = ABCD$.

Bisgaard (1994) also illustrated blocking a 2_{IV}^{4-1} design. Of course the fraction would be constructed using $D = ABC$, so the defining relation is $I = ABCD$. If this design is run in two blocks, what should be the generator for the blocking factor? Clearly it cannot be another three-factor interaction, because the generalized interaction with ABCD would be a two-factor interaction (e.g., if $\mathsf{B} = BCD$, then $I = ABCD = BCD\mathsf{B} = AB$ so that A be confounded with the blocking factor B). Thus, a two-factor interaction (any one will do) would have to be used, so that the blocking factor would thus be confounded with this interaction and the resolution would be reduced from IV to III. That is unavoidable.

Example 5.2   Case Study

Chokshi (2000) gave a case study in which a 2_{IV}^{4-1} design was run in two blocks. This was a rocket engine fabrication experiment. The components of the rocket thrust chamber are brazed together to form the rocket thrust chamber. There are several thousand braze points and the objective is to come as close as possible to 100 percent of the maximum feasible number of braze points. The percent coverage was considered unsatisfactory, which was the motivation for the experiment.

Six factors were identified from engineering considerations and the use of Pareto chart methods, but the team decided to initially study only four of these factors. Each chamber could handle only four test combinations, so two chambers were used, with chamber thus being the blocking factor. The defining relation for the 2_{IV}^{4-1} design before the blocking was $I = ABCD$ and the blocking factor was confounded with the BC interaction, so that the full defining relation with the blocking was $I = ABCD = \mathsf{B}BC = \mathsf{B}AB$. Since the defining relation shows that the blocking factor is confounded with 2 two-factor interactions, the resolution drops from IV to III.

Curiously, there was one percentage that was far above 100 (115.46) among the eight data points and there was no explanation for this. Each percentage was computed as the average braze length per tube in inches divided by the available braze length per tube in inches, with each average computed from either 180 or 360 measurements. The only way the percentage could exceed 100 would be for at least one braze length to exceed the available braze length. The braze coverage measurements were obtained by X-ray. One set of 180 measurements was listed in the paper and there appears to be at least one other error as one of the listed coverage figures is .225, whereas almost all the other numbers exceed .80 and the next smallest value is .65.

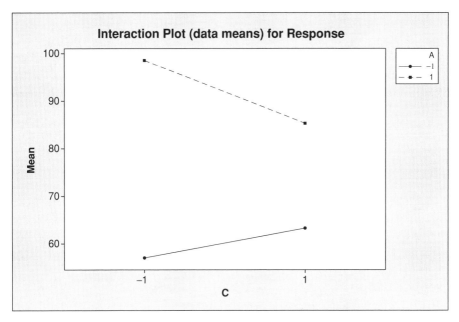

Figure 5.7   *AC* interaction plot for braze coverage data.

Nevertheless, we will use the numbers as given since we have no way of correcting any errors. The numbers for each treatment combination are as follows.

Treatment Combination	Response Value(%)
(1)	46.42
bc	34.94
ad	115.46
abcd	78.01
cd	91.60
bd	67.70
ac	92.64
ab	81.64

There was interest in examining the *AB* and *AC* interactions, in addition to the main effects. These interactions were declared not significant, however, as was the *C* main effect, and these effects were used to create an error term with three degrees of freedom for the ANOVA that was used. It is not clear how this was determined since this is an unreplicated fractional factorial. The *C* factor (which was the number of wires, 0 and 2) does have a noticeable effect at the high level of factor *A*, as can be seen from Figure 5.7.

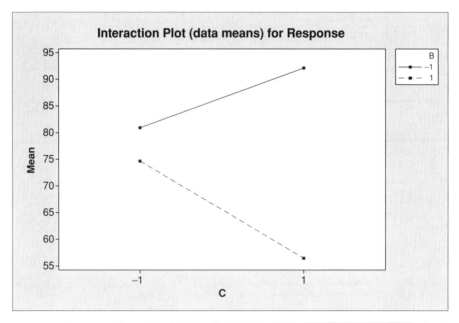

Figure 5.8 *BC* interaction plot for braze coverage data.

Since the slope of one line in Figure 5.7 is positive and that of the other is negative and the absolute values of the slopes do not differ greatly, the interaction effect estimate must be close to zero, which obscures the obvious effect of factor *C* at the high level of *A*. The vertical distances between the lines at each level of *C* being considerably different shows that a single main effect number for factor *A* will also be misleading.

The lines in Figure 5.7 were stated as being "nearly parallel" in the paper, but they clearly aren't, and the degree of departure from parallelism necessitates a conditional effects analysis.

The *BC* interaction is even more extreme, as can be seen from Figure 5.8.

This interaction was not discussed in the paper, probably because it was confounded with the block effect by the manner in which the design was constructed. The question that obviously must be addressed is, "Is this the block effect or the *BC* interaction effect, or some of both?" If there is no reason to suspect that there should be a test chamber effect—and on the surface that would seem to be a reasonable assumption—then this would seemingly be the *BC* interaction effect. In any event, this is something that should have been resolved with additional runs because the conclusion was that there was no *C* effect, whereas there is very clearly a *C* effect at each level of *B*. Since the slopes of the line segments in Figure 5.8 have opposite signs and the absolute values are almost equal, the conditional effects are almost offsetting, resulting in a very small *C* main effect, which was -3.51. The conditional main effects of *C* are -18.20 and 11.18. The former was not much less than the *B* effect estimate, -20.96, that was declared significant, although the numbers are not directly comparable since the estimators have different variances.

Nevertheless a conditional effects analysis should have been performed before declaring that the C effect was not significant. Although the results of the experiment led to changes that resulted in annual savings of more than \$40,000 and management became convinced of the power of experimental design, additional work should have been performed. In particular, the magnitude of the (block $+ BC$) interaction effect cannot be ignored if there is no reason to suspect a sizable block effect, so additional runs should be performed to unconfound these two effects. (*Note*: It is not possible to perform a conditional BC effects analysis because B and C are confounded within each block, so an interaction plot would contain only two points instead of four. Furthermore, even if this were not the problem, there would not be enough data to perform the analysis since each block has only four observations. If the experimenters concluded that there should not have been a block effect, one obvious way to perform additional runs would be to fix the level of the blocking factor and use a replicated 2^2 design for factors B and C, using enough replicates to be able to detect a significant interaction.)

If further experimentation revealed that there really is a (sizable) BC interaction, a conditional effects analysis would have to be performed for factors B and C. There is also evidence of a failure to perform cleaning of the data that were used to compute the coverage percentages used as the response values.

Example 5.3   Case Study

Another case study that involved blocking a fractional factorial was given by Ledolter (2002–2003), which has some very interesting features. As in Example 5.2, there was natural blocking as the experiment had to be carried out over several days.

The experimental setting involved a viscose fiber that was being produced and interest centered on fiber strength and fiber elongation. Ten factors were investigated and a 2_{IV}^{10-5} design was used. Of course, only main effects are estimable with this design, as two-factor interactions are confounded amongst themselves. The experiment was performed over a period of 10 days, so these are the natural blocks. Ignoring the latter could inflate the error term and cause significant effects to not be identified as such.

In addition to the 32 factorial design points, there were two centerpoints, which were run on days 5 and 10. In addition to providing some evidence of experimental error, the placement of these centerpoints could provide a check on possible out-of-control process conditions that could contaminate the results. For each of the two centerpoints, the middle value was used for nine of the factors, but three levels (not two) of the other factor (stretch) were varied. Only one of the three levels was used in the other design points, however, that being 20 percent, with the other levels at the two centerpoints being 15 and 25 percent. Although two levels of the Stretch factor were used at each design point, the levels differed as 10 percent was used in combination with either 20 or 30 percent at some design points.

As pointed out by Ledolter (2002–2003), the three observations cannot be considered to be independent because the measurements were made on the same material. Therefore, this is not true experimental error.

A decision must of course be made regarding the composition of the blocks, which determines the effects that are not estimable. The days with the centerpoint runs became blocks 5 and 10 and these blocks contained only those runs. This means that the other 32 runs were split into eight blocks, with four runs per block. The composition of the viscose was determined by three factors, and the eight combinations of levels of those factors determined the eight blocks. This means that the effects of those factors and their interactions could not be estimated if individual block effects are to be determined because all those effects have the same level within each block. One benefit from this, however, is that the two-factor interactions of the remaining seven factors all become estimable, along with the main effects, of course. This essentially has the effect of "splitting up" the two-factor interactions into a group of estimable interactions and a group of nonestimable interactions, whereas with the 2_{IV}^{10-5} design no such grouping is formed, directly or indirectly, with certain two-factor interactions estimable only if the two-factor interactions with which they are confounded do not exist.

One shortcoming of this approach is that the main effects that are likely to be real generally aren't known before the experiment is conducted; to determine this is the objective of the experiment.

The data from the experiment are given in Ledolter (2002–2003, pp. 315–316). Readers may wish to perform their own analyses of the data and compare their results with those given by Ledolter (2002–2003).

Loeppky and Sitter (2002) give additional information about blocked fractional factorial designs and discuss the analysis of data from such designs.

5.8.1.1 Blocks of Size 2

A block size of 2 is often appropriate, such as when human subjects are involved (two eyes, two hands, two feet, etc.). When this block size is used for either a full or a fractional factorial, the usual approach is to have the set of second observations for the blocks be a foldover of the set of first observations. (One type of foldover design is a mirror image of the original design, which is then used in combination with that design. Foldover designs are discussed in Section 5.9.) This permits the estimation of all main effects free of block effects, but only main effects are estimable and all the factor levels must be changed to produce the mirror image. This will clearly be a serious disadvantage if there are any hard-to-change factors.

Draper and Guttman (1997) proposed a design strategy with more runs ($(k - p) \times 2^{k-p}$ runs for a 2^{k-p} design) that allows all effects to be estimated free of block effects. For example, in the simple case of a 2^2 design, four blocks, with two observations per block, are needed to estimate the A and B main effects and the AB interaction. Note that this is twice as many observations that are needed without blocking, but this is unavoidable if blocking is necessary and all effects are to be estimated. The following statement is especially important if either A or B or both are hard-to-change factors: "One choice, for example, would be to use pairings that require a change of only one factor level per pair." That is, only one factor-level change would be required within each block.

These designs might be cost prohibitive if $k - p$ is not small, requiring, for example, 64 runs for four factors. Consequently, these designs will be most useful when $k - p$ is small, such as in a highly fractionated factorial when k is not small. Draper and Guttman (1997) do not discuss such designs, however. They did illustrate how a 2^{4-1} design would be blocked, noting that there are 28 possible blocks of size 2 to select from, since $\binom{8}{2} = 28$ with $2^{4-1} = 8$. They noted that 12 of these are needed to estimate all effects, so 24 runs are necessary. They did not give the 12 pairs (blocks) and clearly such enumeration would be laborious, in general, since as Draper and Guttman (1997) pointed out, the 12 cannot be arbitrarily chosen. Consequently, it is almost imperative that software be used to construct the blocks.

As might be expected, however, software is not generally available for this purpose. For example, MINITAB cannot be used to block a 2^{k-p} design in such a way that the number of runs will exceed 2^{k-p}, and this is also true of JMP and Design-Expert.

5.9 FOLDOVER DESIGNS

It is often necessary to make additional experimental runs in order to resolve ambiguities. For example, assume that a 2_{IV}^{6-2} design was used and the AB interaction effect was judged significant. Since it is aliased with the CE interaction plus two four-factor interactions with the particular fraction that was used (defining relation of $I = ABCE = BCDF = ADEF$), this raises the question of whether the AB interaction or the CE interaction is a real effect, or are they both real effects.

Assume that a priori there is no reason to believe that any particular two-factor interaction is apt to be significant or not significant. Therefore, there is a need to make additional runs to dealias these two interactions, and of course the objective should be to make the minimum number of runs necessary to accomplish this.

We will return to this example near the end of this section after first discussing foldover designs in general.

A standard *foldover design* (also called a mirror image foldover design; see, e.g., http://www.itl.nist.gov/div898/handbook/pri/section3/pri3381.htm) results when a 2^{k-1} design is "folded over"; that is, all of the signs in the fraction are reversed and a new fraction of the same size is combined with the original fraction. Obviously this must produce the full 2^k design because by reversing all of the signs we create 2^{k-1} additional treatment combinations, and since there is a total of 2^k treatment combinations in the full factorial, the combination of the treatment combinations in the two half fractions must constitute the full 2^k design.

Although this should be apparent, another way to view this, which will be helpful when we consider foldovers of 2^{k-2} designs, is to recall that a 2^{k-1} design is a full factorial in $k - 1$ factors with the signs of the kth factor obtained as the product of the signs of the $k - 1$ factors in one fraction, and the negative of that product in the other fraction. For example, if we constructed a 2_V^{5-1} design we have $E = ABCD$ in one fraction and $E = -ABCD$ in the other fraction. If we combine the two fractions we have the full 2^5 design.

Note also the following. If we write out a 2^3 design, we notice that the second half of the treatment combinations (when written in Yates order) are the sign reversals of the treatment combinations in the first half, with the pairing of the treatment combinations being (4, 5), (3, 6), (2, 7), and (1, 8). Of course these halves are not the usual half fractions of the 2^3 design, however, which of course are formed by constructing the 2^2 design and letting $C = -AB$ for one fraction and $C = AB$ for the other fraction. If, however, we take one fraction and reverse all the signs, that will produce the other fraction and of course this must be true regadless of how the first fraction is constructed. This same type of relationship will exist when $k > 3$.

This equivalence does not extend to smaller fractions, however, because more than one additional column is constructed. For example, consider a 2_{III}^{5-2} design. The two additional columns beyond the 2^3 design would logically be constructed as $D = AB$ and $E = AC$. Changing all of the signs in this fraction so as to create a second fraction would result in $D = -AB$ and $E = -AC$. Thus, the defining relation for the first fraction is $I = ABD = ACE = BCDE$ and for the second fraction $I = -ABD = -ACE = BCDE$. Thus, $BCDE$ is the only interaction that is common to both defining relations, so not surprisingly the defining relation for the design that results from combining the two fractions is $I = BCDE$. (In essence, we are adding the components of the defining relations with the components that have opposite signs since they add to zero; see John (1971, p. 161) for a related discussion/illustration.) Of course a 2_V^{5-1} design should be constructed so that the defining relation is $I = ABCDE$, so the half fraction that results from folding over the 1/4 fraction is not a useful one as most of the two-factor interactions are confounded with other two-factor interactions, whereas two-factor interactions would be confounded with three-factor interactions if a 1/2 fraction were constructed in the usual way.

Thus, we have seen that combining 1/4 fractions does not produce a useful 1/2 fraction, a result that should not be surprising since none of the words in the defining relations for the two 1/4 fractions were of length 5, whereas the defining relation for the 1/2 fraction consists of a single word of length 5.

Similar problems would be encountered in folding over 1/4 fractions of larger designs, in that we would not obtain one of the 1/2 fractions with the maximum resolution. For example, a 2_{IV}^{6-2} design could be constructed, if done so by hand or with software, by constructing the 2^4 full factorial design and then defining the two additional factors as $E = ABC$ and $F = BCD$. The defining relation is then $I = ABCE = BCDF = ADEF$. One of the other fractions has $E = ABC$ and $F = -BCD$ for a defining relation of $I = ABCE = -BCDF = -ADEF$. The defining relation will contain, in addition to other terms, the interaction that has the same sign in each fraction. Thus, $ABCE$ is part of the defining relation of the combination of the two fractions. The defining relation of the optimal half fraction is $I = ABCDEF$, so we see very quickly that this will not produce the optimal half fraction, nor would this happen if we merged any of the other five pairs of 1/4 fractions, as the reader can easily verify in Exercise 5.1.

This is not to suggest that foldover designs aren't useful, as the purpose is to de-alias effects when at least one effect in a pair is significant. The point to be made is simply that the design that results from a foldover will generally not have the highest resolution possible for that number of design points, and could be considerably less than the highest resolution possible, as in the preceding example. A related disadvantage of a (full) foldover, as pointed out by Wu and Hamada (2000, p. 175), is that the number of effects to de-alias is not in proportion to the size of the original experiment, so a full foldover will be quite inefficient, relative to the optimal design, when the original experiment has a moderate-to-large number of runs. Essentially the same point was made by Mee and Peralta (2000), who stated that a full foldover design following a 2^{k-p} design is degree-of-freedom inefficient.

These shortcomings in foldover designs mandate the consideration of semifoldover designs, which are covered in the next section.

5.9.1 Semifolding

Let's return to the problem posed at the beginning of Section 5.9, with a pair of two-factor interactions to be dealiased. We would like to be able to perform the dealiasing using only half as many observations as would be used in a foldover design, and this can be accomplished with the *semifolding* technique, the term being due to Barnett, Czitrom, John, and León (1997), although the general idea was apparently first proposed by Daniel (1962).

First, let's consider our options for dealiasing the *AB* interaction from the *CE* interaction. Technically, if we wanted to estimate one more effect, we need just one additional run. This approach is illustrated by Montgomery (1997, p. 553). Although this does have the advantage of economy, there is a rather heavy price that must be paid. One disadvantage is the loss of orthogonality, which is also lost when semifolding is used since the fraction that results from the use of semifolding is an irregular fractional factorial (i.e., it is not in the form 2^{k-p}). So it is a question of degree, but we don't want to have a serious departure from orthogonality. In Montgomery's example, the two interactions that he wishes to separate are very highly correlated, as the correlation between the columns in the design is .80. The only difference in the two columns occurs at the additional run and it would be very risky to try to separate them using a single run. Furthermore, with an odd number of observations, each effect estimate would be computed from unbalanced data as each effect estimate will be computed from five observations at one level and four observations at the other level.

With the semifolding technique, eight additional runs would be made rather than the 16 runs that would have been made if a foldover design had been used, assuming that a 16-run fractional factorial had originally been run. There are many possible combinations of those runs that might be selected, as discussed in detail by Mee and Peralta (2000), who provided guidance. They emphasize that it is advantageous to view semifolding as a two-step process with the first step being to select the (full) foldover fraction from which half the runs will be used. Second, a level of a factor (or interaction) is selected (e.g., the high level of factor *E*), which defines the half of the runs in the full foldover fraction that will constitute the foldover. Mee and Peralta (2000) refer to this as "subsetting" on the effect.

How are those runs selected? There are 15 estimable effects for the 2^{6-2}_{IV} design and one of the 15 would be selected to "semifold on," with the treatment combinations selected that have the same sign relative to the selected estimable effect. Since there are two possibilities for each of the 15 estimable effects, there are thus $(15)(2) = 30$ possibilities for each fraction, and thus $(30)(3) = 90$ possibilities altogether since there are three other fractions.

Assume that when the original design was run, the principal fraction was used so that the defining relation was $I = ABCE = BCDF = ADEF$, and a semifold is to be performed using the fraction with the defining relation $I = -ABCE = BCDF = -ADEF$. Operationally, however, it isn't necessary to select the fraction that is to be used, and in fact the user interface of Design-Expert does not permit such a selection, as the user instead is asked to select a factor on which to semifold. If we select factor A to (semi)fold on, then we reverse the sign of A in each of the 16 runs. That is, we indirectly let $A = -BCE$, which algebraically is obviously equivalent to $E = -ABC$, so it is equivalent to using the generator $E = -ABC$, which leads to the desired defining relation. Thus, selecting factor A to fold on is equivalent to specifying the defining relation $I = -ABCE = BCDF = -ADEF$.

Which of the 15 effects should the fold be performed on? The position taken by Mee and Peralta (2000) is that in general it is best not to fold on a two-factor interaction, as doing so will result in fewer estimable two-factor interactions than can be achieved by folding on a main effect. A simplification of their approach, based on the assumptions of effect heredity and three-factor and higher-order interactions not being real, was given by Wang and Lee (2006).

One advantage of folding on a factor for the current example is that all two-factor interactions containing that factor are dealiased. This result holds for 2_{IV}^{k-p} designs in general; Mee and Peralta (2000) sketch a proof of this. Included in their proof is the result that subsetting on a main effect results in the same estimable functions of two-factor interactions as does any full foldover fraction that is obtained by reversing the sign of a single factor. (Of course this is not how a full (mirror image) foldover is performed, however.)

Whereas the recommendation of Mee and Peralta (2000) to fold on a main effect is generally good advice, we don't have a great need to estimate interactions when sets of confounded interactions are not significant in the original analysis. That is, we don't need to separate interactions whose sum is not significant. One problem with folding on a main effect is that we would create a major imbalance in the precision of the estimation of that particular main effect. For example, assume that the fold was performed on factor A. This means that with the eight additional runs, factor A will be at either the high level or at the low level for all of those runs, so that when these runs are combined with the original 16 runs, factor A will be at the high level 16 times or at the low level 16 times, and at the other level only 8 times. This means that the average effect at one level will be estimated with a standard error of $\sigma/4$ and with a standard error of $\sigma/2.8$ at the other level.

Of course this is undesirable but is not a serious problem, whereas some desirable properties, including orthogonality, must be relinquished regardless of how the semifolding is performed. Of course such semifolding could be justified if the factor that is folded on is not significant when the original fraction, in this case the 2_{IV}^{6-2}, is analyzed, but the general recommendation is to fold on a factor whose interactions one wants to estimate, and usually a significant two-factor interaction will be accompanied by at least one of the main effects of those factors being significant. (Of course it was observed in Chapter 4 in the discussions of conditional effects that a large interaction effect can cause main effects to be small, so there are exceptions.)

We don't have an imbalance problem with a main effect if we fold on a two-factor interaction. If we decide to fold on the DF interaction, we would use the eight runs for which DF was either $+1$ or -1. If we select the former, the design is as given below.

Row	A	B	C	D	E	F
1	−1	−1	−1	−1	−1	−1
2	1	−1	−1	−1	1	−1
3	−1	1	−1	−1	1	1
4	1	1	−1	−1	−1	1
5	−1	−1	1	−1	1	1
6	1	−1	1	−1	−1	1
7	−1	1	1	−1	−1	−1
8	1	1	1	−1	1	−1
9	−1	−1	−1	1	−1	1
10	1	−1	−1	1	1	1
11	−1	1	−1	1	1	−1
12	1	1	−1	1	−1	−1
13	−1	−1	1	1	1	−1
14	1	−1	1	1	−1	−1
15	−1	1	1	1	−1	1
16	1	1	1	1	1	1
17	−1	1	−1	−1	−1	−1
18	1	1	−1	−1	1	−1
19	−1	−1	1	−1	−1	−1
20	1	−1	1	−1	1	−1
21	−1	1	−1	1	−1	1
22	1	1	−1	1	1	1
23	−1	−1	1	1	−1	1
24	1	−1	1	1	1	1

The design is not orthogonal as three pairs of main effect estimates are correlated (two with a correlation of .333 and one with a correlation of −.333), but this is part of the price that is paid when less than 16 additional design points are added to the original 16 points. The interaction effect estimates for AB and CE are also correlated with a correlation of .333. All the effects are estimated with 12 runs at the high level and 12 at the low level. That is somewhat offset, however, by the fact that folding on factor A results in only three of the effect estimates being correlated, instead of four when we fold on the DF interaction.

All things considered, there is little to choose for this design between folding on DF and folding on A if our objective is simply to de-alias the AB and CE interactions. If, however, certain pairs of confounded two-factor interactions are close to being judged significant, it would be desirable to disentangle them and take a closer look.

TABLE 5.4   Data from Barnett et al. (1997)

A	B	C	D	E	F	Log (etch uniformity)
−1	−1	−1	−1	−1	−1	2.4
−1	1	−1	−1	1	1	2.31
−1	−1	1	−1	1	1	2.16
−1	1	1	−1	−1	−1	2.22
−1	−1	−1	1	−1	1	1.16
−1	1	−1	1	1	−1	1.59
−1	−1	1	1	1	−1	1.76
−1	1	1	1	−1	1	1.06
1	−1	−1	−1	1	−1	1.13
1	1	−1	−1	−1	1	1.28
1	−1	1	−1	−1	1	1.28
1	1	1	−1	1	−1	2.04
1	−1	−1	1	1	1	−0.22
1	1	−1	1	−1	−1	3.71
1	−1	1	1	−1	−1	4.26
1	1	1	1	1	1	0.41
0	0	0	0	0	0	1.36
0	0	0	0	0	0	1.65

This would favor folding on a main effect since that will permit the estimation of more two-factor interactions, as discussed previously. Of course we would also want to do so if there were multiple confounded interactions that were judged to be significant, as in the following example.

Example 5.4

In the example given by Barnett et al. (1997), the authors analyzed data from an experiment in which a 2_{IV}^{6-2} design was used plus two centerpoints for a total of 18 runs, with the design and the data (on three response variables) given in their Table 17.2, and are given here, in Table 5.4, with the generators for factors E and F being $E = ABC$ and $F = BCD$.

This was a semiconductor etch uniformity experiment and the objective was to certify a new vapor phase etching process for 200-mm wafers. With data available on three response variables, the experimenters elected to use log (etch uniformity) as the response variable in their analyses. The factors were A = revolutions per minute, B = pre-etch total flow, C = pre-etch vapor flow, D = etch total flow, E = etch vapor flow, and F = amount of oxide etched.

There are 13 main effects and two-factor interactions that are estimable with the 2_{IV}^{6-2} design, leaving two degrees of freedom for estimating the error variance. The analysis of Barnett et al. (1997) showed six significant effects (as judged by $p < .05$), four of which were interaction effects. When the two centerpoint runs are used, however, there are only five significant effects, as the AC interaction, which is confounded with the BE interaction, has a p-value of .06. The p-value for the AC interaction doubled when the centerpoints were added because the estimate of σ^2 was approximately four times larger with the centerpoints than without them. This difference resulted from the fact that there was a degree of freedom for curvature when the centerpoints were used and although the curvature was not significant (the p-value was .085), the curvature sum of squares was large enough to inflate the estimate of σ^2. (We may note that the estimate of σ^2 using just the pure error, with one degree of freedom, is close to what it is when all four available degrees of freedom are used for the estimation.)

Even though it isn't necessary to use Lenth's method since there are degrees of freedom available for estimating the error term, it is still of interest to see what results are obtained with this approach. Six effects are declared significant, in agreement with the results given by Barnett et al. (1997). Thus, there is agreement between two of the methods, but data are being discarded when each method is used.

Regardless of the results that one goes by, there are major problems, as for example, the significant AE interaction is confounded with the BC and DF interactions, although the engineer stated before the experiment that any DF interaction would be negligible. There were three other interactions involving factor A such that the sum of the interaction and the two-factor interaction that it was confounded with were significant. Thus, there were 9 two-factor interactions that were potentially significant, so semifolding should be performed in a manner so as to isolate as many of these as possible. By semifolding on factor A, Barnett et al. (1997) were able to isolate seven of the nine.

There is a price for doing so, however. The BC and DF interactions are still confounded, but in this example that is not a problem as long as the engineer is correct in believing that any DF interaction would be negligible. If the engineer's belief were incorrect, however, this would create subsequent analysis problems. A real problem is that there are correlations between some of the effects that were identified in the first stage of the analysis that are .333 in absolute value. This precludes a straightforward analysis using t-statistics and p-values. Barnett et al. (1997) attempted to circumvent this problem by using stepwise regression. The results of the stepwise regression given in their Table 17.5 have many t-statistics that are much less than 1, however, which is not the way stepwise regression is generally used.

If stepwise regression is to be performed, it would logically be done using the main effects that were identified in the first stage of the analysis as candidate factors in addition to the set of seven interactions that can now be estimated separately from the group of nine that were aliased in the first stage. Thus, the candidate predictors would be E, F, AC, BE, AD, EF, AE, BC, AF, and DE, with DF excluded since it is confounded with BC. (Note that four of these interactions involve factor A even though the main effect of A was not significant. We would, however, expect to see some effect of factor A when we look at conditional effects. We address this issue is the second stage of the analysis.)

Variable selection is actually not a good strategy when variables are highly correlated because the results will be very sensitive to small data perturbations, although there aren't any high correlations with these data. Since stepwise regression can easily give misleading results under certain conditions, it is advisable to use more than one variable selection approach, and perhaps use all possible regressions. If we use the latter here we do not have a clear-cut choice for a model. Certainly a good model, though, would be one with the following five terms: F, AF, E, BC, and EF. This produces an R^2 value of 87.2 percent. This differs sharply from the model selected by Barnett et al. (1997), which oddly includes five main effects that have extremely large p-values, including one that is .8683!

One possible criticism of our model with the five terms is that it is nonhierarchical since the main effects of A, B, and C are not included. The fact that interactions involving these factors entered the model when stepwise regression was used suggests a possible strong need for a conditional effects analysis of these data, which is performed in Section 5.9.1.1.

Barnett et al. (1997) included a block term in their model that measures the difference between the first 18 runs and the 8 runs that comprised the semifold. The inclusion of a block term is of dubious value as that corresponds to using a blocking variable in the model when a randomized block design is used, with the blocking variable being an extraneous variable. Furthermore, although there was some evidence of a difference between variables with these data, a blocking variable cannot be "set" the way that a physical variable such as temperature would be set, so it is of no value in future applications of a regression model; it simply helps explain the variability in the response variable for the data that were used in developing the model.

Thus, the function of the block variable in the analysis is to remove from the error term an effect that is significant, and the block variable is selected when stepwise regression is used, with the result that R^2 increases to 90.8 percent when the block variable is included.

There are many applications of the semifoldover technique that are easily accessible. For example, a case study showing the utility of the technique given by Mark Anderson and Pat Whitcomb of Stat-Ease, Inc., can be found at http://www.statease.com/pubs/semifold.pdf, and other case studies are described by Mark Anderson at http://www.statease.com/pubs/breaddoe.pdf and at http://pubs.acs.org/subscribe/journals/tcaw/11/i02/html/02comp.html, with the latter also having appeared in print (Anderson, 2002).

5.9.1.1 Conditional Effects

Many practitioners would object to using a model with E, F, AF, BC, and EF because this is not a hierarchical model, as noted previously, but recall the discussion in Section 4.18 regarding hierarchical and nonhierarchical models. Barnett et al. (1997, pp. 241–242) address this issue, pointing out that the corresponding main effects "have little meaning" when interactions are significant. While this is certainly true, we need to look at conditional effects and in this case we need to look at the conditional main effects of A, B, and C. Since the block effect was significant, we might calculate two conditional main effects: one computed using only the first 16 runs (centerpoints have no effect on effect estimates) and one computed using all 24 runs. The second conditional main effect should not differ greatly from the former; if it does, that would be cause for concern and further investigation.

The A effect would of course be obtained by splitting the data on the F factor since AF is the only interaction involving factor A that is significant. For the first 16 runs, the conditional effects of A are -0.985 at high F and 0.7925 at low F. These are comparatively large effect estimates as they exceed in absolute value all the regular effect estimates that were judged to be significant, although we should keep in mind that these conditional effect estimates have a variance that is twice the variance of the regular effect estimates since they are based on half as many data points. Nevertheless, the magnitude of these conditional effects cannot be ignored, and we should conclude that there really is an A effect. For the 26 points the conditional effects of A are -0.6325 at high F and 0.8575 at low F. Although there isn't much difference between the two sets of conditional effects at low F, there is a considerable difference at high F.

Similarly the B and C conditional effects would be obtained by splitting the data on C to obtain the B conditional effects and on B to obtain the C conditional effects. These are also comparatively large, as we might expect. For example, for the 16 data points the B conditional effects are 1.105 at low C and -0.9325 at high C.

In determining if certain significant effects are real, an *ad hoc* approach that would probably be sufficient in many situations would be to simply compare the conditional effects with the whole effect analysis obtained using a normal probability plot, recognizing that the variances of the effect estimates differ by a factor of 2. Note that this was essentially done in analyzing the data from Daniel (1976). That is, we might say that a conditional effect is "large" if it is at least equal to the smallest whole effect that was significant.

From a modeling perspective, a (regular) regression model could not be constructed using a mixture of conditional effects and regular effects because different amounts of data are used in constructing each. Furthermore, different data (with the same number of observations) are used when the data are split on different factors, so a model could not be constructed using a standard approach even with a constant number of observations. (Recall from Example 4.4, however, that we can approach this in a somewhat indirect way and obtain the appropriate model coefficients.)

Despite problems in trying to model conditional effects in a conventional manner, it is important to compute conditional effects because they will exist when there are large interactions. This is by definition because a first-order (i.e., two-factor) interaction effect estimate is simply the difference of the conditional effect estimates for each factor divided by $2r$, with r denoting the number of replicates. This is, in essence, the definition of a two-factor interaction (see Appendix A to Chapter 4). Thus, the difference in the two conditional effect estimates is the same for each factor.

Of course a large difference does not necessarily mean that each conditional effect is large in absolute value, however, as one of the conditional effects could theoretically be zero.

In general, the computation and consideration of conditional effects is just as important in fractional factorial designs as it is in factorial designs. As another example, Myers and Montgomery (1995, p. 142) analyzed data from a 2^{4-1} design and concluded that factor B is not significant and can be dropped from consideration because the main effect estimate is only 1.50. The conditional effects were 20.5 and -17.5, however, which were not small as all the main effect estimates were less than 20.

5.9.1.2 Semifolding a 2^{k-1} Design

If we semifold a 2^{k-1} design, we will create a 3/4 fraction of the 2^k, with 3/4 fractions of 2^k designs due to John (1961, 1962) and also described in John (1971, p. 163). John used the notation, slightly altered here to accommodate the notation of this chapter, $3(2^{k-2})$ to designate a 3/4 fraction of a 2^k design, the simplest practical design of which would be the $3(2^{4-2})$ design. These designs are discussed in more detail in Section 5.10.

There are many ways to construct a 2^{4-2} design, depending upon the choice of effects in the defining relation, but once those effects are chosen, there are four fractions, defined by the four combinations of plus and minus signs preceding the effects in the defining relation.

For example, one of the four fractions could be defined as $I = AB = ACD = BCD$, with the last effect of course being the generalized interaction of the first two effects. Then, fixing AB and ACD as the effects in the defining relation (and thus BCD also), the other three fractions would be obtained by the other three combinations of signs for the first two effects, with the combinations being $+/-$, $-/+$, and $-/-$. One way to view the treatment combinations that comprise each fraction is to recognize that these sign combinations correspond to even/odd, even/even, odd/odd, and odd/even, as stated previously. Thus, the first fraction would consist of the treatment combinations that have an even number of letters in common with AB and an even number of letters in common with ACD. There are 16 treatment combinations altogether and it should be apparent that from those 16 the following treatment combinations comprise the even/even fraction: ab, c, d, and $abcd$.

Assume, for the sake of the example that we will use, that the 3/4 replicate is constructed by omitting the even/odd fraction. This gives the following treatment combinations, which are listed in the same order that they were given in the case study of Prat and Tort (1989) that will be discussed shortly.

Treatment Combination	A	B	C	D
ac	−	+	−	+
bcd	+	−	−	−
ad	−	+	+	−
b	+	−	+	+
abcd	−	−	−	−
c	+	+	−	+
ab	−	−	+	+
d	+	+	+	−
acd	−	+	−	−
bc	+	−	−	+
a	−	+	+	+
bd	+	−	+	−

We immediately observe that this is not an orthogonal design since the correlation between the first two columns is $-1/3$. This is not a small correlation, but of course factor A might not be significant and the five other pairs of factors are uncorrelated. If we look at two-factor interactions, however, there are nonzero correlations involving all the other factors. For example, the correlation between the BC and D columns is also $-1/3$.

Of course a regression approach can still be used to obtain the coefficients for a fitted model, but a normal probability plot for identifying significant effects is, strictly speaking, inapplicable since the effect estimates are correlated.

Another way to view this design is to recognize that the design is equivalent to first constructing a 2^{4-1} design with $I = BCD$ and then semifolding on the AB interaction. This can be easily seen by recognizing that the first eight treatment combinations all have an odd number of letters in common with BCD and the last four treatment combinations have the same sign (minus) for the AB interaction. Also observe that these last four treatment combinations all have an even number of letters in common with BCD and thus are part of the other half fraction defined by $I = -BCD$.

Of course we always construct a 2^{k-1} design by confounding the highest-order interaction between the two fractions, so the obvious question is whether the design would be better by constructing the $1/2$ fraction in this manner and then semifolding on the AB interaction. The alias structure is actually worse when this approach is used, however, as all four of the factors are involved in nonzero correlations (A and B are correlated and C and D are correlated).

Example 5.5

The case study of Prat and Tort (1989) alluded to previously is an excellent example of some of the problems that are encountered in the statistical design of experiments (DOE). A plant experiment was run with the objective of determining how to improve the process of manufacturing a particular brand of pet food. As is generally the case with designed experiments, there was more than one response variable of interest, and observations were made on four response variables: powder in the product, powder in the process, yield, and consumption.

The experiment was preceded by lengthy discussions extending over a few weeks, which included having to convince the plant manager of the desirability of conducting the experiment. A pet food is of course a mixture, and the possibility of using a mixture design (see Section 13.12) was put off to a later date because the number of runs that such a design would require was considered unacceptably high for the first experiment.

The combination of economical considerations and engineering knowledge of the process resulted in the decision to use four factors, each at two levels. The factors were the formula, conditioning temperature, flow, and compression zone in die.

There were some constraints that served as complicating factors. One constraint was that the experiment had to be run in the real plant (i.e., not a pilot plant), so naturally another requirement was that salable product should be produced. Another requirement was that all of the experimental runs had to be performed in one day, which meant that no more than 13 runs could be made.

If we consider 2^{4-p} designs, the last requirement is a severe restriction as a 2_{III}^{4-1} design does not permit the estimation of two-factor interactions, whereas the objectives of the experiment stipulated that all two-factor interactions should be considered, and there was to be no confounding of main effects and two-factor interactions.

A 12-run Plackett–Burman design (see Section 13.4.1) would not be a possibility since this is a resolution III design and main effects are heavily confounded with two-factor interactions. Consequently, one is led to consider a $3(2^{4-2})$ design and the one that was used in the Prat and Tort (1989) study was given earlier in this section.

A major problem is that the fourth factor, compression zone in die, was hard to change. Prat and Tort presented the engineers with two possibilities, one of which was to use a split-plot design with the fourth factor defining the main plots. The other option was to use complete randomization. As a compromise, the hard-to-change (HTC) factor was changed four times during the day, whereas seven changes would have been necessary if the experiment had been run in accordance with the design layout given earlier in the section.

There is no indication of whether the particular run sequence was randomly selected from among all run sequences that require four changes of the HTC factor. There is a large number of ways in which the four changes could occur. The most obvious general arrangement of the runs would be to alternate the level in such a way that four groups of 3 have the same level. For example, the ordering that was actually used was $---+++---+++$. If we fix the treatment combinations that are to be in each group of three, there are $(3!)^4 = 1,296$ possible orderings of treatment combinations that have this pattern, and the same number that start with $+++$ and end with $---$, again fixing the treatment combinations in each group. Of course there are $\binom{6}{3} = 20$ ways of choosing the three for each group that have a plus sign, and the same number for a minus sign. Thus, for these two logical patterns there are $(20)(20)(2)(1296) = 1,136,800$ possible orderings of four groups of three. Randomly selecting one of these would be reasonable, but it seems unlikely that this occurred since there was no discussion in Prat and Tort (1989) of the sequence being determined randomly.

Prat and Tort (1989) proved that the design is resolution V by determining the alias structure, but we should keep in mind that this is a nonorthogonal design and correlations that are $1/3$ in absolute value are not small. Consequently, the results of a normal probability plot analysis of estimated effects should not necessarily be strictly followed since the plot is for independent effects.

Prat and Tort concluded that only the second and fourth factors "seem to have a significant effect on the mean" (for the first response variable), with no interactions standing out in their normal probability plot of effect estimates. We should note that their plot was the type of plot that was typically performed during that era, with no decision rule that selects significant effects, as in Lenth (1989), for example. When we employ the latter, however, no effects are identified as being significant. This is obviously due to the fact that the effect sparsity requirement is not met as six of the ten estimable effects have t-statistics ranging in absolute value from 1.94 to 4.53. The effect estimate of the BC interaction is exceeded in absolute value only by the effect estimate of D, so an examination of the conditional effects of B and C should be performed, and especially for factor C since it was judged to be nonsignificant.

One problem with a nonorthogonal design is that the effect estimates will not be equal to twice the regression coefficients *if* we computed each effect estimate using all the observations. That isn't done, however, which undoubtedly seems counterintuitive. John (1971, p. 161) indicated that A is estimable from the second and third fractions (i.e., the even/even and odd/odd fractions), and *not* from using all 12 observations. This can be seen by adding the defining relations together for the three quarters of the 2^4 that are used. Doing so produces $I = -AB = -ACD = -BCD$. This means that A is confounded with B, CD, and $ABCD$. Confounding with $ABCD$ is of no concern, but confounding with B and CD is of concern. This "defining relation" for the design allows us to see which effects are partially aliased, with the correlations between the partially aliased effects through first-order interactions all being $-1/3$ because of the minus signs in the "defining relation" and because we are summing the defining relations over three fractions to obtain the minus sign.

We don't need to literally use only parts of the data for estimating each effect, however, as John (1971, p. 161) showed that the least squares estimates for the effects using all the data—and hence the effects estimates being twice those coefficients—are the same effect estimates that would be obtained by selecting the appropriate data subsets, computing the effect estimates one at a time and averaging the estimates. (See John, 1971, pp. 161–162 for an explanation of why this occurs.) Thus, the effect estimates are the "correct" ones from uncorrelated effect estimates.

Regarding which effects are judged significant, there is an important point to be made that is undoubtedly better understood today than it was when Prat and Tort (1989) analyzed these data. Specifically, the line on a normal probability plot of effect estimates must go through the point $(0, 0)$ if these correspond to (normal score, effect estimate), or through (50 percent, 0) if the vertical axis is the normal cumulative probability scale; the line cannot be drawn in such a way as to connect or nearly connect the majority of points that are plotted. Theoretically, it is possible, although unlikely, that the majority of the effect estimates could be large and be such that they line up as almost a straight line on the plot. Then if a line were drawn through these points, we would indirectly be declaring the most significant effects not significant.

Notice that no (standardized) effect is declared significant (i.e., no point is labeled) even though one point in particular is very far from the line. (The plotted points are just t-statistics instead of effect estimates so as to make a point.) The reason for no effect being declared significant can be seen from the fact that only two points are close to the line, signifying that effect sparsity does not exist. Stated differently, only four of the t-statistics are less than 1.94 in absolute value. Thus, the approach breaks down with these data as there are not enough small effects to combine to create a pseudo-error term. Of course if we tilted the line in Figure 5.9, the line would almost pass through all but two of the points, but then the line would not come close to going through the point (50 percent, 0).

The obvious question to ask then is "Which effects are significant?" If we tried to answer this question by constructing a Pareto chart of effect estimates (as in Section 5.1.4, for example), we would have the same problem relative to the dotted line that is used for determining effect significance, as there are not enough small effect estimates to provide a reliable estimate of σ. Stated differently, the Pareto chart shown in Figure 5.10 shows no evidence of "the vital few and the trivial many." Rather the chart comes close to exhibiting almost perfect stair steps.

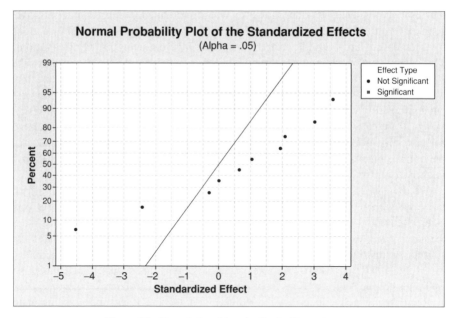

Figure 5.9 Normal plot of (standardized) effect estimates.

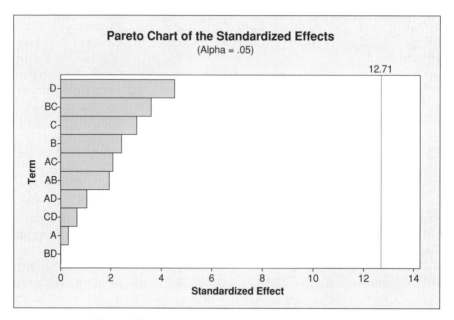

Figure 5.10 Pareto chart of the standardized effect estimates.

Furthermore, the analysis is complicated by the fact that the *BC* interaction is large and the *AC* interaction is large relative to the *C* effect, and that interaction dwarfs the *A* effect. This means that an analysis of conditional main effects should be performed, and we can see from Figure 5.10 that one of the conditional main effects of *A* will be moderately large. The *BC* interaction being larger than the *B* and *C* main effects means that there will be a conditional main effect for each of the factors *B* and *C* that will be large, and the largest conditional main effect of *C* will be slightly larger than the largest conditional main effect of factor *D* since there is no large interaction involving factor *D*.

We miss this if we look only at main effects, and indeed Prat and Tort (1989) stated "The conclusion is then that only the compression zone of the die has an important effect. . . ."

This is a difficult dataset to analyze because of the absence of effect sparsity, the presence of large interactions, and the apparent absence of randomization in the determination of the run sequence. These problems render a traditional analysis implausible and the large interactions necessitate the use of conditional main effect estimates.

Should a different design have been used? A split-plot design, with the hard-to-change factor being the main plot factor, was considered, but was not used, apparently because of concern that a lurking variable could be confounded with the hard-to-change factor. Although that could certainly happen, there is always this danger when a split-plot design is used to accommodate a hard-to-change factor, but is probably not a sufficient reason for ruling out a split-plot design unless there is strong reason to suspect a particular lurking variable.

The problems involved in this application of experimental design are by no means atypical, and illustrate that the statistical DOE and the analysis of the resultant data are by no means routine.

5.9.1.3 General Strategy?

It is inadvisable to try to formulate a general strategy for semifolding because so many factors must be considered. Clearly the advantage of folding on a factor instead of an interaction may be slight at best if suspected interactions do not involve a single factor but rather are spread over the factors. The level of the factor that is used in selecting the semifold should generally be the "best" level in the sense that this will aid in the maximization or minimization, whichever is desired, of the response and allow the variance of the response at that level to be minimized.

5.9.1.4 Semifolding with Software

Although semifolding is reasonably simple and straightforward, manual semifolding would be tedious and impractical for a large number of factors, and we would also like to avoid it for even a moderate number of factors.

Design-Expert 7.0 has considerable semifolding capability, which can be employed with a Plackett–Burman design as well as with fractional factorial designs. Indeed, semifolding a Plackett–Burman design will result if, for example, the user specifies a minimum run resolution IV design and that design turns out to be a Plackett–Burman design. If a resolution V design has been constructed, semifolding is not an option and only a full foldover can be used.

Semifolding can also be used with a resolution III design, which of course confounds main effects with two-factor interactions. Such designs are generally used as screening designs; so one possible use of a semifoldover would be to increase the resolution of the design so that main effects can be de-aliased from two-factor interactions. This may not be easily done, however. For example, Mee and Peralta (2000) state that only one of the 31 foldover fractions of the 2_{III}^{9-5} design used by Hsieh and Goodwin (1986) would increase the resolution.

5.10   JOHN'S 3/4 DESIGNS

Designs that are a 3/4 fraction of a 2^k design were mentioned briefly in Section 5.9.1.2. These designs, which were proposed by John (1961, 1962) and also presented in John (1971), deserve consideration because they are efficient designs in that the available degrees of freedom is not much greater than the number of main effects and two-factor interactions to be investigated. John's work was motivated by a practical problem. As described in John (2003), he had a client for whom he recommended a 2^3 design, but when the experimental material arrived, there was enough material only for six runs. In another application, an experiment had to be planned for four factors, each at two levels, but there was enough raw material only for 12 runs. So the 3/4 fraction was born. Because these are nonorthogonal designs, the degree of correlation between the effect estimates must be considered.

Design-Expert constructs a 3/4 fraction when the Irregular Fraction option is selected and the number of factors is 4, 5, or 6. For 7 factors the fraction is 3/8; for 8 factors the fraction is 3/16; and for 9, 10, and 11 factors the fraction is 1/8, 1/16, and 3/64, respectively.

The 3/4 fraction of a 2^4 design that the software produces is given below.

A	B	C	D
1.00	1.00	−1.00	−1.00
1.00	−1.00	−1.00	1.00
1.00	1.00	1.00	−1.00
−1.00	−1.00	1.00	−1.00
1.00	1.00	−1.00	1.00
−1.00	−1.00	−1.00	−1.00
−1.00	1.00	−1.00	1.00
−1.00	1.00	1.00	1.00
−1.00	−1.00	−1.00	1.00
−1.00	1.00	1.00	−1.00
1.00	−1.00	1.00	1.00
1.00	−1.00	1.00	−1.00

Design-Expert gives the alias structure for this design, which is given below.

```
                4 Factors: A, B, C, D

           Design Matrix Evaluation for Factorial 2FI Model

      Factorial Effects Aliases
 [Est. Terms] Aliased Terms
      [Intercept]= Intercept - 0.5 * ABC - 0.5 * ABD
      [A]  = A - ACD
      [B]  = B - BCD
      [C]  = C
      [D]  = D
      [AB] = AB
      [AC] = AC - BCD
      [AD] = AD - BCD
      [BC] = BC - ACD
      [BD] = BD - ACD
      [CD] = CD - 0.5 * ABC - 0.5 * ABD
```

Of course with 12 runs we can estimate 11 effects. Design-Expert lists 10 effects on the left, with the *ABCD* interaction not listed in the alias set. This is because interactions beyond second order are not displayed with the default, so the feature that displays them would have to be selected.

This is a reasonable design because two of the main effects are unaliased and the two that are aliased are each aliased with a three-factor interaction. Similarly, one two-factor interaction is unaliased and the others that are aliased are each aliased with a three-factor interaction. It may look odd that the intercept is part of the alias structure. This occurs because the columns for *ABC* and *ABD* do not sum to zero, as the reader can observe by constructing them or by simply performing the appropriate multiplication and summation with the design matrix above. In general, the effects that are aliased with the intercept are those whose column does not sum to zero.

Mee (2004) recommends the use of John's $(3/4)2^{8-2}$ design for $k = 8$ (48 points) and the $(3/4)2^{11-4}$ design for $k = 11$ (96 points). (Note that this notation differs from the notation used by John, 1971.) Both of these designs can be generated using Design-Expert.

The alias structure for the $(3/4)2^{8-2}$ design given by Design-Expert is listed below.

```
           8 Factors: A, B, C, D, E, F, G, H

           Design Matrix Evaluation for Factorial 2FI Model

      Factorial Effects Aliases
 [Est. Terms]  Aliased Terms
      [Intercept] = Intercept - 0.333 * ACF - 0.333 * BEG -
 0.333 * CEH - 0.333 * DFG
      [A]  = A - 0.5 * BDH - 0.5 * CDG - 0.5 * EFH
      [B]  = B - 0.5 * ADH - 0.5 * CGH - 0.5 * DEF
      [C]  = C - ADG
```

```
[D]  = D - 0.5 * ACG - 0.5 * BEF
[E]  = E - BDF
[F]  = F - AEH
[G]  = G - BCH
[H]  = H - 0.5 * AEF - 0.5 * BCG
[AB] = AB - 0.5 * AEG - 0.5 * BCF + CDE + FGH
[AC] = AC - AEH + BDE
[AD] = AD + BCE - CDF - EGH
[AE] = AE - 0.5 * ACH + BCD - 0.5 * CEF - 0.5 * DGH
[AF] = AF - ADG + BGH
[AG] = AG - 0.5 * ADF + BFH - 0.5 * CFG - 0.5 * DEH
[AH] = AH + BFG - CFH - DEG
[BC] = BC + ADE - 0.5 * BEH - 0.5 * CEG - 0.5 * DFH
[BD] = BD + ACE - CFH - DEG
[BE] = BE + ACD - BCH
[BF] = BF + AGH - 0.5 * BDG - 0.5 * CDH - 0.5 * EFG
[BG] = BG + AFH - BDF
[BH] = BH + AFG - CDF - EGH
[CD] = CD + ABE - 0.5 * ADF - 0.5 * CFG - 0.5 * DEH
[CE] = CE + ABD - 0.5 * AEF - 0.5 * BCG
[CF] = CF - 0.5 * BDH - 0.5 * CDG - 0.5 * EFH
[CG] = CG - CDF - EGH
[CH] = CH - BDF
[DE] = DE + ABC - 0.5 * BDG - 0.5 * CDH - 0.5 * EFG
[DF] = DF - BCH
[DG] = DG - AEH
[DH] = DH - 0.5 * AEG - 0.5 * BCF
[EF] = EF - CFH - DEG
[EG] = EG - 0.5 * ADH - 0.5 * CGH - 0.5 * DEF
[EH] = EH - ADG
[FG] = FG + ABH - 0.5 * ACG - 0.5 * BEF
[FH] = FH + ABG - 0.5 * ACH - 0.5 * CEF - 0.5 * DGH
[GH] = GH + ABF - 0.5 * BEH - 0.5 * CEG - 0.5 * DFH
```

We observe that all the main effects are aliased with two-factor interactions, and the two-factor interactions are aliased with three-factor interactions. Despite the complicated alias structure, the design is resolution V and obviously has good properties. (We may note that the 2^{8-2} design is also resolution V, so there is no loss in resolution in using 3/4 of this number of design points.)

The $(3/4)2^{11-4}$ design with 96 points, which is described in detail in the next section, has a similar alias structure in the sense that main effects and two-factor interactions are aliased with three-factor interactions.

If appropriate software is unavailable, there are various ways in which these designs can be constructed, one way being the semifolding technique as was discussed in Section 5.9.1.2. The 96-point design for $k = 11$ recommended by Mee (2004) is obviously not a 3/4 fraction of a 2^{11} design, as such a design would have over 1000 points. The obvious question then is of what design is this a 3/4 fraction? John (1971) describes how the design is constructed, stating that it is obtained by thinking of the four quarters of a 2^7 design and dropping the quarter defined by $I = AB = CDEFG = ABCDEFG$, with those treatment combinations eliminated by John (1969) for which $AB = CDEFG = -1$. (Note of course that this is not the way that a 2^{7-2} design would be constructed as the main effects of A and B would be confounded with this design.) Then four additional factors are added to the 3/4 of a 2^7 design by defining them as $H = ABDEF$, $J = ABCEG$, $K = AEFG$, and $L = BCDE$.

How good is this design, remembering that the design is nonorthogonal? One obvious problem is that A and B will not be orthogonal because of the way that the design was constructed. The dot product of A and B is -32 because 32 points have been excluded, for which the dot product is $+32$. Since there are 96 design points and the sum of all columns is zero, it follows that the correlation between the A and B columns is $-1/3$. The correlation between all other pairs of main effect estimates is zero. Thus, although John (1971) describes this as a resolution V design, it is a nonorthogonal resolution V design. The complete alias sets for this design are given in John (1969).

The obvious question to ask at this point is could the partial aliasing of main effects be avoided by, for example, omitting the quarter defined by $I = ABC = DEFG = ABCDEFG$ and defining the four additional factors the same as before? Although this might seem to be an improvement, Mee (2004) concluded that when the partial aliasing of all effects is considered, the design given by John (1969) is the best possible design.

The notation used by Mee (2004) is preferable to the notation used by John since $(3/4)2^{11-4}$ clearly indicates that this is 3/4 of a 2^{11-4} design, which is 3/64 of a 2^{11} design. Furthermore, it makes no difference whether we first construct a 2^{11-4} and then delete a specified 1/4 of the points, or following the implied steps given by John (1971, p. 167), delete 1/4 of the 2^7 and then construct the four additional factors.

Readers interested in learning more about the various types of foldover designs are referred to the sources cited previously, in addition to Li and Mee (2002), Li and Lin (2003), and Li, Lin, and Ye (2003). The first two papers are concerned with optimal foldover plans for 2^{k-p} designs with optimality determined by the minimum aberration criterion (see Section 5.7) for the combined designs. The third paper is concerned with optimal foldover plans for nonregular designs, including 12-run and 20-run designs. The authors showed that the full foldover is the optimal design for all 12-run and 20-run orthogonal designs. Other papers on foldover designs and their properties that may be of interest include Ye and Li (2003), Montgomery and Runger (1996), and Miller and Sitter (2001).

5.11 PROJECTIVE PROPERTIES OF 2^{k-p} DESIGNS

Assume that a 2^{6-2}_{IV} design has been used and it was determined that three of the six factors seem to be important. The design contains a replicated full factorial in those factors and thus has the advantage of a replicated full factorial over a fractional factorial. Of course we know that the design is also a full factorial in the first four factors (A–D), as those factors were used to form the full factorial from which the two additional columns were generated. But it is not a full factorial in any other set of four factors, whereas it is a full factorial in *any* set of three factors.

The general result, which is due to Box and Hunter (1961a,b), is that a 2^{k-p} design of resolution R contains a full factorial in $R - 1$ factors. This results applies almost trivially to a 2^{3-1} resolution III design since it is constructed from a 2^2 design, which is virtually of no practical value, but the real applicability of the result is for designs of resolution IV and V.

This is an important result because all the factors that are examined in a design are almost certainly not going to be significant. Of course there is no guarantee that the number of significant factors will turn out to be one less than the resolution of the design, but there is a good chance of this happening when the number of factors is small to moderate. At the other extreme, assume that a 2_V^{5-1} design has been used and only three factors seem to be important. The design in those factors is then a replicated 2^3 design. This can be seen from the following 2_V^{5-1} design. It is easily seen that this is a 2^3 design in factors B, C, and D with two replicates, but note that this is also true for each of the other nine 2^3 designs that could be constructed from these points.

A	B	C	D	E
−1	−1	−1	−1	1
1	−1	−1	−1	−1
−1	1	−1	−1	−1
1	1	−1	−1	1
−1	−1	1	−1	−1
1	−1	1	−1	1
−1	1	1	−1	1
1	1	1	−1	−1
−1	−1	−1	1	−1
1	−1	−1	1	1
−1	1	−1	1	1
1	1	−1	1	−1
−1	−1	1	1	1
1	−1	1	1	−1
−1	1	1	1	−1
1	1	1	1	1

5.12 SMALL FRACTIONS AND IRREGULAR DESIGNS

As stated in Section 5.3, Snee (1985) considered the 2^{5-1}, 2^{6-2}, 2^{7-3}, and 2^{8-4} designs for the study of five, six, seven, and eight factors, respectively, to be the most useful fractional factorial designs, and provided an application of 2^{5-1} and 2^{7-3} designs. There is clearly a need for fractions smaller than $1/2$ and $1/4$ fractions when the number of factors under consideration is at least moderate. For example, a 2^{8-1} design has 127 degrees of freedom available for estimating eight main effects, 28 two-factor interactions, and if desired, 56 three-factor interactions. The latter are seldom significant, but even if we estimated all of the three-factor interactions (3fi's), we would still have 35 degrees of freedom that are unused and thus could be used to estimate the standard deviation of the error term.

This is wasteful of degrees of freedom, so something smaller than a $1/2$ fraction would be a better choice. In general, when a $(1/p)$th fraction of a 2^k design is used, the defining relation will contain p effects and all generalized interactions of those effects. We can use the fact that the sum of binomial coefficients in the binomial expansion is 2^p to arrive at the number of generalized interactions. That number is obtained as $\binom{p}{2} + \binom{p}{3} + \cdots \binom{p}{p}$, which equals $2^p - p - 1$. Therefore, the total number of effects in the defining relation is $(2^p - p - 1) + p = 2^p - 1$. Obviously this number grows rapidly with p, so the alias structure is quite involved when p is at least moderate, but that complexity is a small price to pay for the savings in time and cost that are realized by using a small fraction.

We should be ever mindful of the minimum detectable effects with 16-point designs, as that minimum might be unacceptable. For example, for a 2^{5-1} design, if $\alpha = .05$ and the power is .90, the minimum detectable effect is 2.5σ, assuming that at least three degrees of freedom are used to estimate σ. This result is from Table 3 of Lynch (1993), from which it is apparent that 32-run 2^{k-p} designs have much better detection properties. A quick summary of power for various numbers of factors and numbers of runs is given in Figure 2 of a Stat-Teaser newsletter from Stat-Ease, Inc., that is available at http://www.statease.com/news/news0409.pdf.

Oftentimes we will prefer to use irregular fractions but we still must consider the size of the design relative to the sizes of effects that we would like to detect. For example, assume that there are eight factors to investigate and we want to be able to estimate all two-factor interactions (2fi's). A 2_V^{8-2} design could be constructed, and although this would have good detection properties, 64 runs would be used to estimate eight main effects and 28 2fi's, thus leaving 27 unused degrees of freedom. When we think about powers of 2, we have $2, 4, 8, 16, 32, 64, 128, \ldots$, with the gaps between the numbers becoming greater as we move up the line. Here we need at least 36 degrees of freedom, but we don't need 63 degrees of freedom.

We lose orthogonality when we use irregular fractions such as $3/4$ fractions, but if that loss is small, it can be more than offset by the gain in cost by using far fewer experimental runs.

As an admittedly extreme example, but one that did occur in practice, Mee (2004) refers to a ballistic missile simulation experiment requiring the estimation of all main effects and two-factor interactions for 47 factors. Obviously a design with a very large number of runs is required, which is generally not a major problem with computer/simulation experiments, as there are 1,128 effects to be estimated. If we tried to bracket this number with powers of 2, and thus provide a sufficient number of degrees of freedom, we would need 2,048 runs. Thus, several hundred additional runs would be made beyond what is necessary if a 2^{k-p} design were used.

There are various efficiency measures for judging the efficiency of a design, which are based upon a particular model assumption. As discussed in Section 5.5, following Daniel (1956), Mee (2004) discussed df-efficiency for assessing the efficiency of a design, under the assumption that the true model contains only main effects and 2fi's in the factors that are used in the design. Of course the true model is never known, and in particular, all of these effects won't be significant, but this is a useful measure because it gives an upper bound of the efficiency relative to the effects that are judged significant.

Using the expression for df-efficiency given in Section 5.5, a two-level resolution V design for estimating the 1,363 effects with 2,048 runs that was alluded to earlier would have a df-efficiency of

$$
\text{df-efficiency} = \frac{\text{Number of effects to be estimated}}{\text{Total number of degrees of freedom}}
$$
$$
= \frac{0.5(k)(k+1)}{n-1}
$$
$$
= 1128/2047
$$
$$
= .551
$$

for the 2^{47-36} design. If the df-efficiency were .50, this would mean that the power of 2 in 2^{k-p} was 1 more than necessary, so we can think of this as a lower bound on df-efficiency, with designs that have a df-efficiency value close to this being highly df inefficient.

The researchers initially used a 2_{IV}^{47-38} design and then followed that with 17 groups of runs so as to separate the aliased 2fi's. Due to the complexity of such a sequence of runs for separating the aliased 2fi's, the investigators later ran, under a different scenario, a 2_V^{47-35} design, even though this has 4,096 design points. Since each run took 30 minutes, an extra 2,000 runs, say, would require an additional 1,000 hours! Therefore, a design with fewer design points would obviously be highly desirable.

As shown in Hedayat, Sloane, and Stufken (1999), it is possible to construct a resolution V design with $n = 2048$ and 47 factors, although the design would have to be constructed using nonlinear error-correcting codes. This approach for constructing the 47-factor design is also explained in the appendix of Mee (2004). Since the method is rather involved, it will not be explained here; the interested reader is referred to one of these sources for details. Mee (2004) recommended the use of either this design or a nonorthogonal subset of the design.

Benefits in the form of df-efficiency can also be realized for other numbers of factors by using an irregular design. For example, designs that are smaller than the regular resolution V designs can be constructed for various values of k for $n = 128$, 256, 2048, and 4096 design points, as described in Hedayat et al. (1999).

5.13   AN EXAMPLE OF SEQUENTIAL EXPERIMENTATION

As an example of the sequential use of designs, we will examine the three sequential experiments described by Eibl, Kess, and Pukelsheim (1992). The response variable was the coating thickness resulting from a painting operation, and the objective was to identify the factors that influenced the value of the response variable and to determine the settings of those factors so as to achieve a target value of 0.8 mm. (The thickness had been varying between 2.0 mm and 2.5 mm.)

Interestingly, the first experiment was a *replicated* fractional factorial, with the design consisting of four replications of a 2_{III}^{6-3} design. The reason given for replicating the design was that prior data had indicated that the standard deviation of the error term might be large. Recall from Eq. (1.1) in Section 1.4.3 that the necessary sample size to detect a difference in factor levels of a specified amount Δ is a function of σ, so a large value of σ can dictate the use of a large number of observations for each factor level.

Randomization was not used in the first experiment because the machine operators felt that it would be too time-consuming. They were subsequently convinced that this was not the case, however, so randomization was used in the second and third experiments. This did not include, however, resetting the factor levels after each run, as was done in the first experiment. Paint viscosity was a hard-to-change factor and was not varied randomly in the first experiment. Instead, the low level was used for the first 16 runs and the high level used for the last 16 runs. Those sets of runs were made one week apart but there was no analysis of a possible time effect or apparent consideration of any lurking variables that might be confounded with a possible time effect.

All four observations (replicates) of one treatment combination were much higher than any of the other observations, suggesting the possibility of a lurking variable or at least an interaction effect resulting from the specific combination of factor levels.

Only main effects are estimable with the design that was used in the first experiment, and four of the six main effects were identified as being significant. The target value for the response variable was 0.8 mm, a number that was exceeded by 30 of the 32 observations. Consequently, the experimental region was changed for the second experiment.

The design used for the second experiment was a foldover of the first experiment, except that only four factors were used instead of six, and the levels of three of the four factors were adjusted to force the response values in the direction of the target value. Accordingly, the factor levels in the second experiment were stated relative to the levels of the first experiment (e.g., 0 and -2 rather than the customary -1 and $+1$). Two replications were used since 16 runs could be made in one day, with the first eight runs and the last eight runs each being randomized.

The first factor, belt speed, was adjusted in the direction opposite of that suggested by the coefficient of the factor from the first experiment Based on their knowledge of the belt apparatus, the authors felt that they should reduce the belt speed in the second experiment, rather than increase it. Their conjecture was not supported by the data from the second experiment, however.

The coefficient of the first factor was, oddly, the same in the fitted model from the second experiment as that in the fitted model from the first experiment. The other coefficients differed noticeably from those in the first fitted model, although the data from the first experiment were added to the data from the second experiment in obtaining the fitted model.

The choice of levels for the second experiment turned out to be not much better than the choice for the first experiment as 14 of the 16 observations were above the target value.

For the third experiment, paint viscosity was fixed at its low level, a seemingly reasonable decision since the two observations below the target value occurred at that level, with most observations well above the target value. Therefore, a 2^3 design was used with two replications, so the number of observations was again 16. The factor levels were changed again with an eye toward driving the response values toward the target value. The experimenters overcorrected somewhat, however, because 15 of the observations were below the target value, although 5 were between 0.74 and 0.79.

The coefficients in the final fitted model, using all 64 observations, differed very little from the coefficients in the second fitted model. From the final fitted model, which had an R^2 value of .93, the experimenters determined factor-level combinations that produced a predicted response value equal to the target value and indicated two such combinations that had small standard deviations of the predicted response value.

5.13.1 Critique of Example

The use of three experiments performed in sequence with information obtained from one experiment used in designing the following experiment is highly commendable. There are, however, some important points that should be made. It is questionable how much influence the data from the first two experiments should have on the coefficients of the fitted model since the factor levels were such that many of the response values were more than twice the target value. This is especially problematic since the first experiment had 32 observations, some of which were almost three times the target value, whereas the third experiment had only 16 observations.

The nonorthogonality of the design that results when the three sets of design values are pieced together should also be of concern since the correlations given below are not small enough for the nonorthogonality issue to be dismissed.

```
        Correlations:   A, B, C, D

              A            B            C
   B       -0.187
   C       -0.187       0.342
   D       -0.086       0.281        0.281
```

Strictly speaking, the nonzero correlations mean that the coefficients in the fitted model are not directly interpretable. Consequently it is perhaps not surprising that one of the two sets of factor-level combinations recommended by the authors is outside the experimental region because the suggested level of the first factor is outside the range of levels used in the 64 runs. The other recommended combination was not one of the 24 combinations used in the experiment, and although each factor level was within its range over the 24 combinations, that does not mean that the set of combinations is within the experimental region for a nonorthogonal design. (In general, there is no exact way to determine if design points are inside or outside the experimental region for such designs when there are more than two factors.)

The experimenters found the BC interaction to be significant in the third experiment, but analyzing the data using ANOVA, since the design is replicated, does not show the BC interaction to be significant. The analysis does show, however, that the A effect is not significant. Furthermore, since factor D was not varied in the third experiment, its coefficient and significance or nonsignificance cannot be compared with the results obtained using all the data.

Thus, there are some major concerns about the conclusions that were drawn from the experiment. In particular, questions must be raised about the wisdom of combining data from experiments in which the response values were not close to the target value. Furthermore, there are problems in trying to use the numerical values of the coefficients in determining optimum operating conditions when there is more than slight nonorthogonality that results from combining the factor levels for the three experiments. Finally, extrapolating beyond the experimental region can be risky and it can be difficult to determine when this has occurred with a nonorthogonal design.

Despite these problems, the authors reported that the information obtained from the experiments resulted in the target value of 0.8 mm being "achieved" with a standard deviation of 0.1 mm. There is no discussion of whether or not 0.8 is the average of the post-experimentation thickness values, although the authors may have assumed that their wording implies this. If 0.8 was the approximate average, then the experimentation was successful in leading to the distribution of thickness values being centered at the target value, which was clearly a vast improvement over the range of 2.0–2.5 mm for the thickness that existed before the experimentation.

We might prefer the post-experimentation standard deviation to be somewhat smaller, and perhaps the standard deviation could be reduced by seeking to identify potentially influential factors that were not included in the experimentation.

The results from this experiment were quite useful in helping to achieve the desired target value when a new paint was introduced. (The standard deviation of the thickness values for the new paint was not given.)

5.14   INADVERTENT NONORTHOGONALITY—CASE STUDY

Nonorthogonal designs, such as John's 3/4 fractions, can be quite useful. Users must be careful not to inadvertently create nonorthogonality, however. This can easily happen, as was illustrated by Ryan (2003). In a lead recovery experiment performed by the NIST in 2001, the objective was to identify the factors that influence the extent of lead recovery from paint. There were 112 paint specimens available for the experiment and 7 specimens could be handled by the sonicator on each run. Since $112/7 = 16$ and five factors were investigated, it follows that a 2^{5-1} would be one possible choice for the design.

The design that was run, however, was one for which the 2^5 design was split into two sets of 16 design points, with one set replicated three times and the other set replicated four times. Despite this imbalance between replicates, there is no problem with orthogonality as long as the 2^5 design is split into two 2^{5-1} designs. That was not what was done, however, so there was a small departure from orthogonality, but it was not enough to cause any major problem with the analysis of the data. This does illustrate, however, that orthogonality can be "lost" through design if care is not exercised, and the loss could be consequential.

5.15 FRACTIONAL FACTORIAL DESIGNS FOR NATURAL SUBSETS OF FACTORS

In recent years there has been interest in developing designs in recognition of the fact that many experiments are used for factors that would be logically grouped into subsets. Yates and Mee (2000) proposed two methods of constructing such designs, which would have application in assembly experiments, in particular. Bisgaard (1997) also proposed an approach for constructing such designs for tolerancing assembly products.

For example, several factors might be identified that probably affect one part of a machine assembly and different factors might be identified that likely affect a second part of the assembly. It should be apparent that constructing a design for subsets of factors will restrict the number of feasible treatment combinations. For example, if there are 10 factors in two subsets of 5 factors each, no treatment combination can contain more than five letters. The method proposed by Bisgaard (1997) stipulates that a design be selected for each subset of factors, and then the product array of the designs is formed, analogous to Taguchi's idea of a product array in robust parameter design (see Section 8.4). Since a product array generally results in an unnecessarily large design (see, e.g., Ryan, 2000, p. 447), it is desirable to use only some portion of the treatment combinations in the product array. Bisgaard (1997) referred to this process as *post-fractionation* and the process of obtaining the fraction for each subset as *pre-fractionation*. Yates and Mee (2000) provided tables of designs for 16 runs and 32 runs that gave the pre- and post-fractionation for the number of factors ranging from 6 to 15 and up to five subsets. Each design is either a resolution III design or a resolution IV design. Although the designs actually aren't listed, they are designated by the plan number in the catalog of designs given by Chen et al. (1993).

For example, their Table 1 shows that if eight factors are grouped into two subsets of four factors, a 2^{4-1} design would be used for each design, and then a 1/4 fraction of the product array would be constructed, for a total of 16 runs. The design is resolution III. If a resolution IV design were desired, a 1/2 fraction of the product array would be constructed, giving 32 runs. (Of course this is the same number of design points in a 2^{8-3}_{IV} design when there is no restriction on treatment combinations.)

To illustrate the construction of these designs, assume that there are three factors of interest for each of two stages of an assembly operation and a 2^6 design would result in two many runs. A 2^{3-1} design could be used in each stage. This would require 16 runs, which would generally not be an excessive number. Therefore, assume that there is no post-fractionation. With A–C denoting the factors in the first stage and D–F the factors in the second stage, the first fraction would logically be constructed using $I = ABC$ and the second fraction would be constructed using $I = DEF$. The product array, which would be a 2^{6-2}, would then have a defining relation obtained from the product of the defining relations for each fraction. That is $(I + ABC)(I + DEF) = I + ABC + DEF + ABCDEF$. Notice that this is a resolution III design whereas it is possible to construct a 2_{IV}^{6-2} design. Thus, there is a loss of resolution but that is unavoidable because the treatment combinations that would be used when a 2_{IV}^{6-2} design is employed are not available when only three factors are varied in each experiment. This is not to suggest that inefficiency results when fractions are used in stages, as the post-fractionization will generally result in a saving.

To illustrate the latter, assume that there are four factors in each of two stages with a 2^{4-1} design used in each stage, with the resultant 64-run product array considered to have an unsatisfactorily high number of runs so that post–fractionation will be necessary. With the letters A–D denoting the first subset of factors and E–H denoting the second subset, the 2^{8-2} product array has 64 runs, so a 1/4 fraction of this is needed to reduce the design to 16 runs. Before the post-fractionation, the defining relation would be $(I + ABCD)(I + EFGH) = I + ABCD + EFGH + ABCDEFGH$. This is obviously a resolution IV design, which is the highest resolution possible. We need a 1/4 fraction of this design, so the result will be a 2^{8-4} design, with resolution IV being possible for a single-stage 2^{8-4} design. We will see if that resolution is possible for the two-stage design. We cannot use $I = ABCD = EFGH = ABCDEFGH$ as the defining relation for the 1/4 fraction since the treatment combinations have already been split on $ABCD$ and $EFGH$, but after some trial and error we see that we can construct a defining relation I + $ACEF$ + $BCEH$ + $ABFH$ such that a three-factor interaction is not involved in the product of this with I + $ABCD$ + $EFGH$ + $ABCDEFGH$. Thus, the full defining relation written in "alphabetical order" is $(I + ABCD + EFGH + ABCDEFGH)(I + ACEF + BCEH + ABFH) = I + ABCD + ABEG + ABFH + ACEF + ACGH + ADEH + ADFG + BCEH + BCFG + BDEF + BDGH + CDEG + CDFH + EFGH + ABCDEFGH$. (We can see that the product should have 15 interactions because it has to be one less than 4×4 since $I \times I = I$, with 4 being the number of components in each defining relation.)

The treatment combinations for the design are given in Table 5.5.

Looking at the treatment combinations and remembering that factors A–D will be studied in the first stage and then factors E–H in the second stage, we see that there are eight distinct treatment combinations for each set of four factors, which of course is the way that it should be, since a 2^{4-1} design was used for each set in the pre-fractionation stage. We can also note that for each treatment combination in either set, the pair of corresponding treatment combinations in the other set (i.e., in the same row of the overall design) form a foldover pair (e.g., note the first two rows of the design).

Before the post-fractionation there are of course eight treatment combinations in the second group for each treatment combination in the first group since it is a product array. Then a 1/4 fraction is taken so it stands to reason that there should be two treatment combinations in the second group for each treatment combination in the first group after the fractionation.

These designs would be extremely laborious to construct by hand and they apparently cannot be constructed with widely available software, so practitioners should rely on the tables of Yates and Mee (2000) and the designs of Chen et al. (1993) as

TABLE 5.5 $(2^{4-1} \times 2^{4-1})/2^2$ **Design**

Row	A	B	C	D	E	F	G	H
1	−1	−1	−1	−1	−1	−1	−1	−1
2	1	1	1	1	−1	−1	−1	−1
3	−1	1	1	−1	1	−1	−1	1
4	1	−1	−1	1	1	−1	−1	1
5	1	1	−1	−1	−1	1	−1	1
6	−1	−1	1	1	−1	1	−1	1
7	1	−1	1	−1	1	1	−1	−1
8	−1	1	−1	1	1	1	−1	−1
9	1	−1	1	−1	−1	−1	1	1
10	−1	1	−1	1	−1	−1	1	1
11	1	1	−1	−1	1	−1	1	−1
12	−1	−1	1	1	1	−1	1	−1
13	−1	1	1	−1	−1	1	1	−1
14	1	−1	−1	1	−1	1	1	−1
15	−1	−1	−1	−1	1	1	1	1
16	1	1	1	1	1	1	1	1

much as possible, as well as the tables of Block and Mee (2005), who enumerated all 128-run resolution IV designs.

5.16   RELATIONSHIP BETWEEN FRACTIONAL FACTORIALS AND LATIN SQUARES

Latin square designs were covered in Section 3.3. There are relationships between Latin square designs and fractional factorial designs that are well known. For example, Kempthorne (1973, pp. 279–283) showed that a replicated 2^3 design can be arranged in two or three replicated 4×4 Latin squares. These cannot be constructed in such a way that an entire replicate is blocked, however, so such designs might be best viewed as being appropriate for field experiments with the need to block on soil fertility in each direction.

Similarly, Wu and Hamada (2000, p. 280) discuss the relationship between 5^{k-p} designs and 5×5 Latin squares and their variations. (Fractional factorial designs with more than two levels are discussed in Chapter 6.)

A less-known relationship between fractional factorials and Latin squares was given by Copeland and Nelson (2000), who showed that a standard $t \times t$ Latin square with $t = 2^k$ corresponds to a 2^{3k-k} fractional factorial. For example, a standard 4×4 Latin square corresponds to a 2^{6-2} fractional factorial. Such a relationship is by no means obvious since a Latin square is for one factor of interest whereas the 2^{6-2} design is obviously for six factors.

Copeland and Nelson (2000) stated that a Latin square configuration is easier for a nonstatistically oriented experimenter to understand, and obviously that is the case since fractional factorials are not intuitive. Apart from simplicity, however, there is an obvious question of what can be gained by knowing this relationship.

This raises the question of what would happen if we tried to analyze a Latin square as a fractional factorial, rather than going in the other direction and simply using a Latin square to represent a fractional factorial. That won't work in general, as Copeland and Nelson (2000) point out that there is only one such standard Latin square of each order that has this correspondence.

This creates somewhat of a problem relative to randomization as ideally a Latin square would be selected at random.

5.17 ALTERNATIVES TO FRACTIONAL FACTORIALS

People in many different fields have used statistically designed experiments and many people outside the statistics profession have written about design and attempted to contribute to the state of the art, sometimes with poor results.

Over the years there have been various types of experimental designs presented in books and journal articles that have poor properties compared to fractional factorials and other well-accepted designs, but unfortunately these alternative designs have gained acceptance in industry to a considerable extent, simply due to the industrial stature of the developers of these designs.

Included in this group of designs are those credited to Dorian Shainin. Undoubtedly many people in industry have used Shainin's variable search technique, which was recommended by Shainin when there are more than four factors. These are tiny designs (relative to the number of factors), and if they were fractional factorials they would be called highly fractionated designs. The methodology is not based on sound principles, however, and in their study of the designs, Ledolter and Swersey (1997) concluded, "We find that there is little reason to abandon the traditional and well-studied fractional factorial designs in favor of the Variables Search design strategy." More generally, Nelson (2004) stated, "The inclusion of Shainin requires a rather broad definition of DOE," with "DOE" representing design of experiments. See also the discussion of Shainin's methods in Woodall (2000).

5.17.1 Designs Attributed to Genichi Taguchi

The statistical designs of Genichi Taguchi have been much debated, as have the statistical methods that he proposed. Ryan (1988) was apparently the first to point out that some of Taguchi's designs correspond to suboptimal fractional factorials (suboptimal in the sense that the resolution of the design is not optimized). Only a few of the designs that have been attributed to Taguchi were actually invented by Taguchi, and some misinformed writers have even stated that Taguchi invented design of experiments!

The designs advocated by Taguchi are commonly known as orthogonal arrays, some of which correspond to fractional factorials. As stated at the beginning of this chapter, all fractional factorials are orthogonal arrays but not all orthogonal arrays are fractional factorials. Taguchi's experimental design contributions are discussed in detail in Chapter 8 in this book and also in Chapter 14 of Ryan (2000).

One highly beneficial result of Taguchi's foray into experimental design has been that he has sparked a considerable amount of research to develop better methods. The most striking of these developments is that his work has motivated researchers to work on designs for identifying factors that influence variability rather than just the mean, and on robust designs that allow the inclusion of noise factors. (These are not model-robust designs, but rather the objective is to use designs that enable the identification of factor levels such that performance of what is being measured will not vary greatly under variations in the levels of noise factors that are uncontrollable during regular production.) These designs and design issues are discussed in detail in Chapter 8.

5.18   MISSING AND BAD DATA

Missing data in full factorial designs were discussed in Section 4.12 and the effects of bad data were illustrated in Section 4.10 and discussed further in Sections 4.12.1 and 4.12.2. Missing data and bad data will have the same deleterious effects in frac-tional factorial designs: loss of orthogonality and can cause erroneous decisions to be reached, as was illustrated for a full factorial design in Section 4.10.

Srivastava, Gupta, and Dey (1991) examined the robustness of resolution III frac-tional factorial designs to one missing observation, in addition to doing this for other designs. (Their definition of *robustness* is that a design is robust if all parameters are still estimable for a given model in the presence of a missing observation.) They found that the consequence of a missing observation varied, depending on which observation was missing (see also Srivastava, Gupta, and Dey, 1990).

5.19   PLACKETT–BURMAN DESIGNS

It should be noted that a Plackett–Burman design, which is a two-level design, is a useful alternative to a 2^{k-p} design, especially when there is a desire to use something between 16 and 32 design points, which of course cannot be done with 2^{k-p} designs. Within this interval, Plackett–Burman designs can be constructed for 20 and 24 design points, as well as for larger and smaller numbers of points. The designs are discussed in detail in Section 13.4.1. (They are not discussed in this chapter in part due to the length of the chapter.)

5.20   SOFTWARE

Virtually all statistical software packages have the capability for fractional factorial designs, including SAS, MINITAB, JMP, D. o. E. Fusion Pro, and Design-Expert. The latter has a nice interface that shows, in particular, a grid of the 2^{k-p} designs that are available with bright color shading to indicate the resolution of each design and the grid constructed using the number of factors and the number of design points. MINITAB has a similarly informative interface that shows the same information and additionally shows the full factorial designs that are available, with the available Plackett–Burman designs (covered in Section 13.4.1) shown on the same display. JMP shows available designs once the user indicates the number of continuous and categorical factors that are to be used in the design.

Books on experimental design also give tables of 2^{k-p} designs, including the design generators. Robinson (2003) pointed out that there is a lot of redundant information in those tables, however, and provided the same information in a more concise manner in his Table 2.

Not all statistical software packages have the capability for all the design methods presented in the chapter. For example, Design-Expert 7.0 has the capability for semifolding, but MINITAB Release 14 does not, nor does most statistical software, in general. (It should be noted, however, that the former does not permit the foldover to be selected by doing the foldover on an interaction, although the instances in which there will be a need for doing so seem to be rather limited, following the discussion in Mee and Peralta, 2000.)

Design-Expert also has the capability for John's 3/4 designs discussed in Section 5.10 and given in John (1961), including the two designs recommended by Mee (2004). The irregular designs that can be produced by Design-Expert are those given by Addelman (1961). These designs cannot be produced using either MINITAB or JMP. Specifically, JMP gives many options for a screening design with 11 factors at two levels, but none has 96 runs, which is the number of runs for a $(3/4)2^{11-4}$ design. JMP also cannot produce a design for 11 two-level factors with 96 runs when its Custom design option is used.

Design-Expert 7.0 can produce minimum run resolution IV designs for 5-50 factors, and resolution V designs for estimating main effects and two-factor interactions for 6-30 factors. Some care must be exercised in using these designs, however, because they are not orthogonal designs and thus are not fractional factorials.

For example, the minimum run resolution IV design for five factors has pairwise correlations of $-.33$ and $.33$ for 4 of the 10 pairwise correlations. These are not small correlations. (For six factors, in 14 runs, all of the correlations are $-.143$. For seven factors in 16 runs all of the correlations are 0 because this is a 2_{IV}^{7-3} design with defining relation $I = ABDF = ABEG = ACDG$. For eight factors in 18 runs, the correlations are all $-.111$ and $.111$, etc.)

The pairwise correlations for the minimum run resolution V designs are generally lower because the minimum number of runs is higher. For example, the design for six factors has 22 runs and the correlations between the columns for the factors are $.091$ and $-.091$. These of course are small correlations, however. The correlation structure for the design for seven factors in 30 runs is disturbing, however, because five of the correlations are either $.2$ or $-.2$ and one correlation is $-.33$. For eight factors in 38 runs, three of the correlations are $.158$ in absolute value; the others are $.053$ in absolute value. For nine factors in 46 runs, the correlations are lower, and so on.

Although numerical output is both useful and standard, and Design-Expert has a unique feature of permitting the selection of backward elimination, a variable selection technique, to arrive at the terms to use in the model as an alternative to a normal probability plot, graphical methods can often be more informative. For example, a Pareto chart of effect estimates, as in Figure 5.3, is more informative than having the same information displayed numerically without the effect estimates being ordered by magnitude. That graph was produced using MINITAB; a Pareto chart of effect estimates can also be produced with Design-Expert.

One unique feature of Design-Expert is that it does a Box–Cox transformation analysis on the response variable, providing a confidence interval on the power transformation parameter, λ, and on the basis of that result, making a recommendation as to whether or not a transformation is needed. This is part of the expert systems software flavor that Design-Expert possesses, and it may be the only commercial software that warns users of the poor effect detection capability of eight-run fractional factorials, for example. (The warning is also given for the 2^3 design, as was stated in Chapter 4.) Similarly, it may be the only commercial software (at least among well-known software) that gives advice to the user (provided that the Annotated ANOVA option is selected), as was illustrated in Section 4.18. Given below is advice when a 2^{5-2} design was constructed:

> Aliases are calculated based on your response selection, taking into account missing datapoints, if necessary. Watch for aliases among terms you need to estimate.

Then, when hypothetical data were analyzed for that design, the following messages resulted:

> Proceed to Diagnostic Plots (the next icon in progression). Be sure to look at the:
>
> 1) Normal probability plot of the studentized residuals to check for normality of residuals.
> 2) Studentized residuals versus predicted values to check for constant error.
> 3) Externally Studentized Residuals to look for outliers, i.e., influential values.
> 4) Box-Cox plot for power transformations.
>
> If all the model statistics and diagnostic plots are OK, finish up with the Model Graphs icon.

D. o. E. Fusion Pro also gives advice and has more of an expert systems flavor than Design-Expert in the sense that users are guided toward an appropriate path leading to design selection based on the answers they provide to questions that they are asked when the Design Navigator Wizard option is followed.

The extensive comparison study of leading software with experimental design capability by Reece (2003) was mentioned in Section 2.3 and also in Section 4.18. It is mentioned again here because the software in the study is rated on its capability regarding full factorial and fractional factorial designs, as well as foldover designs and design augmentation, in general. Regarding the latter, Design-Expert and JMP received the highest possible score on design augmentation among the 11 software packages in the comparison study, in addition to D. o. E. Fusion Pro. The aforementioned features of

Design-Expert in addition to various types of graphs, including diagnostic plots with some unusual features that can be produced dynamically, seem to justify its rating.

The software packages that received the highest rating in the Reece (2003) study in the category of "Factorial amd Fractional Factorial Designs" were, in alphabetical order, D. o. E. Fusion Pro, Design-Expert, JMP, MINITAB, Statgraphics Plus, and Statistica.

5.21 SUMMARY

Just as factorial designs are useful in a wide variety of application fields, so are fractional factorial designs. Although we associate their use with manufacturing, industry, and the physical sciences in general, they have also been used extensively in the social sciences, with examples given by Stolle, Robbennolt, Patry, and Penrod (2002).

The primary advantage of fractional factorial designs is they permit effects that may be significant to be estimated with a reasonable number of runs. However, since fractional factorial designs that are used most often in practice have a small number of runs, such as 16, it is important to consider the minimum detectable difference in factor level means for a selected design before the design is used. It is also important to use designs that are degrees-of-freedom efficient so that degrees of freedom are not wasted.

Resolution III designs are frequently used in the first stage of experimentation for the purpose of identifying factors that seem important. Such an approach will work well, provided that interactions are small. A resolution IV or resolution V design might be used in the second stage. If the former is used, a foldover design might be needed to de-alias certain two-factor interactions. There are various types of foldover designs, including a mirror image foldover. Semifoldover designs, as discussed in Section 5.9.1, are useful for de-aliasing two-factor interactions, although it appears as though virtually no statistical software or DOE software has full semifoldover capability.

Although Lenth's (1989) normal probability plot method will generally be a good choice, it won't always be the best choice, especially if a ratio of real effects to total estimable effects exceeds 0.4. Therefore, it would be useful if the user had at least a rough idea of the number of real effects to anticipate.

Missing and/or bad data can occur when any design is used. Although imputing missing data simplifies computation, the use of artificial data would be unsettling to many experimenters.

Even when there are sufficient resources and time to conduct a full factorial with at least a moderate number of runs, a fractional factorial would be the preferred choice if there was reason to believe that problems might ensue, such as mechanical failures, that would prevent all of the runs in the full factorial from being performed. If there were no problems, the additional runs that would comprise the full factorial could be made (such as folding over a half fraction to produce the full factorial). If problems ensue during the experiment, however, such that, for example, only half of the planned runs could be made, orthogonality will almost certainly be lost and the "design" for the runs that are actually made may be very poor. Recall the discussion in Section 1.4.1 and the quote from Peter John regarding an experience early in his career when he was working as an industrial statistician. In that example, a 2^4 design was planned but only eight runs could be made because of machinery failure. Thus, half of the data were "missing." If the runs had been ordered such that the first eight runs constituted a half fraction, a meaningful analysis could be performed. If there were no problems after the first eight runs, the other eight runs might then be made, or perhaps, as John suggested, four of the runs could be made to form a 3/4 fraction, as discussed in Section 5.10.

Conditional effects analyses are just as important when fractional factorials are used as when full factorials are used, as large interactions can also render main effect estimates meaningless for fractional factorial designs. The situation can be more complicated, as large interactions may be confounded with main effects that are estimated, with the interactions not estimated and thus not detected.

Finally, although the use of two-level fractional factorial designs does predominate in industry, if there are very many factors, there will generally be at least one or two factors for which there will be interest in three or more levels. Then mixed factorials will have to be considered, and these designs are covered in Section 6.4, with mixed fractional factorial designs covered in Section 6.5. (Note that many designs that are referred to as "Taguchi designs" are mixed factorials, as discussed in Chapter 8.)

REFERENCES

Addelman, S. (1961). Irregular fractions of the 2^n factorial experiments. *Technometrics*, **3**, 479–496.

Anderson, M. (2002). The knead for speed. *Today's Chemist at Work*, **11**(2), 21, 23–24.

Barnett, J., V. Czitrom, P. W. M. John, and R. V. León (1997). Using fewer wafers to resolve confounding in screening experiments. In *Statistical Case Studies for Industrial Process Improvement*, Chap. 17 (V. Czitrom and P. D. Spagon, eds.). Philadelphia: Society for Industrial and Applied Mathematics.

Berk, K. N. and R. R. Picard (1991). Significance tests for saturated orthogonal arrays. *Journal of Quality Technology*, **23**, 79–89.

Bisgaard, S. (1993). A method for identifying defining contrasts for 2^{k-p} experiments. *Journal of Quality Technology*, **25**(1), 28–35.

Bisgaard, S. (1994). Blocking generators for small 2^{k-p} designs. *Journal of Quality Technology*, **26**(4), 288–296.

Bisgaard, S. (1997). Designing experiments for tolerancing assembled products. *Technometrics*, **39**, 142–152.

Block, R. and R. W. Mee (2005). Resolution IV designs with 128 runs. *Journal of Quality Technology*, **37**(4), 282–293.

Box, G. E. P. and J. S. Hunter (1961a). The 2^{k-p} fractional factorial designs, Part I. *Technometrics*, **3**, 311–351.

Box, G. E. P. and J. S. Hunter (1961b). The 2^{k-p} fractional factorial designs, Part II. *Technometrics*, **3**, 449–458.

Box, G. E. P., W. G. Hunter, and J. S. Hunter (1978). *Statistics for Experimenters*. New York: Wiley.

Box, G. E. P. and K. B. Wilson (1951). On the experimental attainment of optimum conditions. *Journal of the Royal Statistical Society, Series B*, **13**, 1–45.

Butler, N. A. (2005). Classification of efficient two-level fractional factorial designs of Resolution IV or more. *Journal of Statistical Planning and Inference*, **131**, 145–159.

Carroll, R. J. and D. B. H. Cline (1988). An asymptotic theory for weighted least squares with weights estimated by replication. *Biometrika*, **75**, 35–43.

Chen, J., D. X. Sun, and C. F. J. Wu (1993). A catalog of two-level and three-level fractional factorial designs with small runs. *International Statistical Review*, **61**, 131–145.

Chokshi, D. (2000). Design of experiments in rocket engine fabrication. In *Proceedings of the 54th Annual Quality Congress*, American Society for Quality, Milwaukee, WI. (http://www.asq.org/members/news/aqc/54_2000/14037.pdf)

Cochran, W. G. and G. M. Cox (1958). *Experimental Designs*. New York: Wiley.

Copeland, K. A. F. and P. R. Nelson (2000). Latin squares and two-level fractional factorial designs. *Journal of Quality Technology*, **32**(4), 432–439.

Daniel, C. (1956). Fractional replication in industrial research. In *Proceedings of the Third Berkeley Symposium on Mathematical Statistics and Probability*, Vol. 8, pp. 87–98.

Daniel, C. (1959). Use of half normal plots in interpreting factorial two-level experiments. *Technometrics*, **1**, 311–341.

Daniel, C. (1962). Sequences of fractional replicates in the 2^{p-q} series. *Journal of the American Statistical Association*, **67**, 403–429.

Daniel, C. (1976). *Applications of Statistics to Industrial Experimentation*. New York: Wiley.

Dong, F. (1993). On the identification of active contrasts in unreplicated fractional factorials. *Statistica Sinica*, **3**, 209–217.

Draper. N. R. and I. Guttman (1997). Two-level factorial and fractional factorial designs in blocks of size two. *Journal of Quality Technology*, **29**(1), 71–75.

Eibl, S., V. Kess, and F. Pukelsheim (1992). Achieving a target value for a manufacturing process: A case study. *Journal of Quality Technology*, **24**, 22–26.

Finney, D. J. (1945). Fractional replication of factorial arrangements. *Annals of Eugenics*, **12**, 291–301.

Fries, A. and W. G. Hunter (1980). Minimum aberration 2^{k-p} designs. *Technometrics*, **22**, 601–608.

Giesbrecht, F. G. and M. L. Gumpertz (2004). *Planning, Construction and Statistical Analysis of Comparative Experiments*. Hoboken, NJ: Wiley.

Haaland, P. D. and M. A. O'Connell (1995). Inference for contrast-saturated fractional factorials. *Technometrics*, **37**, 82–93.

Hamada, M. and N. Balakrishnan (1998). Analyzing unreplicated factorial experiments: A review with some new proposals. *Statistica Sinica*, **8**(1), 31–35.

Hedayat, A. S., N. J. A. Sloane, and J. Stufken (1999). *Orthogonal Arrays*. New York: Springer.

Hill, W. J. and R. A. Wiles (1975). Plant experimentation (PLEX). *Journal of Quality Technology*, **7**, 115–122.

Hsieh, P. and W. Goodwin (1986). Sheet molded compound process improvement. In *Fourth Symposium on Taguchi Methods*. Dearborn, MI: American Supplier Institute, pp.13–21.

Jacroux, M. (2003). A modified minimum aberration criterion for selecting regular 2^{m-k} fractional factorial designs. *Journal of Statistical Planning and Inference*, **126**(1), 325–336.

John, P. W. M. (1961). Three-quarter replicates of 2^4 and 2^5 designs. *Biometrics*, **17**, 319–321.

John, P. W. M. (1962). Three-quarter replicates of 2^n designs. *Biometrics*, **18**, 172–184.

John, P. W. M. (1969). Some non-orthogonal fractions of 2^n designs. *Journal of the Royal Statistical Society*, Series B, **31**, 270–275.

John, P. W. M. (1971). *Statistical Design and Analysis of Experiments*. New York: Macmillan.

John, P. W. M. (2003). Plenary presentation at *the 2003 Quality and Productivity Research Conference*, IBM T. J. Watson Research Center, Yorktown Heights, NY, May 21–23, 2003. (The talk is available at http://www.research.ibm.com/stat/qprc/papers/Peter_John.pdf.)

Kempthorne, O. (1973). *Design and Analysis of Experiments*. New York: Robert E. Krieger Publishing Co. (copyright held by John Wiley & Sons, Inc.)

Ledolter, J. (2002–2003). A case study in design of experiments: Improving the manufacture of viscose fiber. *Quality Engineering*, **15**(2), 311–322.

Ledolter, J. and A. Swersey (1997). Dorian Shainin's variables search procedure: A critical assessment. *Journal of Quality Technology*, **29**(3), 237–247.

Lenth, R. V. (1989). Quick and easy analysis of unreplicated factorials. *Technometrics*, **31**, 469–473.

Li, W. and D. K. J. Lin (2003). Optimal foldover plans for two-level fractional factorial designs. *Technometrics*, **45**(2), 141–149.

Li, W., D. K. J. Lin, and K. Q. Ye (2003). Optimal foldover plans for two-level nonregular orthogonal designs. *Technometrics*, **45**(4), 347–351.

Li, W. and R. W. Mee (2002). Better foldover fractions for Resolution III 2^{k-p} designs. *Technometrics*, **44**(3), 278–283.

Loeppky, J. L. and R. R. Sitter (2002). Analyzing unreplicated blocked or split-plot fractional factorial designs. *Journal of Quality Technology*, **34**(3), 229–243.

Loh, W. Y. (1992). Identification of active contrasts in unreplicated factorial experiments. *Computational Statistics and Data Analysis*, **14**, 135–148.

Lynch, R. O. (1993). Minimum detectable effects for 2^{k-p} experimental plans. *Journal of Quality Technology*, **25**(1), 12–17.

Margolin, B. H. (1967). Systematic methods of analyzing $2^m 3^n$ factorial experiments with applications. *Technometrics*, **9**, 245–260.

Mee, R. W. (2004). Efficient two-level designs for estimating all main effects and two-factor interactions. *Journal of Quality Technology*, **36**, 400–412.

Mee, R. W. and R. M. Block (2006). Constructing regular 2^{k-p} designs by bounding length of alias chains. Technical Report No. 2006-1, Department of Statistics, Operations, and Management Science, University of Tennessee. (http://stat.bus.utk.edu/techrpts)

Mee, R. W. and M. Peralta (2000). Semifolding 2^{k-p} designs. *Technometrics*, **42**, 122–134.

Miller, A. (2005). The analysis of unreplicated factorial experiments using all possible comparisons. *Technometrics*, **47**(1), 51–63.

Miller, A. and R. R. Sitter (2001). Using the folded-over 12-run Plackett–Burman design to consider interactions. *Technometrics*, **43**(1), 44–55.

Montgomery, D. C. (1997). *Design of Experiments*, 4th ed. New York: Wiley.

Montgomery, D. C. and G. C. Runger (1996). Foldovers of 2^{k-p} Resolution IV experimental designs. *Journal of Quality Technology*, **28**, 446–450.

Myers, R. H. and D. C. Montgomery (1995). *Response Surface Methodology: Process and Product Optimization using Designed Experiments* (2nd ed., 2002). New York: Wiley.

Nelson, L. S. (2004). *Review of Quality Control*, 7th ed. by D. H. Besterfield. *Journal of Quality Technology*, **36**(2), 241–242.

Prat, A. and X. Tort (1989). Case study: Experimental design in a pet food manufacturing company. Report No. 37. Center for Quality Improvement, University of Wisconsin-Madison.

Reece, J. E. (2003). Software to support manufacturing systems. In *Handbook of Statistics*, Vol. 22, Chap. 9 (R. Khattree and C. R. Rao, eds.). Amsterdam: Elsevier Science B. V.

Robinson, L. W. (2003). Concise experimental designs. *Quality Engineering*, **15**(3), 403–406.

Ryan, T. P. (1988). Taguchi's approach to experimental design: Some concerns. *Quality Progress*, **21**, 34–36 (May).

Ryan, T. P. (2000). *Statistical Methods for Quality Improvement*, 2nd ed. New York: Wiley.

Ryan T. P. (2003). *Analysis of Lead Recovery Data*. Gaithersburg, MD: Statistical Engineering Division, National Institute of Standards and Technology. (Case study available at http://www.itl.nist.gov/div898/casestud/casest3f.pdf)

Schneider, H., W. J. Kasperski, and L. Weissfeld (1993). Finding significant effects for unreplicated fractional factorials using the n smallest contrasts. *Journal of Quality Technology*, **25**, 18–27.

Snee, R. D. (1985). Experimenting with a large number of variables. In *Experiments in Industry: Design, Analysis, and Interpretation of Results*, pp. 25–35 (R. D. Snee, L. B. Hare, and J. R. Trout, eds.). Milwaukee, WI: Quality Press.

Srivastava, R., V. K. Gupta, and A. Dey (1990). Robustness of some designs against missing observations. *Communications in Statistics, Theory and Methods*, **19**(1), 121–126.

Srivastava, R., V. K. Gupta, and A. Dey (1991). Robustness of some designs against missing data. *Journal of Applied Statistics*, **18**(3), 313–318.

Steinberg, D. M. and W. G. Hunter (1984). Experimental design: Review and Comment. *Technometrics*, **26**, 71–97 (discussion: pp. 98–130).

Stolle, D. P., J. K. Robbennolt, M. Patry, and S. D. Penrod (2002). Fractional factorial designs for legal psychology. *Behavioral Sciences and the Law*, **20**(1–2), 5–17.

Wang, P. C. and C. Lee (2006). Strategies for semi-folding fractional factorial designs. *Quality and Reliability Engineering International*. **22**(3), 265–273.

Woodall, W. H. (2000). Controversies and contradictions in statistical process control. *Journal of Quality Technology*, **42**, 341–350; discussion: 351–378.

Wu, C. F. J. and M. Hamada (2000). *Experiments: Planning, Analysis, and Parameter Design Optimization*. New York: Wiley.

Yates, P. and R. W. Mee (2000). Fractional factorial designs that restrict the number of treatment combinations for factor subsets. *Quality and Reliability Engineering International*, **16**, 343–354.

Ye, K. Q. and W. Li (2003). Some properties of blocked and unblocked foldovers of 2^{k-p} designs. *Statistica Sinica*, **13**(2), 403–408.

Zahn, D. A. (1975). Modifications of and revised critical values for the half-normal plot. *Technometrics*, **17**, 189–200.

EXERCISES

5.1 Show that no combination of pairs of 2^{6-2} designs produces a 2^{6-1} design with the maximum resolution.

5.2 Explain why the variances of effect estimates in 2^{k-p} designs are equal regardless of the number of observations that are missing.

5.3 Show that the alias structure of the 2^{3-1} design in Section 5.1 can be obtained using Yates' algorithm in the manner described in that section by using zeros for the treatment combinations that are not in the fraction that is used.

5.4 There are 16 treatment combinations used when a 2^{5-1} design is run with $I = ABCDE$. Which one of the following treatment combinations is used in error with the other four (correct) treatment combinations: ab, $bcde$, acd, bc, and $abce$?

5.5 Explain why a 2^{5-1} design should not be run with $ABDE$ confounded with the difference between the two fractions. What would you do instead?

5.6 An example of sequential experimentation was given in Section 5.9. A follow-up experiment is often necessary to dealias effects, as was discussed in Section 5.7 and subsequent sections. Snee (1985, references) described an experiment conducted to determine the factors affecting viscosity measurements in an analytical laboratory, as the variation in those measurements was thought to be too great. The chemists felt that five factors should be used in the experiment, but were less enthusiastic about two other factors that were under consideration. Snee stated, "They agreed to the seven variable test when it was pointed out to them that these two variables (spindle and protective lid) would not involve any additional runs beyond the sixteen required to test the first five variables." That is, a 2^{7-3} design was used whereas the chemists might have preferred a 2^{5-1} design. Do you believe that the former is a suitable substitute for the latter? What would you say if there was a strong a priori belief that the two additional factors would not likely be significant?

When the experiment was run, spindle had the largest effect (!). There was also a large interaction component (which represented $X_1X_4 + X_3X_6 + X_5X_7$). Since X_1 and X_7 had very small effects, it was believed that X_1X_4 and X_5X_7 might not be significant. Do you agree with that assessment, especially when viewed in the context of conditional effects? Accordingly, a 2^2 design was run using only X_3 and X_6 in an effort to assess the magnitude of the X_3X_6 interaction. A moderate interaction was observed and the predicted response values of the four additional runs were close to the predicted values from the model based on the first 16 runs. The interaction graph was interpreted in terms of the best combination of levels of X_3 and X_6 and this is where the analysis ended. Would you have proceeded further? Explain.

5.7 Assume that a 2^{4-1} design has been run using the treatment combinations (1), ab, ac, bc, ad, bd, cd, and $abcd$.

(a) What was the defining contrast?

(b) Was it necessary to use the methodology given in Section 5.4 in answering the question? Why, or why not?

(c) What would you recommend if the treatment combination bc is an impossible combination of factor levels?

5.8 Assume that four factors, each at two levels, are studied with the design points given by the following treatment combinations: (1), ab, bc, abd, acd, bcd, d, and ac.

(a) What is the defining contrast?

(b) Could the design be improved using the same number of design points?

(c) In particular, which three main effects are confounded with two-factor interactions?

5.9 Assume that a 2^{5-2} design with three replications has been used.

(a) What is the smallest possible value of the F-statistic for testing the significance of the A effect and when will that occur? What is the largest possible value?

(b) Construct an example with six observations at each level that will produce this minimum value.

(c) Fractional factorial designs are generally not replicated. Assume that you have pointed this out to the experimenter, who responds that he was not willing to trust the result of a normal probability plot analysis because he expected more than a few effects to be significant. Is this a reasonable reason for using a replicated fractional factorial? Explain.

5.10 A design with six factors was run and the alias structure involving factor E is $E = ABCDF$. How many design points were there (assuming no replication)?

5.11 Critique the following statement: "I don't understand why textbooks give alias structures such as $AB = DE$. The effects are almost certainly not equal, and no hypothesis test of their equality is being performed. Therefore, I don't see any point in listing these effects as being equal and believe that this is misleading. Instead, this should be written as $AB + DE$ since that is what is being estimated, and similarly for other aliased effects."

5.12 Given below is the alias structure for a fractional factorial design with seven factors, each at two levels.

Alias Structure

$$I + ABCE + ABFG + ACDG + ADEF + BCDF + BDEG + CEFG$$

$$A + BCE + BFG + CDG + DEF + ABCDF + ABDEG + ACEFG$$
$$B + ACE + AFG + CDF + DEG + ABCDG + ABDEF + BCEFG$$
$$C + ABE + ADG + BDF + EFG + ABCFG + ACDEF + BCDEG$$
$$D + ACG + AEF + BCF + BEG + ABCDE + ABDFG + CDEFG$$
$$E + ABC + ADF + BDG + CFG + ABEFG + ACDEG + BCDEF$$
$$F + ABG + ADE + BCD + CEG + ABCEF + ACDFG + BDEFG$$
$$G + ABF + ACD + BDE + CEF + ABCEG + ADEFG + BCDFG$$
$$AB + CE + FG + ACDF + ADEG + BCDG + BDEF + ABCEFG$$
$$AC + BE + DG + ABDF + AEFG + BCFG + CDEF + ABCDEG$$
$$AD + CG + EF + ABCF + ABEG + BCDE + BDFG + ACDEFG$$
$$AE + BC + DF + ABDG + ACFG + BEFG + CDEG + ABCDEF$$
$$AF + BG + DE + ABCD + ACEG + BCEF + CDFG + ABDEFG$$
$$AG + BF + CD + ABDE + ACEF + BCEG + DEFG + ABCDFG$$
$$BD + CF + EG + ABCG + ABEF + ACDE + ADFG + BCDEFG$$
$$ABD + ACF + AEG + BCG + BEF + CDE + DFG + ABCDEFG$$

(a) What is the order of the fraction that was used?

(b) How would you explain to a manager who knows very little about experimental design what this alias structure means in practical terms?

5.13 Consider a 2^{4-1} design constructed so as to maximize the resolution. If the AB interaction is very close to zero, what will be the relationship between the conditional effects of factor D when the data are split on factor C?

5.14 Explain the condition(s) under which the selection of a one-half fraction of a full factorial design should not be chosen randomly.

5.15 Explain why a minimum aberration design isn't necessarily the best design to use.

5.16 What is the resolution of any fractional factorial design that has k factors and $k + 1$ design points?

5.17 Consider a 2^{8-4} design.

(a) How many words are in the defining relation?

(b) What is the maximum resolution of the design?

(c) Explain how a novice might erroneously construct the design such that the resolution is one less than the maximum?

(d) Explain to that person why it is desirable to construct a design so that it has maximum resolution. Are there any conditions under which you would recommend the use of a design that did not have maximum resolution?

5.18   Assume that an engineer constructed the following design, with "−" and "+" used in place of −1 and +1 for the sake of brevity.

−	−	−	+	−
+	−	−	−	−
−	+	−	−	+
+	+	−	+	+
−	−	+	−	+
+	−	+	+	+
−	+	+	+	−
+	+	+	−	−

What is the name of the design, in the form 2^{k-p}? Would you have constructed the design differently? If so, give the design.

5.19   What are the dangers of using highly fractionated designs?

5.20   Explain how you would construct a 2^{6-2} design to be run in two blocks.

5.21   Would you analyze with a normal probability plot the data from an experiment that used a 2^{5-2} design? Why, or why not? If not, how would you analyze the data?

5.22   For the data given in Example 5.2, construct the AC interaction plot using factor A on the horizontal axis. Does this plot suggest a weaker or stronger interaction than the interaction plot given in Figure 5.7? Explain.

5.23   Assume that an experimenter decides to use all four main effects in a model for the data in Example 5.2. The R^2 value is 85.9 percent for that model. Would you recommend the use of R^2 for that scenario, considering the number of data points and the number of terms in the model? Explain. Would you recommend that all four of those terms be used in the model? Why, or why not?

5.24   In his paper "Establishing optimum process levels of suspending agents for a suspension product" (*Quality Engineering*, **10**(2), 347–350), 1997–1998) A. Gupta used a design that was stated as "the L_{16} orthogonal-array design." (This is Taguchi-type terminology; see Chapter 8.) The design for five factors, each at two levels, is given below in the original units.

A	B	C	D	E
8	50	0.2	0.4	Usual
8	50	0.4	0.4	Modified
8	60	0.4	0.4	Usual
8	60	0.2	0.4	Modified
16	50	0.4	0.4	Usual

```
16   50   0.2   0.4   Modified
16   60   0.2   0.4   Usual
16   60   0.4   0.4   Modified
 8   50   0.2   0.6   Usual
 8   50   0.4   0.6   Modified
 8   60   0.4   0.6   Usual
 8   60   0.2   0.6   Modified
16   50   0.4   0.6   Usual
16   50   0.2   0.6   Modified
16   60   0.2   0.6   Usual
16   60   0.4   0.6   Modified
```

Since there are 16 runs, this must be "some" 2^{5-1} design. What is the resolution of the design and why is this a suboptimal fractional factorial? Convert this to a 2^{5-1} design with the maximum possible resolution by changing the numbers in one of the columns.

5.25   Given below are the runs for a two-level design with five factors. What was the defining relation?

A	B	C	D	E
−	−	−	−	−
−	−	−	+	+
−	−	+	−	+
−	−	+	+	−
−	+	−	−	+
−	+	−	+	−
−	+	+	−	−
−	+	+	+	+
+	−	−	−	−
+	−	−	+	+
+	−	+	−	+
+	−	+	+	−
+	+	−	−	+
+	+	−	+	−
+	+	+	−	−
+	+	+	+	+

5.26   Explain the difference between complete confounding and partial aliasing. Assume the use of a fractional factorial design that is not run in blocks.

5.27   What is the maximum possible resolution of a 2^{8-3} design? Give one possible defining relation that would produce that resolution. What is the *minimum* possible resolution? Give one possible defining relation that would give this resolution and explain why the design would never be used.

5.28 Consider the design given in Exercise 5.18. Let the columns represent the factors A, B, C, D, and E (as usual), and assume that when the design was run, the values of the response variable were, in order of the rows in the design, 16.2, 18.4, 17.3, 19.4, 15.9, 17.1, 18.0, and 16.8. Analyze the data and present your conclusion. Do you believe that at least one follow-up experiment is necessary? Explain.

5.29 Referring to Yates and Mee (2000), construct a design with a reasonable number of treatment combinations for which there are seven two-level factors, with four factors in the first subset and three factors in the second subset.

5.30 The fact that an interaction that is the product of the two selected interactions which are selected to be part of the defining relation when a 1/4 fraction is used is also part of the defining relation is not necessarily intuitive. To see that this does in fact occur, proceed as follows for the following example. Construct a 2^{8-1} design, using of course $I = ABCDEFGH$. Then split this design in half by using only those treatment combinations for which $ABCD = 1$. (This can be easily done using MINITAB, for example.). Look at the value of $EFGH$ for each of those 64 treatment combinations. What must that value be? Is this what happens? Explain.

5.31 In his article "In the soup: A case study to identify contributors to filling variability" (*Journal of Quality Technology*, **20**(1), 1988), L. B. Hare describes an experiment performed with the objective of minimizing the filling variation among pockets of dry soup mix. The use of control charts had shown that the variation was unacceptable. Five factors were examined in a 2^{5-1} design: number of ports, temperature, mixing time, batch weight, and delay. (The latter is the delay in days between mixing and packaging. For temperature, "C" means that cool water was used and "A" means that no water was running through the mixing jacket.) The design and the values of the response variable are given below.

Row	A	B	C	D	E	Y
1	1	C	60	2000	7	0.78
2	3	C	80	2000	7	1.10
3	3	A	60	1500	1	1.70
4	3	C	80	1500	1	1.28
5	1	A	60	1500	7	0.97
6	1	C	80	1500	7	1.47
7	1	A	60	2000	1	1.85
8	3	A	80	2000	1	2.10
9	1	A	80	2000	7	0.76
10	3	A	60	2000	7	0.62
11	1	C	80	2000	1	1.09
12	1	C	60	1500	1	1.13

13	3	C	60	1500	7	1.25
14	3	A	80	1500	7	0.98
15	3	C	60	2000	1	1.36
16	1	A	80	1500	1	1.18

   (a) Analyze the data by using appropriate computer software. Is a conditional effects analysis necessary? (Notice which effects are the three largest effects.)

   (b) Consider your analysis versus the hand analysis given in the paper. Are most of your effect estimates a constant multiple of the effect estimates given in the paper? What does this suggest about one error that was made in the paper (several errors were made in the analysis)?

5.32 Explain or show why a $(2^{3-1} \times 2^{4-1})/2^1$ assembly design could not be resolution IV.

5.33 There is an upper limit on the number of experimental runs in a design constructed using various statistical software (e.g., with MINITAB the limit is 128 runs). Assume that you are using MINITAB, for which operations on columns can be easily performed, and you need to construct a 2^{9-1} design, as you need the design for an application in which experimental runs are very inexpensive. If you construct a 2^{9-2} design (i.e., 128 runs) and then fold it over using appropriate column operations in MINITAB, you would obtain a 2^{9-1} design. Would you want to use this design, however? Explain.

5.34 Assume that your boss is distrustful of fractional factorials, preferring instead to see "the whole thing, not a fraction." You ran a 2^{5-1} design last week but now he insists that the full factorial be used. You could run the other 1/2 fraction and add it to the results from the first fraction, or you could start over and run the 2^5 design. Assume that it will be two weeks before the other 1/2 fraction can be run and that certain conditions could change by then. How will you proceed?

5.35 The projective properties of a 2^{k-p} design were of relevance in a study described by T. Jørgensen and T. Naes in their paper "A design and analysis strategy for situations with uncontrolled raw material variation" (*Journal of Chemometrics*, **18**, 45–52, 2004). A cheese experiment was conducted that involved 6 two-level factors in a 2^{6-1} design with eight raw material blocks. Only the first four factors were analyzed in the paper, however, because the response variable is measured before the final process steps are performed, which involve the other two factors. *If* blocking had not been used, what effects would be estimable and would be unconfounded when the first four factors are analyzed? How would you describe the design when used for only these factors? Now, given your answers to these questions, how should the blocks be viewed and indeed what is one way that the blocks could have been constructed so as to minimize the confounding of effects with blocks that might be significant? Read the paper, if desired, to learn more of this somewhat unusual use of a fractional factorial.

5.36 (Harder problem) Using a table of the signs for a 2^4 design, construct a 2^{4-1} design with 12 blocks of size 2 such that the same effects are estimable that are estimable with the 2^{4-1} design when run without blocking.

5.37 J. V. Stephenson and D. Drabenstadt (*Surface Mount Technology*, November 1999) described sequential experimentation relative to a solder paste printing process. Eleven factors were used in the first experiment and it was stated that the design was resolution IV. The 32 treatment combinations (so the design is a 2^{11-6}) are given at the following Web site: http://smt.pennnet.com/Articles/Article_Display.cfm?Section=Archives&Subsection=Display&ARTICLE_ID=111949&KEYWORD=printing, and the defining relation was not given. Use the method of Bisgaard (1993) or any other method you prefer and determine the defining relation.

5.38 Continuing Exercise 5.37, a second experiment was performed with five factors that were identified as important in the first experiment. A 2^{5-1} design was used so that the design was resolution V. The data for the second experiment (the data were not given for the first experiment) were given in the article at the URL stated in Exercise 5.37. The authors stated that one main effect and 2 two-factor interactions were significant, with print pressure being the significant main effect and the interactions were between print pressure and print speed and print pressure and snap off. Do you agree? If so, are you concerned about the fact that there are significant interactions involving the factor that is significant since a regression equation was developed for predictive purposes in the optimization stage? If so, what would you do differently, if anything. If you are not concerned, explain why.

5.39 B. N. Gawande and A. Y. Patkar (*Biotechnology and Bioengineering*, **64**(2), 168–173) described two experiments in which the effect of dextrin, peptone, yeast extract, ammonium dihydrogen orthophosphate, and magnesium sulfate on enzyme production was investigated in two fractional factorial experiments. In the first experiment, six observations were made at the center of the design. The observations, which were not given in the article, produced a standard error of the response of 0.144. The centerpoint replicates of course permit the construction of significance tests and it was found that all of the main effects and one interaction effect were significant when the data from the 2^{5-1} design were analyzed. (The numbers suggest that there should have been two significant interaction effects, contrary to the impression created by the article.) Significance was declared when an effect estimate exceeded twice its standard error. The design and the data are given below.

Row	Detrin	Pepton	Yeast Extract	$NH_4H_2PO_4$	$MgSO_4 \cdot 7H_2O$	Activity
1	-1	-1	-1	-1	-1	3.97
2	-1	-1	-1	+1	+1	5.99
3	-1	-1	+1	-1	+1	4.13
4	-1	-1	+1	+1	-1	5.59
5	-1	+1	-1	-1	+1	5.18
6	-1	+1	-1	+1	-1	6.47
7	-1	+1	+1	-1	-1	5.12
8	-1	+1	+1	+1	+1	6.53
9	+1	-1	-1	-1	-1	5.39
10	+1	-1	-1	+1	+1	5.25
11	+1	-1	+1	-1	+1	5.39
12	+1	-1	+1	+1	-1	6.06
13	+1	+1	-1	-1	+1	4.98
14	+1	+1	-1	+1	-1	6.74
15	+1	+1	+1	-1	-1	5.66
16	+1	+1	+1	+1	+1	8.42

(a) Analyze the data. In particular, note that the response values when the first, second, and fourth factors are at the low level are considerably smaller than the other values. Is this possibly suggestive of anything in particular?

(b) Compare the number of significant effects that you obtained with the number obtained by the authors of the article. What do you think accounts for the discrepancy? Which set of results would you be inclined to use?

(c) Of the 15 estimable effects, about how many effects, recalling the discussion in Section 4.11, would you expect to be significant without even looking at the data?

A second fractional factorial was used (actually the mirror image foldover of the first design), with the design points centered at the best design point from the runs that were made along the path of steepest ascent (see Section 10.4). The method of steepest ascent was also used after the second design. We return to this study in Exercise 10.12 in Chapter 10 and critique the response surface analysis.

5.40 In referring to the 2_{III}^{9-5} design given by Hsieh and Goodwin (1986), mentioned in Section 5.9.1.4, with generators $E = -ABCD$, $F = -CD$, $G = ACD$, $H = -BC$, and $J = ABC$, Mee and Peralta (2000, references) stated in part " ... a better design exists...." Explain why they would make such a statement. If you agree, give an example of a superior design.

5.41 (Requires knowledge of regression) It was stated in Section 5.20 that Design-Expert allows the user the option of selecting the model terms to use by using backward elimination, a variable selection technique that is used in regression analysis. Is this of any value when the design is orthogonal, however (as are most of the designs in the chapter), or will the model be the same as when the results of t-tests are used to arrive at the model?

5.42 Consider Example 5.3. If there are no restrictions on effects to be used in constructing a 2^{10-5} design to be run in eight blocks of size 4, is it possible to construct the design so that no main effect is confounded with a block estimate? Use appropriate software to construct the design and describe, in general, the alias structure.

CHAPTER 6

Designs with More Than Two Levels

Although two-level designs are the most popular and simplest designs to use, experimenters will oftentimes be interested in using more than two levels. In this chapter we consider full and fractional factorials with more than two levels, including full and incomplete mixed factorials. We do not consider Box–Behnken designs, which have three levels. These designs are intended to be used as response surface designs, so they are covered in Section 10.8.

6.1 3^k DESIGNS

Designs with three levels (a middle level in addition to low and high) permit the investigation of quadratic effects, although 3^k or 3^{k-p} designs might require too many runs, depending on the values of k and p.

Designs in the 3^k series can be generated rather easily using most statistical software, but not so with 3^{k-p} designs. This is discussed later. Given below is a 3^2 design that was created using the general full factorial design capability in MINITAB, with the runs randomized.

Row	A	B
1	1	1
2	3	2
3	3	1
4	3	3
5	1	2
6	2	3
7	1	3
8	2	1
9	2	2

Modern Experimental Design   By Thomas P. Ryan

(*Note*: There are various ways of denoting the three levels. One alternative way is to use −1, 0, and 1 to denote the three levels and another way is to use the numbers 0, 1, and 2, which is what is used later in the chapter, beginning with Section 6.1.1. The reason for this choice should become apparent in this section.) The following is MINITAB output for hypothetical response data from using this 3^2 design with the data simulated as $Y = A + e/20$, with e denoting a random error term that is $N(0, 1)$.

```
General Linear Model: Response versus

Factor Type Levels Values

Analysis of Variance for Response, using Adjusted SS

for Tests
```

Source	DF	Seq SS	Adj SS	Adj MS	F	P
A	1	5.8535	0.1219	0.1219	26.55	0.014
A*A	1	0.0015	0.0015	0.0015	0.34	0.603
B	1	0.0002	0.0047	0.0047	1.02	0.388
B*B	1	0.0081	0.0081	0.0081	1.76	0.277
A*B	1	0.0017	0.0017	0.0017	0.38	0.582
Error	3	0.0138	0.0138	0.0046		
Total	8	5.8788				

```
S = 0.0677626    R-Sq = 99.77%    R-Sq(adj) = 99.38%
```

Term	Coef	SE Coef	T	P
Constant	-0.1591	0.2757	-0.58	0.604
A	1.0572	0.2052	5.15	0.014
A*A	-0.02779	0.04792	-0.58	0.603
B	0.2067	0.2052	1.01	0.388
B*B	-0.06351	0.04792	-1.33	0.277
A*B	0.02083	0.03388	0.61	0.582

Here A^*A denotes the quadratic component of factor A and similarly for B, and A^*B is the interaction term. Note that in this chapter we will denote interactions in this way rather than the more customary AB notation because one way to decompose the A^*B interaction is to decompose it into components that are designated as AB and AB^2, each with two degrees of freedom when A and B each have two levels. These components don't have any physical meaning but under certain conditions they can be related to another method of decomposing the interaction that does have physical meaning. This is discussed in Section 6.1.1.

The "adjusted" sums of squares and adjusted mean squares are just the usual sums of squares and mean squares. That is, the sums of squares are the values for each of the effects when the other effects are in the model. The sequential sums of squares are the sums of squares for the effects when the terms are entered in the order in which the effects are listed in the table. These two sets of columns are different because the model components are not orthogonal, as they are when two-level designs are used, which obviates the use of both sequential and adjusted sums of squares. In particular, notice that the two sum of squares for the A effect differ greatly, due to the correlation between A and A^*A, in particular. (*Note*: These numbers are obtained without any "orthogonal polynomial" approach. That is, the columns for the quadratic terms are not transformed, such as by centering, in any way. In general, some type of centering should be used, however (see Ryan, 1997, p. 251).)

Another way to decompose the A^*B interaction is in terms of the trend components. Since each main effect has two degrees of freedom, which allows the linear and quadratic effects to be estimated, with one degree of freedom for each, it logically follows that the interaction of two 3-level factors would have four components with one degree of freedom each that results from "crossing" the linear and quadratic components, just as one obtains the four treatment combinations in a 2^2 design. That is, the four components are linear $\times$ linear, linear $\times$ quadratic, quadratic $\times$ linear, and quadratic $\times$ quadratic. These are referred to as the trend components of interaction.

Of course the (linear) main effect of A is significant, as expected, and no other terms are significant, which is also as expected since we know the true model because these are simulated data.

In the output above, each factor has two df because of the three levels, with, as stated, one df used to estimate the linear effect and the other used to estimate the quadratic effect. Then why does the A^*B interaction have only one df instead of four since the df for the interaction term must be the product of the df for each factor that comprises the interaction? The different components of the A^*B interaction are explained in detail in the next section, but for the moment we will note that the interaction cannot be broken down using MINITAB unless the interaction components are entered physically when the model is specified, and only the trend components can be entered (i.e., AB and AB^2 cannot be entered). The same is true of the linear and quadratic effects of a factor. In general, any term that is to appear in the ANOVA table must be specified in the model, which would obviously be cumbersome if there were more than a few factors. Design-Expert will display the quadratic components in the ANOVA table and they can also be specified in the model. Confidence intervals are given for all the parameters corresponding to the nonlinear components if there is at least one df for the error term and the fitted model with the parameter estimates is also given. MINITAB will also give the confidence intervals and parameter estimates. MINITAB does not provide the capability for a normal probability plot for effect estimates with its general linear model (GLM) capability, which would be used for the types of designs covered in this chapter, nor does Design-Expert.

This is unfortunate because a normal probability plot can be constructed for *any* design, provided that the plotted effect estimates are uncorrelated and have the same variance. This is discussed further in Section 6.1.2.

Even when quadratic effects are displayed, the advantage of being able to investigate quadratic effects is offset somewhat by the complexity of interactions and the difficulty in interpreting them. When all factors have only two levels, each interaction has one degree of freedom, regardless of the order of the interaction. The number of degrees of freedom for an interaction term when all factors have three levels is 2^c, with c denoting the number of factors involved in the interaction. The interpretation of even a simple interaction such as the A^*B interaction in a 3^2 design is somewhat involved, partly because of the various possible ways of decomposing the four degrees of freedom for the interaction.

6.1.1 Decomposing the A^*B Interaction

As stated, one way to decompose the A^*B interaction is to decompose it into AB and AB^2. These have also been referred to as the I–J components. To understand these components, it is helpful to understand modular arithmetic, which, recall, was used in Section 5.1 for two-level fractional factorials.

To facilitate this, we will use 0, 1, and 2 to represent the three levels. (In principle, any set of three consecutive integers could be used.) With this designation, the column for AB is $A + B$, modulo 3, and AB^2 is $A + 2B$, modulo 3. (Modulo 3 means that $1 + 2 = 0$ and $2 + 2 = 1$, which is analogous to our base 10 (i.e., "modulo 10") number system if we consider only the units digit, since the latter is 0 for $1 + 9$, 1 for $2 + 9$, etc.) Thus, we have the following:

A	B	AB	AB^2
0	0	0	0
1	0	1	1
2	0	2	2
0	1	1	2
1	1	2	0
2	1	0	1
0	2	2	1
1	2	0	2
2	2	1	0

Orthogonality of these columns could be verified by subtracting 1 from each number (so as to make the column sums zero, as is the case with two-level designs) and showing that all column dot products are zero, as the reader is asked to show in Exercise 6.9. With column sums of zero, the columns for A^2 and B^2 would each be 1−2 1 1−2 1 1−2 1. The reader can verify that each of these two columns is orthogonal to what the four columns above become after 1 is subtracted from each number. As is probably self-evident, the columns for A^2 and B^2 cannot be obtained by transformation of the columns for A and B (using modular arithmetic or otherwise) since those columns measure the *linear* effect of A and B, respectively. That is, the numbers 1, 2, 3 form a linear progression, whereas a graph of $Y = 1, 2, 1$ against a sequence index would show quadrature. Similarly, if the value of the response variable at the middle level of a factor is quite different from the values at the low level and the high level, respectively, the linear combination $1(Y_{\text{low}}) - 2(Y_{\text{middle}}) + 1(Y_{\text{high}})$ will be quite different from zero.

Of course we should note that referring to the linear effect of a factor makes sense only if the factor is quantitative. It wouldn't make any sense to speak of the linear effect of a qualitative factor because the level designations are arbitrary. Thus, the terms low, middle, and high in referring to the levels would not make any sense unless there was an ordinal scale involved (such as army ranks).

As stated, the components AB and AB^2 have no physical meaning but there is a connection between these components and the trend components. Specifically, Ryan (1981) showed that the sum of squares for the AB and AB^2 components are equal when exactly one of the trend components of the interaction is nonzero. There is a related discussion in Sun and Wu (1994).

The interaction plots have a definite type of configuration for each of these four cases, and prior information that one of the four cases may be very likely to occur, such as the interaction being strictly linear × linear, may be available. When a 3^2 design is run in blocks of size 3, the usual approach is to confound either AB or AB^2 with the difference between blocks. If AB^2 were large and AB quite small, a serious mistake would be to confound the larger component, as that could lead to the false conclusion that there is no interaction. It would be comforting to know that nothing is lost when one of these two components is selected to be confounded. Further research is needed in this area for other 3-level designs and for mixed factorials.

6.1.2 Inference with Unreplicated 3^k Designs

Unreplicated 3^k designs, like unreplicated 2^k and 2^{k-p} designs, do not have any degrees of freedom for an error term unless some interaction components are not estimated and instead are used to create an error term. Unlike 2^k and 2^{k-p} designs, however, software developers have not provided the capability for normal plots for use with factorial designs with three (or more) levels, as indicated previously. There is no reason why such plots cannot be constructed, however. Indeed, a normal probability plot could be constructed for *any* set of continuous data or any set of statistics computed from such data, provided the assumptions of independence and equal variances are met. Similarly, the method of Lenth (1989) could be applied to any unreplicated factorial, but this has generally not been done by software developers. Indeed, both MINITAB and Design-Expert utilize Lenth's method only for two-level designs.

Wu and Hamada (2000, pp. 222, 226) do give two half-normal plots for effect estimates from a 3^{4-1} design with three replications and conclude for one of them that the linear effect of the first factor may be significant, the conclusion being confirmable by using Lenth's method, which the authors leave as an exercise. Of course we much prefer to see normal probability plots of effect estimates identifying significant effects for us, rather than having to use hand computations to assess significance. (Designs in the 3^{k-p} series are discussed in Section 6.3.) Although it is, of course, easier to restrict consideration to two-level designs, the absence of normal probability plot methods for general factorial designs seems to be a common shortcoming of statistical software and software specifically for design of experiments (DOEs). Interestingly, this shortcoming was not mentioned in the software survey of Reece (2003).

Example 6.1

An application in which a 3^2 design with three replicates of the centerpoint (which unlike a 2^k design happens to be one of the points in the design) was given by Vázquez and Martin (1998). The objective was to "optimize the growth of the yeast *Phaffia rhodozyma* in continuous fermentation using peat hydrolysates as substrate." The two factors in the experiment were dilution rate at levels of 0.13, 0.23, and 0.33, and pH at levels of 5, 7, and 9. There were four response variables: biomass concentration at steady-state conditions (g/L), substrate concentration at steady-state conditions (g/L), biomass yield (g/g), and biomass volumetric productivity [g/(L z h)].

We will use the first of these response variables for illustration, with the data given below. (Vázquez and Martin (1998) stated that the response values given below are the averages of three numbers; the 33 original observations were not given.)

A	B	Y
-1	-1	3.8
-1	0	4.0
-1	+1	1.4
0	-1	5.1
0	0	4.8
0	0	4.9
0	0	4.9
0	+1	1.9
+1	-1	4.5
+1	0	3.8
+1	+1	0.8

(Note that *A* and *B* are orthogonal, as must be the case, just as columns representing the levels of two-level factors are orthogonal.)

We saw in Chapter 4 that the definition of a main effect for two-level factorials was quite intuitive as we simply calculate the average response value at the high level of a factor and subtract from that average the average response value at the low level of the factor.

The way to measure the effect of a factor may not be obvious when the factor has three levels, however, and indeed various alternative definitions have been proposed and can be proposed. We need an estimate of the *linear effect* and the *quadratic effect* of each factor. The middle level can be viewed as the "additional" level that is used to detect quadrature, so it would be ignored when the linear effect is estimated. Thus, *one way* to define the linear effect of a 3-level factor is as the average response at the high level of the factor minus the average response at the low level. An alternative definition, which will give one-half of the value obtained using the intuitive definition, is to bring the middle level into play and define the linear effect as the average of two linear effects—middle minus low and high minus middle. This gives (average at high level–average at low level)/2, since the middle level drops out in averaging the two linear effects. We could make a good argument for each definition but it is the second definition that is generally used.

A similar debate could occur regarding the definition of the quadratic effect of a factor since all three levels must be used directly. In order for the effect of a 3-level factor to be 100 percent linear, the difference between the average response at the high level and the average response at the middle level must be the same as the difference between the average response at the middle level and the average response at the low level. If not, the line segment that connects the three averages will not be a straight line and there will thus be evidence of quadrature. Therefore, the quadratic effect should seemingly be measured by the difference of these two differences, but instead it is defined as the (less intuitive) difference of the differences divided by 2.

To summarize, let $\overline{A_0}$, $\overline{A_1}$, and $\overline{A_2}$ denote the average response value at the low, middle, and high levels, respectively, for factor A, and let A_L and A_Q denote the linear and quadratic effects, respectively, of factor A. Then

$$A_L = \frac{\overline{A_2} - \overline{A_0}}{2}$$

and

$$A_Q = \frac{\overline{A_2} - 2\overline{A_1} + \overline{A_0}}{2}$$

and similarly for factor B.

For the current example, $A_L = -0.017$, $A_Q = -2.54$, $B_L = -1.55$, and $B_Q = -1.40$. We could similarly define four interaction effect estimates, $A_L B_L$, $A_L B_Q$, $A_Q B_L$, and $A_Q B_Q$, which would be computed using coefficients that are products of the component coefficients.

The sum of squares for the terms in the model can, for a balanced design, be obtained simply using products of the columns for A and B times Y and divided by the sum of the squares of the numbers in the column that is multiplied times Y. It is more complicated for unbalanced data, however, as in this example, and a complete discussion would be beyond the intended scope of this text.

Vázquez and Martin (1998) analyzed these data by fitting a regression model with linear and quadratic terms in A and B and the interaction term. All terms except the linear term in A are (highly) significant when this regression is run.

Critique

It is not clear why an unbalanced design was used; it appears, though, that this may have been done to facilitate a lack-of-fit test from the repeated runs. This was unnecessary, however, as the 33 original data points could, assuming they were available, be used and that would permit a lack-of-fit test. Furthermore, the design would be balanced as it would be a 3^2 design with three replicates, thus avoiding the headaches that are caused by unbalanced data.

Example 6.2

Clark (1999) described an experiment in which a 3^3 design with three replicates was used in a resistance welding operation experiment. The three factors studied were current, weld time, and pressure. The data were not given so we cannot analyze it, but we can critique the design and the methods of analysis. The three-factor interaction was not investigated, despite the fact that there were plenty of degrees of freedom for doing so. Eighteen degrees of freedom are needed to estimate the main effects and two-factor interactions, so with an unreplicated 3^3 design there would be eight degrees of freedom for estimating the error term, which are the eight degrees of freedom for the A^*B^*C interaction.

Only three degrees of freedom were listed for "interaction" in the ANOVA table, however, whereas the number should have been 12 since there are four degrees of freedom for each of the three interactions. This analysis was performed in MINITAB and it seems likely that only the linear × linear component of the interaction was broken out in the ANOVA table. Ignoring most of the degrees of freedom for the interaction is very risky. Two of the three factors were significant at the .01 level in terms of the linear effect and one quadratic component was also significant, as well as one of the two-factor interactions. The latter should necessitate a conditional effects approach (see Section 6.2).

The R^2 value was only .806 with all nine fitted terms in the model, suggesting that the 81 observations might have been used more efficiently to look at additional factors, especially since one of the three linear effects and two of the three quadratic effects were significant. There were two huge standardized residuals (-5.12 and 5.13) on consecutive observations, but this was not noted in the paper. It is interesting that these observations are in opposite directions relative to their respective fitted values. A possible explanation is that the tooling process was badly out of control at those two observations. If so, then all the effect estimates are invalid and some of them might be off by a considerable margin.

A Pareto chart effect analysis was performed and the model was reduced to six terms, which was then analyzed to determine optimum operating conditions. Those conditions soon ceased to be optimum because variation in tip height and taper created problems with the process. This underscores the importance of maintaining processes in a state of statistical quality control, as was emphasized in Section 1.7 for this particular experiment.

After a state of statistical quality control has been reached and maintained, future experimentation might best be performed with fewer observations, which might not only greatly reduce the cost of experimentation, but the use of 81 observations could result in tests of significance that will detect significant effects that do not have practical significance. A better choice for a design might be either a response surface design (Chapter 10), a 3^{k-p} design if more factors are to be examined and all have three levels, or perhaps a mixed fractional factorial (Section 6.5) if differing numbers of levels are to be used for the factors. Indeed, Bisgaard (1997) in a nontechnical article expressed a preference for response surface designs over three-level designs for technological applications.

6.2   CONDITIONAL EFFECTS

We need to examine conditional effects for 3-level designs just as we do for 2-level designs. Assume that an experiment with a 3^2 design has been run and the results are

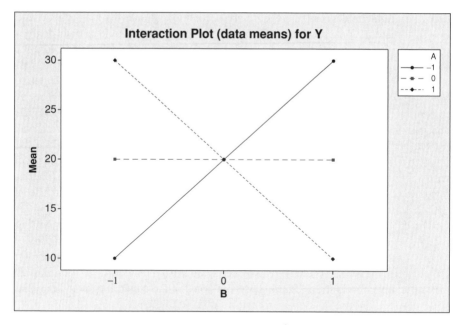

Figure 6.1 Interaction plot for 3^2 design.

as follows:

A	B	Y
-1	-1	10
-1	0	20
-1	1	30
0	-1	20
0	0	20
0	1	20
1	-1	30
1	0	20
1	1	10

The data are plotted in Figure 6.1.

If we computed A_L, A_Q, B_L, and B_Q, we would find that each one is zero. (Figure 6.1 is the counterpart to Fig. 4.2 that was for a 2^2 design.) Thus, if we ran these data through software that produced an ANOVA table, we would find that the sum of squares for each effect is zero.

Given below is the MINITAB output for these data (using the GLM command and using only A and B as the model terms). Even though the linear and quadratic sum of squares are not shown separately, in this case it is unnecessary because if the sum is zero, each of the two component parts must be zero.

```
General Linear Model: Y versus A, B

Factor   Type    Levels   Values
A        fixed    3       -1, 0, 1
B        fixed    3       -1, 0, 1

Analysis of Variance for Y, using Adjusted SS for Tests

Source   DF   Seq SS   Adj SS   Adj MS     F       P
A        2       0.0      0.0      0.0   0.00   1.000
B        2       0.0      0.0      0.0   0.00   1.000
Error    4     400.0    400.0    100.0
Total    8     400.0

S = 10    R-Sq = 0.00%    R-Sq(adj) = 0.00%
```

The $A*B$ interaction is not displayed in the table because it was not specified in the model. Since this is an unreplicated factorial, it is confounded with the error term.

It is clear from Figure 6.1 that A and B each have a strong linear effect when viewed in the context of conditional effects; the conditional main effects simply add to zero as one is -20 and the other is 20. (These numbers result from conditioning, in turn, on the low and the high levels of each factor.)

Of course this is the most extreme case, and the zero sums of squares, which correspond to the main effect estimates being zero, are a sizable red flag. The point, however, is that the use of conditional effects when factors have more than two levels should match the use of conditional effects for two-level designs.

In general, the need to consider conditional effects applies to *every* type of factorial design, and also to designs with blocking since there is no guarantee that the assumption of no block $\times$ treatments interaction in a randomized complete block design will be met, and if the assumption is not met, then the rank ordering of the treatments within a block might differ considerably over the blocks. Then the conditional effects, which might be defined as the deviations of treatment i within each block from the average response for the observations in that block, could differ considerably over the blocks.

6.3 3^{k-p} DESIGNS

Although a 3^k design generally requires too many runs, a 3^{k-p} design reduces that number by a factor of $1/3$—a sizable reduction. Nevertheless, historically these designs have not been viewed favorably, especially relative to other designs, covered in Chapter 10, for estimating second-order effects because the 3^{k-p} designs have been assumed to be relatively inefficient. In particular, Box and Wilson (1951) made this argument.

Recall from Section 5.9.1.4, however, that 2_{III}^{k-p} designs are used as screening designs (i.e., the focal point is identifying significant main effects). There is no reason why 3^{k-p} designs cannot be used for the same purpose in a two-stage procedure, with possible interaction effects investigated in the second stage, just as is done when a 2_{III}^{k-p} design is used in the first stage. This is essentially the point that was made by Cheng and Wu (2001), who pointed out, for example, that up to 13 factors could be studied with a 3^{k-p} design with 27 runs. This is apparent because each factor would have 2 df and $13(2) = 26$, the available number of degrees of freedom. Thus, the design is saturated relative to the estimation of single-factor effects. It should be noted that Cheng and Wu (2001) proposed a method that permits factor screening and model fitting in the same stage, thus reducing the two-stage procedure to a single stage. This is discussed further in Section 10.1.

As discussed in Section 6.7, statistical software generally does not provide the capability for generating many of the designs discussed in this chapter, especially a 3^{13-10}.

It is also worth noting that many 3^{k-p} designs will not have enough df for fitting the full second-order model, anyway. For example, there are eight df for estimating effects with a 3^{3-1} design, but since each two-factor interaction has four df, there are only enough df to estimate the linear and quadratic effects of each factor, with two df left. Thus, a full 3^3 factorial would have to be used to fit the full second-order model: 6 df for the main effects (linear and quadratic) and 12 df for the 3 two-factor interactions, and there would be 8 df for the three-factor interaction if that were fit.

A 3^{4-1} design will *not* work, however, as 8 df are needed to estimate the main effects and 24 df are needed for all of the two-factor interactions. A 3^{5-1} design will work, as 80 df far exceeds the 10 that are needed for main effects and the 40 that are needed for the two-factor interactions. A 3^{5-2} design will obviously not have enough degrees of freedom, however. Similarly, a 3^{6-1} design will work, as will a 3^{6-2}, but not a 3^{6-3}. It should be clear that as we let the number of factors increase, we will not be able to fit a full second-order model with a reasonable number of design points.

Of course designs with three-level factors are used when there are three levels of interest. Three levels don't necessarily have to be used at the screening stage, however, so we have to think about what the justification would be for doing so. The general belief is that when quadrature is present, it is generally not present in the absence of linearity, and indeed models with polynomial terms are built on this assumption.

Assume that we have a temperature factor and we are primarily interested in 300° and 500°F. If we elect to additionally use 400°F at the screening stage, we are implicitly assuming that if the factor has an effect, the quadrature will dominate the linearity, as would occur, for example, if the plot of the response values against temperature has a V-shape configuration.

On the other hand, if 300°F and 400°F are the primary levels of interest but 500°F is chosen for use as a "test for linearity," the three levels would be quite revealing if the response value for the middle temperature is slightly higher than that for the lowest temperature, but the response is much lower at 500°F (as in the plot of factor D in Figure 5.1 in Wu and Hamada, 2000). In this case the linear effect will dominate the quadratic effect, but we wouldn't have been able to detect the linear effect without the use of the third level.

So the justification for using a 3^{k-p} design as a screening design will hinge largely upon whether or not a possible quadratic effect might be expected to dominate a possible linear effect, and whether or not the third level is within the range of the two levels if a 2^{k-p} design had been used instead.

Certain concepts that were discussed in Chapter 5 for 2^{k-p} designs also apply to the 3^{k-p} series. Specifically, the resolution of a 3^{k-p} design is determined in the same way as for a 2^{k-p} design, and we may also speak of minimum aberration 3^{k-p} designs, which are discussed in Section 6.7.

Of course this implies that the 3^{k-p} designs have the same type of projective properties that the 2^{k-p} designs possess; for example, a 3^{4-1} design would project into a full 3^3 design in any combinations of three factors, regardless of whether $D = ABC$ or $D = ABC^2$, as the reader is asked to show in Exercise 6.10.

Projective properties for 3^{k-p} designs are somewhat complicated in general, however. The interested reader is referred to Theorem 1 of Cheng and Wu (2001). Part of that theorem states that if we start with a 3^{k-p} design and project it onto q factors, then the projected design is either a fractional factorial design or a 3^q design with $3^{k-p}/3^q$ replicates.

6.3.1 Understanding 3^{k-p} Designs

To try to simplify a complex subject as much as possible, let's consider a 3^{2-1} design. This would not be a practical design because with only two degrees of freedom only one of the main effects could be estimated, but we need to start with a very simple design because things can become complicated in a hurry. There are two degrees of freedom associated with the three possible fractions of the 3^2 design, so we need to confound either AB or AB^2 with the estimate of the mean. Since we must relinquish information on one of the two components, it would be nice to know when these will be approximately equal in magnitude. As stated in Section 6.1.1, they will be equal when only one of the four trend components of the interaction is nonzero. Since the interaction plots for these four scenarios have distinct forms, as the reader is asked to show in Exercise 6.14, the experimenters might provide input as to whether or not any of these scenarios would be likely.

Clearly, we don't want to confound a large component and then conclude from the estimable small component that the interaction is not significant, nor do we want to do the reverse and conclude that an interaction is significant when it may be borderline at best.

Assume for the sake of illustration that we elect to confound AB^2. Recall that for the 2^{k-p} series the arithmetic was performed modulo 2 because the designs had two levels. For the 3^{k-p} series the arithmetic is modulo 3. For that arithmetic, the AB^2 interaction is represented by $x_1 + 2x_2 = 0, 1$, and 2 for the three fractions, with x_1

representing the first factor and x_2 representing the second factor. Unlike the 2^{k-p} series, we cannot use the presence or absence of a factor to represent the level of a factor since there are three levels. Therefore, *for the sake of constructing the fractions*, the levels are denoted as 0, 1, and 2.

Thus, with AB^2 confounded with the estimate of the mean, the three fractions, which would be the three blocks if a 3^2 design were run in blocks of size 3, are

$$
\begin{array}{ccc}
00 & 02 & 01 \\
11 & 10 & 12 \\
22 & 21 & 20
\end{array}
$$

with the fractions corresponding to, in order, $x_1 + 2x_2 = 0$, 1, and 2. If AB had been confounded with the estimate of the mean, the fractions would have been

$$
\begin{array}{ccc}
00 & 01 & 02 \\
12 & 10 & 11 \\
21 & 22 & 20
\end{array}
$$

Since an exponent can be a 2, the reader may wonder why A^2B and A^2B^2 are not being considered. The algebraic representation of the first is $2x_1 + x_2$ and $2x_1 + 2x_2$ for the second. The reader can observe that the first expression will produce the same three fractions as was produced by the expression $x_1 + 2x_2$, and $2x_1 + 2x_2$ produces the same set of fractions as was produced by $x_1 + x_2$. Thus, there is a reason for not using expressions for which the exponent of the first factor is not a 1. Furthermore, the squaring of such expressions ceases to be a "black box operation" when we see what underlies such actions.

It is important to note that, as was implied earlier, we have information on one component of the A^*B interaction—the one that is not selected for the defining contrast. That is, we have lost half of the information about the interaction. When we have more than two factors and a $1/3$ fraction is used, we retain information on interaction components that are not confounded, and for the 3^{k-p} series we will have some information on components of interactions that are selected to be confounded with the mean.

There is an analogous interaction breakdown when there are three factors. Since the A^*B^*C interaction has $(2)(2)(2) = 8$ df, there are other components of the interaction, with the full set being ABC, ABC^2, AB^2C, and AB^2C^2, each of which has 2 df. Notice that the exponent of A is one, by convention, and the combinations of exponents of B and C are all possible combinations of each exponent being either 1 or 2.

6.3.2 Constructing 3^{k-p} Designs

The construction and understanding of these designs is complicated by the fact that the factors involved in interactions can have an exponent of 2. For example, as stated in Section 6.1, the A^*B interaction has four degrees of freedom, two each for AB and AB^2.

These designs are constructed analogous to the way 2^{k-p} designs are constructed. That is, a full factorial is constructed in $k - p$ factors, with one or more generators used to construct the additional column(s). For example, a 3^{4-1} design would be constructed by first constructing the 3^3 design and then defining the fourth factor, D, to be equal to one of the components of the A^*B^*C interaction, such as $D = ABC^2$. More specifically, the level of D in each experimental run would be determined as $D = x_1 + x_2 + 2x_3$ (mod 3), with x_1, x_2, and x_3 denoting the levels of A, B, and C, respectively in the 3^3 design, and "mod 3" indicating that modulo 3 arithmetic is performed, as stated previously.

Of course there are three possible fractions for a given choice of the effect to use in constructing the factor D, just as there are two possible fractions in the 2^{k-1} series. The three fractions are defined by $x_1 + x_2 + 2x_3 = 0$, 1, and 2, mod 3. Unlike the 2^{k-1} series, however, the use of, say, $D = ABC$ to generate the additional factor does not mean that the defining relation is $I = ABCD$.

This type of lack of correspondence will be explained shortly, but first we will apply the modular arithmetic approach to a 2^{4-1} design. Letting $D = ABC$ means $x_4 = x_1 + x_2 + x_3$, so $x_1 + x_2 + x_3 - x_4 = 0$. We need to remove the $-x_4$ because we need a positive coefficient for that term, which can be accomplished by adding $2x_4$ to each side of the equation. This is equivalent to adding zero to each side because $2x_4 = 0$ (mod 2), regardless of whether $x_4 = 0$ or 1. This produces $x_1 + x_2 + x_3 + x_4 = 0 = I$, so $I = ABCD$.

Modular arithmetic was not used in explaining the 2^{k-p} designs in Chapter 5 because it wasn't needed there. This can be done here, however, to serve as a bridge that may make certain aspects of the 3^{k-p} series easier to understand. Consider the simple case of a 2^{3-1} design constructed by first constructing the 2^2 design and then letting $C = AB$. Using 0 and 1 to denote the two levels since modular arithmetic is being used, the design is thus represented as

A	B	C
0	0	0
1	0	1
0	1	1
1	1	0

Note that $x_3 = x_1 + x_2$ and that $x_1 + x_2 + x_3 = 0$ (mod 2). (Note that if we wanted to translate this design back to the $(+1, -1)$ level designations, we would have to let $0 = 1$ and $1 = -1$. This substitution would produce the first of the two fractions in Figure 5.1 in Section 5.1, with the treatment combinations in the reverse order.

Returning to the 3^{3-1} and 3^{4-1} design explanations, for the former $C = AB$ so $x_3 = x_1 + x_2$ and $x_1 + x_2 - x_3 = 0$. Notice that in this instance adding $2x_3$ to each side of the equation won't work because, for example, $2(1) = 2$ (mod 2), not zero. Therefore, we must add $3x_3$ to each side of the equation. This works because $3(0) = 3(1) = 3(2) = 0$ (mod 2). Therefore, $x_1 + x_2 + 2x_3 = 0$, so $I = ABC^2$, even though $C = AB$. The design is thus as follows:

A	B	C
0	0	0
0	1	1
0	2	2
1	0	1
1	1	2
1	2	0
2	0	2
2	1	0
2	2	1

Notice that $ABC^2 = 0(= I)$. We have thus illustrated that we cannot obtain the defining relation directly from the expression for the additional factor for three-level designs. The 3^{4-1} design can be similarly constructed, as the reader is asked to show in Exercise 6.22.

6.3.3 Alias Structure

The alias structure for a 3^{k-1} design is determined differently from the way it is determined for the 2^{k-1} series, and similarly for 3^{k-p} designs in general. Specifically, for a 3^{k-1} design the alias structure is determined by multiplying each effect by the effect that is confounded with the mean and the square of the effect that is confounded with the mean.

Therefore, for the 3^{2-1} design in the current example, if we use AB^2 as the effect that is confounded with the mean and pick one of those three fractions (preferably at random), we know that A, B, and AB must share the two degrees of freedom for estimating effects, and thus be confounded. Thus, $A(AB^2) = A^2B^2 = AB$ and $A(AB^2)^2 = B^4 = B$, so that $A = AB = B$. Similarly, $B(AB^2) = A$ and $B(AB^2)^2 = A^2B^2 = AB$, so that $B = A = AB$, as of course determined with the first set of multiplications.

Again, this is not a practical design; the intent was only to illustrate the construction of a 3^{k-p} design with the smallest number of factors possible for simplicity.

As an aside, it is worth noting the expression given by Wu and Hamada (2000, p. 216) for the maximum number of factors that can be studied with a 3^{k-p} design. The expression is $(3^{k-p} - 1)/2$, which does give 1 for the previous example as there are two degrees of freedom for estimating effects, and A and B each have two degrees of freedom. The way this result should be viewed, so that it doesn't seem nonsensical since k appears to be fixed in $k - p$, is to view $k - p$ as the fixed quantity, not k.

6.3.4 Constructing a 3^{3-1} Design

As a second example, we consider the construction of a 3^{3-1} design for which we know that we *could* estimate four factor main effects with that number of design points, but of course we have only three factors. With three factors, we would logically select one of the four components of the A^*B^*C interaction (ABC, AB^2C, ABC^2, and

AB^2C^2) to confound with the estimate of the mean. Note that with four choices for the interaction and three choices for the fraction since this is a $1/3$ fraction, there are 12 fractions from which one is selected.

Assume that ABC is selected. Then obviously each factor effect will be confounded with a two-factor interaction component and part of the three-factor interaction component since, for example, $A(ABC) = AB^2C^2$ and $A(ABC)^2 = B^2C^2 = BC$.

Of course, there are eight degrees of freedom and A, B, and C account for six of them. The other two degrees of freedom are represented by the alias string $AB^2 = AC^2 = BC^2$, as the reader is asked to show in Exercise 6.3.

Giesbrecht and Gumpertz (2004, p. 384) comment on the 3^{3-1} design, stating "We must warn the reader that while this fractional plan using 9 of 27 treatment combinations has proved to be quite popular in some circles, it is not really a good experimental plan. The problem is that the main effects are all aliased with two-factor interactions." Although the warning does seem to be appropriate, the statement is somewhat misleading. The problem here is somewhat analogous to running a 3^2 design in blocks of size 3. If we confound AB with the block differences, we still have information on AB^2. Specifically, the eight df are broken down as blocks (2), A(2), B(2), and AB^2(2).

For the 3^{3-1} design with ABC confounded, the BC component of the B^*C interaction might be much smaller than the BC^2 component, in which case the fact that A is confounded with BC might not be a major problem, and similarly for the other two factors. Although there are at present no theoretical results to provide the basis for a strategic approach, if the sum of squares associated with an alias string of two-factor interaction components is large, it might be desirable to look at the trend components of the interactions and look at the relative magnitude of those components. There is a need for research in this area, but at present it would be best to be cautious and heed the warning of Giesbrecht and Gumpertz (2004) while recognizing that the aliasing of main effects with two-factor interaction components won't always be a problem.

The construction of the design can proceed analogous to the way that a 2^{k-p} design is constructed; that is, a 3^2 design could first be constructed with the additional factor created as $C = AB$ or $C = AB^2$. Assume that the first option is used. If this were to be a 2^{3-1} design, then $I = ABC$. This does not apply to the 3^{k-p} series, however, as was illustrated.

6.3.5 Need for Mixed Number of Levels

Of course one problem with the 3^{k-p} designs is the large jumps in run size for the various plans, from 9 to 27 to 81 and beyond. Eighty-one runs will be viewed by most experimenters as too many runs, yet the tables of Wu and Hamada (2000, pp. 250–251) show that there is only one 27-run 3^{k-p} design, the 3^{4-1}, that is resolution IV. (See Chen, Sun, and Wu (1993) for the complete enumeration of 27-run 3^{k-p} designs, upon which Table 5a.2 in Wu and Hamada (2000) is based. This work has been extended by Xu (2005), who gives a large number of 3^{k-p} designs for up to 729 runs.)

There shouldn't be a need, however, to use three levels for every factor when there is at least a moderate number of factors, as some factors may be qualitative with only two levels of interest. There may also be a strong belief that a given quantitative factor has a linear effect over the range of levels that are of interest and are considered feasible. Therefore, a fraction of a design with a mixture of the number of levels for the factors will often be the appropriate choice. Such designs are discussed in Section 6.5.

6.3.6 Replication of 3^{k-p} Designs?

While we normally don't think about replicating a 2^{k-p} design, although an example of one was given in Section 5.13, replicating a 3^{k-p} design is actually not a bad idea. One reason for doing so is the lack of a normal probability plot capability; of course that problem can be avoided simply by replicating the design. Although this is probably the most compelling reason for the replication, we should note that we have partial information on interactions whenever a 3^{k-p} design is used since not all components are confounded, as stated previously. If we are fortunate enough to have the "right" (i.e., largest) components unconfounded, then by using replication we obtain more precise estimates of those components. The downside is that the number of runs is of course $c(3^{k-p})$, with c denoting the number of replicates. This number will grow rapidly with c if 3^{k-p} is not small.

6.4 MIXED FACTORIALS

Frequently there is interest in using designs for which the number of levels for the factors is not the same, and as noted in Section 6.3.3, the number of runs with a 3^{k-p} design will be large when $k - p$ is not very small. A design with differing numbers of levels of the factors is called a *mixed factorial*.

A mixed (full) factorial (also called an asymmetrical factorial) in two factors is of the general form $a^{k_1} b^{k_2}$, with k_1 and k_2 greater than or equal to 1, and $2 \leq a < b$. Mixed factorial designs have been discussed, in particular, by Addelman (1962) and in Chapter 18 of Kempthorne (1973), although the treatment in these sources is somewhat mathematical. A more recent source, albeit with somewhat limited treatment, is Giesbrecht and Gumpertz (2004). Hinkelmann and Kempthorne (2005) devote a chapter to confounding in mixed factorial designs, which they term asymmetrical factorial designs.

Example 6.3

The simplest example of a mixed factorial is a 2×3 design. Sahin and Örnderci (2002) investigated the effect of Vitamin C (two levels) and chromium picolinate (three levels) and concluded that the high level of Vitamin C (250 mg) combined with either the high or middle level of chromium "can positively influence the performance of laying eggs reared under low ambient temperature." There were 180 laying hens, all 32 weeks old, which were used in the experiment, and they were divided into six groups of 30 hens each. This provides a considerable amount of power for detecting effects—which would be too much power in a typical application. In many if not most applications, it would be impractical, impossible, or simply too costly to obtain this many observations for each treatment combination. There were actually three response variables, so this example is revisited in Exercise 12.10 of Chapter 12. Experiments can be run inexpensively in many fields of application, however, and in such cases a full mixed factorial may be affordable.

In this application, 48 design points could be afforded, and in fact all of the runs could be made in one day. One problem with mixed factorials is that in many, if not most, applications, if k_2 in the general expression for a mixed factorial given earlier is at least 2, the number of design points may be prohibitive. We can use a fractional mixed factorial design, but the design won't necessarily be orthogonal.

Consider the 3×2^4 design and assume that only 24 runs can be afforded. The necessary fractionation must obviously be in regard to the two-level design in order to produce half the original points, and also because one cannot fractionate on a single 3-level factor, as multiple 3-level factors would have to be involved. Clearly, a $3 \times 2^{4-1}$ design would produce the desired number of points and would be an orthogonal design. If, however, something other than 6, 12, or 24 points were desired, the design would not be orthogonal. This becomes clear if we recognize that a design with any other number of points would not be of the form $3 \times 2^{4-p}$. For example, a design with 14 points wouldn't work because 14 is not a multiple of 3. Similarly, 15 would not work because $2^{4-p} \neq 5$, for any integral value of p, and so on.

We encounter the same type of analysis problem with mixed factorials as we do with three-level designs in that we can't construct a normal probability plot for a mixed factorial with commonly used statistical software.

6.4.1 Constructing Mixed Factorials

A standard way of constructing mixed factorials is by using the method of collapsing levels, due to Addelman (1962), such as collapsing a 3-level factor into a 2-level factor. Giesbrecht and Gumpertz (2004) discuss how a $2^2 3^2$ design would be constructed by using this approach, which starts with a 3^4 design and ends with a $2^2 3^2$ design with 81 runs, so that some of the runs are repeated. Of course the "extra" runs would have to be extracted if the experimenter wanted to use only 36 runs.

For a relatively small mixed factorial such as this, we could simply construct a 3^2 design and then construct a 2^2 design at each point in the 3^2 design. Of course the easiest approach would be to use software. The following $2^2 3^2$ design was created using the "full factorial" DOE option in JMP, with the runs not randomized. It is easy to see the 2^2 design at each point in the 3^2 design.

```
11--   1   1   -1   -1
11-+   1   1   -1    1
11+-   1   1    1   -1
11++   1   1    1    1
12--   1   2   -1   -1
12-+   1   2   -1    1
12+-   1   2    1   -1
12++   1   2    1    1
```

```
13--   1   3   -1   -1
13-+   1   3   -1    1
13+-   1   3    1   -1
13++   1   3    1    1
21--   2   1   -1   -1
21-+   2   1   -1    1
21+-   2   1    1   -1
21++   2   1    1    1
22--   2   2   -1   -1
22-+   2   2   -1    1
22+-   2   2    1   -1
22++   2   2    1    1
23--   2   3   -1   -1
23-+   2   3   -1    1
23+-   2   3    1   -1
23++   2   3    1    1
31--   3   1   -1   -1
31-+   3   1   -1    1
31+-   3   1    1   -1
31++   3   1    1    1
32--   3   2   -1   -1
32-+   3   2   -1    1
32+-   3   2    1   -1
32++   3   2    1    1
33--   3   3   -1   -1
33-+   3   3   -1    1
33+-   3   3    1   -1
33++   3   3    1    1
```

6.4.2 Additional Examples

A $3^2 2^3$ design was used in one of the two experiments described by Inman, Ledolter, Lenth, and Niemi (1992). The experiment involved a Baird spectrometer. Control charts of spectrometer readings had exhibited instability, so an experiment was conducted to try to identify the causes of the instability. Spectrometers must be frequently calibrated, especially when there is evidence of instability, but if the causes of the instability could be identified and removed, many of the time-consuming recalibrations could be eliminated.

It is worth noting in this experiment that the response variable is the same variable that is out of control. This is different from the situation where factors that are extraneous to the experiment are out of control and might be affecting the measurements of the response variable. When the response variable and the out-of-control variable are the same, there is the possibility of the results of the experiment being misleading if the factors that affect the response variable have not been identified. For example, what if the experiment is performed and the R^2 value is only .62? This means that 38 percent of the variability in the response measurements has not been explained. The question then must be addressed as to whether this is common cause variability or is there (unacceptable) variability due to assignable causes that have not been accounted for by the factors in the experiment. In general, if a model is wrong, the expected value of the effect estimates will generally not be equal to the effect that we are trying to estimate.

It is possible, although unlikely, that variability due to extraneous factors that are out of control could even go completely undetected. For example, process shifts could conceivably exactly coincide with changes in factor levels, so that changes in the values of the response variable could be erroneously attributed to changes in the levels of the factors in the experiment. Of course this is very unlikely to occur exactly, but something approximating this scenario could occur and ruin an experiment.

There were five factors in the experiment under discussion: room temperature of the laboratory (three levels), sharpness of the counterelectrode tip (two levels), the boron nitride disk (two levels—new and used), cleanliness of the entrance window (clean and in use for a week), and placement of the sample with respect to the boron nitride disk.

The 72 runs in the experiment were blocked on temperature, with temperature changes made 8 hours before the experimental runs were made at each level of temperature. The 24 runs made within each block were completely randomized. It was necessary to block on temperature because it was not possible to change temperature between individual runs; that is, temperature was a "hard-to-change" (actually, impossible to change) factor, which occurs much more frequently than is probably realized.

Wang (1999) gave an application of a 2×3^7 design in 18 runs to Poisson data. This was a blackening experiment conducted in an electric company using a three-layer oven. Thirty masks from each layer in the oven were collected for examining the number of defects in each mask. The response variable was the number of defects summed over the 30 masks in each layer and Wang (1999) gave the data for the upper layer.

It would be reasonable to assume a Poisson distribution for the defects, although all assumptions should be tested. Using the average number of defects per mask would be one possible approach, assuming that the average is approximately normally distributed. This would not work very well, however, if the probability of a defect was extremely small. Another approach would be to transform the data, such as by using some form of a square root transformation.

Example 6.4 — Case Study

We consider the second experiment described by Sheesley (1985) as this has some interesting features. The experiment involved lead wires for incandescent lamps. The objective was to determine if lead wires produced by a new process would perform in a superior way, relative to the current process, during the lamp-making process in terms of feeding into the automatic equipment.

The experiment involved five factors as Sheesley (1985) considered replications to be the fifth factor. We don't normally consider replications to be a factor, but in this case that does seem appropriate because the replications were made over days,

TABLE 6.1 Lead Wire Experiment Data

Lead Type	Plant	Machine	Shift		Replications		
C	A	S	1	24.6	38.5	34.1	16.4
C	A	S	2	35.8	20.1	44.1	47.1
C	A	S	3	35.3	18.1	15.3	44.2
C	A	H	1	24.4	29.2	39.8	28.2
C	A	H	2	17.0	33.5	22.3	29.2
C	A	H	3	35.2	25.4	24.2	19.0
C	B	S	1	17.6	19.2	21.4	22.7
C	B	S	2	18.3	18.0	19.9	23.2
C	B	S	3	10.8	39.4	23.7	19.6
C	B	H	1	40.5	37.8	25.1	49.4
C	B	H	2	18.9	24.7	19.6	24.9
C	B	H	3	6.5	16.6	67.8	30.3
T	A	S	1	17.0	20.8	26.0	23.7
T	A	S	2	11.8	22.0	22.4	23.5
T	A	S	3	21.0	52.4	20.6	12.4
T	A	H	1	35.5	13.1	21.7	30.1
T	A	H	2	12.1	30.5	21.3	22.1
T	A	H	3	17.4	23.4	28.8	18.0
T	B	S	1	12.4	7.8	25.6	11.2
T	B	S	2	28.1	16.7	23.7	21.5
T	B	S	3	11.5	16.8	26.9	18.9
T	B	H	1	6.7	18.6	12.7	13.1
T	B	H	2	8.7	19.0	34.4	27.7
T	B	H	3	12.7	9.1	31.2	17.7

C and T denote the control and test lead types; A and B are the two plant types; S and T are the
standard and high-speed machines, and 1, 2, 3 denote the three shifts, with the replications listed
in order within the treatment combinations for the other four factors.

so we might view replications as being a proxy for days. The other four factors were
shift (three different shifts), lead type (current process, new process), plant (A, B),
and machine (standard, high speed). Thus, this is a $2^3 \times 3 \times 4$ design, with all factors
fixed except replications. The response variable was the average number of leads
missed per running hour. The data, 96 observations, are given in Table 6.1.

Of course the first step that we would take in analyzing such data would be to
look for bad data since this occurs with almost all experiments. A dotplot of the data
reveals one extreme data point, 67.8, which is probably a bad data point since this
value is considerably greater than the sum of the other three replications in the 12th
row in Table 6.1. Sheesley (1985) did not do any preliminary data analysis, so there is
no discussion of this data point. Consequently, we will keep an eye on the influence
of the data point during the analyses and, if necessary, suggest corrective action, that
being the best that we can do at this point in time. (The 52.4 in the 15th row of
Table 6.1 also looks somewhat suspicious since it is only slightly less than the sum
of the other three numbers in that row.)

The output using the GLM command in MINITAB is given below, with the fitted model containing the largest three-factor interaction but no other three-factor interaction. (In the expected mean squares section, Q [number] denotes the fixed effect of the corresponding numbered term.)

```
General Linear Model: Y versus T, P, M, S, R

Factor   Type      Levels     Values
T        fixed        2       C, T
P        fixed        2       A, B
M        fixed        2       H, S
S        fixed        3       1, 2, 3
R        random       4       1, 2, 3, 4
```

Analysis of Variance for Y, using Adjusted SS for Tests

Source	DF	Seq SS	Adj SS	Adj MS	F	P
T	1	1201.3	1201.3	1201.3	47.70	0.006
P	1	404.3	404.3	404.3	4.03	0.138
M	1	32.7	32.7	32.7	1.06	0.378
S	2	1.9	1.9	1.0	0.02	0.977
R	3	639.8	639.8	213.3	**	
T*P	1	7.5	7.5	7.5	0.07	0.796
T*M	1	48.5	48.5	48.5	0.44	0.511
T*S	2	172.4	172.4	86.2	0.78	0.464
T*R	3	75.6	75.6	25.2	0.23	0.877
P*M	1	208.9	208.9	208.9	1.88	0.175
P*S	2	14.2	14.2	7.1	0.06	0.938
P*R	3	300.9	300.9	100.3	0.90	0.445
M*S	2	227.0	227.0	113.5	1.02	0.366
M*R	3	92.2	92.2	30.7	0.28	0.842
S*R	6	251.6	251.6	41.9	0.38	0.890
T*P*M	1	291.9	291.9	291.9	2.63	0.110
Error	62	6880.8	6880.8	111.0		
Total	95	10851.3				

** Denominator of F-test is zero.

S = 10.5347   R-Sq = 36.59%   R-Sq(adj) = 2.84%

Unusual Observations for Y

Obs	Y	Fit	SE Fit	Residual	St Resid
33	39.4000	21.8333	6.2694	17.5667	2.07 R
39	52.4000	27.6708	6.2694	24.7292	2.92 R
60	67.8000	38.0750	6.2694	29.7250	3.51 R

R denotes an observation with a large standardized residual.

Expected Mean Squares, using Adjusted SS

```
  Source Expected Mean Square for Each Term

1   T          (17) + 12.0000 (9) + Q[1, 6 , 7 , 8 , 16]
2   P          (17) + 12.0000 (12) + Q[2, 6 , 10 , 11 , 16]
```

```
 3    M         (17) + 12.0000 (14) + Q[3, 7 , 10 , 13 , 16]
 4    S         (17) +  8.0000 (15) + Q[4, 8 , 11 , 13]
 5    R         (17) +  8.0000 (15)+ 12.0000 (14) + 12.0000 (12)
                     + 12.0000 (9)+ 24.0000 (5)
 6    T*P       (17) + Q[6, 16]
 7    T*M       (17) + Q[7, 16]
 8    T*S       (17) + Q[8]
 9    T*R       (17) + 12.0000 (9)
10    P*M       (17) + Q[10, 16]
11    P*S       (17) + Q[11]
12    P*R       (17) + 12.0000 (12)
13    M*S       (17) + Q[13]
14    M*R       (17) + 12.0000 (14)
15    S*R       (17) +  8.0000 (15)
16    T*P*M     (17) + Q[16]
17    Error     (17)
```

Error Terms for Tests, using Adjusted SS

```
Source Error DF Error MS Synthesis of Error MS
 1   T        3.00     25.2   (9)
 2   P        3.00    100.3   (12)
 3   M        3.00     30.7   (14)
 4   S        6.00     41.9   (15)
 5   R        3.05      *     (9) + (12) + (14)
                             + (15) - 3.0000 (17)

 6   T*P     62.00    111.0   (17)
 7   T*M     62.00    111.0   (17)
 8   T*S     62.00    111.0   (17)
 9   T*R     62.00    111.0   (17)
10   P*M     62.00    111.0   (17)
11   P*S     62.00    111.0   (17)
12   P*R     62.00    111.0   (17)
13   M*S     62.00    111.0   (17)
14   M*R     62.00    111.0   (17)
15   S*R     62.00    111.0   (17)
16   T*P*M   62.00    111.0   (17)
```

Variance Components, using Adjusted SS

```
           Estimated
Source     Value
R           14.502
T*R         -7.150
P*R         -0.891
M*R         -6.687
S*R         -8.631
Error      110.981
```

There are some important observations that should be made from these results:

(1) The Lead Type (T) factor is the only one that is close to being significant, and it is highly significant.

(2) The denominator of the F-test for testing the significance of the Replications factor is given as zero. (This will be explained shortly.)

(3) The R^2 for the model is low since Lead Type is the only significant factor.

(4) Four estimated variance components are negative. A variance component cannot, by definition, be negative. Special estimation methods must be employed to prevent negative estimates, which will otherwise frequently occur when effects are far from being significant.

(5) The two suspicious observations that were mentioned previously have very large standardized residuals, 3.51 and 2.92. Of course a data point is an outlier relative to the model that is used and here the model has many non-significant terms. The standardized residuals are even larger, however (4.03 and 3.20, respectively), when the model contains only Lead Type. Action would normally be taken regarding such extreme points. In this case, however, there are 96 data points so one or two extreme points are not likely to have much effect, especially since none of the effects are close to the dividing line between significance and non-significance, although two of the three-factor interactions are larger than we might prefer.

The message from the computer output that the denominator degrees of freedom is zero for testing the Replications factor undoubtedly looks strange. This can be explained as follows. There are not always exact tests for effects; oftentimes a pseudo-error term must be created by using Satterthwaite's (1946) procedure, which provides an approximate test rather than one based on distribution theory, and sometimes strange things can happen when this is done.

Satterthwaite's procedure uses linear combinations of mean squares. To understand the mechanics, consider the discussion of expected mean squares (EMSs) for a 2^2 design in Appendix C of Chapter 4. It was stated that $E(\mathrm{MS}_{AB}) = \sigma^2 + 2\sigma_{AB}^2$ and $E(\mathrm{MS_E}) = \sigma^2$. To test the hypothesis that $\sigma_{AB}^2 = 0$, we thus construct an F-test as $F = \mathrm{MS}_{AB}/\mathrm{MS_E}$, so that if $\sigma_{AB}^2 = 0$, the numerator and denominator are estimating the same thing. In general, the expected value of the numerator minus the expected value of the denominator should be equal to a multiple of what is being tested. Now assume that we have a design for which it is not possible to accomplish this using one mean square in the numerator and one in the denominator, but rather a linear combination is necessary for both the numerator and the denominator. These linear combinations should be constructed using the guiding principle that the expected value of the numerator minus the expected value of the denominator is equal to a multiple of what is to be tested, as stated.

It will generally give a fractional degrees of freedom, which in this case is 0.49, and is treated as zero by MINITAB. PROC GLM in SAS also uses Satterthwaite's procedure.

Since there is obviously no shift effect and the Replications factor is not significant, we would be tempted to analyze the other three factors as having come from a 2^3 design. Although such ad hocery is not recommended, in general, in this case the change in the conclusion is that the plant effect is borderline significant with such an analysis.

We will illustrate Satterthwaite's procedure by using an example that is simpler than the one just used. Specifically, one of the sample datasets that comes with MINITAB is EXH_AOV, which contains sets of columns of data of unequal length that are used in the ANOVA online help. Four of those columns constitute a useful example for illustrating Satterthwaite's procedure, however. There are two factors at three levels (one fixed and one random) and one fixed factor at two levels. So it is a 2×3^2 design with Thickness being the response variable and Time, Operator, and Setting the factors. The MINITAB output is given below.

```
ANOVA: Thickness versus Time, Operator, Setting

Factor      Type      Levels   Values
Time        fixed     2        1, 2
Operator    random    3        1, 2, 3
Setting     fixed     3        35, 44, 52

Analysis of Variance for Thickness

Source                  DF       SS        MS        F      P
Time                    1        9.0       9.0       0.29   0.644
Operator                2        1120.9    560.4     4.91   0.090 x
Setting                 2        15676.4   7838.2    73.18  0.001
Time*Operator           2        62.0      31.0      1.29   0.369
Time*Setting            2        114.5     57.3      2.39   0.208
Operator*Setting        4        428.4     107.1     4.46   0.088
Time*Operator*Setting   4        96.0      24.0      7.08   0.001
Error                   18       61.0      3.4
Total                   35       17568.2

x Not an exact F-test.

S = 1.84089  R-Sq = 99.65%  R-Sq(adj) = 99.32%

                               Variance    Error   Expected Mean Square for Each
   Source                      component    term   Term (using unrestricted model)
1  Time                                     4      (8) + 2 (7) + 6 (4) + Q [1,5]
2  Operator                    37.194       *      (8) + 2 (7) + 4 (6) + 6 (4)+12(2)
3  Setting                                  6      (8) + 2 (7) + 4 (6) + Q[3,5]
4  Time*Operator               1.167        7      (8) + 2 (7) + 6 (4)
5  Time*Setting                             7      (8) + 2 (7) + Q[5]
6  Operator*Setting            20.778       7      (8) + 2 (7) + 4 (6)
7  Time*Operator*Setting       10.306       8      (8) + 2 (7)
8  Error                       3.389               (8)

*Synthesized Test.

Error Terms for Synthesized Tests
                                      Synthesis of
Source          Error DF   Error MS   Error MS
2 Operator      3.73       114.1      (4) + (6) - (7)
```

Consider the EMSs. Notice that for each effect except the second one (i.e., Operator), it is possible to find an effect whose EMS is a subset of the EMS of the effect that is to be tested, with the only difference being a term in the effect that is being tested. For example, we can see that Time should be tested against Time*Operator; the latter should be tested against Time*Operator*Setting, and so on. We need to construct a denominator that contains every term in EMS(Operator) except $\sigma^2_{operator}$. We see that we can accomplish this by using Time*Operator + Operator*Setting − Time*Operator*Setting. Thus, this combination of mean squares forms the denominator for testing the Operator effect, realizing that this is just an approximate test.

Example 6.5

One of the sample datasets, PANCAKE.MTW, which comes with the MINITAB software is data from a 2×4 experiment with three replications conducted to study the effect of two factors on the quality of pancakes. The two factors are supplement (present or absent) and four levels of whey. Obviously, the first factor is fixed and we will also assume that the second factor is fixed. Three experts were asked to rate the quality of the pancake and the average of the three ratings was used as the value of the response variable. This process was performed three times for each treatment combination, providing a total of 24 observations.

The analysis of the data is as follows

```
ANOVA: Quality versus Supplement, Whey

Factor       Type    Levels   Values
Supplement   fixed     2      1, 2
Whey         fixed     4      0, 10, 20, 30

Analysis of Variance for Quality

Source        DF        SS    MS          F        P
Supplement     1    0.5104    0.5104     2.31    0.145
Whey           3    6.6912    2.2304    10.08    0.000
Error         19    4.2046    0.2213
Total         23   11.4062

S = 0.470419    R-Sq = 63.14%    R-Sq(adj) = 55.38%
```

The four levels of the whey factor permit the fitting of second- and higher-order effects, but the plot of the response variable against whey in Figure 6.2 shows no evidence of a nonlinear effect.

Although the scatterplot does not show any evidence of a nonlinear relationship between the response variable and whey, the clear separation that results in two distinct groups of three values at the first two values of whey is something for which an explanation should be sought. This separation tends to reduce the R^2 value, which is low partly because of the separation and partly because the supplement factor is not significant.

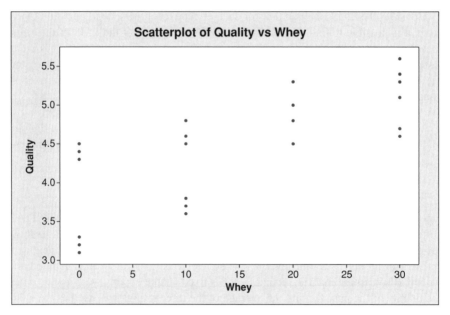

Figure 6.2   Scatterplot of the pancake experiment data.

6.5   MIXED FRACTIONAL FACTORIALS

Because of the number of runs required for a mixed full factorial, a mixed fractional factorial is often used. For example, a $2^{3-1}3^{3-1}$ design was used in the case study of Hale-Bennett and Lin (1997). (We won't analyze the data from that study here because the individual data values were not given.)

Such a design can of course be viewed analogous to the way that a $2^2 3^2$ is viewed; that is, the $2^{3-1}3^{3-1}$ design can be viewed as a 2^{3-1} design at each point of a 3^{3-1} design (or the reverse). Notice that 216 design points would be required if fractionation were not used, whereas only 36 points are needed for the mixed fractional factorial.

Although tables and software are of course available for constructing such designs, the $2^{3-1}3^{3-1}$ design could be easily constructed by hand. First, a 3^{3-1} design would be constructed by confounding one component of the $A*B*C$ interaction. Of course we know that main effects are confounded with two-factor interactions in the 2^{3-1} design since $I = ABC$.

When we concatenate the two designs we can think of the combined design as a replicated 3^{3-1} design and as a replicated 2^{3-1} design. Replication has no effect on estimability, so the estimability for the separate designs is the same as that for the combined design, and the alias structure cannot be more palatable than it is for the separate designs.

The bottom line is that 36 design points provide for the estimation of the six main effects plus certain two-factor interactions, and the main effects are estimable only if two-factor interactions for the two-level factors and the appropriate two-factor interaction components for the three-level factors do not exist. In general, fractions of three-level designs and designs that include fractions of three-level designs are not particularly good designs. (A related discussion of fractions of the three-level factorials is given by Giesbrecht and Gumpertz (2004, pp. 384–385)).

One problem with designs like the $2^{3-1}3^{3-1}$ design is that the interaction components eat up multiple degrees of freedom (as contrasted with two-level designs for which each interaction has one degree of freedom). When these degrees of freedom are added, the total and hence the run size become at least moderate in size.

In addition to the case study of Hale-Bennett and Lin (1997), a detailed analysis of the design and resultant data are given by Wu and Hamada (2000, pp. 271–278), and the reader is referred to these sources for further details.

Designs with differing numbers of levels are generally presented in the literature as orthogonal arrays (OAs), and many tables of OAs are available. This of course encourages the blind use of these designs, which is certainly undesirable, although the alias structure for incomplete factorial designs with more than two levels is generally complex. Consequently, most practitioners would probably be better off avoiding it, but not completely avoiding an understanding of OAs and their strengths and weaknesses.

Some tables of 4^m2^{k-p} designs are given at http://iems.northwestern.edu/~bea/articles/Appendix.PDF.

6.6 ORTHOGONAL ARRAYS WITH MIXED LEVELS

It was mentioned at the beginning of Chapter 5 that fractional factorial designs are OAs, but not all OAs are fractional factorials. Specifically, not all OAs are regular designs, meaning that not all effects to be estimated can be estimated independently.

Xu, Cheng, and Wu (2004) give an application of an OA with mixed levels that is not a regular design, using experimental data originally given by King and Allen (1987). The objective of the experiment described in the latter had radio frequency choke as the response variable and the objectives were to identify the factors that affected the choke readings and to identify the best settings of those factors. (Recall the definitions of a "regular design" and a nonregular design given at the start of Chapter 5. For each pair of factorial effects in a regular design, the effects are either independent or completely aliased. Designs for which this condition is not met are nonregular designs.)

There was one 2-level factor and seven 3-level factors in 18 runs, with each run replicated twice. The design is given below. (*Note*: This design can be generated with various statistical software packages. MINITAB was used to create this design, creating it as an L18 Taguchi design. Such designs are discussed in Chapter 8.)

A	B	C	D	E	F	G	H
0	0	0	0	0	0	0	0
0	0	1	1	1	1	1	1
0	0	2	2	2	2	2	2
0	1	0	0	1	1	2	2
0	1	1	1	2	2	0	0
0	1	2	2	0	0	1	1
0	2	0	1	0	2	1	2
0	2	1	2	1	0	2	0
0	2	2	0	2	1	0	1
1	0	0	2	2	1	1	0
1	0	1	0	0	2	2	1
1	0	2	1	1	0	0	2
1	1	0	1	2	0	2	1
1	1	1	2	0	1	0	2
1	1	2	0	1	2	1	0
1	2	0	2	1	2	0	1
1	2	1	0	2	0	1	2
1	2	2	1	0	1	2	0

How do we know this could not be a $2 \times 3^{6-4}$ design and thus be a regular design? In order for this to be true, a 3^{6-4} design in the 6 three-level factors would have to be crossed with the two levels of the other factor, which means that there would have to be replicate rows for those six factors. We can see that there are no such replicate rows, so it could not be this design. Since there is no other way to form 18 runs with one 2-level factor and six 3-level factors using a factorial/fractional factorial representation, the design cannot be a mixed fractional factorial or a combination of a factor with a fractional factorial.

We can also verify that the design is not a regular design by computing correlations between effect estimates. The 17 degrees of freedom for estimating effects could be used to estimate the linear effect of the two-level factor and the linear and quadratic effects of the three-level factors, with four df left, which could be used to estimate one two-factor interaction involving three-level factors. Or, interactions could be estimated among the factors in lieu of estimating quadratic effects. The correlation between C and the GH interaction is .50 and there are many other correlations between 0 and 1 since this is a nonregular design.

Since interactions are not orthogonal to main effects with this design but interaction terms might be more likely to be real than pure quadratic effects for these data, the question arises as to how to proceed. As when a supersaturated design (Section 13.4.2) is used, a logical strategy is to use some variable selection approach such as stepwise regression to identify what seem to be the most important effects and then estimate those effects.

Xu et al. (2004) used an alternative two-stage approach, however, first performing only a main effects analysis and identifying four factors, B, E, G, and H, as being significant at the .01 level. At this point, stepwise regression might have been applied to all possible terms in those four factors, with the available degrees of freedom used to estimate the effects that were selected.

Xu et al. (2004) did not use that approach, however, but rather fit the full second-order model in the four factors, although there would not be enough degrees of freedom to do that if an ANOVA approach were taken. There *are* enough degrees of freedom, however, if only the linear × linear component of each interaction is estimated. Significant effects were then identified through the use of t-tests and the effects that were identified were $B, E, G, H, BE, EG, EH, GH$, and E^2.

This approach is quite similar to the two-stage approach that is used in response surface methodology (Chapter 10) for model identification, except that in the latter two designs (i.e., two experiments) are used instead of one. Cheng and Wu (2001) argued that their two-stage analysis is a useful alternative to this response surface methodology approach. This is discussed further in Section 10.1.

There are potential problems with either the stepwise regression or the t-test approach, however, as stepwise regression can be undermined by correlations between variables, as discussed in Section 13.4.2, and correlations can also cause t-test results to be misleading.

Since first-order effects generally dominate second-order effects in the same factors, the Cheng and Wu strategy may be expected to work well, in general. There could be problems, however, when main effects are moderately correlated with two-factor interactions. With this example, the GH interaction was found by Xu et al. (2004) to be significant at the .01 level. The correlation between that effect estimate and the C effect estimate is 0.5, however, which raises the question of how much of the GH interaction is due to the C effect. We cannot answer questions of this type without running another experiment unless we employ methods such as those of Chevan and Sutherland (1991) to obtain a measure of the "independent effects" of effects that are correlated.

The data for this experiment are given in Exercise 6.25 and the reader is asked to analyze it there. An algorithm for constructing orthogonal and nearly orthogonal arrays with mixed levels was given by Xu (2002).

6.7 MINIMUM ABERRATION DESIGNS AND MINIMUM CONFOUNDED EFFECTS DESIGNS

Minimum aberration designs and minimum confounded effects designs were discussed in Section 5.7 in the context of 2^{k-p} designs. We can also discuss such designs when there are more than two levels and when the number of levels is mixed. The discussion must necessarily be different, however, for the following reason. With 2^{k-p} designs, each interaction has one degree of freedom, so the interaction is either estimable or not estimable, whereas, as has been explained, this is not the case with a 3^{k-p} design. Just as with the 2^{k-p} series, a minimum aberration 3^{k-p} design isn't necessarily the best design to use. In particular, Sun and Wu (1994) explain why designs that are not minimum aberration may be preferred for $k = 6, 7$, and 8.

6.8   FOUR OR MORE LEVELS

Not infrequently, a multiple-factor design is used in which at least one of the factors has four levels. This will generally occur when the four-level factor(s) are qualitative. Of course the required number of runs of a full factorial may be prohibitive if there are very many factors and some of them have four levels. Consequently, a fractional mixed factorial design or an orthogonal main effect design may have to be used. Lorenzen (1993) showed how a variety of orthogonal main effect plans could be easily constructed, including those with four or more levels. By using the catalog of 33 designs combined with some simple rules, a total of 7172 distinct orthogonal main effect plans could be created that have up to six levels and 50 runs.

As discussed by Wu and Hamada (2000, p. 258), the simplest way to construct a design with a mixture of two-level and four-level factors that is not a full factorial is to start with a saturated 2^{k-p} design and use the *method of replacement*. For each four-level factor this entails replacing three of the two-level factors by using the four combinations to designate the levels of the four-level factor. (The third factor is a product of the first two in their example; hence there are four combinations, not eight. Viewed differently, a 2^{k-p} design contains a replicated 2^{k-p-1} design.) If the construction began with the 2^{7-4} design, the resultant design is a 4-level factor combined with a 2^{4-1}. This method has been referred to as using *pseudofactors* by Giesbrecht and Gumpertz (2004).

Full factorial designs in four or more factors can be easily created with almost all statistical software packages; fractional factorials are another matter. The construction of fractional factorials and various types of custom designs by JMP are discussed in the next section.

Assume that we wish to construct a 4^{3-1} design, which of course would have 16 runs. JMP produces the following design when the user selects 16 runs:

A	B	C
0	0	0
0	1	2
0	2	1
0	3	3
1	0	3
1	1	0
1	2	2
1	3	1
2	0	1
2	1	0
2	2	2
2	3	3
3	0	2
3	1	1
3	2	3
3	3	0

As we would expect, there is a full factorial in A and B. We note that the levels for C cannot be obtained from the levels of A and B, however, unlike the case with two-level and three-level fractional factorials. That is, we cannot use $C = AB^s$ for $s = 1, 2,$ or 3 and obtain the levels of C given in the design above. Does this mean that a 4^{3-1} design cannot be constructed? It can't be done with JMP. The default option for three four-level factors is 16 design points. When that option is used, JMP generates a D-optimal design (see Section 13.7 for a discussion of various types of optimal designs). A similar problem is encountered when Design-Expert is used, as no 4^{3-1} design can be constructed, but a D-optimal design with 41 runs is constructed when the D-optimal design option is selected.

It is possible to construct a 4^{3-1} design, however, using the same general approach as is used in constructing 2^{k-p} and 3^{k-p} designs. If we let $C = AB$, we obtain the following design:

A	B	C
0	0	0
0	1	1
0	2	2
0	3	3
1	0	1
1	1	2
1	2	3
1	3	0
2	0	2
2	1	3
2	2	0
2	3	1
3	0	3
3	1	0
3	2	1
3	3	2

It can be observed that $A + B + 3C = 0$ (mod 4) for every combination so this is one of the four fractions (the others summing to 1, 2, and 3). Multiplying both sides of $C = AB$ by C^3, we obtain $I = ABC^3$ as *part* of the defining relation. From this point the development does not parallel the development for 2^{k-p} and 3^{k-p} designs because 4 is not a prime number. For notational simplification, let $A + B + 3C = 0$ (mod 4) be denoted by (113). Then the rest of the defining relation is obtained as 2(113), mod 4, and 3(113), mod 4. This produces the set (113), (222), and (331), so the entire defining relation is $I = ABC^3 = A^2B^2C^2 = A^3B^3C$. (Note that the exponent of A is allowed to be something other than 1 resulting from the fact that 4 is not a prime number.) The reader can verify that each of the last two components of the defining relation also give a sum of zero, mod 4, when applied to each of the 16 treatment combinations.

The (complicated) alias structure is then easily obtained from this defining relation and part of it is as follows:

$$A = A^2BC^3 = A^3B^2C^2 = B^3C$$
$$A^2 = A^3BC^3 = B^2C^2 = AB^3C$$
$$A^3 = BC^3 = AB^2C^2 = A^2B^3C$$
$$B = AB^2C^3 = A^2B^3C^2 = A^3C$$
$$B^2 = AB^3C^3 = A^2C^2 = A^3BC$$
$$B^3 = A^2C^3 = A^2BC^2 = A^3B^2C$$
$$C = AB = A^2B^2C^3 = A^3B^3C^2$$
$$C^2 = ABC = A^2B^2 = A^3B^3C^3$$
$$C^3 = ABC^2 = A^2B^2C = A^3B^3$$
$$AC = A^2B = A^3B^2C^3 = B^3C^2$$
$$BC = AB^2 = A^2B^3C^3 = A^3C^2$$

With 16 runs we, of course, have 15 degrees of freedom for estimating effects. Each main effect has three degrees of freedom and there are three factors. This leaves six df for estimating two-factor interactions, but the components of these interactions are so badly confounded that estimating them would generally be unwise. Although the main effects can be estimated, this is contingent upon the interaction components with which they are confounded not being significant.

It is also possible, of course, to construct designs for factors with five or more levels. Fractional factorials with five levels are easier to deal with than are fractional factorials with six levels because 5 is a prime number whereas 6 is not. The construction of an r^{k-p} design with r a prime number of at least 5 is discussed by Giesbrecht and Gumpertz (2004, p. 282).

6.9 SOFTWARE

Virtually all software packages with experimental design capability can be used to construct designs with more than two levels, including mixed factorials. For example, the "General Full Factorial Design" capability in MINITAB can be used to create full factorial designs with any number of levels up to 100, provided that the number of factors does not exceed 15 and the number of runs does not exceed 100,000. Similarly, the "General Factorial Design" capability in Design-Expert can be used to create a design with as many as 999 levels but the number of factors cannot exceed 12.

Neither MINITAB nor Design-Expert can be used to generate a fraction of a design with more than two levels, however, which means that neither a 3^{k-p} design nor a fraction of a mixed factorial can be generated. Design-Expert will, however, create a design that is D-optimal. For example, if five 3-level factors and two 2-level factors are specified, Design-Expert creates a design with 79 points. Of course there is no way to obtain that number of design points by fractionation since 79 is a prime number. The alias structure for the design is, as one might guess, extremely complex. Correlations between effect estimates can be obtained if that option is selected.

JMP has limited capability for generating 3^{k-p} designs, with only designs that are Taguchi OAs being available. (The latter are discussed in Chapter 8.) For example, a 3^{3-1} design can be selected, which is listed as an L_9 Taguchi design. If we convert the levels given in the JMP output to the level designation used in Section 6.1.1, we have the design as follows:

A	B	C
0	0	0
0	1	1
0	2	2
1	0	1
1	1	2
1	2	0
2	0	2
2	1	0
2	2	1

Using modulo arithmetic as in Section 6.1.1, we can see that $C = AB$ since $A + B = C$, mod 3, with AB being one of the two components of the $A*B$ interaction. The eight degrees of freedom that are available for estimating effects can thus be used to estimate A (2 df), B (2 df), C (2 df), and AB^2 (2 df).

The defining contrast is $I = ABC^2$, which can be obtained by multiplying each side of the equation $C = AB$ by C^2, and it can be observed that $A + B + 2C = 0$ (mod 3), for each of the nine treatment combinations. The design is thus resolution III, with main effects confounded with two-factor interaction components. The full alias structure (not given by JMP) is thus as follows:

$$A = ABC^2 = BC^2$$

$$B = AB^2C^2 = AC^2$$

$$C = AB = ABC$$

$$AC = AB^2 = BC$$

There would have been 26 df for estimating effects if the 3^3 design had been used; we can see how the 26 df are broken down when the 3^{3-1} design is used. Two df are of course confounded with the mean, with the remaining 24 df confounded in four sets of three, as shown above, with each component in each set of three having 2 df.

Thus, one component of the $A*B*C$ interaction is confounded with the estimate of the mean, and the other three components are confounded with main effect estimates, which are each confounded with a component of a two-factor interaction.

A 3^{4-1} design and a 3^{4-2} design can also be constructed by JMP. These are listed as the L_{27} and the L_9, respectively, but only the L_{27} is available when there are more than four factors. Somewhat similarly, a 3^{4-2} design can also be constructed as an L_9, using Design-Expert.

There are many 3^{k-p} designs in existence, as listed in Wu and Hamada (2000, pp. 250–255); see also Chen et al. (1993). Design-Expert can generate two other 3^{k-p} designs as OAs, which we will write in OA notation as the $L_{27}(3^{13})$ and $L_{27}(3^{22})$.

The 3^{4-2} design is as follows, again converting the JMP and Design-Expert notation into the notation used in the preceding example:

A	B	C	D
0	0	0	0
0	1	1	1
0	2	2	2
1	0	1	2
1	1	2	0
1	2	0	1
2	0	2	1
2	1	0	2
2	2	1	0

As in the construction of 2^{k-p} designs, we can see that there is a full 3^2 factorial in the first two factors. We can also observe that $C = AB$ and $D = AC$. Therefore, $I = ABC^2 = ACD^2 = AB^2D = BCD$, with the third component obtained as $(ABC^2)(ACD^2)$ and the last component obtained as $(ABC^2)(ACD^2)^2$. From this defining relation, the full alias structure can be obtained, as the reader is asked to obtain in Exercise 6.17.

JMP also has very limited capabilities for fractions of mixed factorials that are cataloged as screening designs. Specifically, only four fractions of mixed factorials are available as screening designs in JMP, and these are listed as OAs: (1) the L_{18} array due to Peter John that can be used for at most one 2-level factor and up to seven 3-level factors, (2) the L_{18} array due to Chakravarty that can be used for up to three 2-level factors and six 3-level factors, (3) the L_{18} array due to Hunter that can be used for eight 2-level factors and four 3-level factors, and (4) the L_{36} array that can be used for eleven 2-level factors and twelve 3-level factors. These designs are all orthogonal, although all of them are not balanced. For example, an L_{18} array for five 3-level factors and two 2-level factors cannot be balanced because the four combinations of the two-level factors cannot all occur an equal number of times because $18/4$ is not an integer.

(It is possible to create balanced designs in JMP, however, such as by using the "Custom Design" option and selecting "Default.") It is also possible to use the custom design capability in JMP to create D-optimal designs (covered in Section 13.7) for certain numbers of design points.

Design-Expert has greater capability than JMP for constructing fractions of mixed factorials. For example, for a mixture of two-level and three-level factors, Design-

Expert will also construct a $L_{36}(2^{11} \times 3^{12})$, $L_{36}(2^3 \times 3^{13})$, and $L_{54}(2 \times 3^{25})$. There is also a $L_{32}(2 \times 4^9)$ design that can be used for one factor at two levels and up to nine factors at four levels, and an $L_{50}(2 \times 5^{11})$ that can be used for one 2-level factor and up to eleven 5-level factors.

D. o. E. Fusion Pro has some capability for three-level and mixed level designs that are not full factorials, although in the case of the latter the properties of the design are not given. More specifically, only 3^k and mixed full factorial designs are available unless one selects the "Design Navigator Wizard" option and lets the design be generated automatically—without even specifying the number of design points. When the user lists five categorical factors, three at three levels and two at two levels, the following design is generated when no points are to be replicated and there is to be no blocking.

```
        Experiment Design - Experiment 1
             Var. A   Var. B   Var. C   Var. D   Var. E
   Run No.
   1           A        A        C        B        A
   2           B        A        A        C        B
   3           B        A        B        B        A
   4           B        B        C        A        A
   5           A        B        A        B        B
   6           C        B        B        B        A
   7           A        B        B        C        A
   8           A        A        A        A        A
   9           C        A        A        A        A
   10          C        B        C        C        B
   11          A        A        B        A        B
```

The User's Guide gives the following explanation as to how the design points are selected: "D.o.E. FUSION generates model-robust designs based on a combination of D-, A-, G-, and V-Optimality. Its algorithms and approach result in a design that is not strictly optimal according to any one letter goal, but meets the requirements of (1) good coverage of the design space, including the interior, (2) low predictive variances of the design points, and (3) low model term coefficient estimation errors." Thus, although the G-efficiency of the generated designs is given, and indeed is given when the user selects "Design Menu Wizard" and then selects "mixed level," this is *not* the (sole) criterion under which a design is constructed. (Of course the G-efficiency is given as 100 when a mixed full factorial is generated.)

Although it is nice to have designs generated automatically, the user needs to know more than the G-efficiency. Specifically, the alias structure should be known, at least generally since alias structures for three-level and mixed-level designs are not particularly tidy, so viewing an entire alias set when there are more than a few factors could be more confusing than enlightening.

Perhaps there is some commercial software that will generate 4^{k-p} designs, but if so, I am not aware of it.

6.10   CATALOG OF DESIGNS

Since not all statistical software packages have the capability for 3^{k-p} designs and fractions of mixed factorials, some users may have to resort to catalogs of these designs. The downside of doing so, however, is that the catalogs are not constructed in such a way that all of the factor-level combinations are listed, so that cutting and pasting won't work. Consequently, the designs would have to be entered manually, thus leading to the possibility of errors.

Nevertheless, the catalogs can be quite useful. One such catalog, by Xu (2005), lists three-level fractional factorial designs (available at http://www.stat.ucla.edu/~hqxu/pub/ffd/ffd3a.pdf). Designs with 27, 81, 243, and 729 runs are given. This is a more extensive list of designs than were given in Wu and Hamada (2000) and Chen et al. (1993). In particular, the designs listed in Wu and Hamada (2000) do not go beyond 81 runs.

6.11   SUMMARY

It will frequently be necessary to use a design with more than two levels, especially when at least one of the factors is qualitative rather than quantitative. One impediment to the use of designs given in this chapter and elsewhere is that software for constructing many of the designs is not readily available. In particular, there is hardly any statistical software that will generate a wide variety of 3^{n-k} designs. Consequently, catalogs of designs such as were mentioned in Section 6.10 take on greater importance than would be the case if software packages were readily available.

The interpretation of computer output for designs with more than two levels is complicated by the fact that interaction components can be decomposed into either trend components or *I–J* components, and the latter have no physical meaning. Alias structures are also rather complicated for fractional factorial designs with three or more levels.

A conditional effects analysis should be performed when there are large interactions, just as when a 2^{k-p} design is used.

REFERENCES

Addelman, S. (1962). Orthogonal main effect plans for asymmetrical fractional factorial experiments. *Technometrics*, **4**(1), 47–58.

Bisgaard, S. (1997). Why three-level designs are not so useful for technological applications. *Quality Engineering*, **9**(3), 545–550.

Box, G. E. P. and K. B. Wilson (1951). On the experimental attainment of optimum conditions (with discussion). *Journal of the Royal Statistical Society, Series B*, **13**, 1–45.

Buckner, J., B. L. Chin, and J. Henri (1997). Prometrix RS35e gauge study in five two-level factors and one three-level factor. In *Statistical Case Studies for Industrial Process Improvement*, Chap. 2 (V. Czitrom and P. D. Spagon, eds.). Philadelphia:

Society for Industrial and Applied Mathematics, and Alexandria, VA: American Statistical Association.

Chen, J., D. X. Sun, and C. F. J. Wu (1993). A catalog of two-level and three- level fractional factorial designs with small runs. *International Statistical Review*, **61**, 131–145.

Cheng, S.-W. and C. F. J. Wu (2001). Factor screening and response surface exploration. *Statistica Sinica*, **11**, 553–580; discussion: 581–604. (available online at http://www3.stat.sinica.edu.tw/statistica/oldpdf/A11n31.pdf)

Chevan, A. and M. Sutherland (1991). Hierarchical partitioning. *The American Statistician*, **45**, 90–96.

Clark, J. B. (1999). Response surface modeling for resistance welding. In *Annual Quality Congress Proceedings*, American Society for Quality, Milwaukee, WI.

Giesbrecht, F. G. and M. L. Gumpertz (2004). *Planning, Construction, and Statistical Analysis of Comparative Experiments*. Hoboken, NJ: Wiley.

Gupte, M. and P. Kulkarni (2003). A study of antifungal antibiotic production by *Thermomonospora* sp MTCC 3340 using full factorial design. *Journal of Chemical Technology and Biotechnology*, **78**(6), 605–610.

Hale-Bennett, C. and D. K. J. Lin (1997). From SPC to DOE: A case study at Meco, Inc. *Quality Engineering*, **9**(3), 489–502.

Heaney, M. D., W. A. Lidy, E. G. Rightor, and C. G. Barnes (2000). Analysis of a difficult factorial designed experiment from polyurethane product research. *Quality Engineering*, **12**(3), 425–438.

Hinkelmann, K. and O. Kempthorne (2005). *Design and Analysis of Experiments, Volume 2: Advanced Experimental Design*. Hoboken, NJ: Wiley.

Inman, J., J, Ledolter, R. V. Lenth, and L. Niemi (1992). Two case studies involving an optical emission spectrometer. *Journal of Quality Technology*, **24**(1), 27–36.

Kempthorne, O. (1973). *Design and Analysis of Experiments*. New York: Krieger (copyright held by John Wiley & Sons, Inc.).

King, C. and L. Allen (1987). Optimization of winding operation for radio frequency chokes. In *Fifth Symposium on Taguchi Methods*, Dearborn, MI: American Supplier Institute, pp. 67–80.

Lenth, R. V. (1989). Quick and easy analysis of unreplicated factorials. *Technometrics*, **31**, 469–473.

Lorenzen, T. J. (1993). Making orthogonal main effect designs useful. General Motors Research Report 8025. General Motors Research and Development Center, Warren, MI.

Peyton, B. M. and W. G. Characklis (1993). A statistical analysis of the effect of substrate utilization and shear stress on the kinetics of biofilm detachment. *Biotechnology and Bioengineering*, **41**(7), 728–735.

Reece, J. E. (2003). Software to support manufacturing systems. Chapter 9 in *Handbook of Statistics* 22, Chap. 9 (R. Khattree and C. R. Rao, eds.). Amsterdam: Elsevier Science B. V.

Ryan, T. P. (1981). Relationships between the trend components and I-J components of interaction in a 3^2 design. Unpublished manuscript.

Ryan, T. P. (1997). *Modern Regression Methods*. New York: Wiley.

Sahin, K. and M. Örnderci (2002). Optimal dietary concentrations of Vitamin C and chromium for alleviating the effect of low ambient temperature on serum insulin, corticosterone, and some blood metabolites in laying hens. *The Journal of Trace Elements in Experimental Medicine*, **15**(3), 153–161.

Satterthwaite, F. E. (1946). An approximate distribution of estimates of variance components. *Biometrics Bulletin*, **2**, 110–114.

Sheesley, J. H. (1985). Use of factorial designs in the development of lighting products. In *Experiments in Industry: Design, Analysis, and Interpretation of Results*, pp. 47–57. (R. D. Snee, L. B. Hare, and J. R. Trout, eds.). Milwaukee, WI: Quality Press.

Sun, D. X. and C. F. J. Wu (1994). Interaction graphs for three-level fractional factorial designs. *Journal of Quality Technology*, **26**, 297–307.

Taguchi, G. (1987). *System of Experimental Design*. White Plains, NY: UNIPUB/Kraus International Publications.

Vázquez, M. and A. M. Martin (1998). Optimization of Phaffia rhodozyma continuous culture through response surface methodology. *Biotechnology and Bioengineering*, **57**(3), 314–320.

Wang, P. C. (1999). Comparisons of the analysis of Poisson data. *Quality and Reliability Engineering International*, **15**, 379–383.

Wu, C. F. J and M. Hamada (2000). *Experiments*: *Planning, Analysis, and Parameter Design Optimization*. New York: Wiley.

Xu, H. (2002). An algorithm for constructing orthogonal and nearly-orthogonal arrays with mixed levels and a small number of runs. *Technometrics*, **44**, 356–368.

Xu, H. (2005). A catalogue of three-level fractional factorial designs. *Metrika*, **62**, 259–281. (preprint available online at http://www.stat.ucla.edu/~hqxu/pub/ffd/ffd3a.pdf and see the online supplement (catalogs) at http://www.stat.ucla.edu/~hqxu/pub/ffd)

Xu, H., S. W. Cheng, and C. F. J. Wu (2004). Optimal projective three-level designs for factor screening and interaction detection. *Technometrics*, **46**, 280– 292.

EXERCISES

6.1   After reading a few articles on experimental design, but not really being knowledgeable in the subject, a scientist proclaims that he is going to use a mixed factorial: a 1×3 design. Explain why this is really not a mixed factorial. What type of design is it? Would it be of any practical value if the design were unreplicated? Explain. Would your answer be the same if the design were replicated? Explain.

6.2   Buckner, Chin, and Henri (1997, references) described an experiment that investigated the effects of one 3-level factor and five 2-level factors in 16 runs. The design was a modification of a 2^{6-2} design to accommodate the three-level factor and would be properly called an "Addelman design." It is interesting to note that the number of design points was determined from a calculation such as is performed using Eq. (1.1), although the authors used a different value for α and a different value for the power of the test.

(a) What can be said about the levels of the three-level factor?

(b) Two response variables were examined in an ANOVA analysis of the data and none of the six factors was even close to being significant for either response variable. Regarding this, the authors stated, "The Addleman [sic] design has a more complex confounding scheme than ordinary fractional factorials, but since no factors appeared important, confounding was not considered in this application." Do you agree with that position?

(c) For the response variable, Uniformity, the sum of squares for one of the factors, Prior Film, was given as 0.000000000. Since the values for this response variable were all two-digit numbers, what is the only way that this number could be correct? Would you be inclined to suspect that it may be incorrect? Explain.

6.3 Derive the alias string for the 3^{3-1} design that was discussed in Section 6.3.4, with part of the alias string given in that section.

6.4 It was shown in Section 4.2.1 that it will generally be a good idea to construct interaction plots for two-level factors in such a way that each of two factors involved in a two-factor interaction is, in turn, the horizontal axis label, since the message of the extent of the interaction can be quite different for the two plots. This is also a good idea for three-level factors. For the data shown in Figure 6.1, will the plot have a different configuration if factor A is plotted on the horizontal axis? If so, construct the plot. If not, explain why not.

6.5 The sample dataset PENDULUM.MTW that comes with the MINITAB software contains data from a 2×4 experiment for which an estimate of gravity is the response variable. A weight (heavy or light) is suspended at the end of a string. The time (T) required for a single cycle of the pendulum is related to the length (L) of the string by the equation $T = (2\pi/g)(\sqrt{L})$, so g can be solved for a given combination of T and L. Four different lengths were used: 60, 70, 80, and 90 cm. Three replicates were used so that there were 24 observations altogether. Analyze the data and draw conclusions.

6.6 Consider the 3^{9-6} design and accompanying response values in Table 9 in Cheng and Wu (2001, references), for the electric wire experiment given by Taguchi (1987, p. 423, references) experiment that they described. (This is available online at http://www3.stat.sinica.edu.tw/statistica/oldpdf/A11n31.pdf.) Perform your analysis of the data, including computing conditional effects if necessary, and compare your results with the results given by the authors. Do the results agree? If not, justify your results and approach. Does the two-stage approach with a single design seem appropriate for this example? Explain.

6.7 Peyton and Characklis (1993, references) examined the effects of substrate utilization and shear stress on the rate of biofilm detachment, using three levels for each factor. Read the article, analyze the data, and draw appropriate conclusions.

6.8 Gupte and Kalkarni (2003, references) used a 3^3 design to investigate the effects of three factors, with three levels used for each factor. The three factors used in the experiment were selected as a result of previous one-factor-at-a-time experimentation. Does this two-stage approach seem reasonable? Read the article and determine whether or not you agree with the authors' conclusions and state whether you would have used a different design strategy.

6.9 Show that all the dot products alluded to in the discussion in Section 6.1.1 are zero.

6.10 Verify the statement made in Section 6.3 that a 3^{4-1} design projects into a full 3^3 for any combination of three factors, with either $D = ABC$ or $D = ABC^2$.

6.11 It was stated in Section 6.3 that it is conventional to list the components of an interaction, such as the A^*B^*C interaction, by listing the components with the first factor in the interaction always having an exponent of 1. Assume that someone decides to break with convention and creates a 3^{4-1} design by letting $D = A^2B^2C$ instead of $D = ABC^2$. Will the designs be different? If so, are there any conditions under which one of the designs might be preferred over the other one? Explain.

6.12 Consider a 3^2 design that is run in blocks of size 3, with the AB component of the A^*B interaction confounded with the difference between blocks. Even though it is unconventional, assume that the blocks are constructed using A^2B^2 rather than AB. Will this result in the same set of three blocks being constructed as when AB is confounded with the difference between blocks?

6.13 Consider Exercise 6.11. Does it follow from the result in that exercise that the three 3^{4-1} fractions that result from using $D = A^2B^2C$ to construct the fractions are the same set of three fractions that result when $D = ABC^2$ is used?

6.14 Using the data in Example 6.1, compute the four interaction graphs for a 3^2 design that correspond to $A_{\text{linear}}B_{\text{linear}}$, $A_{\text{linear}}B_{\text{quadratic}}$, $A_{\text{quadratic}}B_{\text{linear}}$, and $A_{\text{quadratic}}B_{\text{quadratic}}$, and comment on the relationship between them.

6.15 Can a 3^{7-p} design be constructed with a reasonable number of runs for fitting a full second-order model? If so, construct the design. If not, explain why it can't be done.

6.16 Give the design columns for the $4 \times 2^{4-1}$ design that was alluded to in Section 6.8, using the method for constructing the design that was discussed therein.

6.17 Obtain the alias structure for the 3^{4-2} design whose defining relation was given in Section 6.9.

6.18 Use JMP or other software to construct a 3^{4-1} design and determine the defining relation and full alias structure.

6.19 Assume that you have three quantitative variables and an experimenter tells you that you should use a 4^3 design. Would this be a reasonable choice? Why, or why not? Assume that you start to use JMP but you discover that the software will permit you to construct the design only if the factors are categorical (i.e., qualitative). Why do you think this is the case?

6.20 Assume that an unreplicated 4^2 design is used. Give the degrees of freedom breakdown.

6.21 An experimenter has four factors to be investigated and wants to use three levels for three of the factors and two levels for the other factor. Economic considerations restrict the design to at most 48 experimental runs, however. The experimenter believes it is highly probable that some interactions will exist but isn't sure which interactions could be real. What design would you suggest?

6.22 Construct a 3^{5-1} design using the same general approach that was used for constructing the 3^{3-1} design in Section 6.3.2.

6.23 What is the resolution of a 3^{10-5} design? Is this a practical design for fitting a second-order model? Explain.

6.24 An application of a 3^2 design is given at http://www.smta.org/files/SMTAI02-Wang_Paul.pdf. The three levels of one factor, pressure, were 10, 14, and 18 (Kgf), and the levels of the other factor, speed, were 15, 25, and 50 (mm/s). Thus, the factor levels were not equally spaced. Would this have created a problem when the data were analyzed? Explain.

6.25 The data for the experiment described in Section 6.6 are as follows, with the first pair of observations for the first treatment combination, the next pair for the second treatment combination, and so on: 106.20, 107.70, 104.20, 102.35, 85.90, 85.90, 101.15, 104.96, 109.92, 110.47, 108.91, 108.91, 109.76, 112.66, 97.20, 94.51, 112.77, 113.03, 93.15, 92.83, 97.25, 100.60, 109.51, 113.28, 85.63, 86.91, 113.17, 113.45, 104.85, 98.87, 113.14, 113.78, 103.19, 106.46, 95.70, and 97.93.

    (a) Do any of the data look suspicious? (*Hint*: Look at the data in the appropriate pairs.)

    (b) If you identified one or more data points that you believe are probably in error, will the apparent error(s) likely affect the analysis? Explain. What would be a possible explanation for any suspected errors, recalling what has been discussed about factor level resettings in previous chapters?

(c) Analyze the data using whatever methods you prefer and compare the effects you select with the effects selected by Xu et al. (2004, references), as given in Section 6.6.

6.26 Heaney, Lidy, Rightor, and Barnes (2000, references) gave an example of a $3 \times 4 \times 5$ experiment that was questionably designed with the subsequent deleterious effects of that compounded by problems that occurred when the experiment was conducted. Indeed the design used was questioned by the authors in their "Conclusions" section. Read the article as well as the first referenced article if you find that necessary and can gain access to it.

(a) Would you have designed the experiment differently? If so, how?

(b) Would you have used a different data analysis approach in an effort to recover from the problems that occurred when the experiment was performed? If so, what would you have done differently? If not, why not?

CHAPTER 7

Nested Designs

Nested designs and the analysis of data from experiments that use these designs are considered to be complicated. Indeed, Reece (2003, p. 384) stated the following regarding nested designs and their analysis using Design-Expert, Version 6.0 (DX6): "The analyses of these designs are complicated at best, but are possible with DX6."

The problem is compounded somewhat by the fact that most books on experimental design devote comparatively little space to the subject, and hardly any of the leading design books have a chapter devoted exclusively to nested designs. Exceptions include Mason, Gunst, and Hess (2003) and Yandell (1997), each of which has two chapters on nested designs. Many design books barely even mention nested designs, so it can be somewhat difficult to gain an in-depth knowledge of the subject without consulting a small number of good sources of information.

One example of the confusion that exists regarding nested designs is the following statement, which can be found at www.sas.org/E-Bulletin/2002-03-08/features/body.html : "Nested designs are not an exclusive category, because a full 2^2 factorial design is also a nested design." As was shown in Section 4.1, in a 2^2 factorial design each level of each factor occurs with each level of the other factor in the design. This is not the case when a nested design is used. That is, the levels of the factors are not crossed. Therefore, the statement is in error.

Nested designs are also called hierarchical designs, with the term emanating from the hierarchical (i.e., "nesting") structure between the factors. There can be either strict nesting of factors or a combination of nesting and a factorial arrangement. When the latter exists, the design is called a *nested factorial design*.

Assume that we have a manufacturing experiment in which temperature at some stage of a process is varied in an experiment to see if it has a significant effect on process yield. Three temperatures are used and each is used for two consecutive weeks, so the experiment runs for six weeks. The factor of interest is temperature, but there might be a week effect (we hope not), which would be a nuisance factor.

Modern Experimental Design   By Thomas P. Ryan
Copyright © 2007 John Wiley & Sons, Inc.

Assume that we are interested in looking at both weeks and temperature. What type of design is this? In a factorial design each level of every factor is "crossed" with each level of every other factor. Thus, if there are a levels of one factor and b levels of a second factor, there are then ab combinations of factor levels. Since there are actually six weeks involved in the temperature-setting experiment, there would have to be $6 \times 3 = 18$ combinations for it to be a cross-classified design. Since there are obviously only six combinations of weeks and temperatures, it clearly cannot be a cross-classified (factorial) design.

Then what type of design is this? It is actually a *nested factor design* as weeks are "nested" within temperature. The corresponding model is

$$Y_{ijk} = \mu + A_i + B_{j(i)} + \varepsilon_{k(ij)} \qquad i = 1, 2, 3 \quad j = 1, 2 \quad k = 1, 2, 3, 4, 5 \qquad (7.1)$$

where i designates the temperature, j the week, and k the replicate factor (days in this case). Further, $j(i)$ indicates that weeks are nested within temperature, and $k(ij)$ indicates that the replicate factor is nested within each (i, j) combination.

Notice that the model does not contain an interaction term. There can be no interaction terms in the model when a strictly nested design is used since an interaction cannot be computed unless all possible combinations of the levels of factors that comprise the interaction are used. Even though interactions cannot be computed, this doesn't mean that they don't exist. The users of these designs simply must assume that they don't exist, but assuming that interactions don't exist certainly doesn't cause them to not exist. It should be noted that it is possible to have a design with a combination of nesting and a factorial structure, as is explained in Section 7.4.

In some applications there can, by definition, be no interaction. For example, assume we have two different types of machines and three types of heads that would fit on one machine but not on the other one. There can be no interaction between heads and machines because the heads cannot be swapped between machines. So for this type of application the assumption of no interaction is certainly true. (There is more discussion of the machines–heads scenario in later sections and in Exercise 7.12.)

On the other hand, assume that a drug-testing experiment is conducted using two hospitals and four drugs. Assume that two drugs are used at one hospital and the other two drugs used at the other hospital. Here we do have to assume no interaction because the drugs should be interchangeable between the hospitals. Assume that the variation in response values would have been much less (for whatever reasons) if the first two drugs had been used at the second hospital instead of the first hospital. Then the assumption of no interaction would not be valid and the results of the experiment could be misleading.

Nested designs are used in a variety of applications, including the chemical industry (see, e.g., Lamar and Zirk, 1991 who gave illustrative case studies of a three-level nested design and a four-level nested design) and manufacturing (see, e.g., Liu and Batson, 2003). More specifically, the use of these designs in process control and process variation studies has been discussed by Bainbridge (1965), Sinibaldi (1983), Snee (1983), and Pignatiello (1984). The designs are also used in agricultural and psychological research, in addition to many applications in biology. Vander Heyden,

De Braekeleer, Zhu, Roets, Hoogmartens, De Beer, and Massart (1999) described the use of nested designs in ruggedness testing in pharmaceutical work and proposed an alternative analysis method for use in ruggedness testing.

One of the primary objectives in the use of nested designs is to remove uncontrolled variation due to a priori differences in primary sampling units from the experimental error. Another reason is that in biological experiments, in particular, there is often a desire to be able to make inferences regarding hierarchically arranged environments, habitats, or species.

Often there is no choice between a nested design and a crossed design because circumstances dictate use of the former. (For example, a head might be usable on only one particular machine, the Florida Everglades can obviously be found only in Florida and not in any other state, etc.)

As is discussed in Chapter 9, it is important to recognize a split-plot structure when it exists, since analyzing the data as if this structure does not exist can cause misleading results. Similarly, it is important to distinguish a nested design structure from other structures. We consider this issue with the following example.

Example 7.1

Pignatiello (1984) discussed a scenario in which an engineer was working for a military contractor and sampling from lots was performed. Specifically, five parts were randomly sampled from each of two lots, and two measurements were made on each part since measurement error was known to exist. The engineer apparently decided to use the average measurement on each part if there was no difference in the variances between the two lots. In other words, measurement error would be removed from the analysis. But equality of variances, if it exists, does not obviate an analysis that uses the measurement error and the fact that there are repeat readings on each part. Furthermore, this raises the question of whether or not this is really a repeated measures design (Chapter 11) and whether the data should be analyzed as such. With a repeated measures design each experimental unit receives more than one treatment. That doesn't happen here, as measuring a part twice does not mean that two treatments are being applied. Therefore, this is not a repeated measures design. There might seem to be a type of nesting involved here, however, since measurement error on each part is nested within that part, but that is hardly different from saying that the variability within each factor level in a one-factor design is nested within each level.

We may note, however, that what the engineer wanted to do is an example of a practitioner trying to collapse data for simplicity, which is generally not a good idea. There is information in numbers, just as there is more information in a cholesterol reading of 217 than simply reporting "over 200." There were no data reported in Pignatiello (1984) since the author simply reported an inquiry, and so no analysis can be given here.

As stated above, each experimental unit receives more than one treatment when a repeated measures design is used, whereas each experimental unit receives only one treatment when a nested design is used.

Example 7.2

An example of a purported nested design appears on the Internet (http://www. psych-stat.missouristate.edu/introbook/sbk23m.htm). The experiment consisted of six males and nine females and a t-test was performed to test for the equality of means, with the response variable being "finger-tapping speed" (no other details were given). The statement is made that "The design is necessarily nested because each subject has only one score and appears in a single treatment condition." There is no treatment, however, at least not in the usual sense. To state that this is an example of a nested design would be similar to saying that one-factor ANOVA is a nested design because each experimental unit receives only one treatment. Furthermore, this scenario is really more of an observational study than a designed experiment. Obviously there is no randomization involved because if we think of sex as the factor and the person as the experimental unit, we obviously can't assign sex to the person.

7.1 VARIOUS EXAMPLES

There are many good examples of nested designs. An experiment might be conducted with the objective of improving process yield, with the experiment involving five machines and five operators. Unless each operator uses each machine during the experiment (which might be both impractical and cost prohibitive) the operator effect will be nested under the machine effect.

As another example, consider an experiment to study the strain readings of glass cathode using four different machines. Assume that there are different heads that will fit on each machine, but will not fit on any of the other three machines. One head is to be randomly selected for each machine. Since each head that is selected cannot be used on any of the other machines, heads are thus nested within machines, so the head factor is a nested factor.

Another example, which has been found on the Internet, is that of a forestry experiment that involves three forests, five trees that are sampled within each forest, and five seedlings that are sampled for each tree. Thus, the tree effect, if it exists, is nested within forests and if a seedling effect exists, it is nested within trees. The ANOVA table setup would be as follows.

```
Source                            df

Forests                            2
Trees within forests              12
Seedlings within trees            60
TOTAL                             74
```

Hypothesis testing is performed by constructing F-tests using the rules that apply to factorial designs regarding fixed and random factors. That is, Forests are tested against Seedlings if both Forests and Trees are fixed; tested against Trees if the latter is random and Forests are fixed, or both factors are random; and Trees are of course tested against Seedlings regardless of whether the two factors are fixed or random.

7.2 SOFTWARE SHORTCOMINGS

Whereas data from experiments with nested designs can be analyzed using SAS Software, trying to do the analysis with most other software packages is either difficult or impossible. For example, Reece (2003) points out, in referring to Design-Expert: "By manipulating it's analysis capabilities, a user can analyze nested designs, but it does not support generating them directly." (This comment was in regard to Version 6.0.5.) Similarly, there is no mention of nested designs in D. o. E. Fusion Pro. Data may be analyzed as having come from a nested design in JMP, however, by using the FitModel dialog with the "Nest" button. Users of nested designs may wish to consider RS/Discover, as Reece (2003, p. 364) stated, "RS/Discover is one of the few products tested that provides support for nested designs involving a number of random effects."

Moderately extensive nested design analysis capability is available in MINITAB, but the key word here, as it is in JMP, is *analysis*. That is, data from the use of a nested design can be analyzed, both for a fully nested design and for an unbalanced nested design. *However*, the "fully nested ANOVA" routine in MINITAB is only for random factors and thus cannot be used if there is at least one fixed factor. When the latter is true, the general linear model (GLM) routine must be used, and the model and the nesting structure must be specified. That is, if the two factors used in an experiment are Machines and Operators and the latter is nested under the former, the terms in the model would have to be specified as "Machines Operators (Machines)," regardless of whether command mode or menu mode is used. Since it is not uncommon for one of the factors in a nested experiment to be fixed (such as the first factor, under which the other factors are nested), many MINITAB users will have to use the GLM capability for analysis.

Data from a gauge $R \& R$ (reproducibility and repeatability) study can also be analyzed with nested factors, as will often occur, such as operators being nested under machines rather than being crossed with machines. (It isn't necessary, in MINITAB, to specify the nesting structure for this routine as that is implied by the designation of the appropriate columns for "Operators" and "Part numbers.")

The situation regarding software isn't quite as bleak as it may seem, however, as is explained in the next section.

7.2.1 A Workaround

Experimenters and others who are using something other than JMP, SAS, or MINITAB can still use their software, although some of the work will have to be performed manually. Because of this, we need to look at some sum of squares results in certain detail.

Accordingly, assume that there are two factors and factor B is nested under factor A. The latter has two levels and factor B is at two levels for each level of factor A.

Since the levels are different, this means that factor B has four levels. Thus there are four combinations of levels of A and B. With a factorial arrangement there would have been eight combinations so this obviously isn't a factorial arrangement. Assume further that there is only one observation per treatment combination, so there are four observations altogether.

We can use the computations that would be used *if* this were a 2^2 design to arrive at the sums of squares for the nested design. That is, we will pretend that the third and fourth levels of factor B are the same as the first and second levels. Recall the formulas for estimating the B effect and the $A \times B$ interaction effect that were given in Section 4.3, and also recall from Section 4.5 that for an unreplicated 2^2 design the sum of squares for a particular effect is equal to the square of the effect estimate.

Squaring the effect estimates for B and the $A \times B$ interaction and adding them together produces $[(A_1 B_1)^2 + (A_1 B_2)^2 + (A_2 B_1)^2 + (A_2 B_2)^2]/2 - A_1 B_1 A_1 B_2 - A_2 B_1 A_2 B_2$.

Now let's consider how we would compute the sum of squares for a nested factor, and specifically B nested within A. We are interested in the variability of the response using the two levels of B at each level of A. Therefore, we want to compute $\sum_{i=1}^{2} \sum_{j=1}^{2} (A_i B_j - \overline{A}_i)^2$, with $\overline{A}_i$ denoting the average response at the ith level of A. Expanding and simplifying this expression produces the same expression given above for the sum of B and $A \times B$, as the reader is asked to show in Exercise 7.2. This means that software that will handle a 2^2 design (as virtually all general purpose statistical software packages will) can be used to produce the sum of square components that are needed for the analysis of the two-factor nested design. Of course it isn't particularly satisfying to have to use software to produce components that then must be combined manually to provide the ANOVA table, but this approach will have to be used by many practitioners, depending upon the software to which they have access.

We will illustrate the necessary computations with the following example.

Example 7.3

Assume that the two major hospitals in a small city are selected for an experiment involving four drugs, with these drugs being the only ones of interest. For the sake of illustration we will assume that the 12 patients who will be involved in the experiment (3 for each hospital–drug combination) are essentially homogenous in regard to any physical characteristics that could affect the response values. The layout and the (coded) response values are given below, with H_i denoting the ith hospital, $i = 1, 2$; and D_i denotes the ith drug, $i = 1, 2, 3, 4$.

	H_1		H_2	
D_1	D_2	D_3	D_4
(5, 7, 4)	(6, 8, 5)	(8, 9, 6)	(9, 8, 10)

If we analyze the data as having come from an experiment with a replicated 2^2 design, we obtain the sum of squares given in the following ANOVA table.

```
Source   DF    SS

H         1    18.750
D         1     4.083
H*D       1     0.083
Error     8    16.000
Total    11    38.917
```

It follows that the ANOVA table for the nested design is

```
Source   DF    SS

H         1    18.750
D(H)      2     4.166
Error     8    16.000
Total    11    38.917
```

Notice that the degrees of freedom for $D(H)$ after the pooling is as it should be because D has one degree for each level of H since there are two levels of D at each level of H. If we calculated the sum of squares for $D(H)$ directly, we would have done so by computing $3 \sum_{i=1}^{2} \sum_{j=1}^{2} (\overline{H_i D_j} - \overline{H_i})^2$, with the "3" resulting from the fact that there are three observations for each hospital–drug combination. The reader is asked to show in Exercise 7.3 that this produces the sum of squares for $D(H)$ shown in the ANOVA table.

```
Source   DF    SS        MS       F       p

H         1    18.750    18.750   9.375   0.016
D(H)      2     4.166     2.083   1.04    0.396
Error     8    16.000     2.000
Total    11    38.917
```

The F-statistics are computed using the fact that both factor H and factor D are fixed. We see that the hospital effect is significant at the .01 level, but the drug effect is not significant, which is the reverse of what an experimenter would want to see for this experiment since the conclusion is that the drugs do not differ in effectiveness.

Of course it is bothersome to have to obtain the F-statistics and p-values manually. We can avoid the manual construction of F-tests by using JMP or SAS. The output from JMP for this example is as follows.

```
Summary of Fit

RSquare          0.588865
RSquare Adj         0.43469
Root Mean Square Error        1.414214
Mean of Response        7.083333
Observations (or Sum Wgts) 12
```

Analysis of Variance

Source	DF	Sum of Squares	Mean Square	F Ratio	Prob > F
Model	3	22.916667	7.63889	3.8194	0.0575
Error	8	16.000000	2.00000		
C.Total	11	38.916667			

Effect Tests

Source	Nparm	DF	Sum of Squares	F Ratio	Prob > F
Hospital	1	1	18.750000	9.3750	0.0155
Drugs [Hospital]	2	2	4.166667	1.0417	0.3962

The output using MINITAB is as follows.

General Linear Model: Response versus Hospital, Drug

Factor	Type	Levels	Values
Hospital	fixed	2	1, 2
Drug(Hospital)	fixed	4	1, 2, 3, 4

Analysis of Variance for Response, using Sequential SS for Tests

Source	DF	Seq SS	Adj SS	Seq MS	F	P
Hospital	1	18.750	18.750	18.750	9.38	0.016
Drug(Hospital)	2	4.167	4.167	2.083	1.04	0.396
Error	8	16.000	16.000	2.000		
Total	11	38.917				

S = 1.41421 R-Sq = 58.89% R-Sq(adj) = 43.47%

Expected Mean Squares, using Sequential SS

	Source	Expected Mean Square for Each Term
1	Hospital	(3) + Q[1, 2]
2	Drug(Hospital)	(3) + Q[2]
3	Error	(3)

Error Terms for Tests, using Sequential SS

	Source	Error DF	Error MS	Synthesis of Error MS
1	Hospital	8.00	2.000	(3)
2	Drug(Hospital)	8.00	2.000	(3)

7.3   STAGGERED NESTED DESIGNS

Even though a nested design does not have a factorial structure, the degrees of freedom does have a "product structure" that can result in a large number of degrees

of freedom—more than is necessary—for effects at the bottom of the hierarchical structure. This is analogous to what happens with interactions in factorial designs with more than two levels, as for example, the ABC interaction in a 4^3 design has 27 degrees of freedom, far more than is necessary to estimate an effect that probably isn't real. Although there are of course no interactions in nested designs, the degrees of freedom for effects in nested designs has a somewhat similar product structure. Specifically, the degrees of freedom for a given effect at least two levels from the top is the product of the number of levels of the factors above it in the hierarchical structure multiplied times one less than the number of levels of that effect. For example, if factor C is nested under factor B, which in turn is nested under factor A and all factors have four levels, factor C has $(4)(4)(4 - 1) = 48$ degrees of freedom and the error degrees of freedom would be much larger than this, resulting in a sample size that may be cost prohibitive. Thus, nested designs can be inefficient, as discussed, for example, by Khuri and Sahai (1985).

The largest reductions in this number would occur when the levels of factor A and/or factor B are reduced, but this might not be acceptable if an experimenter has a certain number of levels in mind to be investigated.

Various alternatives to a nested design have been discussed in the literature, some of which are covered by Bainbridge (1965). A *staggered nested design* is one practical alternative, which is recommended by Bainbridge (1965). With this type of design, there is not full nesting. For example, consider the following staggered design layout given by Smith and Beverly (1981).

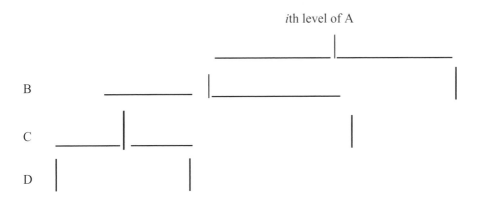

*i*th level of A

B

C

D

Factors A, B, C, and D might represent raw material lots, batches, samples, and measurements, respectively. Notice that two samples are obtained from only one of the two batches (with one sample taken from the other batch), with one of the two samples having a repeat measurement. Thus, there are only four observations for the *i*th level of A with this staggered nested design, whereas there would be eight observations with a regular nested design.

Assume that there are three levels of A. The degrees of freedom breakdown is then

Source	df
A	2
B (within A)	3
C (within B)	3
D (error)	3
Total	11

Since a staggered nested design is an unbalanced design, it can be analyzed by software that handles unbalanced designs, such as the GLM capability in MINITAB. Although these designs are rarely used in biology, Cole (2001) gave an example of their use in such an application.

7.4 NESTED AND STAGGERED NESTED DESIGNS WITH FACTORIAL STRUCTURE

It is possible to construct designs that have a factorial structure for some factors and a nested structure for other factors, as stated previously. Similarly, it is also possible to construct a design that has a staggered nested structure for some factors and a factorial structure for other factors. Smith and Beverly (1981) discuss these designs and the models to which the designs correspond. Khattree and Naik (1995) gave statistical tests for random effects for these designs; Khattree, Naik, and Mason (1997) discussed the estimation of variance components for the designs; and Naik and Khattree (1998) gave a computer program for estimating the variance components. Ojima (1998) gave certain theoretical results.

As in Eq. (7.1), the way in which the model is written gives the nesting structure. For example, the model

$$Y_{ijkl} = \mu + A_i + B_j + C_{k(ij)} + \varepsilon_{l(ijk)} \tag{7.2}$$

shows that the effect of factor C is nested within the AB combinations and that the error term also has a nested structure, as in Eq. (7.1). Thus, the model in Eq. (7.2) has both a factorial structure and a nested structure. Although it was stated at the beginning of the chapter that the model for a nested design cannot contain any interaction terms, this is assuming that all of the factors are nested. It is possible to have interaction terms involving factors that have factorial structure, so an AB_{ij} term in Eq. (7.2) would be permissible since factors A and B have factorial structure.

7.5 ESTIMATING VARIANCE COMPONENTS

To this point in the chapter, only inference using ANOVA has been presented. Random factors generally occur when nested designs are used, as factors that are nested are generally random, and sometimes all factors are random, as in Exercise 7.9. (When all factors are random, the NESTED command in MINITAB can be used for the analysis.)

Since variance reduction is important in quality improvement work, for example, having variance component estimates for factors in an experiment can be quite useful. Estimates will frequently be negative, however, which is a problem since variances are of course never negative.

The following output will be used to illustrate how this can occur.

```
              Analysis of Variance for Y

Source    DF        SS             MS          F        P
A          2    30236.7222    15118.3611    166.696   0.001
B          3      272.0833       90.6944      0.694   0.573
C         12     1569.0000      130.7500      7.679   0.000
D         18      306.5000       17.0278
Total     35    32384.3056
```

```
               Variance Components

                             % of
   Source    Var Comp.      Total     StDev
      A       1252.306      94.43     35.388
      B         -6.676*      0.00      0.000
      C         56.861       4.29      7.541
      D         17.028       1.28      4.126
   Total      1326.194                36.417
```

```
* Value is negative, and is estimated by zero.
```

```
              Expected Mean Squares

   1    A    1.00(4) + 2.00(3) + 6.00(2) + 12.00(1)
   2    B    1.00(4) + 2.00(3) + 6.00(2)
   3    C    1.00(4) + 2.00(3)
   4    D    1.00(4)
```

The last part of the output is used in combination with the first section to obtain the variance component estimates that are shown in the middle section. To illustrate, $E(\text{MS}_D) = \sigma_D^2$ so $\widehat{\sigma}_D^2 = \text{MS}_D = 17.028$. $E(\text{MS}_C) = \sigma_D^2 + 2\sigma_C^2$. Using this last equation as the basis for solving for $\widehat{\sigma}_C^2$ produces $\widehat{\sigma}_C^2 = (130.75 - 17.028)/2 = 56.861$. We encounter a problem when we try to solve for $\widehat{\sigma}_B^2$, however. $E(\text{MS}_B) = \sigma_D^2 + 2\sigma_C^2 + 6\sigma_B^2$, as shown. Using the estimates for σ_D^2 and σ_C^2, we obtain $\widehat{\sigma}_B^2 = (90.6944 - 17.028 - 2(56.861)/6 = -6.676$. Notice that the output sets the estimate to zero. The negative estimate is caused by the large value of $\widehat{\sigma}_C^2$, especially relative to the value of MS_B. The large value should be investigated and in fact was investigated, as is explained in the particular chapter exercise for which the analysis of the data results in this computer output.

Readers interested in hand computation of variance component estimates are referred to Nelson (1995a), with confidence intervals for the corresponding standard deviations given in Nelson (1995b). Variance component estimation for up to three-level nesting is covered by Searle, Casella, and McCulloch (1992), and for confidence interval construction of variance components readers are referred to Burdick and Gray-bill (1992). For applications of variance components estimation with nested designs in the semiconductor industry, see Jensen (2002); see also Ankenman, Liu, Karr, and Picka (2002). Another useful reference, but now out of print, is Rao and Kleffe (1988), and Vardeman and Wendelberger (2005) provide some instruction in the estimation of variance components for an unbalanced two-factor nested design.

7.6   ANOM FOR NESTED DESIGNS?

Analysis of Means (ANOM) was used initially in Section 2.2 and subsequently used to analyze data from various types of designs. As was stated in Section 3.1.5 and also stated by Nelson (1993), ANOM can be used with any complete design. The factors must be fixed, however, and with a nested design this requirement will hardly ever be met.

Gonzalez-de la Parra and Rodriguez-Loaiza (2003) proposed a ANOM procedure for nested designs with random factors. The proposed method should be viewed as only an ad hoc procedure, however, for the following reasons. The authors simply applied ANOM for a single factor to each of the factors separately and then proceeded as one would do if the factors were fixed. They interpreted the charts by looking for "...patterns (random or systematic) of the subgroup means across the factors...." Thus, this is not a well-defined statistical procedure, nor can it be since ANOM is not applicable to random factors. So even though the use of these charts was beneficial in the application that was described, such an analysis would be best used only as a graphical supplement to accepted procedures.

7.7   SUMMARY

Nested designs have been used in industrial applications for decades, and medical, agricultural, and biological applications are also natural, as are applications in psycho-logical research. Applications in other areas, such as aquatics, have been slow coming, however. Staggered nested designs are a practical alternative to nested designs, as the former provides the opportunity to have a reasonable number of degrees of freedom for factors at the top of the hierarchical structure without having a high number of degrees of freedom at the bottom. Unfortunately, there are some software shortcom-ings, detailed in Section 7.2, that can hinder the effective use of nested designs and analysis of the data from experiments in which these designs are used. Consequently, experimenters need to be aware of the limitations of specific statistical software.

REFERENCES

Ankenman, B. E., H. Liu, A. F. Karr, and J. D. Picka (2002). A class of experimental designs for estimating a response surface and variance components. *Technometrics*, **44**(1), 45–54.

Bainbridge, T. R. (1965). Staggered nested designs for estimating variance components. *Industrial Quality Control*, **22**, 12–20.

Burdick, R. K. and F. A. Graybill (1992). *Confidence Intervals on Variance Components*. New York: Marcel Dekker.

Cole, R. G. (2001). Staggered nested designs to identify hierarchical scales of variability. *New Zealand Journal of Marine and Freshwater Research*, **35**, 891–896. (This article is available at http://www.rsnz.org/publish/nzjmfr/2001/78.pdf.)

Croarkin, C. and P. Tobias, eds. (2002). *NIST/SEMATECH e-Handbook of Statistical Methods* (http://www.itl.nist.gov/div898/handbook), a joint effort of the National Institute of Standards and Technology and International SEMATECH.

Gonzalez-de la Parra, M. and P. Rodriguez-Loaiza (2003). Application of analysis of means (ANOM) to nested designs for improving the visualization and understanding of the sources of variation of chemical and pharmaceutical processes. *Quality Engineering*, **15**(4), 663–670.

Gruska, G. F. and M. S. Heaphy (2000). Designing experiments—an Overview. *ASQC Statistics Division Newsletter*, **19**(2), 8–13.

Hicks, C. R. (1956). Fundamentals of Analysis of Variance, Part III: Nested designs in analysis of variance. *Industrial Quality Control*, **13**(4), 13–16.

Jensen, C. R. (2002). Variance component calculations: Common methods and misapplications in the semiconductor industry. *Quality Engineering*, **14**, 645–657.

Khattree, R. and D. N. Naik (1995). Statistical tests for random effects in staggered nested designs. *Journal of Applied Statistics*, **22**, 495–505.

Khattree, R., D. N. Naik, and R. L. Mason (1997). Estimation of variance components in staggered nested designs. *Journal of Applied Statistics*, **24**, 395–408.

Khuri, A. I. and H. Sahai (1985). Variance components analysis: A selective literature review. *International Statistical Review*, **53**, 279–300.

Lamar, J. M. and W. E. Zirk (1991). Nested designs in the chemical industry. In *Annual Quality Congress Transactions*, American Society for Quality Control, Milwaukee, WI. pp. 615–622.

Liu, S. and R. G. Batson (2003). A nested experiment design for gauge gain in steel tube manufacturing. *Quality Engineering*, **16**(2), 269–282.

Mason, R. L., R. F. Gunst, and J. L. Hess (2003). *Statistical Design and Analysis of Experiments: With Applications to Engineering and Science*, 2nd ed. New York: Wiley.

Naik, D. N. and R. Khattree (1998). A computer program to estimate variance components in staggered nested designs. *Journal of Quality Technology*, **30**, 292–297.

Nelson, P. R. (1993). Additional uses for the analysis of means and extended tables of critical values. *Technometrics*, **35**(1), 61–71.

Nelson, L. S. (1995a). Using nested designs: I. Estimation of standard deviations. *Journal of Quality Technology*, **27**(2), 169–171.

Nelson, L. S. (1995b). Using nested designs: II. Confidence limits for standard deviations. *Journal of Quality Technology*, **27**(3), 265–277.

Ojima, Y. (1998). General formulae for expectations, variances and covariances of the mean squares for staggered nested designs. *Journal of Applied Statistics*, **25**, 785–799.

Pignatiello, J. (1984). Two-stage nested designs. *ASQC Statistics Division Newsletter*, **6**(1), September. (This can be accessed by ASQ members at http://www.asq.org/forums/statistics/newsletters/vol_6_no_1_sept1984.pdf.)

Rao, C. R. and J. Kleffe (1988). *Estimation of Variance Components and Applications*. Amsterdam: North-Holland.

Reece, J. E. (2003). Software to support manufacturing systems. In *Handbook of Statistics*, Vol. 22, Chapter 9 (R. Khattree and C. R. Rao, eds.). Amsterdam: Elsevier Science B. V.

Salvia, A. A. and W. C. Lasher (1989). The power of the *F*-test in two-factor nested experiments. *Journal of Quality Technology*, **21**(1), 20–23.

Searle, S. R., G. Casella, and C. E. McCulloch (1992). *Variance Components*. New York: Wiley.

Sinibaldi, F. J. (1983). Nested designs in process variation studies. In *ASQC Annual Quality Congress Transactions*, pp. 503–508. American Society for Quality Control, Milwaukee, WI.

Smith, J. R. and J. M. Beverly (1981). The use and analysis of staggered nested factorial designs. *Journal of Quality Technology*, **13**(3), 166–173.

Snee, R. D. (1983). Graphical analysis of process variation studies. *Journal of Quality Technology*, **15**(2), 76–88.

Vander Heyden, Y., K. De Braekeleer, Y. Zhu, E. Roets, J. Hoogmartens, J. De Beer, and D. L. Massart (1999). Nested designs in ruggedness testing. *Journal of Pharmaceutical and Biomedical Analysis*, **20**(6), 875–887.

Vardeman, S. B. and J. Wendelberger (2005). The expected sample variance of uncorrelated random variables with a common mean and some applications in random unbalanced effects models. *Journal of Statistical Education*, **13**(1). (This is an online journal available at http://www.amstat.org/publications/jse.)

Yandell, B. (1997). *Practical Data Analysis for Designed Experiments*. London: Chapman & Hall.

EXERCISES

7.1 Describe an experiment from your field of study for which the use of a nested design would be appropriate.

7.2 Derive the expression for $SS_{B(A)}$ that was given in Section 7.2.1 and similarly derive the expression for $SS_{A \times B} + SS_B$ and show their equality.

7.3 Compute directly the sum of squares for $D(H)$ in Example 7.3.

7.4 Consider the schematic diagram of a nested design given in Vardeman and Wendelberger (2005, references). Determine the number of degrees of freedom for error.

7.5 In listing advantages and disadvantages of nested designs Gruska and Heaphy (2000, references) stated the following: "DISADVANTAGE—Nesting is a specialized design appropriate only when some hierarchical structure is present." Respond to that statement.

7.6 Salvia and Lasher (1989, references) gave the following example of data from a two-factor nested design. A food processor produces 12 batches of strawberry jam per day. Each batch contains 100 cases of 24 jars. The response variable is the sugar content of the product. Four batches are randomly selected from the 12; three cases are then randomly selected, and then two jars are sampled. The data are given below.

Batch

Case	1	2	3	4
1	5.821	5.991	6.123	6.215
	5.815	5.997	6.133	6.203
2	5.772	5.973	6.028	6.143
	5.778	5.947	6.063	6.158
3	5.763	5.924	6.025	6.045
	5.746	5.937	6.018	6.039

Perform an appropriate analysis and state your findings to the head statistician and to the chief operating officer.

7.7 Smith and Beverly (1981, references) gave the following example. A product in pellet form is received in bulk form in hopper-type trailers. An experiment was performed to estimate the between-trailers (long term) variability and the within-trailer (short term) variability of a quality characteristic, impurities. Ten trailers were used. Measurement precision was also to be assessed and the two laboratories used in the experiment were to be compared. Four samples were obtained from each trailer, with two assigned to one laboratory and two assigned to the other. Each laboratory made three tests—two replicate measurements on the same sample and a single measurement on a second sample. Within this restriction, samples were assigned randomly to the laboratories. Thus, there are 60 observations: 10*2*3. The data are given below.

Trailer	Lab	Sample	Measurement	Trailer	Lab	Sample	Measurement
1	1	1	47.06	6	1	1	46.99
1	1	1	44.37	6	1	1	50.87
1	1	2	49.30	6	1	2	51.87
1	2	1	47.40	6	2	1	52.14
1	2	1	47.80	6	2	1	49.56
1	2	2	50.43	6	2	2	48.03
2	1	1	47.43	7	1	1	47.49
2	1	1	50.35	7	1	1	51.55
2	1	2	50.42	7	1	2	58.57
2	2	1	50.43	7	2	1	51.61
2	2	1	53.07	7	2	1	49.86
2	2	2	49.18	7	2	2	46.32
3	1	1	48.90	8	1	1	47.41
3	1	1	48.05	8	1	1	47.63

3	1	2	50.64	8	1	2	48.63
3	2	1	52.52	8	2	1	48.46
3	2	1	50.38	8	2	1	46.14
3	2	2	47.64	8	2	2	47.41
4	1	1	52.32	9	1	1	48.37
4	1	1	52.26	9	1	1	51.03
4	1	2	53.47	9	1	2	50.15
4	2	1	47.39	9	2	1	50.53
4	2	1	50.73	9	2	1	47.82
4	2	2	54.49	9	2	2	49.37
5	1	1	46.53	10	1	1	54.80
5	1	1	45.60	10	1	1	51.57
5	1	2	53.98	10	1	2	54.52
5	2	1	48.07	10	2	1	53.02
5	2	1	47.59	10	2	1	51.95
5	2	2	46.50	10	2	2	50.50

Perform an appropriate analysis and state your conclusions.

7.8 Assume that there are five levels of factor A for the schematic diagram in Section 7.3, instead of three as was assumed in the example. What will be the number of degrees of freedom for error?

7.9 Snee (1983, references) gave an example of a nested design with four factors, with the factors being Operator, Specimen, Run, and Analysis. There were three operators, two specimens that were nested under each level of operator, three levels of run that were nested under each level of specimen, and two levels of analysis that were nested under each level of run. All factors were random. Since the description was sketchy, this may not have been an actual experiment, or perhaps the description was sketchy for proprietary reasons. The data are given below.

Operator	Specimen	Run	Analysis	Response Value
1	1	1	1	156
1	1	1	2	154
1	1	2	1	151
1	1	2	2	154
1	1	3	1	154
1	1	3	2	160
1	2	4	1	148
1	2	4	2	150
1	2	5	1	154
1	2	5	2	157
1	2	6	1	147
1	2	6	2	149
2	3	7	1	125
2	3	7	2	125
2	3	8	1	94

2	3	8	2	95
2	3	9	1	98
2	3	9	2	102
2	4	10	1	118
2	4	10	2	124
2	4	11	1	112
2	4	11	2	117
2	4	12	1	98
2	4	12	2	110
3	5	13	1	184
3	5	13	2	184
3	5	14	1	172
3	5	14	2	186
3	5	15	1	181
3	5	15	2	191
3	6	16	1	172
3	6	16	2	176
3	6	17	1	181
3	6	17	2	184
3	6	18	1	175
3	6	18	2	177

(a) Since all factors are random, interest would be focused on variance component estimation. Snee (1983) makes the point that, in general, atypical observations can inflate variances. Do any of these observations seem unusual?

(b) Perform appropriate analyses and draw conclusions regarding the sources of variation.

(c) You will notice a problem that occurs with a variance component estimate. Is the cause of the problem apparent from inspection of the data? If not, perform appropriate analyses (perhaps both graphical and numerical) to try to identify the cause, then read the appropriate portion of Snee (1983) and compare your answer with that given by Snee. Do you agree with the conclusion given by the latter? Explain.

7.10 Gonzalez-de la Parra and Rodriguez-Loaiza (2003, references) gave an example of a nested design with three factors, to which they applied a proposed ANOM procedure that was described in Section 7.6. The objective of the study was to identify the primary source(s) of the overall variability of a synthetic process of a drug substance, as the operators believed that the quality of the product was related to the quality of the starting material. Two suppliers were randomly selected from the list of approved suppliers, three lots of starting material were selected at random from each supplier, and four containers from each lot were selected at random. Three assay determinations were then made on each container for the purpose of estimating the experimental error.

(a) There are some issues here relating to estimating the experimental error, including whether or not the same experimental unit is being measured.

Repeated measures designs are covered in Chapter 11, in which the same experimental unit is measured under different conditions. Is that the case here or, as discussed in Section 1.4.2, is it a matter of distinguishing between multiple readings and replications? In any event, based on the description of the study, do you believe that a reasonable estimate of the experimental error would result?

(b) What will be the degrees of freedom for containers?

7.11 Assume that an experiment in which a fully nested design was used had five factors, all of which were random. How would the divisor for the F-test be constructed for testing for the significance of the only factor that is not nested under another factor?

7.12 An example of a nested design that is often given is one that has been discussed throughout this chapter; namely, an experiment is performed to determine if there is any difference between machine performance in a manufacturing setting. Each machine has, say, four different heads (in four different positions on a machine) and it is felt that heads could influence the readings that are obtained from each machine. Each set of four heads can fit only a given machine, however, so heads are nested within machines. Hicks (1956, references) may have been the first to describe such an experiment with an illustrative example. Four observations were made on each head and the response variable was strain readings. The observations are given below.

	Machine																			
	1				2				3				4				5			
Head	1	2	3	4	5	6	7	8	9	10	11	12	13	14	15	16	17	18	19	20
Readings	6	13	1	7	10	2	4	0	0	10	8	7	11	5	1	0	1	6	3	3
	2	3	10	4	9	1	1	3	0	11	5	2	0	10	8	8	4	7	0	7
	0	9	0	7	7	1	7	4	5	6	0	5	6	8	9	6	7	0	2	4
	8	8	6	9	12	10	9	1	5	7	7	4	4	3	4	5	9	3	2	0

Analyze the data and determine if there is a head effect or a machine effect (or both).

7.13 A survey is conducted in four classrooms in each of two schools, with each classroom–school combination thus constituting a group. Give the breakdown of the degrees of freedom for groups.

7.14 Liu and Batson (2003, references) described an interesting experiment that involved both control factors and noise factors, with the latter mentioned in Section 5.17.1. The experiment is somewhat unique in the sense that noise factors are generally regarded as being controllable under test conditions but not during normal plant operations, as is discussed in Chapter 8, but in this experiment they were not controllable. This means that the noise factors obviously could not be part of the experimental design. Therefore, regression analysis had to be used to determine if the noise factors were related to the response variable. That variable was "gauge gain," which refers to the phenomenon of the gauge of a fabricated steel tube being thicker than that of the coil of steel that is used to produce the steel tube. The problem affects small-dimension tubes manufactured by cold-working.

The design was nested in the following way. There were three factors: gauge size (0.25 in., 0.188 in., and 0.12 in.), gauge shape (rectangular and square), and the size of tube for each shape in inches (3 × 4 and 2 × 5 for rectangular and 4 × 4 and 1.5 × 1.5 for square). From the latter, it is obvious that size of tube is nested under shape, this being the only nesting that occurs. Two replicates were used, so there were 24 observations. The data were as follows, with 11, 7, and 4 used below, in accordance with industry terminology, to represent the three gauge sizes given above; "1" denotes the rectangular shape and "−1" the square shape, with "1" denoting 3 × 4 and "2" denoting 2 × 5 for the rectangular shape, and "1" denoting 4 × 4 and "2" denoting 1.5 × 1.5 for the square shape.

Gauge Gain	Gauge	Shape	Size
0.004	11	1	1
0.004	11	1	1
0.002	7	1	1
0.003	7	1	1
0.004	4	1	1
0.003	4	1	1
0.004	11	1	2
0.005	11	1	2
0.010	7	1	2
0.012	7	1	2
0.007	4	1	2
0.011	4	1	2
-0.001	11	-1	1
0.000	11	-1	1
0.000	7	-1	1
0.001	7	-1	1
0.004	4	-1	1
0.002	4	-1	1
0.001	11	-1	2
0.002	11	-1	2
0.002	7	-1	2
0.000	7	-1	2
0.003	4	-1	2
0.001	4	-1	2

Analyze the data and determine significant effects, if any, recognizing that certain interactions are estimable even though there is nesting.

7.15 (Croarkin and Tobias, 2002, references) gave the following example in Section 3.2.3.3. Pin diameters are studied to see if there is any difference due to machines and operators. There are five machines that each have a day operator and a night operator. Five samples are taken from each machine and for each operator. The data, on pin diameters, are as follows.

	Machine				
	1	2	3	4	5
Operator (Day)	0.125	0.118	0.123	0.126	0.118
	0.127	0.122	0.125	0.128	0.129
	0.125	0.120	0.125	0.126	0.127
	0.126	0.124	0.124	0.127	0.120
	0.128	0.119	0.126	0.129	0.121
Operator (Night)	0.124	0.116	0.122	0.126	0.125
	0.128	0.125	0.121	0.129	0.123
	0.127	0.119	0.124	0.125	0.114
	0.120	0.125	0.126	0.130	0.124
	0.129	0.120	0.125	0.124	0.117

Analyze the data and draw appropriate conclusions.

CHAPTER 8

Robust Designs

We begin at the outset by pointing out that model-robust designs are not discussed in this chapter, as that is a type of robustness that is different from the type with which we are concerned in the following sections. Model-robust designs are robust under various models that contain all the available main effects and some of the interactions. They are discussed in Section 13.19. Similarly, we will not be particularly concerned with designs that are robust to errors in the settings of factors, although this is certainly not unimportant. Box and Draper (1987) stated that "good behavior when errors in settings of input variables occur" is a desirable property of a response surface design and is obviously a desirable property of any type of design. (Some of the other desirable properties of designs were discussed in Section 1.2.)

Box (1994) made the important point that for a robust design (as discussed in the next paragraph) to be effective, the functional relationship between the response variable and the factors needs to be known, at least approximately, as well as "the standard deviations of all of the error variables and design variables, and the *dependence of these standard deviations on the nominal levels of the variables*." This is certainly important, just as good statistical models are always important, in general, but the focus in this chapter is on robust designs as have been given in the literature.

Here is what we *will* discuss. In the 1980s, the concept of a robust design was popularized by Genichi Taguchi, and there have been many refinements since then. If a designed experiment is used to select optimal settings of the factors under consideration so as to optimize (maximize or minimize) the response, it is certainly desirable that when these settings are used in production, the response values are not greatly affected by particular levels of *noise factors*. The latter are variables, such as environmental variables, which can be fixed in the controlled setting used for the experiment, but cannot otherwise be set at any specific values. Companies also want the performance of their products to be relatively insensitive to variations in intended customer usage, such as toys that might be banged around excessively by children.

We will use the term *control factor* to indicate a factor that is of interest, with a robust design consisting of noise factors and control factors. Of course when noise

Modern Experimental Design   By Thomas P. Ryan
Copyright © 2007 John Wiley & Sons, Inc.

factors are used in a design, a decision must be made as to the levels of the noise factors to use, just as this decision must be made for control factors (recall the discussion of this in Section 1.6.2.2). Steinberg and Bursztyn (1998) recommended that the level of each noise factor be set at $\pm 1.5\sigma_N$, with σ_N denoting the standard deviation of the noise factor. Of course this requires an estimate of σ_N.

Another one of Taguchi's major contributions during the 1980s and thereafter was to motivate statisticians and experimenters to separate factors that affect the average value of the response (i.e., location effects) from factors that have dispersion effects (i.e., affect the variability of the response variable). This is easier said than done, however, as location and dispersion effects will generally be confounded in fractional factorial experiments and very often a complete separation will not exist even when there are enough observations to allow the effects to be estimated independently.

Consequently, an experimenter will frequently have to choose between a desirable combination of levels from a location standpoint (i.e., maximizing or minimizing the response), and minimum dispersion. For example, in analyzing Taguchi's arc welding experiment, Bisgaard and Pinho (2003–2004) pointed out that if high tensile strength was to be achieved, the material that exhibited the largest dispersion had to be used, whereas the material that produced the least amount of dispersion seemed to have inferior breaking strength. Occasionally (and perhaps, rarely), it will be possible to accomplish both objectives. For example, Ankenman and Dean (2003, p. 299) described an experiment, concluding "...and so the response variability and the average response be minimized simultaneously in this experiment."

8.1 "TAGUCHI DESIGNS?"

There has been a vast amount of misinformation regarding the field of experimental design and Taguchi's contribution to it. Some sources have even given Taguchi credit for inventing design of experiments. (See, for example, Sharma (2003) who stated "...DOE was invented in the 1960s and developed by people like Taguchi and the military.") Similarly, the promotional material for *Taguchi's Quality Engineering Handbook* by Taguchi, Chowdhury, and Wu (2005) reads "Design of Experiments (known as the Taguchi method)." Gell, Xie, Ma, Jordan, and Padture (2004) stated, "In this study, a Taguchi design of experiments was employed...." Later in this article the authors stated "The selected Taguchi orthogonal array is $L_{27}(3^{13})$." This is another misconception as only a few of the orthogonal arrays that have been attributed to Taguchi were actually invented by Taguchi. (The notation $L_{27}(3^{13})$, as used in Chapter 6, means that 13 three-level factors are examined in 27 runs. Addelman (1962, p. 38) presented a 3^{13-10} design that differs only slightly from the L_{27} array.) Note that full and fractional factorials are orthogonal arrays, but not every fractional factorial is an orthogonal array, as stated in earlier chapters (see Raktoe, Hedayat, and Federer (1981) for details).

Taguchi's approach to experimental design was laid out in Taguchi (1987), a two-volume set. Volume 1 was reviewed by Senturia (1989) and Volume 2 was reviewed by Bisgaard (1989a). The reader may wish to read these reviews, especially Bisgaard (1989a), which is an appropriately critical review and which also draws a conclusion about Volume 1, as well as Box, Bisgaard, and Fung (1988) and Bisgaard (1989b), which is a nontechnical article. Bisgaard (1989a) stated we would be regressing to a level that even the books on experimental design in the 1940s would exceed if Volume 1 were to become "a standard course on the design of experiments" and pointed out that one often encounters methods in Volume 2 that are "...overly complicated, inefficient, and sometimes simply wrong."

The orthogonal arrays that Taguchi has advocated have received some criticism. Ryan (1988) was apparently the first to point out in print that some of these orthogonal arrays are equivalent to suboptimal fraction factorials, suboptimal in the sense that the resolution of the design is not maximized. This is also discussed in Ryan (2000, p. 442).

When these arrays are generated with software, the low level is denoted by "1" and the high level by "2." For example, given below is the $L_8(2^7)$ produced by the Design-Expert software, with the labeling of the columns being that given by Taguchi and Wu (1979), for example, under the assumption that there are four factors and two of the interactions are of interest.

Std. Order	Run Order	B	C	BC	D	BD	A	e
1	5	1	1	1	1	1	1	1
2	8	1	1	1	2	2	2	2
3	4	1	2	2	1	1	2	2
4	2	1	2	2	2	2	1	1
5	6	2	1	2	1	2	1	2
6	7	2	1	2	2	1	2	1
7	3	2	2	1	1	2	2	1
8	1	2	2	1	2	1	1	2

It is not obvious that the column labeled BC is the product of the columns labeled B and C, but this does become apparent if the (1, 2) level designation is replaced by (0, 1), so as to produce the following configuration:

Std. Order	Run Order	B	C	BC	D	BD	A	e
1	5	0	0	0	0	0	0	0
2	8	0	0	0	1	1	1	1
3	4	0	1	1	0	0	1	1
4	2	0	1	1	1	1	0	0
5	6	1	0	1	0	1	0	1
6	7	1	0	1	1	0	1	0
7	3	1	1	0	0	1	1	0
8	1	1	1	0	1	0	0	1

Then the BC interaction levels are the sum of the B and C levels, mod 2. (Recall this use of modular arithmetic in Section 6.1.1, for example.) With this change, the levels of both of the interaction columns are easy to verify, using modular arithmetic.

With this designation we can easily see, for example, that $A = CD$, and also that $C = AD$ and $D = AC$. Thus, the design could not be resolution IV, whereas a 2_{IV}^{4-1} design could be constructed. The L_8 when used with four factors *must* be equivalent to "some" 2^{4-1} design and from the alias structure we know that this is the half fraction with the defining relation $I = ACD$, whereas the defining relation of the 2_{IV}^{4-1} design is of course $I = ABCD$.

8.2   IDENTIFICATION OF DISPERSION EFFECTS

Methods for checking on homogeneity of variance in linear models can be adapted to the detection of dispersion effects. As discussed by, for example, Bisgaard and Pinho (2003–2004), this can be accomplished by computing the residuals from an appropriate model for location. If location were the sole interest, the residuals would be used to test the model assumptions and to check for bad data. Certainly this should still be done, but in searching for dispersion effects the residuals would be additionally used as the response and a normal probability plot analysis performed. It is necessary for the important effects to be used in the first stage in the model that produces the residuals, however, because if this isn't done, important location effects will be spotlighted by the normal probability plot because this is what happens when necessary terms are left out of the model.

Even if all important terms have been included, however, the plot can only be suggestive of possible dispersion effects since replication is necessary to identify such effects, an important but often overlooked point that was originally made by Box and Meyer (1986). It is becoming better known that certain tests for dispersion effects, such as F-tests, are undermined by the failure to identify significant location effects. In particular, Schoen (2004) stated, "Recent literature shows a severe sensitivity of the dispersion F-test to unidentified location effects, to the link function for the variance, and to the presence of other dispersion effects."

The following examples illustrate these ideas.

Example 8.1

Steinberg and Bursztyn (1994) showed with two examples how the failure to model noise factors can cause errors in the identification of dispersion effects. The first example utilized data from Engel (1992), who described an experiment designed to improve an injection molding process. The goal was to determine factor settings for which the amount of percent shrinkage (the response variable) would be close to the target value. The latter was not specified, however.

Seven control factors and three noise factors were used in the experiment. A 2^{7-4} design was used for the control factors, which is referred to as the *inner array*. This design was crossed with a 2^{3-1} design that was used for the noise factors, that being the *outer array*. Thus, there were 32 design points and the use of an inner array and an outer array means that a *product array* was used. In a *combined array*, there is only one array, which contains the columns for both the controllable and noise variables.

TABLE 8.1   Data from Injection Molding Experiment

Control Factors							Noise Factors (M, N, O)			
A	B	C	D	E	F	G	$(-1, -1, -1)$	$(-1, 1, 1)$	$(1, -1, 1)$	$(1, 1, -1)$
-1	-1	-1	-1	-1	-1	-1	2.2	2.1	2.3	2.3
-1	-1	-1	1	1	1	1	0.3	2.5	2.7	0.3
-1	1	1	-1	-1	1	1	0.5	3.1	0.4	2.8
-1	1	1	1	1	-1	-1	2.0	1.9	1.8	2.0
1	-1	1	-1	1	-1	1	3.0	3.1	3.0	3.0
1	-1	1	1	-1	1	-1	2.1	4.2	1.0	3.1
1	1	-1	-1	1	1	-1	4.0	1.9	4.6	2.2
1	1	-1	1	-1	-1	1	2.0	1.9	1.9	1.8

(Early work on constructing outer array points was given by Wang, Lin, and Fang, 1995.)

The design(s) and the observed percentage shrinkages as given in Steinberg and Bursztyn (1994) were as follows, with the triplets in parentheses denoting the levels of the three noise factors (Table 8.1).

We can see by the way that the design in the control factors is listed that although this is equivalent to a 2^{7-4} design, it was probably not constructed as a fractional factorial, but rather a Taguchi orthogonal array was used, specifically the L_8. This seems apparent because the first three columns do not constitute a full factorial design. Columns A, B, and D do constitute a 2^3 design, however, with $C = -AB$, $E = -AD$, $F = -BD$, and $G = ABD$. Similarly, the design in the noise factors has $O = -MN$.

Engel (1992) analyzed the data by computing an average at each design point, averaging over the noise factors, and obtained a model for the mean with terms that were the main effects of factors A, D, and E. Thus, the noise factors were not analyzed and, in particular, no control $\times$ noise interactions could be computed. Steinberg and Bursztyn (1994) analyzed the data as having come from a 2^{10-5} design, so that the noise factors are analyzed as factors and all 32 observations are used. Although this can certainly be done and is clearly preferred over the analysis method of Engel (1992), the "product design" so obtained is not necessarily a good design. The product of the two designs cannot have a higher resolution than the smaller resolution of the two designs that are used to form the product (see, e.g., Ryan, 2000, p. 449). Therefore, the product design cannot be greater than resolution III, which is also obvious from the fact that $C = -AB$, for example. A 2^{10-5}_{IV} design can be constructed, so if it were known in advance that all 10 factors were to be analyzed, then a 2^{10-5}_{IV} could have been used.

In this case Steinberg and Bursztyn (1994) were simply trying to salvage what they could from the experiment. As they pointed out, however, the product design does permit the estimation of all of the control factor $\times$ noise factor interactions, but this comes at the cost of having the main effects of the control factors confounded with two-factor interactions of the control factors. The advantages and disadvantages of a product array are discussed in detail in Section 8.3. With 32 observations, 31 effects could be estimated; the 10 main effects and $7 \times 3 = 21$ two-factor interactions of the control factors and noise factors use all of those degrees of freedom.

One important finding of the experiment, which should be apparent from Table 8.1, is that the variability in the response is much less when factor F is at the low level than when the factor is at the high level. This would suggest using the low level of the factor since the average response for the two levels does not differ greatly. This conclusion, however, would be based on the standard deviations at each design point, which are computed from only four numbers. Furthermore, there are two suspicious data points that have a considerable effect on the conclusions, as noted by Steinberg and Bursztyn (1994). The reader is asked to pursue this line of analysis in Exercise 8.3.

The variability might have been modeled, such as by using $\log(s)$ as the response variable, with s being the standard deviation. Only main effects of the control factors could be estimated, however, as that would use up all of the degrees of freedom, but the control factors that affect the variability can be seen by looking at Table 8.1.

8.3 DESIGNS WITH NOISE FACTORS

There are two major ways in which noise factors can be used in an experimental design: (1) by constructing a separate design for the noise factors and using the resultant number of design points at *each* treatment combination for the main design of the factors of interest, or (2) by constructing a single design and simply having the noise factors be part of the design. The former is called a product array, which will usually be too expensive to run, and the latter is a combined array, as indicated previously.

We hope that noise factors are not significant because since they are uncontrollable in practice, we cannot set a noise factor at what might seem to be the most desirable level. Similarly, noise $\times$ noise interactions that are significant are also of no value to us, but control factor $\times$ noise factor interactions are important and provide useful information. Accordingly, some work has been directed at identifying these interactions, including Russell, Lewis, and Dean (2004). See also Vine, Lewis, and Dean (2005) who proposed a two-stage group screening procedure that was based upon subjective probabilities of the various effects, including control $\times$ noise interactions, being real effects. Bingham and Li (2002) stated that the models of primary interest are those that contain at least one control factor $\times$ noise factor interaction and introduced the model-ordering principle with models being ranked by their order of importance. Bingham and Sitter (2003) considered the use of fractional factorial split-plot designs in robust parameter design work. Other work includes Kuhn (2003), Kuhn, Carter, and Myers (2000), and Bisgaard and Ankenman (1995), with the latter formulating the parameter design problem as a constrained optimization problem.

As a simple illustration of this, assume that we have only a single control factor and a single noise factor and their interaction graph that results from a particular experiment is given in Figure 8.1.

If the objective is to minimize the variability of the response, the high $(+1)$ level of the control factor should be used since the response is constant over the two noise levels for that level of the control factor. Interaction plots such as the idealized, in terms

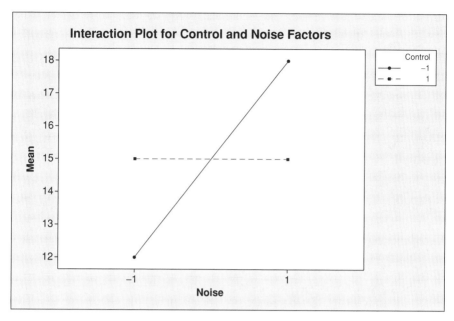

Figure 8.1   Control × noise interaction.

of robustness, plot in Figure 8.1 are potentially very helpful and should be routinely constructed, just as we construct interaction plots when factorial designs are used. We have to look at the plots and not just go by the magnitude of the interaction; the reason for this is explained in Section 8.4.

Certainly these plots are more useful than the results that are obtained by modeling location and dispersion separately, which is the preferred alternative to the use of signal-to-noise (S/N) ratios that were debunked many years ago (see, e.g., Box, 1988; Nair, 1992). (We do not discuss S/N ratios in this chapter, as the emphasis is on experimental designs. Readers interested in the subject can find these ratios discussed in books and articles on the Taguchi methods.)

What should be done if the objective is to maximize the average response? In the absence of any information regarding which noise level should predominate in practice, nothing should be done.

Of course, ideally, we would like to be able to identify control factors that affect dispersion but do not affect location, and then use a set of factors that affect location, but not dispersion, to arrive at the factor settings that maximize or minimize the average response.

It seems unlikely that such a scenario will be encountered very often, however, when we view the problem in terms of conditional effects. The main effect estimate for the control factor shown in Figure 8.1 is zero because the average response is the same at each level of the control factor. The conditional effects are 3 and −3, however, and this is 20 percent of the average response. Such a difference may or may not be deemed significant, but clearly we should not say that the factor has no effect. (Note that if the line for the low level of the control factor were twisted so that the slope increased, the absolute value of the conditional effects would increase but the main effect estimate would remain at zero. Thus, it would be easy to envision a scenario for which the conditional effects would undoubtedly be significant.)

Stated generally, if a control factor has a dispersion effect, then the lines in the interaction graph cannot be parallel, nor can one slope of a line be the negative of the slope of the other line. In the latter case the main effect estimate would be zero and we might erroneously conclude that a factor has neither a location effect nor a dispersion effect even though the conditional effects could be quite large.

Thus, we must consider conditional effects when robust designs are used, just as we must do so when factorial designs are used without noise factors.

8.4   PRODUCT ARRAY, COMBINED ARRAY, OR COMPOUND ARRAY?

An experimenter who decides to use a robust design can select a design from certain types of robust designs, the best known of which are product arrays and combined arrays.

With the product array approach, a separate factorial or fractional factorial design or orthogonal array is constructed for the control factors and the noise factors (termed the inner array and outer array, respectively, as stated in Example 8.1). The product array is simply the product of the inner array and the outer array, as indicated previously.

As a simple example, assume that there are two control factors and a 2^2 design is constructed. Similarly, a 2^2 design is also constructed for the noise factors. When we take the product of the two designs, we will have a design with 16 points because the outer array with the noise factors is at each of the four design points of the inner array. So the design has $4 \times 4 = 16$ points. What is the nature of this particular product array? That is, how can the resultant array (design) be described? As the reader is asked to show in Exercise 8.1, the result is a 2^4 design.

Shoemaker, Tsui, and Wu (1991) proved that the effects that are estimable in the product array are the effects that are estimable in the inner and outer arrays, respectively, plus all generalized interactions of those estimable effects. For the present example, let A and B denote the control factors and let C and D denote the noise factors. The effects that are estimable in the inner array are thus A, B, and AB, with C, D, and CD estimable from the outer array. The combination of these estimable effects with all of their cross products obviously constitutes the effects that are estimable with a 2^4 design.

This is clearly not a large design, but if either the number of control factors or the number of noise factors is larger than 2 (or both), the number of design points could be too large if a full factorial were used for each array. For example, 128 runs would be required if the inner array had four factors at two levels and the outer array had three factors at two levels. Consequently, a fractional factorial would often have to be used, but another disadvantage of the product array approach is the inflexibility regarding the effects that can be estimated. As stated previously, the effects that are estimable beyond those that are estimable in each array are the generalized interactions of those estimable effects. Those generalized interactions will often include effects that the experimenter would not want to estimate, however. With the example given previously, an experimenter would generally not want to estimate the ABCD interaction as four-factor interactions rarely exist.

A combined array has the advantage of economy, but care should be exercised to ensure that certain categories of effects are not confounded. In particular, we don't want to confound control × noise interactions with anything that is likely to be significant, as these interactions are the key to robust design. Therefore, we would want to confound those interactions with noise × noise interactions or with high-order interactions among the control factors.

A compromise between a product array and a combined array is a *compound array*, as introduced by Rosenbaum (1994, 1996). Such a design has a specified number of treatment combinations of noise factors for each treatment combination of control factors, with the number being less than all possible combinations (as would be used in a product array). Rosenbaum (1996) supports the use of a (4, 3, 4) design, with $(\gamma, \lambda, \alpha)$ denoting a design that is of resolution α for the control and noise factors combined, of resolution γ for the control factors only, and of resolution λ for the noise factors only. The advantage of a compound array for the designs considered by Rosenbaum (1996) is that by using a combined array rather than a product array, α will be 4 instead of 3. This is desirable because, as stated previously, it is important to be able to estimate the control × noise interactions. Of course with a resolution IV design the two-factor interactions will be confounded among themselves, so having $\alpha = 4$ is helpful only if the control × noise interactions are confounded with noise × noise interactions, *and* the noise × noise interactions can be assumed to be quite small. That is, each control × noise interaction must not be confounded with a main effect of a control factor nor confounded with another control × noise interaction.

Certainly that is necessary for Figure 8.1 to point the direction toward desired results, as we want to be able to select the value of the control factor so as to reduce variability, but we won't necessarily be able to do that if that particular interaction is confounded with another two-factor interaction that is significant, or with a significant main effect.

It is important to look at graphs such as Figure 8.1 and not simply go by the magnitude of a control × noise interaction, as it is the *shape* of the interaction profile that is important. More specifically, a control × interaction that graphs as an "X" will have a larger sum of squares than the interaction depicted in Figure 8.1, but an interaction that graphs as an "X" is of no value in selecting control factor settings so as to minimize dispersion since with such a configuration the two levels of the control factor (assuming there are two) result in equal variability. So we can't just go by sums of squares and F-statistics. Shoemaker et al. (1991) were apparently the first to point this out.

This begs for an analysis using graphs rather than a strictly numerical approach (ANOVA tables, etc.) for identifying important control × noise interactions. Indeed, Ryan (2000, p. 453) showed that for a 2^2 design, the A × B interaction must be almost to the point of being of practically no value from a robustification standpoint before it is declared significant using a method for analyzing unreplicated factorials such as the one given by Lenth (1989).

The important control × noise interactions are those for which the absolute values of the slopes of the lines that connect the points in the graph differ more than slightly. Consequently, software for robust design and analysis might be constructed so as to rank order the interactions by this criterion, but I am not aware of any software that does so. Of course with some software packages this could be easily done as the effect estimates and the corresponding names could be stored and then sorted into descending order.

Example 8.2

Brennerman and Myers (2003) described a robust parameter designed experiment that utilized a combined array. Engineers in the packaging development department at Procter and Gamble wanted to develop optimum and robust settings for the control variables for a new sealing process so as to hit a target value for the response variable, which was maximum peel strength. This is the maximum amount of strength, measured in pounds of force, which is required to open a package. There were three control factors: temperature, pressure, and speed. The packaging materials were furnished by different suppliers and the engineers did not want to have a single set of manufacturing conditions for each packaging material, so it was important to have settings for the control factors that would be robust to supplier-to-supplier variation. Supplier was the categorical noise variable and there were three suppliers. A 37-run D-efficient design was used with 37 being the maximum number of runs that could be made because of constraints on resources and time. A D-efficient design is (as the name implies) a design that is efficient relative to a D-optimal design. A formal definition of D-efficiency is given on, for example, page 9 of Waterhouse (2005).

The authors did not give the design, nor did they give the number of levels used for each of the control factors. Since 37 is a prime number, it is obvious that the design had to be lacking in balance. That is, for whatever number of levels used (less than 37, presumably), the levels could not occur an equal number of times. The design may have had good properties, however, as it is possible, for example, to construct a design with 37 runs that has two levels for each of the control factors and three levels for the noise factor with very small pairwise correlations, so that the design is near-orthogonal. No other properties of the design were stated, however, but one obvious weakness of the design was that the standard error of the point estimate of the parameters for the terms in the fitted model varied greatly, ranging from 0.021 to 1.344. One problem with such a design is that it is difficult to apply any rules of thumb regarding the magnitude of interactions relative to the magnitude of main effects (as discussed, for example, in Section 4.2), when the standard errors differ by orders of magnitude, as obviously we would strongly prefer that the standard errors be the same in applying any such rules.

8.5 SOFTWARE

Although software that has the capability for Taguchi's orthogonal arrays is plentiful, the designs should be used with caution, if at all. At the very least, the user should know the properties of a chosen design before using it in an experiment.

Unfortunately, Reece's (2003) very comprehensive study of software with experimental design capabilities did not include Taguchi designs, so some guidance will be given in this section.

Release 14 of MINITAB generates Taguchi (orthogonal array) designs, as this is one of the pull-down menu options. Many different designs can be generated: two-level designs for up to 31 factors, three-level designs for up to 13 factors, four-level designs for up to 5 factors, five-level designs for up to 6 factors, plus mixed-level designs. The available mixed-level designs are 2-3, 2-4, 2-8, and 3-6, with, for example, "2-4" referring to a mixture of two-level factors and four-level factors. (Only one 2-8 and one 3-6 are available.) These arrays have 8, 16, 18, 32, 36, and 54 design points.

Design-Expert 7.0 has capabilities for more designs than does MINITAB, with the number of design points for Taguchi orthogonal array designs being 4, 8, 9, 12, 16, 18, 25, 27, 32, 36, 50, 54, and 64. These are for factors with 2, 3, 4, or 5 levels, with seven of the listed designs being mixed-level designs.

Of course, since these are orthogonal arrays, each array could be used with fewer than the maximum number of factors. Thus, there is a moderately large number of possible designs.

The screen display in Design-Expert 7.0 that lists the available designs also states the following.

Use these designs with caution. Always use the design evaluation to examine aliasing before running any experiments and again when analyzing and interpreting the results.

JMP 5.1 also has the capability for Taguchi designs but unlike MINITAB and Design-Expert there is no list of available designs. Rather, the user simply specifies each factor as a two-level or three-level control factor, or as a noise factor. If a moderate number of control factors are specified, design options are listed. For example, the L_{16}, L_{20}, L_{24}, L_{32}, L_{64}, and L_{128} arrays are listed as available designs for the inner array when 15 two-level control factors and two noise factors are indicated (the noise factors must have two levels), with an L_4 array listed as the only available design for the noise factors in the outer array. If too many control factors are specified, however, such as 38 two-level control factors and two noise factors, the program crashes and an "unknown error" message is displayed.

When three-level control factors are specified, different arrays are listed as possible designs, such as the L_{36} when there are 11 three-level control factors and 2 noise factors and the L_{27} when there are 7 three-level factors and 2 noise factors. For a small number of factors, the options could include a full factorial for the inner array, such as when only 3 three-level control factors are indicated.

Although the options (and there might be only one option) are indicated, once the control factors and noise factors are specified, it would be better to know what is available before the latter is done.

Because of the popularity of Taguchi methods/designs in certain quarters, there are many other software packages that will generate these designs. Among the other statistical software packages studied by Reece (2003), D. o. E. Fusion (previously known as CARD) does not have such capability, however, and allows a maximum of only 10 factors. Thus, the software could not be used for screening with a large number of candidate variables, such as is sometimes the case when Taguchi designs are used.

8.6 FURTHER READING

There is much useful information in the statistical literature on critiques of Taguchi's design methods and superior alternatives, especially in Steinberg (1996), Steinberg and Bursztyn (1994, 1998), and Tsui (1994, 1996a,b, 1998). A review of parameter design was given by Robinson, Borror, and Myers (2003) and application articles and case studies were given by Chen, Allen, Tsui, and Mistree (1996), Czitrom, Mohammadi, Flemming, and Dyas (1998), and Muzammil, Singh, and Talib (2003). An interesting feature of the experiment described by Czitrom et al. (1998) is that control charts were used *after* the experiment but there is no mention of them being used during the experiment. Runs at the standard operating conditions were made at the beginning and the end of the experiment, however, which serves a similar purpose and has been recommended as a control procedure, as discussed in Section 1.7. Control charts were used after the experiment to show the reduction in variability that was achieved.

Also listed in the references are articles on various related topics, such as robust designs with cost considerations (Morehead and Wu, 1998), analysis of certain types of robust design experiments (McCaskey and Tsui, 1999; Miller, 2002), and combined arrays with a minimum number of runs (Evangelaras and Koukouvinos, 2004), and an improved dual response method given by Miro-Quesada and Del Castillo (2004). There have also been standard techniques such as split-plot experiments presented as applicable to robust design and discussed by Kowalski (2002), as was the possible use of generalized linear modeling methods in robust design, as discussed by Engel and Huele (1996) and Lesperance and Park (2003). See also Lawson and Helps (1996), Tsui (1999) and Joseph (2003).

Section 14.9 of Ryan (2000) gives certain detailed information regarding Taguchi methods of design that are covered only generally in this chapter, and also contains a very detailed analysis of a designed experiment that was presented by Lewis, Hutchens, and Smith (1997), which is mentioned in Exercise 8.4.

8.7 SUMMARY

The emphasis in this chapter was on certain important aspects of experimental design as related to making products that are robust to variations in manufacturing conditions (noise factors). Readers interested in S/N ratios and other statistical methods used by G. Taguchi will find information about them in many sources, including Fowlkes and Creveling (1995), but those methods are not advocated here, just as Taguchi's design methods were not advocated in sources such as Bisgaard (1989a). Similarly, Anderson and Kraber (2003) gave an example that illustrated the superiority of two-level fractional factorials over Taguchi designs. See also the critique of Taguchi methods in Ramberg, Pignatiello, and Sanchez (1992).

Taguchi's engineering ideas are quite important, however, and have led researchers and practitioners to focus their attention on noise factors and control $\times$ noise interactions.

REFERENCES

Addelman, S. (1962). Orthogonal main-effect plans for asymmetrical factorial experiments. *Technometrics*, **4**(1), 21–46.

Anderson, M. J. and S. L. Kraber (2003). Using design of experiments to make processes more robust to environmental and input variations. *Paint and Coating Industry*, February, 1–7.

Ankenman, B. E. and A. M. Dean (2003). Quality improvement and robustness via Design of Experiments. In *Handbook of Statistics*, Vol. 22, Chap. 8 (R. Khattree and C. R. Rao, eds.). Amsterdam: Elsevier Science B. V.

Bingham, D. and W. Li (2002). A class of optimal robust designs. *Journal of Quality Technology*, **34**(3), 244–259.

Bingham, D. and R. Sitter (2003). Fractional factorial split-plot designs for robust parameter experiments. *Technometrics*, **45**(1), 80–89.

Bisgaard, S. (1989a). Review of *System of Experimental Design, Vol. 2* by G. Taguchi. *Technometrics*, **31**(2), 257–260.

Bisgaard, S. (1989b). Quality engineering and Taguchi methods: A perspective. *Target*, October. (This is also available as Report No. 40, Center for Productivity and Quality Improvement, University of Wisconsin-Madison and may be downloaded at http://www.engr.wisc.edu/centers/cqpi/reports/pdfs/r040.pdf.)

Bisgaard, S. and B. Ankenman (1995). Analytic parameter design. *Quality Engineering*, **8**(1), 75–91. (This article is available as Report No. 103, Center for Productivity and Quality Improvement, University of Wisconsin-Madison and may be downloaded at http://www.engr.wisc.edu/centers/cqpi/reports/pdfs/r103.pdf.)

Bisgaard, S. and A. Pinho (2003–2004). Follow-up experiments to verify dispersion effects: Taguchi's welding experiment. *Quality Engineering*, **16**, 335–343.

Box, G. E. P. (1988). Signal-to-noise ratios, performance criteria, and transformations. *Technometrics*, **30**, 1–40 (including discussion).

Box, G. E. P. (1994). Is your robust design product robust? *Quality Engineering*, **6**(3), 503–514.

Box, G. E. P. and N. R. Draper (1987). *Empirical Model Building and Response Surfaces*. New York: Wiley.

Box, G. E. P. and R. D. Meyer (1986). Dispersion effects from fractional designs. *Technometrics*, **28**(1), 19–27.

Box, G. E. P., S. Bisgaard, and C. Fung (1988). An explanation and critique of Taguchi's contributions to quality engineering. *Quality and Reliability Engineering International*, **4**(2), 123–131.

Brennerman, W. A. and W. R. Myers (2003). Robust parameter design with categorical noise variables. *Journal of Quality Technology*, **35**(4), 335–341.

Chen, W., J. K. Allen, K.-L. Tsui, and F. Mistree (1996). A procedure for robust design: Minimizing variations caused by noise factors and control factors. *ASME Journal of Mechanical Design*, **118**, 478–485.

Czitrom, V., P. Mohammadi, M. Flemming, and B. Dyas (1998). Robust design experiment to reduce variance components. *Quality Engineering*, **10**(4), 645–655.

Engel, J. (1992). Modelling variation in industrial experiments. *Applied Statistics*, **41**, 579–593.

Engel, J. and F. A. Huele (1996). A generalized linear modeling approach to robust design. *Technometrics*, **38**(4), 365–373.

Evangelaras, H. and C. Koukouvinos (2004). Combined arrays with minimum number of runs and maximum estimation efficiency. *Communications in Statistics, Theory and Methods*, **33**, 1621–1628.

Fowlkes, W. Y. and C. M. Creveling (1995). *Engineering Methods for Robust Product Design: Using Taguchi Methods in Technology and Product Development*. Englewood Cliffs, NJ: Prentice Hall.

Gell, M., L. Xie, X. Ma, E. H. Jordan, and N. P. Padture (2004). Highly durable thermal barrier coatings made by the solution precursor plasma spray process. *Surface and Coatings Technology*, **177–178**, 97–102. (The article is available at http://www.matsceng. ohio-state.edu/fac_staff/faculty/padture/padturewebpage/padture/PadturePapers/Padture SurfCoat2.pdf.)

Joseph, V. R. (2003). Robust parameter design with feed-forward control. *Technometrics*, **45**(4), 282–292.

Kowalski, S. M. (2002). 24 run split-plot experiments for robust parameter design. *Journal of Quality Technology*, **34**, 399–410.

Kuhn, A. (2003). Optimizing response surface experiments with noise factors using confidence regions. *Quality Engineering*, **15**, 419–426.

Kuhn, A. M., W. H. Carter, and R. H. Myers (2000). Incorporating noise factors into experiments with censored data. *Technometrics*, **42**, 376–383.

Lawson, J. and R. Helps (1996). Detecting undesirable interactions in robust designed experiments. *Quality Engineering*, **8**(3), 465–473.

Lenth, R. V. (1989). Quick and easy analysis of unreplicated factorials. *Technometrics*, **31**, 469–473.

Lesperance, M. L. and S.-M. Park (2003). GLMs for the analysis of robust designs with dynamic characteristics. *Journal of Quality Technology*, **35**(3), 253–263.

Lewis, D. K., C. Hutchens and J. M. Smith (1997). Experimentation for equipment reliability improvement. In *Statistical Case Studies for Industrial Process Improvement*, Chap. 27. (V. Czitrom and P. D. Spagon, eds.) Philadelphia, PA: Society of Industrial and Applied Mathematics; Alexandria, VA: American Statistical Association.

McCaskey, S. D. and K.-L. Tsui (1999). Analysis of dynamic robust design experiments. *International Journal of Production Research*, **35**(6), 1561–1574.

Miller, A. (2002). Analysis of parameter design experiments for signal-response systems. *Journal of Quality Technology*, **34**(2), 139–151.

Miro-Quesada, G. and E. Del Castillo (2004). Two approaches for improving the dual response method in robust design. *Journal of Quality Technology*, **36**(2), 154–168.

Morehead, P. R. and C. F. J. Wu (1998). Cost-driven parameter design. *Technometrics*, **40**(2), 111–119.

Muzammil, M., P. P. Singh, and F. Talib (2003). Optimization of gear blank casting process using Taguchi's robust design technique. *Quality Engineering*, **15**(3), 351–359.

Nair, V., ed. (1992). Taguchi's parameter design: A panel discussion. *Technometrics*, **34**(2), 127–161.

Raktoe, B. L., A. Hedayat, and W. T. Federer (1981). *Factorial Designs*. New York: Wiley.

Ramberg, J. S., J. J. Pignatiello, Jr., and S. M. Sanchez (1992). A critique and enhancement of the Taguchi method. *Annual Quality Congress Transactions*, **46**, 491–498.

Reece, J. E. (2003). Software to support manufacturing systems. In *Handbook of Statistics*, Vol. 22, Chap. 9 (R. Khattree and C. R. Rao, eds.). Amsterdam: Elsevier Science B.V.

Robinson, T. J., C. M. Borror, and R. H. Myers (2003). Robust parameter design: A review. *Quality and Reliability Engineering International*, **20**(1), 81–101.

Rosenbaum, P. R. (1994). Dispersion effects from fractional factorials in Taguchi's method of quality design. *Journal of the Royal Statistical Society, Series B*, **56**, 641–652.

Rosenbaum, P. R. (1996). Some useful compound dispersion experiments in quality design. *Technometrics*, **38**, 354–364.

Rowlands, H., J. Antony, and G. Knowles (2000). An application of experimental design for process optimisation. *The TQM Magazine*, **12**(2), 78–84.

Russell, K. G., S. M. Lewis, and A. M. Dean (2004). Fractional factorial designs for the detection of interactions between design and noise factors. *Journal of Applied Statistics*, **36**, 545–552.

Ryan, T. P. (1988). Taguchi's approach to experimental design: Some concerns. *Quality Progress*, **21**(5), 34–36.

Ryan, T. P. (2000). *Statistical Methods for Quality Improvement*, 2nd ed. New York: Wiley.

Schoen, E. D. (2004). Dispersion-effects detection after screening for location effects in unreplicated two-level experiments. *Journal of Statistical Planning and Inference*, **126**, 289–304.

Senturia, J. (1989). Review of *System of Experimental Design, Vol. 1* by G. Taguchi. *Technometrics*, **31**(2), 256–257.

Sharma, V. (2003). Six Sigma: A dissenting opinion. *Manufacturing Engineering*, October. (The article is available at http://www.findarticles.com/p/articles/mi_qa3618/is_200310/ai_n9341480.)

Shoemaker, A. C., K.-L. Tsui, and C. F. J. Wu (1991). Economical experimentation methods for robust design. *Technometrics*, **33**, 415–427.

Steinberg, D. M. (1996). Robust design: Experiments for improving quality. In *Handbook of Statistics. Vol. 13: Design and Analysis of Experiments* (S. Ghosh and C. R. Rao, eds.). Amsterdam: Elsevier/North Holland.

Steinberg, D. M. and D. Bursztyn (1994). Dispersion effects in robust-design experiments with noise factors. *Journal of Quality Technology*, **26**, 12–20.

Steinberg, D. M. and D. Bursztyn (1998). Noise factors, dispersion effects, and robust design. *Statistica Sinica*, **8**(1), 67–85. (This article is available online at http://www3.stat.sinica.edu.tw/statistica/oldpdf/A8n13.pdf .)

Taguchi, G. (1987). *System of Experimental Design*, Vols. 1 and 2. White Plains, NY: UNIPUB/Kraus International Publications.

Taguchi, G. and Y. Wu (1979). *Introduction to Off-Line Quality Control*. Central Japan Quality Control Association, Nagaya.

Taguchi, G., S. Chowdhury, and Y. Wu (2005). *Taguchi's Quality Engineering Handbook*. Hoboken, NJ: Wiley.

Tsui, K.-L. (1994). Avoiding unnecessary bias in robust design analysis. *Computational Statistics and Data Analysis*, **18**(5), 535–546.

Tsui, K.-L. (1996a). A multi-step analysis procedure for robust design. *Statistica Sinica*, **6**, 631–648.

Tsui, K.-L. (1996b). A critical look at Taguchi's modeling approach for experimental design. *Journal of Applied Statistics*, **23**(1), 81–95.

Tsui, K.-L. (1998). Alternatives of Taguchi modeling approach for dynamic robust design problems. *International Journal of Reliability, Quality and Safety Engineering*, **5**(2), 115–131.

Tsui, K.-L. (1999). Modeling and analysis of dynamic robust design experiments. *IIE Transactions*, **31**, 1113–1122.

Vine, A. E., S. M. Lewis, and A. M. Dean (2005). Two-stage group screening in the presence of noise factors and unequal probabilities of active effects. *Statistica Sinica*, **15**, 871–888.

Wang, Y., D. K. J. Lin, and K.-T. Fang (1995). Designing outer array points. *Journal of Quality Technology*, **27**(3), 226–241.

Waterhouse, T. H. (2005). *Optimal Experimental Design for Nonlinear and Generalised Linear Models*. Ph.D. thesis, School of Physical Sciences, University of Queensland. (This is available at http://www.maths.uq.edu.au/~thw/research/thesis.pdf.)

EXERCISES

8.1 Show that a product array that results from a 2^2 inner array and a 2^2 outer array is a 2^4 design.

8.2 What are the advantages and disadvantages of a product array relative to a combined array?

8.3 Consider Example 8.1. Steinberg and Bursztyn (1994) discovered that there are two suspicious data points and that the predicted value for one data point is equal to the observed value of the other data point, suggesting that the points may have been accidentally switched. This illustrates the need for a careful analysis, which of course is true for any dataset. Identify the two points, switch them, and reanalyze the data. Does this have a material effect on the results of the analysis? Explain. Assuming the two data points indeed should be switched, what do you conclude and what would you recommend?

8.4 Lewis et al. (1997, references) described an experiment performed by Electro-Scientific-Industries (ESI) that was motivated by the dissatisfaction expressed by a company's customer in Japan. The customer was dissatisfied with the mean-time-between failures (MTBF) performance of one of ESI's products. The company had six months to find a solution to the problem; otherwise, the Japanese company would switch to one of ESI's competitors.

Eight control factors and three noise factors were studied with the inner array for the control factors being a 2^{8-4}_{IV} design and the outer array being a 2^{3-1}_{III} design.

  (a) Critique the design that was used. In particular, what was the design in terms of the $(\gamma, \lambda, \alpha)$ notation that was used in Section 8.4? Would you have recommended that a different design be used? If so, which design? If not, explain why you believe that the product array used was a good design.
  (b) The dataset is rather large since there are 64 design points and 11 factors, so it will not be given here. In addition to Lewis et al. (1997), the dataset can also be found in Ryan (2000). Obtain the data and perform an appropriate analysis. Do you believe that a follow-up experiment is needed to resolve any ambiguities? Why, or why not? What do you conclude from your analysis?
  (c) Compare your analysis to that given by Ryan (2000, pp. 455–462) and comment.

8.5 Consider the following control × noise interaction plot, which is a variation of Figure 8.1. Is this configuration of points helpful in terms of selecting a level of the control factor? Why, or why not?

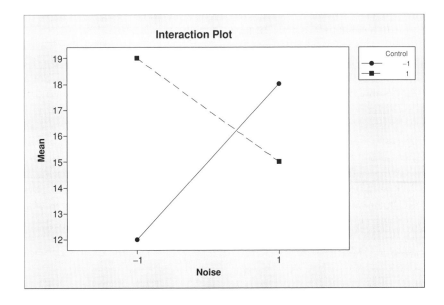

8.6 Which would we rather see, a control × noise interaction profile in which both slopes (assuming two levels of the control factor) are close to zero, or a profile in which one slope is close to +1 and the other slope is close to −1?

8.7 Critique the following statement: "Why should I include noise factors in an experimental design since they can't be controlled during production?"

8.8 Consider the caution given by the Design-Expert software that was quoted in Section 8.5 relative to the following orthogonal array, which was produced by Design-Expert. Under what conditions, if any, would you recommend that this design be used? In particular, could this design be used as a combined array to investigate control × noise interactions. Explain.

1	1	1	1	1	1	1	1	1	1	1
1	1	1	1	1	2	2	2	2	2	2
1	1	2	2	2	1	1	1	2	2	2
1	2	1	2	2	1	2	2	1	1	2
1	2	2	1	2	2	1	2	1	2	1
1	2	2	2	1	2	2	1	2	1	1
2	1	2	2	1	1	2	2	1	2	1
2	1	2	1	2	2	2	1	1	1	2
2	1	1	2	2	2	1	2	2	1	1
2	2	2	1	1	1	1	2	2	1	2
2	2	1	2	1	2	1	1	1	2	2
2	2	1	1	2	1	2	1	2	2	1

8.9 It was shown in Section 8.1 that the L_8 design given in that section is equivalent to a suboptimal fractional factorial. Does the design in Exercise 8.8 have the same weakness? Why, or why not?

8.10 Case 35 in Taguchi et al. (2005, references) is a typical application of an orthogonal array design in the Taguchi literature in which there is no discussion of the properties of the design that is used, which is the $L_{18}(2^2 \times 3^6)$ that can be generated by Design-Expert, for example. (Recall that the notation specifies that the design has 2 two-level factors and 6 three-level factors.) The alias structure for this design is very complex but is produced automatically by Design-Expert. Bearing in mind that we use three-level factors for the purpose of investigating quadrature, use Design-Expert or other software and comment on the alias structure. Specifically, under what conditions would you recommend use of the design?

1	1	1	1	1	1	1	1
1	1	2	2	1	2	2	2
1	1	3	3	2	3	3	3
1	2	1	2	2	2	3	1
1	2	2	3	1	3	1	2
1	2	3	1	3	1	2	3
1	2	1	3	3	2	1	3
1	2	2	1	2	3	2	1
1	2	3	2	1	1	3	2
2	1	1	3	2	1	2	2
2	1	2	1	1	2	3	3
2	1	3	2	3	3	1	1

2	2	1	1	3	3	3	2
2	2	2	2	2	1	1	3
2	2	3	3	1	2	2	1
2	2	1	2	1	3	2	3
2	2	2	3	3	1	3	1
2	2	3	1	2	2	1	2

8.11 One of the problems regarding case studies of Taguchi's methods is that the raw data are generally not given. If the data were given, then readers could perform alternative analyses, if desired, and compare those analyses with the analysis given in the case study. For example, Rowlands, Antony, and Knowles (2000, references) described an application but did not give the raw data, although they did give the ANOVA table for the raw data. This article is available at http://www.emeraldinsight.com/Insight/ViewContentServlet?Filename= Published/EmeraldFullTextArticle/Articles/1060120201.html and at http://www.caledonian.ac.uk/crisspi/downloads/publication6.pdf. Read the article and critique the analysis. Do you believe that the conclusions are justified based on the content of the article?

CHAPTER 9

Split-Unit, Split-Lot, and Related Designs

In this chapter we consider designs that do not have a single error structure. We cover a split-unit design (better known as a split-plot design), a split-lot design, and other similar designs such as a strip-plot design. There has been much research activity regarding these various types of designs during the past 10 years. Although designs such as a split-plot design have not been accorded much space in many books on experimental design, Bisgaard (2000) stated that split-plot designs play a key role in the industrial application of factorial experiments and their use is much more prevalent than the literature on design of experiments in engineering would suggest. Box, Hunter, and Hunter (2005, p. 336) quoted Cuthbert Daniel, famous statistician, author, and consultant of some years ago, as having stated "All industrial experiments are split-plot experiments" although they recognize that may be a slight exaggeration. Somewhat similarly, Langhans, Goos, and Vandebroek (2005) stated that split-plot designs and their properties receive much less attention in the chemometrics literature than the designs receive in the general statistical literature. Robust parameter designs were discussed in Chapter 8 and Bingham and Sitter (2003) discussed the use of fractional factorial split-plot (FFSP) designs in robust parameter experiments. (Fractional factorial split-plot designs are discussed in Section 9.1.3.) Emptage, Hudson-Curtis, and Sen (2003) discussed the treatment of microarrays as split-plot experiments, with that application area possibly also underutilized, although this is a relatively new field. (Microarray experiments are discussed in Section 13.21.)

Goos and Vandebroek (2004) showed that split-plot designs will often outperform completely randomized designs in terms of D-efficiency and G-efficiency, and Goos and Vandebroek (2003) discussed the construction of D-optimal split-plot designs. As with other types of designs, the response variable is assumed to have a normal distribution, but Robinson, Myers, and Montgomery (2004) discussed the analysis of data from industrial split-plot experiments when the response variable does not have a normal distribution.

We will use the terms "split unit" and "split plot" interchangeably in the sections that follow, with the choice between them for discussing a particular type of design determined in part by the terminology that has been used in the literature.

9.1 SPLIT-UNIT DESIGN

This design has for decades been referred to as a *split-plot* design; this was the original name given to it as this design along with many designs were originally used in agricultural applications and a "plot" was a plot of land. That is, a small plot of land was literally "split," so that a "whole plot" was split into two or more pieces, which were called subplots. These designs are now used heavily in industry rather than mainly in agriculture, however, sometimes even unknowingly when the error structure is not understood. Therefore, following Mead (1988), Giesbrecht and Gumpertz (2004), and Ramírez (2004), it seems more appropriate to call the design a *split-unit* design, while realizing that there are still agricultural experiments being performed for which the term "split-plot design" is of course appropriate.

Another reason for eschewing the term "plot" in discussing a split-unit design in general and its variations is that a normal probability plot is often used in analyzing the data from experiments using these and other designs, and that plot is a graph, not something that can be "split." Thus, it seems desirable to avoid the use of the word to mean two different things, especially when discussing the analysis of data from a designed experiment.

Despite these presumably persuasive arguments for a change in terminology, we will still refer to a split-plot design at times in this chapter, especially in discussing journal articles in which the term is used in the title of the article and/or in the article contents. Suggestions for permanent names for the designs presented in this chapter are given in Section 9.4. In this chapter we will often use WP to denote "whole plot" and SP to denote "split-plot."

Consider a very simple example of two fixed factors, each at two levels. We will first assume that all four combinations of the factor levels are feasible, and the level changes of each factor can be easily made. If the four combinations are run in random order, we have a 2^2 design. For example, the order in which the treatment combinations might be run could be $A_1 B_2$, $A_2 B_1$, $A_2 B_2$, and $A_1 B_1$.

Now assume that although each factor is suspected of having a significant effect on the response, one of the two factors is definitely of secondary interest. Also, assume that this factor is hard to change. One of the levels of this factor is randomly selected and then used in combination with each of the two levels of the other factor, which are also randomly selected. Then this process is repeated for the second level of the first factor. Thus, the order in which the treatment combinations are run could be as follows, assuming that factor A is the hard-to-change factor: $A_2 B_2$, $A_2 B_1$, $A_1 B_1$, and $A_1 B_2$.

Notice that this last sequence of treatment combinations and the one given previously *could* of course be the same, but the data would still have to be analyzed differently because of the restricted randomization in the second case. That is, there are only eight possible sequences of treatment combinations with the restriction, whereas there are 24 possible sequences without the restriction.

With the restriction, is this a randomized block design (RCB) since we seem to be blocking on the first factor? Recall from Section 3.1 that the experimental units are considered to be more homogeneous within blocks than between blocks for the RCB design. Recall also that a blocking variable is considered to be an extraneous factor and that blocks are generally considered to be random. Thus, although the experimental layout coincides with the layout that would be used in a randomized block design, we clearly cannot analyze the data as having come from such a design because the conditions are not the same.

Because of the restricted randomization, this also is not a 2^2 design. Then what is left? Although the description of the scenario did not suggest that a plot or unit was literally split (as in splitting a plot of land), the data would be logically analyzed in that manner because of the manner in which the experiment was conducted.

Recall from Appendix C to Chapter 4 that in a 2^k design each effect is estimated with the same precision. This does not happen with a split-plot design as sub-plot factors are generally estimated with greater precision than are whole-plot factors. This should be intuitively apparent for an agricultural experiment in which sub-plots would be strips of land within a whole plot, as the sub-plots would certainly be more homogeneous in terms of land fertility than are the whole plots, especially if the whole plots are large and so the distance between the centers of the two plots is large.

In addition to being quite intuitive, at least for agricultural experiments, it can be shown mathematically that the variance of any whole-plot effect estimate must exceed the variance of any subplot effect estimate for any 2^k design. Bisgaard and de Pinho (2004) state that whole-plot factors and their interactions have a variance of $\frac{4}{N}(2^q\sigma_1^2 + \sigma_0^2)$, whereas subplot factors and their interactions have a variance of $\frac{4}{N}(\sigma_0^2)$. Here σ_1^2 and σ_0^2 denote the whole plot and subplot error variance, respectively.

Sub-plot effects will not necessarily be estimated with greater efficiency in industrial experiments or various other types of experiments, however. It is obvious from the expressions of the variances that this could not happen if σ_0^2 and σ_1^2 were estimated individually using the same data, but this does not happen as a sum of squares estimates a mean square, which is a linear combination of variance components for whole-plot factors and interactions. Thus, the data are used in one way to estimate $\frac{4}{N}(\sigma_0^2)$ and in another way to estimate $\frac{4}{N}(2^q\sigma_1^2 + \sigma_0^2)$.

This issue of whole-plot effect and sub-plot effect variance estimates is addressed by Giesbrecht and Gumpertz (2004, p. 169), who point out that the estimate of the split-plot error may exceed the estimate of the whole-plot error, with opinions differing as to how to proceed when that occurs. Their opinion is to proceed as if the assumed model were valid, with that model containing the whole-plot factor(s), the subplot factor(s), the interaction(s) between them, and the two error terms.

To illustrate the analysis for both cases (complete randomization and restricted randomization for the whole-plot factor), we will consider the following simple example.

Example 9.1

We will assume that factor A is the hard-to-change factor and factor B is not hard to change, with the experiment being such that material (e.g., a board) is divided into two pieces and the two levels of factor *A* applied to the two pieces, one level to each piece. Then the pieces are further subdivided and each of the two levels of factor *B* are applied to the subdivided pieces. Three pieces of the original length (e.g., three full boards) are used. The data are given below.

```
  A         B           Observations (replications)

  1         1           2.5        2.4        2.6
            2           2.7        2.6        2.5
  2         1           2.3        2.3        2.4
            2           2.7        2.7        2.8
```

If the data are improperly analyzed as a 2^2 design with three replications, the results are as follows.

```
            Analysis of Variance for Y

    Source   DF    SS       MS       F       P
       A      1   0.0008   0.0008    0.13   0.733
       B      1   0.1875   0.1875   28.13   0.001
     A*B      1   0.0675   0.0675   10.13   0.013
   Error      8   0.0533   0.0067
   Total     11   0.3092
```

The proper analysis of the data as having come from a split-plot design is not easily achieved. The following statement by Box (1995–1996) needs some clarification: "The numerical calculation of the analysis of variance for split-plot experiments—computation of the degrees of freedom, sums of squares and mean squares—is the same as for any other design and can be performed by any of the many computer programs now available." Presumably, what was meant was that the *basic methods* for computing sums of squares, and so on are the same as for other designs. Other designs, such as full factorial and fractional factorial designs, do not have two or more error terms, however, so the *direct production* of ANOVA tables is not possible with most statistical software.

As Potcner and Kowalski (2004) explained, however, statistical packages can be tricked into performing the correct analysis by assuming a nested model and forcing a nested model analysis. This requires, of course, that the software package being used has nested design capability but unfortunately it is either very difficult or impossible to do the analysis with most software packages, as explained in Section 7.2.

Thus, the analysis of data from a split-plot design can be very difficult. Here, we will take a more straightforward but slightly cumbersome approach and discuss how the analysis can be performed by starting with the analysis without the randomization restriction and simply decomposing the error term into the two appropriate error terms: the whole-plot error and the subplot error. The other sums of squares in the ANOVA are the same as those assuming a completely randomized design.

In decomposing the error term given above, we have to make a decision as to whether or not to isolate the replication factor. Montgomery (1996, p. 524), for example, does so as the replicates were run over days, so it would be appropriate to treat these as blocks. Potcner and Kowalski (2004) in a similar example do not do so. We will initially follow the approach of Potcner and Kowalski (2004), especially since replications was not isolated in the above analysis and we will assume that there is no compelling reason for them to be treated as blocks.

This means that the whole-plot error will have degrees of freedom determined as follows. The degrees of freedom will be the interaction degrees of freedom, replicates $\times A$ since A is the whole-plot factor, plus the replicates degrees of freedom. That is, $2 \times 1 + 2 = 4$. The subplot error degrees of freedom will then be $12 - 4 = 8$.

The corresponding error sums of squares are not difficult to compute by hand, especially since only one has to be computed directly, with the other obtained by subtraction after the "wrong analysis" (i.e., the one given above) is computer generated to produce the error sum of squares that is to be decomposed. The sum of squares for replications is computed the same way that the treatment sum of squares is computed in one-way ANOVA and the replicates $\times A$ sum of squares is computed as the sum of squares for the "cells," with each cell formed by taking a replicate and a level of A (so there are six cells), minus the sum of the squares for the replications and the sum of squares of the whole-plot factor, A. Applied to this example we have

$$\text{SS(replicates} \times A) = \text{SS (cells)} - \text{SS(replications)} - \text{SS}(A)$$

with

$$
\begin{aligned}
\text{SS(cells)} = {} & (2.5 + 2.7)^2/2 + (2.4 + 2.6)^2/2 \\
& + (2.6+2.5)^2/2 + (2.3+2.7)^2/2 + (2.3+2.7)^2/2 + (2.4+2.8)^2/2 \\
& - (\text{sum of all observations})^2/12 \\
= {} & 77.5450 - 77.5208 \\
= {} & 0.0242
\end{aligned}
$$

$$
\begin{aligned}
\text{SS(replications)} = {} & (2.5 + 2.7 + 2.3 + 2.7)^2/4 + (2.4 + 2.6 + 2.3 + 2.7)^2/4 \\
& + (2.6 + 2.5 + 2.4 + 2.8)^2/4 - (\text{sum of all observations})^2/12 \\
= {} & 77.5325 - 77.5208 \\
= {} & 0.0117
\end{aligned}
$$

Therefore,

$$
\begin{aligned}
\text{SS(replicates} \times A) & = \text{SS(cells)} - \text{SS(replications)} - \text{SS}(A) \\
& = 0.0242 - 0.0117 - 0.0008 \\
& = 0.0117
\end{aligned}
$$

The latter would be the whole-plot error sum of squares if replications were one of the components of the ANOVA table. Since we are not using that approach for this example, we add back SS(replications) to obtain SS(whole-plot error) = 0.0234.

We now have what we need to begin constructing the ANOVA table, which leads to

```
                    Analysis of Variance for Y

     Source         DF      SS        MS        F         P
     A               1    0.0008    0.0008    0.138     0.729
     Error (WP)      4    0.0234    0.0058      *         *
     B               1    0.1875    0.1875    25.00     0.008
     A*B             1    0.0675    0.0675     9.00     0.040
     Error (SB)      4    0.0299    0.0075      *         *
     Total          11    0.3091
```

A somewhat different picture emerges when the data are analyzed correctly. Whereas the A*B interaction had a p-value of .013 when the randomization restriction was ignored, now the p-value is much closer to .05, so the evidence that the interaction effect exists is not as strong.

The difference in the conclusions drawn with the wrong analysis and the conclusions made with the proper analysis can be much greater than the difference in this example. This was illustrated by Potcner and Kowalski (2004) who showed that a significant main effect in the complete randomization analysis can become a non-significant whole-plot main effect when the split-plot analysis is performed, and a non-significant main effect in the complete randomization analysis can become a significant subplot main effect when the split-plot analysis is performed. This was illustrated by their first example.

Similarly, Lucas and Hazel (1997) compared a completely randomized design and a split-plot design for the same experimental situation. This provides a tutorial on the use of split-plot designs. Lucas (1999) states that in his experience the most common situation that requires the use of a split-plot design is when there is a single hard-to-change factor. Czitrom (1997) gave an example of such an experiment, and the reader is asked to analyze the data in Exercise 9.1. Goos and Vandebroek (2001) gave an algorithm for constructing D-optimal split-plot designs (optimal designs are covered in Section 13.7), and Goos (2002) discussed optimal split-plot designs, including designs with hard-to-change factors.

Hard-to-change factors were discussed extensively in Section 4.19. Even though the industrial use of split-plot designs motivated by the recognition of hard-to-change factors may seem to have been motivated by research articles during the past 10 years, Daniel (1976, p. 270) discussed the use of a split-plot design with one hard-to-change factor and one factor that is not hard to change. Thus, the idea of using a split-plot design when there is at least one hard-to-change factor is not of recent origin.

One of the best known nonagricultural applications of a split-plot design was given by Box and Jones (1992), which is incorrectly described by Miller (1997) as a strip-plot experiment. (In Section 9.3 we try to clarify somewhat the fine difference between the designs presented in this chapter.)

9.1.1   Split-Plot Mirror Image Pairs Designs

There are many ways in which a split-plot design could be constructed because we can view this, analogous to the discussion in Section 8.2, as a "product design." That is, there is a separate design for the whole-plot factors and one for the subplot factors. One possibility is a *split-plot mirror image pairs design* (SPMIP), which is a design such that there are two subplot runs for each whole-plot run that are mirror images. Of course "mirror image" means two, so we are talking about two-level designs, at least for the subplot factors. These designs have been considered by, in particular, Tyssedal, Kulahci, and Bisgaard (2005) and Tyssedal and Kulahci (2005).

Of course the simplest such design would be a design with a single subplot factor and two levels of that factor: -1 and $+1$. For more complex designs, it is useful to view the general form of the design matrix, given by Tyssedal and Kulahci (2005) for example, in partitioned form as

$$\begin{bmatrix} W & S \\ W & -S \end{bmatrix}$$

Clearly, we need only focus on the form of $[W \quad S]$ since the other parts of the full partitioned design matrix are defined from this submatrix. Many possibilities exist for the form of $[W \quad S]$, including a Plackett–Burman design. Tyssedal and Kulahci (2005) used such a design in illustrating their procedure for computing effect estimates for SPMIP designs.

9.1.2   Split-Unit Designs in Industry

Although the split-plot design originated in agriculture, as stated in Section 9.1, it is now commonly used in industry because of the cost savings relative to other types of designs (Bisgaard and Sutherland, 2003–2004).

In this section we look at a couple of industrial applications of split-unit experiments.

Example 9.2

Bisgaard, Fuller, and Barrios (1996) presented a split-plot experiment in which the surface of a security paper was modified via plasma treatment to make it more susceptible to ink. There were four factors that were believed to affect plasma creation—pressure, power, gas flow rate, and type of gas—in addition to the paper type. A 2^5 design was ruled out for the following reason. The creation of plasma requires a vacuum, with this being done by placing a sample of security paper in a reactor and pumping out the air. This process takes approximately half an hour, so running a 2^5 design would require about 16 hours. To reduce the labor and the time required to run the experiment, the decision was made to place two types of papers in the reactor at the same time.

The paper type is then the subplot factor and the other four factors are run as a 2^4 experiment. The data are given below, with factor E of course being the subplot factor.

A	B	C	D	E −	E +
−	−	−	−	48.6	57.0
+	−	−	−	41.2	38.2
−	+	−	−	55.8	62.9
+	+	−	−	53.5	51.3
−	−	+	−	37.6	43.5
+	−	+	−	47.2	44.8
−	+	+	−	47.2	54.6
+	+	+	−	48.7	44.4
−	−	−	+	5.0	18.1
+	−	−	+	56.8	56.2
−	+	−	+	25.6	33.0
+	+	−	+	41.8	37.8
−	−	+	+	13.3	23.7
+	−	+	+	47.5	43.2
−	+	+	+	11.3	23.9
+	+	+	+	49.5	48.2

Since this is an unreplicated design, there is not an error sum of squares to decompose into two components. Instead, two normal probability plots will be used. We will also analyze these data "by hand," since statistical software in general is inadequate to the task. If we erroneously analyze the data as having come from a 2^5 design, we obtain the following results if we assume that all three-factor and higher-order interactions are not real.

```
Estimated Effects and Coefficients for Y (coded units)

Term    Effect    Coef    SE Coef    T        P

A       11.825    5.913   1.088      5.44     0.000
B        4.225    2.112   1.088      1.94     0.070
C       -3.388   -1.694   1.088     -1.56     0.139
D      -15.100   -7.550   1.088     -6.94     0.000
E        3.137    1.569   1.088      1.44     0.169
A*B     -4.212   -2.106   1.088     -1.94     0.071
A*C      2.975    1.488   1.088      1.37     0.190
A*D     16.563    8.281   1.088      7.61     0.000
A*E     -5.900   -2.950   1.088     -2.71     0.015
B*C     -0.850   -0.425   1.088     -0.39     0.701
B*D     -3.313   -1.656   1.088     -1.52     0.147
B*E     -0.300   -0.150   1.088     -0.14     0.892
C*D      1.675    0.837   1.088      0.77     0.453
C*E     -0.138   -0.069   1.088     -0.06     0.950
D*E      1.025    0.512   1.088      0.47     0.644
```

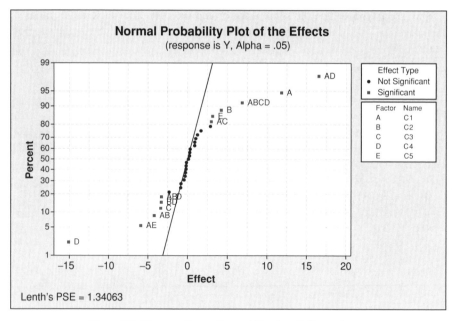

Figure 9.1 "Improper" normal probability plot of effect estimates.

We see that the A, D, and AD effects appear to be real, with the B and AB interactions perhaps deserving some attention. If we construct a normal probability plot of the effect estimates, as shown in Figure 9.1, we obtain a completely different message.

Here we see that 12 of the 31 effects are declared significant using Lenth's procedure with $\alpha = .05$, including a four-factor interaction. The significance of the latter, in particular, should tell us that something is wrong. The problem is that when the data are analyzed improperly in this manner, there is a mixing of the whole-plot error term and the subplot error term. Of course with an unreplicated design, there is a conceptual mixing since the two error terms cannot be computed. Since the subplot factor is estimated with greater precision than the whole-plot factors, too many whole-plot effects may be identified as real when any pseudo-error term is computed in conjunction with the probability plot.

Since we have an unreplicated design, the simplest and best approach would be to assume that any interaction effects could be real and construct the two appropriate normal probability plots, which would be a plot of the whole-plot effects and a plot of the subplot main effect and interactions between the subplot factor and the whole-plot factors. Bisgaard et al. (1996) did just that and obtained rough estimates of the standard errors of a whole-plot effect estimate and a subplot effect estimate from the slopes of the lines fit through the fitted points. The whole-plot standard error was estimated at 7 and the subplot error at 1.

Notice that all but two of the 12 effects labeled significant in Figure 9.1 are whole-plot effects, and also notice that the Lenth estimate of σ is 1.34, which leads to an effect standard error that is far too small for the whole-plot effects. With this in mind and remembering the standard error estimate for subplot effects, we might guess from Figure 9.1 that the real effects are A, D, AD, E, and AE, and in fact this is what Bisgaard et al. (1996) obtained from their two separate plots. (It is worth noting that there must be more than one or two whole-plot factors in order to construct a normal probability plot of the whole-plot effects, as there must be enough plotted points to have a high probability of distinguishing real effects from the effects that are not real. Other methods have been proposed for small designs, but as demonstrated by Goos, Langhans, and Vandebroek (2006), the methods don't work very well.)

There is a large number of published industrial applications of split-plot designs that have been performed; other such examples include Bjerke, Aastveit, Stroup, Kirkhus, and Naes (2004), and Gregory and Taam (1996), with alternative methods of analysis presented for each example.

The Bjerke et al. (2004) article is a case study but there were certain complexities and complications that rendered it somewhat unsuitable as a case study here. It will be discussed briefly, however, and the reader is urged to study the article as an example of how experiments and the subsequent data analysis are not always straightforward. A split-plot structure resulted, according to the authors, because the production samples for a storing experiment were divided into subunits and there were two subplot storing factors: time (7, 14, and 38 weeks) and temperature (4 and 21°C).

This requires some thought, however, because when measurements are made over time, we would generally not regard time as a factor. (Obviously, it is not a factor whose "levels" can be varied at random.) Instead, it would be more appropriate to view the time factor as producing repeated measurements, for which the observations, for a fixed combination of the whole-plot factors, should be autocorrelated. This violates the error structure for a split-plot design, for which the whole-plot errors and split-plot errors are assumed to be independent.

The manufacturer decided to use only the measurements taken at the colder of the two temperatures for analysis because that temperature most closely resembled the real-life storage of mayonnaise.

Therefore, the temperature subplot factor was essentially removed from the analysis and the time subplot factor seems to be actually a repeated measurement. This would mean that there really isn't a split-plot structure. The authors seemed to be aware of this, although they did not state it strongly. They did, however, describe two other analysis approaches: a mixed model approach and a robustness approach. Again, the reader is urged to study the article, which serves as a good example of data analysis complexities at the model-determination stage. Although it is often said that data from a well-designed experiment will practically analyze itself (i.e., be very easy to analyze), that won't always be true.

Gregory and Taam (1996) discussed the use and interpretation of a split-plot experiment to determine significant factors in the development of fracture-resistant automobile windshields. The objective of the experiment was to determine factors that would make the windshields more resistant to fracture caused by the impact of small rocks or pebbles. Three product factors were investigated: glass thickness (two levels), paint type used at the edges of the inner surface (two levels), and windshield coating (three types). All combinations of these factors were tested at eight locations of each windshield produced. Because of manufacturing restrictions, it was necessary to perform certain steps of the process for entire sets of the possible combinations of factors (e.g., all windshields made from the thicker glass were cut at one time). The authors appropriately referred to this as restricted randomization since not all steps in the process were performed in a completely randomized order. Each windshield manufactured was the whole-plot unit and each of the locations tested on each windshield was the subplot unit. Because of the restricted randomization, the authors considered thickness to be a blocking variable, and paint and coating as crossed factors randomized within each glass thickness.

In addition to the analysis of the data as a split-plot experiment, two alternative methods of analysis were suggested. The first option was to use all eight locations as repeated observations for each of the windshields assuming that there is a restriction on randomization of the testing sequence. The second approach was to treat each of the eight observations associated with test sites as a single multivariate response. Results obtained by the first alternative were similar to those obtained by the original interpretation as a split-plot experiment. Those obtained by the multivariate approach were not similar.

These two discussions illustrate that the analysis of data from a split-plot experiment is not necessarily clear-cut.

9.1.3 Split-Unit Designs with Fractional Factorials

The split-unit design concept can be applied when there are more than two factors. Bingham, Schoen, and Sitter (2004), Bingham and Sitter (1999b, 2001), and Huang, Chen, and Voelkel (1998) discussed and illustrated the application of the split-unit design concept to fractional factorials, with Bingham and Sitter (2001) providing an actual industrial example and certain theoretical results (Bingham and Sitter, 1999a) regarding the impact of randomization restrictions on the choice of fractional factorial split-unit designs. Loeppky and Sitter (2002) discussed methods of analyzing data from these designs. Kulahci, Ramírez, and Tobias (2006) question the use of the minimum aberration criterion for these designs and instead suggest the use of the maximum number of clear two-factor interactions criterion recommended for fractional factorials in general by Wu and Hamada (2000).

The notation used in some of the literature for fractional factorial split-unit designs is $2^{(n_1 + n_2)-(k_1 + k_2)}$, with n_1 and n_2 denoting the number of whole-plot and subplot factors, respectively, and k_1 and k_2 denoting the number of whole-plot and subplot fractional generators, respectively (Bingham and Sitter, 1999a, b; Huang et al., 1998). The choice of notation is somewhat unfortunate because (1) here $n_1 + n_2 = n$ denotes the total number of factors, whereas it generally denotes the number of experimental runs, and (2) the standard notation for a two-level fractional factorial design is 2^{k-p}, so if something is to be "split," it would have been better to split k and p than to split n and k. Bisgaard (2000) used different notation as the designs were represented by $2^{(k-p)-(q-r)}$, and gave many examples of such designs. This is better notation than others have used.

Fractional factorial split-plot designs, designated as FFSP designs in the literature, are constructed similar to the way that fractional factorial designs are constructed, as generators are chosen for the whole-plot factors and for the subplot factors. For example, Bingham and Sitter (1999a) illustrated one way to construct a $2^{(3+3)-(1+1)}$ design by letting $C = AB$ be the generator for the third WP factor and $F = ABDE$ be the generator for the third SP factor, with A, B, and C representing the WP factors and D, E, and F denoting the SP factors. (A generator for a WP factor may contain only WP factors, whereas a SP generator may contain both SP and WP factors.) Whole plots are fixed, so using SP factors as WP generators would destroy the split-plot nature of the design.

When we put these together, we have the defining relation given by $I = ABC = ABDEF = CDEF$, with $CDEF$ being the generalized interaction of the first two components of the defining relation, as in a fractional factorial design. This is analogous to how a 2^{6-2} design is constructed as one starts with a 2^4 design and then selects generators for factors E and F. The difference here is that there is a category for WP factors and a category for SP factors and that distinction must be used in constructing the design. For example, we could not use two SP factors as generators. The reader is asked to explain why this is true in Exercise 9.4.

Since factorial effects are estimated with differing precision when a split-unit arrangement is employed, decisions must be made as to which effects are going to be estimated with the higher precision.

Example 9.3

This was illustrated in the example given by Bingham and Sitter (2001). In that example, a wood products company was interested in investigating the factors that affect the swelling properties of a wood product. The goal of the experiment was to determine optimum process settings that would minimize the swelling of the product after it had been saturated with water and allowed to dry. The fabrication process consisted of two stages, with each batch of wood and additives subdivided to form several sub-batches.

There were five factors that were believed to affect the response variable at the first stage, the mixing stage, and three factors that were thought to be influential in the processing stage. Thus, there were eight factors to be examined; a 2^{8-p} design could not be used because the levels of the five factors in the first stage were set when the second stage commenced because of operational restrictions.

If a $2^{(5+3)-(1+1)}$ FFSP design is to be used, it is necessary to choose one WP generator and one SP generator. Letting $A-E$ denote the WP factors and $F-H$ denote the SP factors, an obvious choice for the generator of E is $E = ABCD$. If the generator for H consisted only of SP factors, the generator would be $H = FG$, which, as pointed out by Bingham and Sitter (2001), would be equivalent to constructing a 2^{5-1} design in the WP factors and a 2^{3-1} design in the SP factors. This would confound main effects with interactions among the SP factors, however, so the design would be only resolution III for that group of factors.

We can obviously do much better than this if we let the generator of H be a function of both SP and WP factors. The design recommended by Bingham and Sitter (2001) had $H = ABFG$ so that the defining relation was $I = ABCDE = ABFGH = CDEFGH$, which is obviously resolution V because the shortest word length is 5.

Operationally, the design would be run the same way as any split-plot design, with the runs involving SP factors randomized.

A problem with various methods that had been given for constructing FFSP designs is that they can sometimes result in too many subplots per whole plot or too few whole plots, as noted by Bingham et al. (2004). They gave a method for FFSP designs that is a solution to this problem.

9.1.4 Blocking Split-Plot Designs

Just as it may not be possible to make all the runs in a factorial design or a fractional factorial design under the same set of conditions, it may not be possible to do so with a split-unit design either. McLeod and Brewster (2004) considered the blocking of FFSP designs, with their work motivated by a practical problem.

Specifically, a company was experiencing problems with one of its chrome-plating processes, as excessive pitting and cracking, poor adhesion, and uneven deposition and adhesion across a chrome-plated, complex-shaped part were observed when the part was being plated. A screening experiment was subsequently planned to identify key factors affecting the quality of the process. Six factors were identified, three of which were hard to vary. The latter became the split-plot factors when a split-plot design was used. (There were actually two additional sub-subplot factors that were used in the experiment, but the authors elected not to discuss them so as to focus on the blocking aspect.) Two levels were to be used for each factor and there were multiple responses, including pits and cracks.

The experiment involved a rectifier and different levels of the SP factors could be used with each rectifier. There was only one tank (bath) available for the experiment, and although the tank had four rectifiers, two of them had to be used for another experiment. Consequently, only two parts could be plated each day, suggesting the use of a 2^{3-2} design for the SP factors. Instead, the decision was made to change one of the SP factors to a WP factor so as to simplify the design problem. Of course this meant that one of the hard-to-change factors would not be treated specially. The experiment was to run for 16 days in the authors' modified example (the actual experiment ran for 20 days), which were regarded as four 4-day weeks and it was desirable to block the experiment by week. With four whole-plot factors there was thus a 2^4 design that could be run in four blocks at the whole-plot level, and this is what was done. The design that was used was thus a $2^{(4+2)-(0+1)}$ FFSP design run in four blocks.

As McLeod and Brewster (2004) pointed out, however, blocking a FFSP is not simple because there is fractionation at both the whole-plot and subplot levels, although that was not the case with this design since the whole-plot design was not fractionated. Let A, B, and C designate the original WP factors and let P denote the SP factor that was converted to a WP factor, leaving q and r as the two *SP* factors. The generator $r = ABCPq$ was used to generate the second SP factor. (Recall from Chapter 5 that in two-level fractional factorial designs, factors are created from columns of the base design, which is a full factorial. Here the base design at the subplot level is a 2^1 design.) The blocking generators were, in the authors' notation, $\beta_1 = ABC$ and $\beta_2 = ABP$, which of course also confounds the product, CP.

After the experiment was run, the authors questioned whether or not the design was optimal. They presented three distinct methods of constructing the blocks and provided a lengthy discussion. Their conclusion was that the design used was not at all bad, but they provided what they considered to be a better approach to constructing the design. See their paper for the details.

9.1.5 Split-Unit Plackett–Burman Designs

Split-plot Plackett–Burman designs have been discussed and illustrated to a moderate extent in the literature, including Kulahci and Bisgaard (2005) and Tyssedal and Kulahci (2005). Of course such designs are alternatives to split-plot fractional factorial designs in the same way that Plackett–Burman designs are alternatives to fractional factorial designs.

9.1.6 Examples of Split-Plot Designs for Hard-to-Change Factors

In this section we examine some actual split-plot experiments with hard-to-change factors, with the split-plot arrangement necessitated by the existence of a hard-to-change factor. This is now a common use of split-plot designs and outside of agricultural use is certainly the motivation for the use of these designs.

Example 9.4

An application of this type of design is described at the Stat-Ease Web site: www.statease.com/pubs/morton.pdf. Morton Powder Coatings of Reading, PA, turned to experimental design after some of its customers began detecting defects in one of the company's leading coating products. Consequently, a team of researchers was formed to address the problem. The researchers suspected that the problem was in the formulation of the powder. They had the use of the customer line for only two shifts, and thus needed to collect data on as many different combinations of factor levels as possible within the allotted time. Five factors were selected for an experiment; three of these were components of the powder: level of a catalyst, amount of a certain proprietary additive, and the coarseness of the powder. The other two factors were line speed and oven temperature, which could be changed. The three factors relating to the powder coating formula could not be easily changed, however.

The researchers used a split-plot design, with the hard-to-change factors comprising the whole plots and the easy-to-change factors comprising the subplots. The way that the experiment would be conducted would be to randomly select a level for each of the whole-plot factors and then randomly vary the levels of the subplot factors.

Notice the relationship between this prescription for carrying out the experiment and the way an experiment is performed when a randomized block design is used. With the latter, the levels of a factor are randomly used within each block. With a split-plot design, we might more or less view each set of levels of the whole-plot factors as constituting a "block," with the sub-plot factors randomly varied within each "block." Then the levels of the whole-plot treatments are changed and the process is repeated. (Of course a split-plot design may be blocked, literally, as was discussed in Section 9.1.4.)

Example 9.5

Another experiment that used a split-plot design because there was at least one hard-to-change factor was the experiment described in Kowalski, Landman, and Simpson (2003). The experimental unit was a NASCAR Winston Cup race car. These cars are of course quite expensive, highly valued, and owned by different racing teams, so experimentation with multiple cars would not be feasible.

The overall objective, of course, is to make a car go as fast as possible. In recent years, many race car teams have constructed the layout for experiments, using an ad hoc approach that does not always produce useful results, despite a considerable investment of time and money.

Four factors were used in the experiment and there were four response variables that each measured aerodynamic efficiency. The four factors were: front car height, rear car height, yaw angle (the angle that the car centerline makes with the air stream), and coverage of the radiator grille (tape). The first two factors were hard to change, thus precluding the use of a factorial design in the four factors. Instead, a 2^2 design was constructed for the hard-to-change factors, with the four points replicated and a centerpoint added, for a total of nine points. Another 2^2 design with a centerpoint was constructed and used in conjunction with each of these 9 points, producing a total of 45 points. (Notice that the construction of the design relates to the idea of an assembly experiment, which was discussed in Section 5.15, and it also relates to a product array, which was discussed in Section 8.3.)

In general, factors should be reset before each experimental run, even those factors that are reset to the same level that was used in the previous run (see the discussion of this in Section 4.20), which means that the front car height and rear car height would have to be set 45 times if 45 design points were used. It would take about 35 minutes for the resets for each design point, however, so just the resets would take over 26 hours. Instead, the entire experiment took only 9 hours to perform because the car heights were reset just nine times.

In split-plot terminology, the two car heights are the WP factors and the other two factors are the SP factors.

We won't pursue this further since the data were not given, but we note that the interaction effect of the whole-plot factors was greater than the main effect of the front height factor, with both effects being significant. Thus, there is a problem with the prediction equation that was stated "can be used to predict the front coefficient of lift for various settings of front height, rear height, tape, and yaw." Specifically, the coefficient for front height, which is 0.0059, is not a particularly good representation of the effect of front height at each of the two levels. The coefficient of rear height is also shaky for a similar reason.

Unfortunately, as stated by Reece (2003, p. 327), very few software packages can generate a split-unit (split-plot) design directly.

9.1.7 Split-Split-Plot Designs

A split-plot design could be converted into a split-split-plot design by adding an additional split to accommodate a third factor and such a design could be split further, if desired. Gumpertz and Brownie (1993) are discussed in the literature as having given an example of a split-split-plot design with time as the third factor because repeated measurements were used, but such measurements will be correlated whereas these designs are assumed to have independent errors at each level of splitting. Thus, as in Example 9.2, it might seem inappropriate to say that a split-split-plot design is used when time is claimed to be the third factor with repeated measurements made over time. Milliken (2004) does discuss and illustrate such a design, however, although the repeated measurements aspect does complicate the analysis. The starting point is to initially analyze the data *as if* the split-plot assumptions hold, with those assumptions being that all the effect estimators are identically and independently normally distributed. A covariance structure for the effect associated with the repeated measurements must then be selected, and Milliken (2004) and Milliken and Johnson (2001) suggest starting with a small set of candidate covariance structures and selecting the simplest one.

For an unreplicated design, three normal probability plots would have to be constructed, analogous to the two normal probability plots that are needed for the split-plot design. Therefore, the same consideration must be made as with a split-plot design; namely, is there a sufficient number of effects to permit the construction of each plot? When replications are used, consideration must be given to whether there is a sufficient number of degrees of freedom for each of the error terms to provide sufficient power for each test. Of course, this same consideration must be made for a split-plot design.

9.2 SPLIT-LOT DESIGN

Another type of "split" design is a split-lot design, which is used in manufacturing processes when a product is formed in two or more process stages. These designs were invented by Mee and Bates (1998); a more recent source is Butler (2004).

A unique feature of these designs is that each factor is used in one and only one processing stage, with multiple factors used at each stage and the design at each stage having a split-plot structure. For example, as illustrated by Butler (2004), a fractional factorial design such as a 2^{9-3} design might be used with three of the nine factors used at each of three stages. A somewhat extreme example relative to the number of stages would be nine stages. Mee and Bates (1998) considered a 2^{9-3} split-lot design with one factor used at each of the nine stages. This is not just a matter of splitting the design, however, as a split-plot structure is used *at each stage*, as stated previously. Of course this raises the question of what the design should be called.

Taguchi (1987) has termed the design a multiway split-unit design, which in some ways is a better term as the "split" designs discussed in this chapter are for two groups, such as whole plots and subplots, whereas the type of design discussed in this section can have far more "groups."

9.2.1 Strip-Plot Design

A strip-plot design is one which is applied to a multistage process. It is equivalent to a split-lot design when there are two processing stages, so it is a special case of a split-lot design. Miller (1997) proposed a method of constructing strip-plot designs for fractional factorials and mixed factorials, and gave as an illustrative example a laundry experiment that consisted of washing in the first stage and drying in the second stage. As with a split-plot design, however, we can question whether or not the term "plot" should be used, since technically there is no "plot" that is being "stripped." There is a splitting of sorts that occurs, however, as what would be a single error term if a fractional factorial design were used in a single stage becomes "split" into two or three components.

As Giesbrecht and Gumpertz (2004, p. 176) point out, these designs are also referred to as *strip-block* or *split-block* designs, either of which may be a better term since blocking is involved. A SAS macro for analyzing data from a strip-plot experiment is given at http://64.233.161.104/search?q=cache:-PuFuIroQAoJ:home. nc.rr.com/schabenb/Strip-Plot.html+strip-plot++design&hl=en.

Example 9.6

Miller (1997) states that the primary motivation for using a strip-plot design is that more treatment combinations can be investigated for the same amount of experimental resources and uses a (somewhat disguised) actual laundry experiment for illustration. That experiment was performed to investigate wrinkling of clothes that are washed and dried, and to obviously determine the best way to perform the two-stage operation so as to minimize wrinkling while performing a small experiment in terms of resources expended. Six factors that represented washing conditions and four factors that represented drying conditions were to be used in the experiment, with each factor at two levels.

If this were a single-stage experiment, one obvious possibility would be a 2^{10-p} design for a suitably chosen value of p. It was decided, however, to use two 4×4 strip plots, each arranged as a 4×4 Latin square with columns representing dryers and the rows representing washers. This might at first seem to be a peculiar design layout because the number of washing conditions is not the same as the number of drying conditions. One thing is apparent: There will be 32 treatment combinations used, which is far less than $2^{10} = 1024$, so some fractionization must occur.

The experiment was performed by using a 2^{6-3} design for the washers and a 2^{4-1} design used for the dryers, with the eight treatment combinations for each split up into two blocks of four. When these are crossed in the 4×4 Latin square layout, there are 16 treatment combinations in each block, for a total of 32 treatment combinations.

An obvious question to ask at this point is "Since 32 runs are to be made, why not just use a 2^{10-5} design?" This would be fine if 32 identical washers and 32 identical dryers were available, or if only a single washer and a single dryer were available and the experiment could be conducted over a long period of time. The point is that the use of such a design would ignore the resources that are available—the four washers and the four dryers.

Since a fractional factorial was used for the washers and for the dryers, the fractions should be constructed in an optimal manner. That is, using the notation of Miller (1997) and letting lowercase letters denote the dryer factors and capital letters denote the washer factors, the defining relation for the fraction for the dryers should be $I = abcd$, with various equally good choices available for the defining relation for the washer fraction.

Milliken, Shi, Mendicino, and Vasudev (1998) described the application of a strip-plot design to a two-step process and explained that a mixed model is an appropriate analysis for such a design. They provided SAS code for performing the analysis. That article serves as somewhat of a tutorial for the construction of both split-plot designs and strip-plot designs.

Milliken (2004) also discussed and illustrated a strip-plot design, with PROC Mixed code for SAS used for the analysis since the model contained both fixed effects and random effects. An example of a strip-plot design, including a diagram that shows the layout of the design, is also given in Section 5.5.5 of the *e-Handbook of Statistical Methods* (Croarkin and Tobias, 2002).

Unfortunately, as stated by Reece (2003, p. 327), very few software packages can generate strip-plot designs.

9.2.1.1 *Applications of Strip-Block (Strip-Plot) Designs*

Although a "strip-block" design is the same as a strip-plot design, as stated in the preceding section, in this section we describe applications of the design when it has been described in the literature as a strip-block design.

These designs have been used in agricultural applications since the late 1930s, but there have been relatively few industrial applications. As with the other types of designs discussed in this chapter, the factors to be studied are divided into two (or more) groups so as to enable the experiment to be conducted more efficiently than would be the case if a factorial design without grouping had been used.

An example given by Vivacqua and de Pinho (2004) will illustrate the basic idea. (The same example is given by Vivacqua, Bisgaard, and Steudel (2002), which is available without restrictions at http://qsr.section.informs.org/download/paper1_Carla.pdf.) A battery company, Rayovac, invited a team of consultants to work on a problem with one of its products that was costing the company over $154,000 in losses per year. The first step taken in the project was to construct a detailed flowchart of the specific process involved and of the company in general.

Six factors were selected for use in the experiment; four were associated with the assembly process and two were thought to possibly have an effect on the open-circuit voltage. The ultimate objective was to determine the levels of the process variables that would result in the production of high-quality battery cells. A 2^6 design was out of the question because each curing cycle required 5 days, so 64 runs would require $64 \times 5 = 320$ days. Instead, a strip-block design was used. There were 2000 batteries used for each treatment combination and the experimental design was a 2^4 design in the assembly factors and a 2^2 design in the curing factors. The experiment was performed by first randomizing the 16 runs and the assignment of the 16 treatment combinations to 16 sets of 2000 batteries. Then each lot was split into four sublots and each of these was assigned to one of the four curing conditions. All 16 subplots assigned to the same curing condition were then processed simultaneously.

Vivacqua and de Pinho (2004) make the important point that a normal probability plot of all the effects in a strip-block design cannot be constructed because there are different error terms. (Of course the variance of the plotted effect estimates must be the same when a normal probability plot is constructed, as has been emphasized previously.) Instead, a separate plot would have to be constructed for all effects that have the same error term. Of course this also applies to any design for which there are multiple error terms. In their example there were three separate error terms, which were for the assembly effects, the curing effects, and the interactions between the assembly factors and the curing factors. This would necessitate the use of three distinct normal probability plots, provided that there were enough effect estimates in each of the three groups to justify the construction of each plot. That is obviously not the case in this application, however, because there are only three curing effects: the two curing factors and the interaction between them. Consequently, Vivacqua and de Pinho (2004) used normal probability plots only for the assembly effects and the interaction effects.

One of the results of the experiment was that the defective rate was reduced by 80 percent (of what the old defective rate was) to a defective rate of approximately 1 percent.

Vivacqua et al. (2002) additionally considered a modification of this design scenario, assuming that only eight sublots could be accommodated in the storage room rather than 16. This would necessitate the use of a 2^{4-1} design for the assembly factors. (The use of fractional factorials in strip-block designs was discussed by Miller, 1997.) This design would be of resolution IV because the 2^{4-1} design is resolution IV. Vivacqua et al. (2002) discussed a way of improving the resolution of this design by using a *post-fractionated design* approach. If the interaction of the four assembly factors ($ABCD$) is aliased with the interaction of the curing factors (ED), a resolution VI design results since the defining relation is $I = ABCDEF$. Vivacqua et al. (2002) termed this a "postfractionated design of order 1," since it is a 1/2 fraction of the original design, with again eight sublots used for each of the four curing conditions.

The authors also illustrated the construction of a post-fractionated design of order 2, which would have to be used if the storage room could accommodate only four lots. That results in a design of resolution IV. (The authors presented tables of these various designs and that would be helpful here, but the article is available on the Internet with no restrictions, so readers are referred to those tables of the design layouts.)

The authors also presented a general framework for the construction of post-fractionated strip-block designs and readers are referred to their article for details. They also presented the analysis framework for the analysis of data from such designs, pointing out, in particular, that there are four error strata rather than three with a strip-block design with post-fractionazation, with the fourth stratum being the post-fraction stratum. They provide a rule that allows the effects to be placed in the appropriate categories without the need to compute variances for all of the effect estimates to determine the category into which each effect should be placed, as has been done in other papers. See their paper for details.

9.3 COMMONALITIES AND DIFFERENCES BETWEEN THESE DESIGNS

Until the last 10 years or so, the only "split" design discussed in the literature was a split-plot design. Now, with the various related designs to select from, experimenters need a good understanding of these designs and their relationships and differences in order to select an appropriate design for a given situation. Therefore, in this section we'll examine these relationships.

With what has commonly been referred to as a split-plot design, a whole plot is divided into at least two subplots and subplot treatments are assigned randomly to each subplot for each whole plot, with the whole-plot treatment combinations assigned to the whole plots.

We might think of a strip-block design as somewhat the opposite or reverse of a split-plot design, as with the former the complete set of treatment combinations of the "whole-plot factors" would all be run for each treatment combination of the subplot factors, then all the treatment combinations would be run for the next treatment combination of the subplot factors, and so on until all the runs have been made for all treatment combinations of the whole-plot factors for the last treatment combination of the subplot factors.

A split-lot design is specifically for multiple process stages, with a split-plot design used in each stage. When there are two stages, the design is equivalent to a strip-plot design, which has also been termed a split-block design.

Better understanding as well as more efficient use of these designs might be facilitated by the adoption of names for the designs that allow them to be better distinguished and by using only one name for each design. For example, a split-plot design might be called simply a split-unit design and a split-lot design might be more appropriately termed a multiway split-unit design. A strip-plot design could be called a strip-block design, which of course is one of the other names for the design anyway. See also Guseo (2000) for a review and comparison of these designs.

9.4 SOFTWARE

As one might suspect, the construction and analysis of some of the designs given in this chapter can be somewhat cumbersome and most software do not have this capability. The SAS 9.1 ADX Interface for Design of Experiments *can* be used to construct full factorial and FFSP designs, however, with the analysis of data from such designs performed using a mixed model approach (see Chapter 11 of http://support.sas.com/documentation/onlinedoc/91pdf/sasdoc_91/qc_gs_7304.pdf.)

The list of software that cannot handle split-plot designs is long and includes the following. A split-unit design cannot be constructed and analyzed directly in MINITAB, as discussed in http://www.scimag.com/ShowPR.aspx?PUBCODE= 030&ACCT=3000039460&ISSUE=0306&RELTYPE=PR&ORIGRELTYPE=FE& PRODCODE=00000000&PRODLETT=A. That article does discuss in a general way how MINITAB can be used in a somewhat manual manner to overcome the problem, however, and is an interesting application of a split-plot design.

Similarly, data from a split-plot design cannot be analyzed directly with Design-Expert. The following excerpts from the User's Guide for Design-Expert 6 (the current version is 7) are relevant:

> The analysis of a split-plot design is tricky, even for statisticians. It can be done in Design-Expert by properly designating effects in specific ways for subsequent analysis of variance. Proceed if you dare! (pp. 4–11)

> The analysis of a split-plot design must be done somewhat manually. (pp. 4–13)

Because of the latter, they state that two separate ANOVAs must be run: one for the whole-plot treatment(s) and interactions of those factors, and the other one for the subplot treatment(s) and the interaction(s) between the whole-plot treatment(s) and the subplot treatment(s), as well as interactions between the subplot factors. This is illustrated for Version 6 of their software at http://www.statease.com/x6ug/DX04-Factorial-General.pdf and is also illustrated by Whitcomb and Kraber (manuscript) in their company case study at www.statease.com/pubs/pcr_via_split-plot.pdf.

D. o. E. Fusion Pro provides the capability to design split-plot experiments and analyze the resultant data, and in fact the user is given the option of selecting a split-plot structure on the first screen display when specifying a design. Reece (2003, p. 335) noted in reference to the software's split-plot design capability, "While this investigation didn't formally investigate this function, this ability is unusual among packages such as this." RS/Discover has similar capability as Reece (2003, p. 364) states ". . . it can design split-plot and strip plot designs for the knowledgeable user."

Because split-plot designs and their variations are becoming more popular, motivated to a considerable extent by the work of Box and Jones (1992), it would be helpful if more software companies would give some attention to incorporating split-plot designs into their software, and perhaps include some or all of the other types of designs discussed in this chapter.

9.5 SUMMARY

Basic, introductory information on split-plot designs is given in Box (1995–1996), Box and Jones (2000), Kowalski (2002), Kowalski and Potcner (2003), and Potcner and Kowalski (2004), and the reader is referred to these articles for additional sources that are easily readable. Other introductory-level sources of information on split-plot designs include Gardner and Cawse (2003), with information on the variants presented of this basic design described in the references that have been cited, as well as Vivacqua and Bisgaard (2004). It is best that split-plot designs have more than one or two whole plot factors since the identification of real whole plot effects is difficult when there are few whole plots.

REFERENCES

Bingham, D. R. and R. R. Sitter (1999a). Some theoretical results for fractional factorial split-plot designs. *Annals of Statistics*, **27**, 1240–1255.

Bingham, D. R. and R. R. Sitter (1999b). Minimum aberration fractional factorial split-plot designs. *Technometrics*, **41**, 62–70.

Bingham, D. R. and R. R. Sitter (2001). Design issues in fractional factorial split-plot experiments. *Journal of Quality Technology*, **33**(1), 2–15.

Bingham, D. R. and R. R. Sitter (2003). Fractional factorial split-plot designs for robust parameter experiments. *Technometrics*, **45**, 80–89.

Bingham, D. R., E. D. Schoen, and R. R. Sitter (2004). Designing fractional factorial split-plot experiments with few whole-plot factors. *Applied Statistics*, **53**(2), 325–339.

Bisgaard, S. (2000). The design and analysis of $2^{k-p} \times 2^{q-r}$ split-plot experiments. *Journal of Quality Technology*, **32**, 39–56.

Bisgaard, S. and A. L. S. de Pinho (2004). The error structure of split-plot designs. *Quality Engineering*, **16**(4), 671–675.

Bisgaard, S. and M. Sutherland (2003-4). Split plot experiments: Taguchi's Ina tile experiment reanalyzed. *Quality Engineering*, **16**(1), 157–164.

Bisgaard, S., H. T. Fuller, and E. Barrios (1996). Two level factorials run as split plot experiments. *Quality Engineering*, **8**(4), 705–708.

Bjerke, F., A. H. Aastveit, W. W. Stroup, B. Kirkhus, and T. Naes (2004). Design and analysis of storing experiments: A case study. *Quality Engineering*, **16**(4), 591–611.

Box, G. E. P. (1995–1996). Split-plot experiments. *Quality Engineering*, **8**(3), 515–520.

Box, G. E. P. and S. Jones (2000). Split-plots for robust product and process experimentation. *Quality Engineering*, **13**(1), 127–134.

Box, G. E. P. and S. Jones (1992). Split-plot designs for robust product experimentation. *Journal of Applied Statistics*, **19**, 3–26. (This is available as Report No. 61, Center for Quality and Productivity Improvement, University of Wisconsin-Madison (see http://www.engr.wisc.edu/centers/cqpi).

Box, G. E. P., J. S. Hunter, and W. G. Hunter (2005). *Statistics for Experimenters: Design, Innovation, and Discovery*. Hoboken, NJ: Wiley.

Butler, N. A. (2004). Construction of two-level split-lot fractional factorial designs for multi-stage processes. *Technometrics*, **46**, 445–451.

Croarkin, C. and P. Tobias, eds. (2002). *NIST/SEMATECH e-Handbook of Statistical Methods* (www.itl.nist.gov/div898/handbook), a joint effort of the National Institute of Standards and Technology and International SEMATECH.

Czitrom, V. (1997). Introduction to design of experiments. In *Statistical Case Studies for Industrial Process Improvement*, Chap. 14 (V. Czitrom and P. D. Spagon, eds.). Philadelphia: Society for Industrial and Applied Mathematics; Alexandria, VA: American Statistical Association.

Daniel, C. (1976). *Applications of Statistics to Industrial Experimentation*. New York: Wiley.

Emptage, M. R., B. Hudson-Curtis, and K. Sen (2003). Treatment of microarrays as split-plot experiments. *Journal of Biopharmaceutical Statistics*, **13**, 159–178.

Gardner, M. M. and J. N. Cawse (2003). Split-plot designs. In *Experimental Design for Combinatorial and High Throughput Materials Development*, Chap. 8. Hoboken, NJ: Wiley.

Giesbrecht, F. G. and M. L. Gumpertz (2004). *Planning, Construction, and Statistical Analysis of Comparative Experiments*. Hoboken, NJ: Wiley.

Goos, P. (2002). *The Optimal Design of Blocked and Split-Plot Experiments*. New York: Springer.

Goos, P. and M. Vandebroek (2001). Optimal split-plot designs. *Journal of Quality Technology*, **33**, 436–450.

Goos, P. and M. Vandebroek (2003). D-optimal split-plot designs with given numbers of whole plots. *Technometrics*, **45**, 235–245.

Goos, P. and M. Vandebroek (2004). Outperforming completely randomized designs. *Journal of Quality Technology*, **36**(1), 12–26.

Goos, P., I. Langhans, and M. Vandebroek (2006). Practical inference from industrial split-plot designs. *Journal of Quality Technology*, **38**(2), 162–179.

Gregory, W. L. and W. Taam (1996). A split-plot experiment in an automobile windshield fracture resistance test. *Quality and Reliability Engineering International*, **12**, 79–87.

Gumpertz, M. L. and C. Brownie (1993). Repeated measures in randomized block and split-plot experiments. *Canadian Journal of Forest Research*, **23**, 625–639.

Guseo, R. (2000). Split and strip-plot configurations of two-level fractional factorials: A review. *Journal of the Italian Statistical Society*, **9**, 85–86.

Huang, P., D. Chen, and J. Voelkel (1998). Minimum aberration two-level split-plot designs. *Technometrics*, **40**, 314–326.

Kahn, W. and C. Baczkowski (1997). Factors which affect the number of aerosol particles released by clean room operators. In *Statistical Case Studies for Industrial Process Improvement*, Chap. 10 (V. Czitrom and P. D. Spagon, eds.). Philadelphia: Society for Industrial and Applied Mathematics; Alexandria, VA: American Statistical Association.

Kowalski, S. M. (2002). 24 run split-plot experiments for robust parameter design. *Journal of Quality Technology*, **34**(4), 399–410.

Kowalski, S. M. and K. J. Potcner (2003). How to recognize a split-plot experiment. *Quality Progress*, November, 60–66. (This paper can be read at http://www.minitab.com/resources/Articles/RecognizeSplitPlot.pdf.)

Kowalski, S., D. Landman, and J. Simpson (2003). Design of experiments enhances race car performance. *Scientific Computing and Instrumentation*, June (available at http://www.scamag.com/ShowPR.aspx?PUBCODE=030&ACCT=3000039460&ISSUE=0306&RELTYPE=PR&ORIGRELTYPE=FE&PRODCODE=00000000&PRODLETT=A)

Kulahci, M. and S. Bisgaard (2005). The use of Plackett–Burman designs to construct split-plot designs. *Technometrics*, **47**(4), 495–501.

Kulahci, M., J. G. Ramírez, and R. Tobias (2006). Split-plot fractional designs: Is minimum aberration enough? *Journal of Quality Technology*, **38**, 56–64.

Langhans, I., P. Goos, and M. Vandebroek (2005). Identifying effects under a split-plot design structure. *Journal of Chemometrics*, **19**, 5–15.

Loeppky, J. L. and R. R. Sitter (2002). Analyzing unreplicated blocked or split-plot fractional factorial designs. *Journal of Quality Technology*, **34**(3), 229–243.

Lucas, J. M. and M. C. Hazel (1997). Running experiments with multiple error terms: How an experiment is run is important. In *ASQ Annual Quality Congress Transactions*, pp. 283–296. Milwaukee, WI: American Society for Quality. (This article is available online to ASQ members at www.asq.org/members/news/aqc/51_1997/10549.pdf.)

Lucas, J. M. (1999). Comparing randomization and a random run order in experimental design. *Chemical and Process Industries Division News*, Summer, **7**, 10–14. (This article is available online at http://www.asq.org/forums/cpi/newsletters/summer_1999.pdf.)

McLeod, R. G. and J. F. Brewster (2004). The design of blocked fractional factorial split-plot experiments. *Technometrics*, **46**, 135–146.

Mead, R. (1988). *The Design of Experiments: Statistical Principles for Practical Application*. Cambridge, UK: Cambridge University Press.

Mee, R. W. and R. L. Bates (1998). Split-lot designs: Experiments for multistage batch processes. *Technometrics*, **40**(2), 127–140.

Miller, A. (1997). Strip-plot configurations of fractional factorials. *Technometrics*, **39**(2), 153–160.

Milliken, G. A. (2004). Mixed models and repeated measures: Some illustrative examples. In *Handbook of Statistics*, Vol. 22, Chap. 5 (R. Khattree and C. R. Rao, eds.). Amsterdam, The Netherlands: Elsevier Science B. V.

Milliken, G. A. and D. E. Johnson (2001). *Analysis of Messy Data, Vol. 3: Covariance Models*. London: CRC Press/Chapman-Hall.

Milliken, G. A., X. Shi, M. Mendicino, and P. K. Vasudev (1998). Strip-plot design for two-step processes. *Quality and Reliability Engineering International*, **14**, 197–210.

Montgomery, D. C. (1996). *Design and Analysis of Experiments*, 4th ed. New York: Wiley.

Potcner, K. J. and S. M. Kowalski (2004). How to analyze a split-plot experiment. *Quality Progress*, December, 67–74. (This article can be read at http://www.minitab.com/resources/Articles/AnalyzeSplitPlot.pdf.)

Ramírez, J. G. (2004). To split or not to split: Do we have a choice? *SPES/Q&P Newsletter*, **11**(2), pp. 10–12.

Reece, J. E. (2003). Software to support manufacturing systems. In *Handbook of Statistics*, Vol. 22, Chap. 9 (R. Khattree and C. R. Rao, eds.). Amsterdam: Elsevier Science B.V.

Robinson, T. J., R. H. Myers, and D. C. Montgomery (2004). Analysis considerations in industrial split-plot experiments when the responses are non-normal. *Journal of Quality Technology*, **36**, 180–192.

Taguchi, G. (1987). *System of Experimental Design*, Vols. 1 and 2. White Plains, NY: UNIPUB/Kraus International.

Tyssedal, J., M. Kulahci, and S. Bisgaard (2005). Split-plot designs with mirror image pairs as subplots. (submitted).

Tyssedal, J. and M. Kulahci (2005). Analysis of split-plot designs with mirror image pairs as sub-plots. *Quality and Reliability Engineering International*, **21**, 539–551.

Vivacqua, C. A. and S. Bisgaard (2004). Strip-block experiments for process improvement and robustness. *Quality Engineering*, **16**(3), 495–500.

Vivacqua, C. A. and A. L. S. de Pinho (2004). On the path to Six Sigma through DOE. *Annual Quality Transactions* (available to ASQ members at http://www.asq.org/members/news/aqc/58_2004/20116.pdf)

Vivacqua, C. A., S. Bisgaard, and H. J. Steudel (2002). Using strip-block designs as an alternative to reduce costs of experimentation in robust product design and multi-stage processes. Paper presented at *INFORMS meeting*. Available without restrictions at http://qsr.section.informs.org/download/paper1_Carla.pdf)

Whitcomb, P. and S. Kraber (2005). PCR process optimized via split-plot DOE. Stat-Ease case study. (www.statease.com/pubs/pcr_via_split-plot.pdf.)

Wu, C. F. J. and M. Hamada (2000). *Experiments: Planning, Analysis, and Parameter Design Optimization*. New York: Wiley.

EXERCISES

9.1 Czitrom (1997) gave an example of an experiment for which a split-plot design was used because there was one hard-to-change factor, which was temperature. The data, from a 2^3 design with a centerpoint, are given below.

Run	Temperature (°C)	Pressure	Argon flow	Tungsten deposition rate	Tungsten non-uniformity (%)	Stress
1	440	0.8	0	265	4.44	10.73
2	440	0.8	300	329	8.37	10.55
3	440	4.0	0	989	4.48	9.71
4	440	4.0	300	1019	7.89	9.77
5	500	0.8	0	612	6.39	8.35
6	500	0.8	300	757	8.92	7.73
7	500	4.0	0	2236	4.44	7.55
8	500	4.0	300	2389	7.83	7.48
9	470	2.4	150	1048	7.53	8.59

Notice that this list of the runs, which is in the same order as in Czitrom (1997), is in Yates order. When the experiment was performed, the actual order of the runs was 4, 2, 1, 3, 9, 7, 5, and 8. Analyze the data, using the second response variable (only), and determine the significant effects, if any. If possible, compare your result with that given by Czitrom (1997).

9.2 Kahn and Baczkowski (1997, references) presented a case study that is worth discussing not only in the context of split-plot designs because that is what was used, but also in regard to the interpretation of data from designed experiments, in general. The title of their article "Factors which affect the number of aerosol particles released by clean room operators" explains the objective of the experimentation that is described in their (detailed) case study. The authors stated that previous experimentation had "not achieved adequately reproducible results" because of the failure to identify certain large variance components. The experiment that was described in the case study had eight factors: garment type, garment size, person, day, time, garment, location, and protocol. Five of the six factors whose levels defined the tests that were performed had two levels and the other factor had three levels. Thus, 96 tests were performed and since they were performed at two workstation levels (near hand level and near foot level) in addition to three (protocol) activities (deep knee bends, marching and slapping, and reaching), a total of 576 particle measurements were taken.

This was not a factorial arrangement because the six measurements that resulted from the six combinations of levels of the last two factors were made on each test; that is, the levels of the first six factors were fixed when the six measurements were obtained (together). Because there was no randomization in obtaining the measurements for the combinations of levels of the last two factors, the first six factors constitute WP factors and the last two factors are the SP factors.

The authors initially collapsed the data over the 96 tests and performed a graphical analysis of the two SP factors, which amounts to analyzing data from a replicated 2×3 design. The authors then proceeded to what they termed the "advanced analysis." They pointed out that there are "several possible ways to analyze a split-plot design," and chose to use two separate ANOVA "runs" (their terminology): one for the whole-plot factors and one for the split-plot factors and split-plot-by-whole-plot interactions. (Note that this is contrary to what was discussed in Section 9.4 regarding Design-Expert.)

In their split-plot analysis, the authors' ANOVA table showed 10 of the 12 effects having a p-value of .0001, including a three-factor interaction. Although they did not provide effect estimates, they did state how such estimates would be computed from the ANOVA results and concluded that most of the effects with p-values of .0001 were not of practical interest, apparently reaching this conclusion because certain effects were "... physically small with respect to the other effects...."

(a) Since we would expect most studied effects, and certainly most interactions, to not be significant (see, e.g., the discussion in Section 4.10), what action would you take if you were presented with such results?

(b) Near the end of their case study, the authors pointed out that, for each of the 96 tests, they did not perform each of the three activity protocols twice, once for the workstation-level measurement and once for the hand-level measurement. Rather, each protocol was performed once and then there were two measurements, one for each level. Does this mean that the design was not a split-plot design and could this possibly explain the fact that almost all of the effects were significant in the split-plot analysis? (You are welcome to read the authors' explanations of these matters.)

9.3 Potcner and Kowalski (2004, see references) used an example to illustrate the analysis of data from an experiment in which a split-plot design was used. (The paper is available at http://www.minitab.com/resources/Articles/ AnalyzeSplitPlot.pdf.) Read the article and use appropriate software, such as D. o. E. Fusion Pro to perform your analysis. Does your output agree with the results given by the authors (as it certainly should)? Do you agree with the conclusions drawn by the authors?

9.4 Consider an FFSP design with four WP factors and two SP factors. Let the SP factors be E and F, with A, B, C, and D denoting the WP factors. Show why the design could not be constructed by using the generators $E = ABD$ and $F = BCD$.

9.5 The data from the famous cake mix experiment given by Box and Jones (1992) are given below, using notation somewhat different from that used by Box and Jones, with the response variable being a measure of how good each cake mix tasted.

F	S	E	(1)	a	b	ab
-1	-1	-1	1.1	1.4	1.0	2.9
1	-1	-1	1.8	5.1	2.8	6.1
-1	1	-1	1.7	1.6	1.9	2.1
1	1	-1	3.9	3.7	4.0	4.4
-1	-1	1	1.9	3.8	2.6	4.7
1	-1	1	4.4	6.4	6.2	6.6
-1	1	1	1.6	2.1	2.3	1.9
1	1	1	4.9	5.5	5.2	5.7

The first three factors, Flour (F), Shortening (S), and Egg Powder (E) constitute a 2^3 design in the whole-plot factors; the other two factors (A and B, say) are environmental variables arranged in a subplot whose four treatment combinations are indicated; that is, this is a split-plot design. More specifically, there were eight separate cake mixes and there were four different environmental conditions under which the cakes were made, and these were different baking times and temperatures. Analyze the data and state your conclusions, thinking in part about what one wants to accomplish (as in Chapter 8) regarding robust design and making a product that is robust to the environmental conditions, such as overcooking in this case.

9.6 Box (1995–1996, references) gave the following data from an experiment with a split-plot design, with the data listed as it appears in that publication, which reflects the randomization that was used. (The data and the analysis of it are also in Chapter 9 of Box et al. (2005, references).) The objective of the experiment was to improve the corrisive resistance of steel bars by applying a surface coating and then baking the bars in a furnace for a specified time. Four coatings and three temperatures were used, and each temperature was used twice. The bars with each of the four coatings were randomly positioned in the furnace for each temperature level; the way that the temperatures were run will be discussed later. Here the four coatings are denoted by 1, 2, 3, and 4. The response variable is corrosive resistance and the response values are given in parentheses below.

```
Temperature                      Coatings
   360°          2          3          1          4
               (73)       (83)       (67)       (89)
   370°          1          3          4          2
               (65)       (87)       (86)       (91)
   380°          3          1          2          4
              (147)      (155)      (127)      (212)
   380°          4          3          2          1
              (153)       (90)      (100)      (108)
   370°          4          1          3          2
              (150)      (140)      (121)      (142)
   360°          1          4          2          3
               (33)       (54)        (8)       (46)
```

Notice the progression of temperatures, which would suggest that temperature was not run in random order. Indeed that was the case but we want to compare the coatings, not the temperatures, and the nonrandom temperature sequence does not affect those comparisons.

   Analyze the data and draw a conclusion regarding the coatings. Specifically, is there a coatings effect? If so, what coating would you recommend for use? Is there a specific temperature setting that you would recommend with that coating setting if you feel that one coating is better than the others? Explain. (You may wish to compare your conclusions with those given by the author. In addition to the stated reference, the technical report, which is #131, can be downloaded at www.engr.wisc.edu/centers/cqpi.)

9.7 Explain the difference between a split-plot design and a strip-block design.

9.8 Taguchi (1987, pp. 445–456) gave an application of a $2^{13-9} \times 2^{3-1}$ split-plot design to an experiment on washing and carding of wool. How many effects would be plotted in a normal probability plot analysis of SP factors and their interactions? How many effects would be plotted in the corresponding analysis of WP factors?

9.9 Using the notation of Bisgaard (2000), give three different combinations of values of k, p, q, and r in a $2^{(k-p)-(q-r)}$ design for which you would recommend that a normal probability plot of SP factors and their interactions not be constructed. Would you also suggest that a normal probability plot of WP factors not be constructed for these three combinations? Explain. If you would suggest that the plots be constructed, give three combinations for which you would advise against the construction of the normal plot.

9.10 Consider a split-plot design with one WP factor and one SP factor. There are four levels of the WP factor and two replications of these are used. There are three levels of the SP factor. How many degrees of freedom will there be for the SP error?

9.11 Assume that a split-split-plot design is to be used and there are two WP factors, two first-level SP factors and one second-level SP factor, with all factors having two levels. Would you recommend that a Plackett–Burman design be considered? Why, or why not?

9.12 An inappropriate analysis of data from what was actually a split-plot design with three replications of the whole-plot factor revealed the SP factor to be significant and the WP factor to not be significant, although the factors were simply labeled as *A* and *B*, and an analysis of the "factorial design" was made under the assumption of complete randomization. A proper analysis of the data as having come from a split-plot arrangement revealed that the WP factor was significant. What must have been the general relationship between the magnitude of the error term in the factorial design analysis and the magnitude of the WP and SP error terms in the split-plot analysis?

9.13 It is important to recognize a split-plot data structure when it exists, as was emphasized in the chapter. It is also important, however, although less frequently stressed, to recognize when such a structure does *not* exist. The following is a message posted on an Internet message board several years ago, with some necessary editing.

> We are planning an experiment to test if plants produced by selfing are less fit than those produced by outcrossing. We have plants from three different alpine valleys, picked randomly among all the possible valleys. In each valley, we have a number of individuals, also picked at random. Seeds from these individuals were brought back to the green house and sawned. When they flowered, five of the flowers were selfed and five were outcrossed. The number of seeds produced by each flower was then recorded.
>   Would you all agree that valleys and individuals are random factors? If so, would you also agree that the data should be analysed as a split-plot design?

Explain to this person why the data should not be analyzed as having come from a split-plot design.

9.14 Consider Example 9.1. Assume that after the experiment was run and the data were collected, a discovery was made that suggests the data from the second and third replications should not be used because of problems with the experiment. What will the whole-plot error sum of squares be if computed using just the four numbers from the first replicate? Will it be necessary or at least desirable to estimate the whole-plot standard error from a normal probability plot? Explain.

CHAPTER 10

Response Surface Designs

Response surface methodology (RSM) is, as the name suggests, a collection of tools for fitting a surface to a set of data, and determining optimum factor levels is also part of the methodology. Of course the shape of the surface is determined by the model that is fit and the response values (i.e., the data), and hence the term "response surface." A distinction should of course be made between the fitted surface and the actual surface, which will almost certainly not coincide, just as the estimate of a parameter will rarely be equal to the parameter value. A fitted surface is, in essence, an estimate of the true surface because if an experiment were repeated, the fitted surface would almost certainly differ at least slightly from the first fitted surface.

Typically a full second-order model is fit in trying to determine the optimum combination of factor levels; this is the model with both linear and quadratic terms in each factor and all two-factor interactions. Before that is done, however, it is not uncommon for a first-order model to be fit and then the method of steepest ascent (or descent, see Section 10.4 for both) used to try to zero in on the optimum operating region, and then use a design to accommodate a second-order model for the purpose of characterizing the region.

Thus, the standard approach is to use a three-stage procedure. The procedure can fail, however, if a screening design (i.e., a resolution III design) is used in the first stage in the presence of significant interactions. If interactions are suspected, a resolution IV design should be used in the first stage so that the wrong subset of factors is not identified in that stage. Another possible impediment, as discussed by Steinberg and Bursztyn (2001) and Xu, Cheng, and Wu (2004), is that it is not always possible, or at least practical, to conduct experiments sequentially, although the need to do so has been stressed by G. E. P. Box, in particular, for decades. Thus, experiments beyond the first one may not be possible.

A key component of RSM is response surface designs. Snee (1985) gave a list of desirable properties for response surface designs, the list having first appeared in Box and Draper (1975) and later given in Box and Draper (1987). Most of those properties are also listed online in, for example, the *NIST/SEMATECH e-Handbook of Statistical Methods* (Croarkin and Tobias, 2002). See http://www.itl.nist. gov/div898/handbook/pri/section3/pri3363.htm and scroll to the bottom of the page.

The 14 properties listed (the last source lists 11) were mostly general, however, and could essentially apply to any experimental design. One of those objectives, "sequential construction of higher order designs from simpler designs," will be indirectly debated in Section 10.1, in which the choice between using one design and more than one design will be discussed.

Similarly, Anderson and Whitcomb (2004) gave their list, which included general properties that any type of good experimental design should possess, in addition to some specific properties such as "behave well when errors occur in the settings of factors (the x's)," and "be insensitive to wild observations."

In this chapter we examine both traditional and nontraditional response surface designs and consider the reasons that have been given during the past few years for moving away from traditional designs and considering other types of designs, especially a uniform design (UD), which is discussed and illustrated in Section 10.3. The designs that have traditionally been used, such as central composite designs (CCDs) (discussed in detail in Section 10.5), permit the fitting of second-order models, so we should view response surface designs as an extension of the designs that were presented in previous chapters. We may also regard a CCD as an enlargement of a factorial design since part of the design is a full or fractional factorial.

We also consider a proposal by Cheng and Wu (2001) and others to move away from the traditional manner in which response surface designs have been used. Typically, a two-level screening design has been used to first identify important factors, then those factors are investigated in a response surface design if the experimenter is interested in fitting a second-order model and perhaps also interested in determining the point/region of optimum operating conditions. We discuss this in Section 10.1.

Published applications of RSM have been primarily in the chemical and food industries, although as stated by Myers (1999), interest in RSM has spread to the biological, biomedical, and biopharmaceutical fields. A small sample of recent applications papers is as follows: Ghadge and Raheman (2006) used RSM for process optimization for biodiesel production; Moberg, Markides, and Bylund (2005) used RSM in a tandem mass spectrometry application; Huang, Lu, Yuan, Lü, and Bie (2006) used RSM in a microbiological application to determine the optimum levels of four protective agents; and Kim and Akoh (2005) used RSM in modeling lipase-catalyzed acidolysis in hexane. Other applications papers include Tuck, Lewis, and Cottrell (1993), who applied RSM in the milling industry.

Although there has been a large number of RSM applications, there are not very many researchers who work in the area of RSM, and the few who do so have lamented their small number (e.g., Khuri, 1999). This is paralleled by the relative paucity of books on RSM, the first of which was Myers (1971), with the "second edition" and now the third edition of that book becoming Myers and Montgomery (2002). The original work on RSM, however, was given in the classic paper by Box and Wilson (1951), and Box and Draper (1987) is another well-known book on RSM, as is Khuri and Cornell (1996). More recently, Khuri (2003) surveyed contemporary modeling approaches and design issues in RSM (see also Khuri, 2005).

10.1 RESPONSE SURFACE EXPERIMENTATION: ONE DESIGN OR MORE THAN ONE?

Assume that there are 20 factors that may be related to a response variable and there is a desire to determine the best combination of levels of the factors that are related to the response variable so as to optimize the (value of) response variable. The first step would obviously be to determine which of the 20 factors are really important. Typically this would be performed with some type of screening design that has a small number of runs relative to the number of factors, and for which main effects would be confounded with two-factor interactions. Of course these screening designs work only if main effects dwarf interactions, as is often the case.

If certain two-factor interactions are significant, we may falsely identify the main effects with which they are confounded as being significant. The consequence of this is that too many factors may be examined in a subsequent response surface design, thus resulting in an inefficient design. Clearly it would be much better if factor screening and response surface fitting and optimization could be performed with a single experiment, and have a high probability of identifying the real effects with the factor screening. Such a strategy has been proposed by Cheng and Wu (2001) and Bursztyn and Steinberg (2001), among others.

The use of such a strategy entails projecting a factor space onto a smaller factor space, namely, the space of factors that seem to be important. This means that the design used for factor screening must have at least three levels since the design will be projected onto a smaller factor space *of the same design*.

Box and Wilson (1951) decried the use of 3^{k-p} designs as second-order designs, pointing out that there is no useful 3^{3-p} design, so a 3^3 design would have to be used. They also pointed out that a 3^{5-1} design (81 runs) is needed for fitting a second-order model with 21 parameters. That is, there are degrees of freedom available for estimating higher-order terms that are not being used, and the 59 degrees of freedom that are available for estimating the error variance are far more than would normally be needed. As pointed out by Cheng and Wu (2001), the Box and Wilson claim that a 3^4 design is necessary for fitting a second-order model with four factors and 15 parameters is incorrect as this could be accomplished with any 3^{4-1}_{IV} design. Cheng and Wu (2001) pointed out that the Box and Wilson argument for not using these designs holds only for $k = 3$ and $k = 5$.

Furthermore, as pointed out by Cheng and Wu (2001), the inefficient run size argument for 3^{k-p} designs clearly doesn't hold when such a design is "doing double duty" by serving as both a screening design and a design for fitting a second-order model. For example, assume that a 3^{10-7} design is used and it is found that there are five important factors. The projected design must of course have the same number of design points, so it would be a 3^{5-2}. Cheng and Wu (2001) gave an example of a 3^{6-3} design projected onto a 3^{5-2} design.

We need to keep in mind, however, that designs such as a 3^{10-7} and a 3^{6-3} are resolution III designs, so if interactions exist, they could undermine this approach. Furthermore, a 3^{5-2} design has the same resolution, so interactions could not be estimated; that is, only pure quadratic terms could be fit, not mixed quadratic terms.

There is no 3_V^{k-p} design with 27 runs and the only 3_V^{k-p} design with 81 runs (more than most experimenters would probably want to use) is the 3_V^{5-1} design.

We should also keep in mind that the idea of using a single design with at least three levels will fail if the design does not cover the optimum region, and furthermore cover it in such a way that the fitted surface is essentially the same as the true surface. This would be a very bold assumption.

Cheng and Wu (2001) used the term *eligible projected design* to designate a second-order design. Since the number and the identity of the important factors are of course not known before the experiment is performed, it is thus desirable to use a design that has a relatively large number of eligible projected designs. For example, assume that there are 12 factors to be examined. There are $\binom{12}{4} = 495$ projected designs onto four factors, but not all of these will be eligible designs.

In general, the idea of letting a 3_V^{k-p} design do double duty seems to be not very practical because the design would have to have enough points (at least 81) for the projected design to be resolution V, since all second-order effects should be estimable when a second-order model is fit to characterize the response surface and to seek the optimum combination of factor levels.

However, let's make some comparisons. Assume that there are 10 factors to be screened and we know that five of the factors are important, as well as certain inter-actions among the 5 factors. We could use a 3_{IV}^{10-6} design and project that into a 3_V^{5-1} design, assuming that we have correctly identified the five important factors. We are then able to fit a full second-order model in these five factors with 81 runs, assuming of course that the 3_V^{5-1} design covers the optimum region. If a 2_{IV}^{10-k} design with the fewest number of design points were used, this would be a 2_{IV}^{10-5} design, which would project into a 2^5 design for any five of the factors. Converting this into a CCD would require 10 more runs for the axial points, and let's assume that eight centerpoints were used, for a total of 50 runs, 31 runs fewer than with the 3-level design approach. Thus, the latter is not competitive with the two-level, screening design approach for the first stage.

Of course both approaches would fall apart if the optimum region is not covered by the first design, but the 2-level design approach might still be superior if the method of steepest ascent/descent (Section 10.4) had to be employed in addition to the use of two designs, provided that the region in which the second-order design was to be used was reached quickly without much experimentation performed along the path of steepest ascent/descent. Of course the optimum region will generally be unknown and will not likely be covered by a two-level design unless the factor levels practically extend to the limits of the operability region. Since the number of experiments that are necessary along an assumed path of steepest ascent/descent will vary between experiments and will be unknown before any specific experimentation is undertaken, it is almost impossible to compare the suggested one-design approach with the standard sequential approach.

If the response surface is complex with multiple peaks and/or valleys, it can be difficult to determine optimum operating conditions with either approach. For such surfaces, a method of continuous experimentation such as simplex EVOP (Evolution-ary Operation) might be employed in searching for the optimum region; then use the standard EVOP approach (Box, 1957) or either of the design approaches described here to determine the optimum conditions after the optimum region seems to have been pinpointed. (Simplex EVOP is discussed in Spendley, Hext, and Himsworth (1962), Lowe (1974), and Hahn (1976).)

10.2   WHICH DESIGNS?

The question of which designs to use in RSM work was also raised in the literature by Tang, Chan, and Lin (2004), who reported that the theoretical results of Fang and Mukerjee (2000) showed that the success of factorial designs in the exploration of response surfaces is due to the fact that the designs uniformly cover the entire design space, rather than being due to the combinatorial or orthogonal properties of the design. Tang et al. (2004) described the successful application of a UD, which they believe has considerable potential in certain situations relative to traditional response surface designs. Uniform designs are covered in detail in Section 13.9.1.

In the same general vein of alternative designs, Hardin and Sloane (1991; http://www.research.att.com/~njas/doc/doeh.pdf) discussed the computer generation of response surface designs for spherical regions. They stated, "The best designs found have repeated runs at the center and points well spread out over the surface of the sphere." Similarly, Cox and Reid (2000, p. 181) stated that when the relationship between the response variable and factors is highly nonlinear "a space-filling design is then useful for exploring the nature of the response surface." Mee (2004) recommended the use of D-optimal and I-optimal three-level designs for spherical design regions involving three or more factors.

Furthermore, McDaniel and Ankenman (2000) found that a version of the traditional RSM worked best when the objective is to make small factor changes in search of a maximum or minimum. Of course this is also what is done when EVOP is used when seeking optimum conditions (Box and Draper, 1969).

Motivation for using something other than traditional RSM designs also comes from Giesbrecht and Gumpertz (2004, p. 413), who stated in their chapter on response surface designs that "in practice, it is more common to find severe constraints that limit the investigation to irregular regions."

There is thus evidence that we should move away from traditional RSM designs and use either UDs or designs quite similar to UDs over the acceptability or operability region. Nevertheless, we will first review traditional response surface designs but will later return to a discussion of UDs.

10.3   CLASSICAL RESPONSE SURFACE DESIGNS
VERSUS ALTERNATIVES

With current computing technology, we can easily see how data look in three dimensions. For example, given in Figure 10.1 is the surface plot for two factors using the data from Vázquez and Martin (1998) that was discussed in Section 6.1.

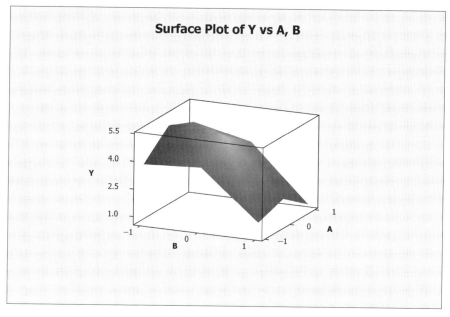

Figure 10.1   Surface plot for the data from Vázquez and Martin (1998).

This surface plot was constructed using Release 14 of MINITAB. The plot was constructed from the data for a 3^2 design, so the nine points were connected to form a surface. The need for a quadratic term in each factor is obvious from the "fold" in the surface on each factor and the interaction is apparent from the bend or kink in the surface. (The surface would have been flat if only main effects were important; the plot suggests that the linear and quadratic terms plus the interaction term explain just about all of the variability in Y, and in fact the model with those terms had an R^2 of .9984.)

We should keep in mind that just as models are unknown, so are response surfaces. The true response surface is almost certainly not as depicted in Figure 10.1, and is very likely irregular with peaks and valleys. We should also keep in mind that the surface in Figure 10.1 is obtained just by connecting points; this is different from a fitted surface that is obtained by fitting a model to data, but such a surface is also random in the sense that the coefficients are realizations of random variables. A fitted surface is discussed and illustrated in Section 10.13.

Obviously the only way we could obtain a very close approximation to the nature of the response surface between the extreme points in the design is to have an extremely large number of points over a fine grid of that region. Unless experimentation is extremely inexpensive, as it is in many computer experiments, the cost of obtaining the necessary number of design points to closely approximate the response surface will generally be prohibitive. The goal then is to approximate the response surface as closely as is needed relative to the objectives of the study, and to do so in an economical manner—not an easy task.

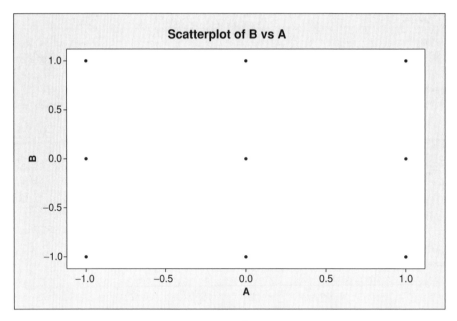

Figure 10.2 3^2 design.

Assume that there are two factors of interest and consider the design layouts depicted in Figures 10.2–10.4.

Figure 10.2 is the design layout for the 3^2 design used by Vázquez and Martin (1998). Although the region appears to be uniformly covered, one weakness is that there is only one point in the interior of the design. This would be an especially poor design if the values of the factors in the original (uncoded) units had been chosen to be extreme values rather than values that were well within the extremes, as with such extreme values the centerpoint might then be the only design point that would be either a point used in practice or in proximity to points used in practice. (Of course here the term "centerpoint" is being used to refer to the design point that is in the center of the design, which of course is different from centerpoints that are added to, say, a 2^k design.)

Figure 10.3 is the layout for a design that is discussed in more detail in Section 10.5: a CCD. This design also has nine unique design points and does not differ greatly from the 3^2 design. As with the latter, the CCD has only one point in the interior of the design. The design was generated by MINITAB using the default values for the axial (star) points, which are the points in proximity to the 2^2 design points. Selection of axial point values is discussed in Section 10.5.

Figure 10.4 is a UD constructed using JMP. (*Note*: This design and the Latin hypercube and sphere-packing designs that are illustrated in Figures 10.6 and 10.7 are random designs in the sense that if JMP, for example, is used to construct the designs, successively generated designs of the same type and with the same design parameters will be different.)

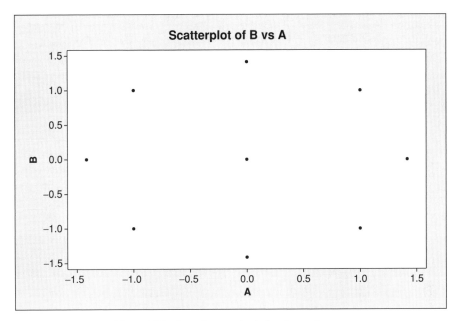

Figure 10.3 Central composite design for two factors.

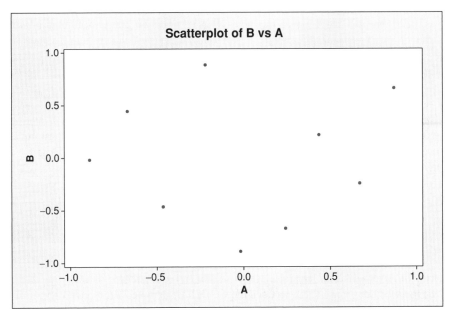

Figure 10.4 Uniform design for two factors.

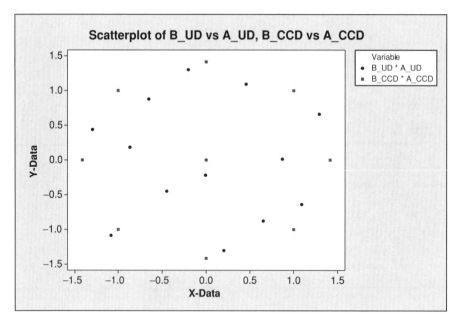

Figure 10.5 Comparison of CCD and UD.

Unlike the first two designs, this design has an irregular configuration, although there is better coverage of the design interior than with the first two designs. Remember that the surface in Figure 10.1 was constructed by connecting the nine design points. Therefore, if serious nonlinearities exist in the interior of the design, a UD would provide a better opportunity for detecting them, although we would probably prefer a few more interior points. That is, if the interior points are not well-fit by the full second-order model, we would like to be able to detect that.

The CCD in Figure 10.3 actually has 13 experimental runs since the centerpoint is replicated five times. Thus, a fairer comparison would be to compare a UD with 13 points to that CCD.

The scaling of the two designs should be the same so as to facilitate a fair comparison, and it would obviously be helpful if the designs could be shown on the same graph. Since the range was $-\sqrt{2}$ to $\sqrt{2}$ for each factor for the CCD (this was not a deliberate choice; this is simply how the design is constructed), this range was also used for the UD.

The comparison is shown in Figure 10.5 and there is obviously no comparison in terms of coverage as a UD provides essentially the same degree of coverage of the periphery of the design space as does the CCD, but has much better coverage of the interior region than the CCD. This of course is due primarily to the fact that there are 13 distinct design points with a UD but only 9 with the CCD, as stated previously.

Some potential UD users might be turned off by the unequal spacing of the design points and by the fact that there is not a unique UD for a given number of factors and

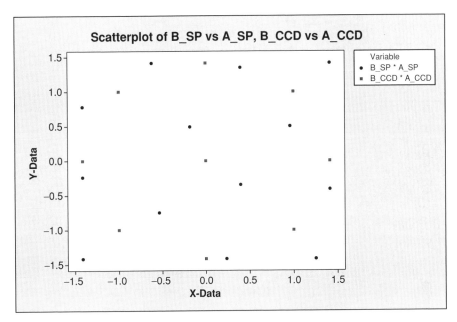

Figure 10.6 Comparison of sphere-packing design and CCD.

design points, as stated previously (see also Section 13.9.1). There are other types of "space-filling" designs, as they are called (Section 13.9), and Figure 10.6 shows the comparison of the sphere-packing design with the CCD.

Finally, Figure 10.7 shows the comparison of a Latin hypercube design with the CCD.

It should be apparent from each of these last three graphs that a space-filling design provides better coverage of the interior part of the design space than does a CCD. There is a price that is paid for this, however, although it is a small one. Specifically, space-filling designs are not orthogonal designs for estimating main effects, although the departure from orthogonality is not great. For example, the correlation between the columns for A and B for the particular Latin hypercube design that was generated is $-.066$, $.064$ for the sphere-packing design, and $-.041$ for the UD. The superior ability to detect nonlinearities more than offsets this slight departure from orthogonality.

In the general spirit of space-filling designs, Design-Expert can generate "distance-based" response surface designs, which spread the points evenly over the experimental region. These are not random designs as only certain interior points are generated randomly, with the other points having fixed coordinate values of 0, 1, and -1.

10.3.1 Effect Estimates?

Effect estimates were discussed for two-level designs, starting with Section 4.1, and for three-level designs in Section 6.1.1. For the uniform and Latin hypercube designs discussed in Section 10.3, the design points all represent different factor levels, and

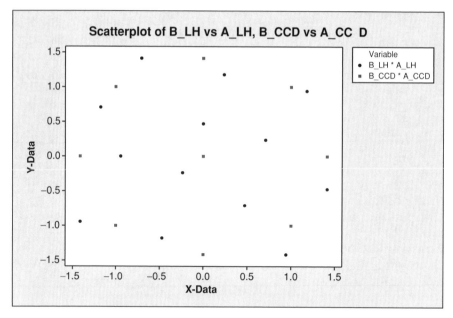

Figure 10.7   Comparison of Latin hypercube with CCD.

indeed the designs are random designs with points that would differ if the designs were successively generated, as was explained in Section 10.3. Therefore, it wouldn't make any sense to think about trying to define effect estimates for such designs, any more than it would make sense to try to do so when a regression model is constructed using a single random regressor (although we do compute regression coefficients, which is different).

We return to UDs and look at some additional statistical properties of them as well as some applications later in the chapter, after first looking at traditional response surface designs. (*Note*: Whereas the designs depicted in Figures 10.5–10.7 were generated using JMP, it isn't possible to use JMP to generate space-filling designs for restricted regions of operability when the region does not form a rectangle. It is, however, possible to do this with SAS Software by using PROC OPTEX, as is illustrated by Giesbrecht and Gumpertz, 2004, p. 415.)

10.4   METHOD OF STEEPEST ASCENT (DESCENT)

This method, due to Box and Wilson (1951), is used to try to identify the region, as narrowly defined as possible, that contains the optimum operating conditions. Methodology for producing a "path of steepest ascent" (or descent) has been available for several decades, but with modern computing capabilities we don't need to think about literally constructing the path.

We will use some simple examples to illustrate the basic idea.

Example 10.1

Assume that there are two factors of interest and for the sake of illustration we will presume that the relationship between them and the response variable is $Y = 2 + 0.4X_1 + 0.3X_2$, with the two factors constrained to lie within $(-1, 1)$, as in a two-level design. The response variable is to be maximized. What is the path of steepest ascent (i.e., path to the maximum), assuming that the current combination of factor levels is $(0, 0)$?

Looking at it intuitively, it should be apparent that the path will be in the quadrant for which both X_1 and X_2 are positive. What might not be obvious, however, is the relationship between X_1 and X_2 along the path. Since the coefficient of X_1 is larger than the coefficient of X_2, the change in X_2 should be expressed in terms of the change in X_1 and will also be a function of the relationship between the coefficients. This relationship can be easily determined if we think about equal increments in Y, starting with $Y = 2$ since that is the value of Y when $(X_1, X_2) = (0, 0)$. We might think about increments of 1.0 for Y, recognizing that Y cannot exceed 2.7 because of the constraints on X_1 and X_2.

If we fix the increment of X_1, ΔX_1, at 0.1, we can then solve for ΔX_2 after determining the direction of steepest ascent, which is simply $\lambda(0.4, 0.3)$, obtained from the coefficients in the assumed equation, and also what we would obtain using calculus directly. That is, with $\Delta X_1 = 0.1$, $\Delta X_2 = (3/4)\Delta X_1 = 0.075$.

Another way of viewing this is to recognize that for a given value of Y such as $Y = 2$, the slope of the line when X_2 is written as a function of X_1 is -1.33. The slope of a line perpendicular to this line is the negative reciprocal of this slope, namely 0.75. A change in X_1 of 0.1 would thus result in a change in X_2 of 0.075.

The line of steepest ascent would, if the scales were the same, thus be perpendicular to the lines in Figure 10.8. Obviously the maximum occurs at $(1, 1)$ and we don't need the method of steepest ascent to tell us that; this was just a simple illustrative example.

Box and Draper (1987, pp. 190–194) presented a method for determining whether the path of steepest ascent has been determined precisely enough (see also Myers and Montgomery, 1995, pp. 194–198). The method involved the construction of a confidence cone about the steepest ascent/descent direction and is described in detail in Section 5.5.1.2 of Croarkin and Tobias (2002). See http://www.itl.nist.gov /div898/handbook/pri/section5/pri5512.htm#Technical%20Appendix.

The cone is given by the inequality

$$\sum_{i=1}^{k} b_i^2 - \frac{\left(\sum_{i=1}^{k} b_i x_i\right)^2}{\sum_{i=1}^{k} x_i^2} \leq (k - 1) s_b^2 F_{\alpha, k-1, n-p}$$

with the b_i denoting the coefficients of the k (linear) model terms and s_b^2 denoting the common variance of the coefficients. Any point with coordinates $(x_1, x_2, \ldots, x_k)$ that

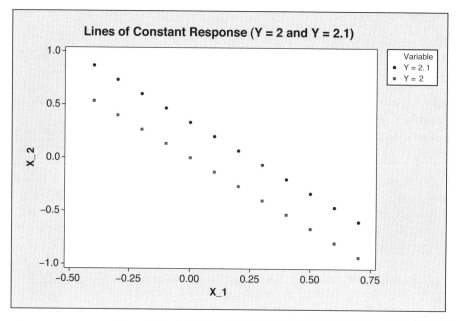

Figure 10.8   Lines of constant response for example.

satisfies this inequality lies within the confidence region provided that $\sum_{i=1}^{k} b_i x_i > 0$
for ascent and $\sum_{i=1}^{k} b_i x_i < 0$ for descent. Box and Draper (1987) gave a measure
of how good a particular search direction is by the fraction of directions that are
excluded. Of course we want that fraction to be as close to 1 as possible. The closer
the fraction is to 1, the greater the confidence in the direction that is used. Of course
we can't see this in high dimensions but Myers and Montgomery (1995, p. 196)
applied the Box and Draper methodology to an example with $k = 2$ to compute the
fraction. Sztendur and Diamond (2002) extended the Box and Draper approach to
cover heterogeneous variances, nonlinear designs, and generalized linear models.
They also used conditional effects in arriving at a path of steepest ascent. This is
discussed in Section 10.9.

   Whether the path is well determined will obviously depend upon the magnitude of
the standard errors of the factor coefficients, which, unlike the example, will generally
be unknown. The value of R^2 can also be used as a general indicator, as a low R^2
value for the fitted model would not provide any confidence that the direction has
been well determined and low-to-moderate R^2 values will generally correspond to a
relatively wide confidence cone.

   One of the criticisms of the method of steepest ascent is that the direction is scale
dependent. To illustrate, consider Example 10.1. If X_1 is in inches and is divided by 12
to convert to feet, the new coefficient would be $0.4 * 12 = 4.8$ and a different direction
would then result since the direction is determined by the relationship between the
model coefficients.

This has motivated recent work on scale-independent steepest ascent/descent methods, as given in Kleijnen, den Hertog, and Angün (2004). Del Castillo (1997) discussed stopping rules for the method of steepest ascent and suggested (see http://www.informs-sim.org/wsc04papers/059.pdf) that the experimentation cease when the value of the selected objective function (which might be R^2) is worse than the preceding value. Thus, experimentation might cease when R^2 exhibits a decline. This could indicate that the first-order model is no longer adequate for the region that has been entered, and so experimentation with a second-order model would be necessary. Of course we would also want to see the average response increase or decrease along the path, depending upon whether maximization or minimization is sought.

Thus, we are looking for two things: the need to switch to a second-order model and an indication that the right path (or at least a good path) is being followed. Nicolai, Dekker, Piersma, and van Oortsmarssen (2004) suggested using a simple t-test on responses from successive experiments to determine if the movement in the average response is in the desired direction.

Of course whether or not the method of steepest ascent is used as part of the standard three-stage approach depends on whether or not identification of the important factors and model fitting are to be performed with the same design, as discussed in Section 10.1. This in turn will depend on several factors, including the costs of running multiple experiments and the practicality of doing so, and whether or not the experimenters believe they know the region of optimal response. If off-line experimentation is deemed too expensive for multiple experiments, the single-design approach might be used followed by an EVOP-type approach (if deemed possible to use), to check on the declared optimal combination of factor levels. That is, small changes would be made in factor levels, starting from the presumed optimal levels, and the changes in the response values noted and a decision made as to whether the changes are statistically significant. If so, further investigation would be warranted.

10.5 CENTRAL COMPOSITE DESIGNS

The most frequently used response surface design is the central composite design, abbreviated as CCD in the beginning of the chapter and abbreviated the same way in this section. There are various forms of the CCD that will be discussed in this section, which is why the word "designs" is in the section title. Readers interested in the motivation for the invention of the design(s) are referred to Box (1999–2000), who invented the design as given in Box and Wilson (1951). Bisgaard (1997) reported that the move away from traditional factorial and fractional factorial designs to a CCD was quite controversial at the time.

The basic design framework consists of three sections: the points of a 2^{k-p} design with, as usual, k denoting the number of factors and p usually but not necessarily equal to zero, axial (star) points, and centerpoints. All the sections are variable in the sense that although a CCD is generally presented with the factorial portion of the design composed of the points in the 2^k design, a fractional factorial could be used. MINITAB, for example, can be used to construct a "full" CCD (i.e., using the 2^k design in the factorial part) for k between 2 and 7, a half CCD for k between 5 and 8, and a quarter CCD for k equal to 8 or 9, with "half" and "quarter" of course being the half and quarter fractions, respectively. Specifically, for $k = 5$ the half CCD would have 32 points that would consist of the 16 factorial points, 10 axial points, and 6 centerpoints. This is a useful alternative to the full CCD for $k = 5$, since the factorial part of the CCD is resolution V. (The CCD has the same resolution as the factorial portion of it for estimating main effects and interactions since the axial points and centerpoints are not involved in the estimation of interactions.)

Whereas the number of factorial points is determined by the selection of a full or fractional factorial, the number of axial points is always twice the number of factors. There is no hard-and-fast rule for selecting the number of centerpoints, but studies have shown that the best response surface designs have replicated centerpoints. Of course centerpoints aid in the investigation of curvature in the region around the center. These points also aid in stabilizing $\mathrm{Var}(\widehat{Y})$ in and around the center, as illustrated by Myers and Montgomery (1995, pp. 310–311).

A CCD in two factors was shown in Figure 10.3 and compared with other designs in a graphical comparison. The design points are listed below.

Number	A	B
1	-1.000	-1.000
2	1.000	-1.000
3	-1.000	1.000
4	1.000	1.000
5	-1.414	0.000
6	1.414	0.000
7	0.000	-1.414
8	0.000	1.414
9	0.000	0.000
10	0.000	0.000
11	0.000	0.000
12	0.000	0.000
13	0.000	0.000

The star points are $(\alpha, 0)$, $(-\alpha, 0)$, $(0, \alpha)$, and $(0, -\alpha)$ for two factors, with the value of α to be selected. The value of α for the design listed above is $\sqrt{2}$, which is the default value in MINITAB. The reason for this choice will be explained shortly. Notice that if $\alpha = 1$ and the set of full factorial points is used, the resultant design is a 3^2 design when these points are combined with the centerpoints. Otherwise, the design has five levels for each factor, more than enough to estimate the linear and quadratic effects, although the quadratic effect columns are not orthogonal to each other. Of course only the factorial points are used in estimating the AB interaction, as only the first four elements in the column obtained from the product of the A and B columns are nonzero.

The objective is generally not to create a 3^2 design, however, but rather to solve for α so as to have a *rotatable* design. The latter is a design for which $\text{Var}(\widehat{Y}_x)$ is the same for all points $\mathbf{x}$ (with $\mathbf{x}$ denoting the set of point coordinates) that are equidistant from the center of the design, which is the point $(0, 0)$ when there are two factors, independent of direction from the center. It is well known that this condition is met when $\alpha = \sqrt[4]{F}$, with F denoting the number of points in the factorial part. Thus, $\alpha = \sqrt[4]{4} = \sqrt{2} = 1.414$ when $k = 2$ and the factorial part contains the full factorial. Notice that all the points beyond the centerpoint are equidistant from the origin with this choice of α. (Note that some software packages, such as JMP, will label a CCD with α determined in this manner a *uniform precision design*.)

Since the axial points (which are also called star points since they form a star configuration when combined with the other points) are orthogonal for any choice of α and of course the factorial points are also orthogonal, the design will be orthogonal for estimating the linear factor effects, regardless of the number of centerpoints that are used. The quadratic factor effects will not be orthogonal to each other because that could not happen unless the columns representing the quadratic effects had some negative numbers, but that can't happen since the square of a positive or negative number is a positive number. The columns for the quadratic effects are orthogonal to the linear effects however, and are also orthogonal to the interaction effects.

Speaking of "effects," which were used in only a general way in the preceding paragraph, we can address the question for a CCD. If we consider the CCD for two factors given earlier in this section, we see that observations at the five levels of each factor occur with different frequencies. That is, there is only one observation at each axial point, two observations at each factorial point level and five centerpoints. Computing conditional effects using only the factorial points is potentially useful, however; this is discussed in Section 10.10.

Myers and Montgomery (1995, p. 311) asked the question: "How important is rotatability?" Their view is that a near-rotatable design should be good enough, and the same should probably be said of orthogonality, thus not ruling out a space-filling design on the grounds of nonorthogonality. It should be noted that the quadratic effect estimators as a group are likely to be less correlated with each other with a space-filling design than with a CCD, although this is offset by the fact that the quadratic effect estimators will be correlated with the interaction effect estimators and with the linear effect estimators, and of course the latter two will also be correlated with each other.

Given below is JMP output that shows the correlation between effect estimates when a UD with 19 design points is used to fit a full quadratic model in three factors. Notice that two of the three correlations between the quadratic terms are small and, ignoring correlations with the intercept term, only three of the correlations exceed .2567 in absolute value, with all but one of the other correlations being less than .20 in absolute value. The user would have to decide whether the magnitude of the correlations involving linear and interaction terms offsets the uniform coverage of the design space, but this is not a bad design in terms of the correlations.

```
Corr
              Intercept       X1       X2       X3    X1*X2    X2*X2    X1*X1    X3*X3    X2*X3    X3*X1
Intercept        1.0000   0.0708  -0.0051  -0.0817  -0.0657  -0.5310  -0.6977  -0.6227  -0.1054   0.0472
X1               0.0708   1.0000   0.0238  -0.0313   0.0440  -0.0895  -0.0108  -0.0501   0.0974   0.0059
X2              -0.0051   0.0238   1.0000  -0.0089  -0.0477   0.0479   0.0382  -0.0723   0.1674   0.1241
X3              -0.0817  -0.0313  -0.0089   1.0000   0.0817   0.1614   0.0086   0.0115  -0.0581  -0.0260
X1*X2           -0.0657   0.0440  -0.0477   0.0817   1.0000  -0.0141   0.0035   0.1592   0.1124   0.2129
X2*X2           -0.5310  -0.0895   0.0479   0.1614  -0.0141   1.0000   0.1679   0.0121   0.0211   0.0886
X1*X1           -0.6977  -0.0108   0.0382   0.0086   0.0035   0.1679   1.0000   0.3239   0.1692  -0.1311
X3*X3           -0.6227  -0.0501  -0.0723   0.0115   0.1592   0.0121   0.3239   1.0000   0.0382  -0.0645
X2*X3           -0.1054   0.0974   0.1674  -0.0581   0.1124   0.0211   0.1692   0.0382   1.0000   0.1851
X3*X1            0.0472   0.0059   0.1241  -0.0260   0.2129   0.0886  -0.1311  -0.0645   0.1851   1.0000
```

Although rotatability is a desirable property of a response surface design, it loses some of its luster when we think about how the fitted equation that results from the experimentation will likely be used. Once the optimum factor levels have been determined, it would undoubtedly be of interest to obtain a prediction interval for the response. There may not often be a need for a second prediction interval for a combination of factor levels such that the point is the same distance from the center as the point representing the optimum conditions. Furthermore, if one or more factor levels are altered slightly, for whatever reason, from the supposed optimum combination of factor levels, the point representing the combination of levels to be used almost certainly won't be the same distance from the center of the design as the optimum point. Even if the distance were the same, there is no guarantee that the two variances of the fitted values will be the same, since rotatability applies only to points used in the design, not to all points in the region covered by the design that are the same distance from the center as the design points.

Before we look at CCD applications in the next section, we should address the question of whether a CCD gives us too many design points, similar to the discussion at the beginning of Chapter 6 regarding three-level full and fractional factorials. For four factors, a CCD will have 25 points if there is a single centerpoint $(16 + 1 + 8)$, whereas a 3^{4-1} design has 27 points—not much difference, and of course CCDs generally have more than one centerpoint. (The effects that can be estimated with the 3^{4-1} design, which is resolution IV, are given by Wu and Hamada, 2000, p. 250.) Wu and Hamada show that all main effects can be estimated, in addition to being able to estimate each two-factor interaction using two degrees of freedom (instead of four df, which each interaction has). Thus, 20 df are used for estimating effects from the 26 df that are available.

Notice that the number of effects to be estimated in a full second-order model is $k + k + \binom{k}{2} = (k^2 + 3k)/2$, whereas the number of points in the CCD is $2^k + 2k + c$, if a full factorial is used, with c denoting the number of centerpoints. This will provide more than enough degrees of freedom for estimating all the effects. In fact, the design is somewhat wasteful in this respect as, for example, there are 20 effects to be estimated for $k = 5$, whereas there would be 43 design points even if only a single centerpoint were used, which is not recommended. Furthermore, we don't need factors with five levels to estimate second-order effects.

We can reduce the number of design points considerably by using either a 2^{k-1} or a 2^{k-2} design in the factorial portion.

10.5.1 CCD Variations

The CCD described in the preceding section is the standard CCD. There are variations of the design that are sometimes useful, however. For example, the region of operability may be defined by the factorial points. If so, no point in the design space could have a coordinate of $\pm\sqrt{2}$, for example. The star points would thus have to be "shortened" to conform to that region, that is, the largest absolute value of any coordinate could not exceed 1. For two factors this means that the factorial and star points constitute a 3^2 design except for the fact that the point $(0, 0)$ is missing, which of course would not be missing from the overall design since the design contains centerpoints. In general, such a design is called a *face center cube* because the axial points occur at the center of the faces of a cuboidal region. We will let this design be designated as CCF so as to distinguish it from the standard CCD.

A face center cube makes sense if the region of interest is cuboidal; if the region of interest is spherical, the standard CCD should be used. The latter should also be used if rotatability is important, since the face center cube is obviously not rotatable since the star points do not have the coordinates that are necessary for rotatability.

If a spherical region is desired but no coordinate can exceed 1 in absolute value, a solution is to "scale down" the design so that the star points fall within the restricted region and the factorial points have smaller coordinates. This can be accomplished by starting with the regular CCD and dividing through by α. For example, for $k = 2$, the nine distinct points would be as follows.

A	B
-0.707	-0.707
0.707	-0.707
-0.707	0.707
0.707	0.707
1	0
-1	0
0	1
0	-1
0	0

This is often called an *inscribed* central composite design, with the notation CCI used in some sources to distinguish this design from the CCD and CCF. Of course the CCI is rotatable since it is just a scaled version of a rotatable design.

A compact summary of the different types of CCDs, including a graphical comparison, is shown in http://www.itl.nist.gov/div898/handbook/pri/section3/pri3361.htm.

10.5.2 Small Composite Designs

There are various ways of constructing composite designs that have fewer runs than a CCD. Such designs date from Hartley (1959), who contended that one could use a resolution III design for the factorial portion, with other methods proposed by Ghosh and Al-Sabah (1996) and Draper and Lin (1990), among others. In addition to being economical, small composite designs have value in that the correlation between the quadratic effect estimates is reduced as the number of factorial points is reduced, as the reader is asked to show in Exercise 10.25. (This follows from the fact that it is the factorial portion of the design that is creating the correlation, which is a perfect correlation for that portion since each column entry is a 1. Therefore, reducing the proportion of factorial-point runs will reduce the correlation.)

10.5.2.1 Draper–Lin Designs

Work on producing composite designs with a fewer number of design points than the composite designs that result from using either a full factorial or a fractional factorial for the factorial portion also includes Westlake (1965), who used irregular fractions of the 2^k system. Draper (1985) used columns of Plackett–Burman designs (see Section 13.4.1) to produce designs for the same number of factors but with fewer design points. Draper and Lin (1990) improved on that set of designs by finding designs for higher values of k and also finding designs that Draper (1985) conjectured did not exist in a singular version. The Draper–Lin designs are not rotatable, however. Another deficiency is that coefficients for terms of the same order, such as linear term coefficients, are not estimated with equal precision. This will be illustrated later with an example. Draper and Lin (1990) did not discuss the choice of α for these designs, but Croarkin and Tobias (2002) indicate that α should be chosen between $(F)^{1/4}$ and $\sqrt{k}$.

Example 10.2

Chapman and Masinda (2003) used what was stated to be a Draper–Lin small composite design with 18 runs for four factors to determine the settings of those factors needed to minimize the effort that is necessary to close a car door on a new prototype design. They also wanted to use the factor settings that were the least sensitive to variations in the process operating conditions and hoped that the same settings would meet both goals.

Although the authors used a randomized run order, the design runs are given, along with the data, in Table 10.1 in an order that allows the nature of the design to be more easily seen.

The authors analyzed these data and arrived at the following fitted equation: $\widehat{Y} = 1.34 - 0.068X_1^2 + 0.021X_2^2 + 0.083X_1X_4 - 0.036X_3X_4$. Notice that there is no main effect in the model, which is as nonhierarchical a model as one could imagine! This would suggest that both the experiment and the data be checked. (Notice also that the sum of the column for A is not zero; the ramifications of this will be discussed later.)

The design given in Table 10.1 is *not* a Draper–Lin design for four factors as described in Draper and Lin (1990). Those designs were constructed by using specific columns from the appropriate Plackett–Burman design, adding to those points the axial points, with centerpoints added if desired. Specifically, for four factors the Plackett–Burman design has eight runs and is a 2_{III}^{4-1} design and they recommend that columns 1, 2, 3, and 6 be chosen so that these runs will constitute the maximum

TABLE 10.1 Data from Chapman and Masinda (1985) Experiment

A	B	C	D	Y
0.5	0	0	0	1.19
0.5	0	0	0	1.18
−0.5	2	2	1	1.33
−0.5	−2	−2	−1	1.31
−0.5	−2	2	−1	1.63
−0.5	2	−2	1	1.40
1.5	2	2	−1	1.14
1.5	−2	−2	1	1.42
1.5	2	−2	−1	1.08
1.5	−2	2	1	1.29
0.5	0	−3.364	0	1.40
0.5	3.364	0	0	1.57
0.5	0	3.364	0	1.29
0.5	−3.364	0	0	1.60
0.5	0	0	−1.682	1.35
0.5	0	0	1.682	1.48
2.182	0	0	0	1.07
−1.182	0	0	0	1.33

"D-value," which they define as the determinant of $X^T X$ divided by n^p. Of course the D-value is computed for the full second-order model that is to be fit. As explained by Draper and Lin (1990), maximizing the D-value causes the design to be the most spread out in the coordinate space.

The 8-run Plackett–Burman (2_{III}^{4-1}) design is as follows.

```
Run   A   B   C   D   E   F   G

1     +   +   +   -   +   -   -
2     -   +   +   +   -   +   -
3     -   -   +   +   +   -   +
4     +   -   -   +   +   +   -
5     -   +   -   -   +   +   +
6     +   -   +   -   -   +   +
7     +   +   -   +   -   -   +
8     -   -   -   -   -   -   -
```

When the eight axial points are added to the appropriate columns, this produces a 16-point design that could be used to estimate the four main effects, four quadratic effects, and six interaction effects, in addition to fitting the constant term for a total of 15 model coefficients, although certainly not all the terms would be significant.

A Draper–Lin design can be generated using Design-Expert by specifying that a "small" CCD is to be constructed. The design is given below, in an "unrandomized" order rather than the randomized order given by Design-Expert.

A	B	C	D
-1.00	-1.00	-1.00	-1.00
-1.00	-1.00	1.00	-1.00
-1.00	1.00	-1.00	1.00
1.00	-1.00	1.00	1.00
-1.00	1.00	1.00	1.00
1.00	1.00	-1.00	-1.00
1.00	1.00	1.00	-1.00
1.00	-1.00	-1.00	1.00
1.68	0.00	0.00	0.00
-1.68	0.00	0.00	0.00
0.00	1.68	0.00	0.00
0.00	-1.68	0.00	0.00
0.00	0.00	1.68	0.00
0.00	0.00	-1.68	0.00
0.00	0.00	0.00	1.68
0.00	0.00	0.00	-1.68

This design was first given by Hartley (1959). Notice that the design is constructed using columns 1, 2, 3, and 6 of the Plackett–Burman design. Thus, the design attains the maximum D-value for a fixed Plackett–Burman design from which the columns are selected, and this is also true for the other Draper–Lin designs that are constructed by Design-Expert.

Since 8 is a power of 2, the factorial portion of the design must be a 2^{4-1} design. We can see that $B = -AD$ so $I = -ABD$. Of course this is not the way we would construct a 2^{4-1} design but such designs cannot be used to estimate quadratic effects, and here we are concerned with the best design for estimating the effects in a full quadratic model.

The design employed by Chapman and Masinda (2003), which should have been termed a modification of a Draper–Lin design, can be converted to a Draper–Lin design by dividing the X_2 and X_3 columns by 2 and subtracting 0.5 from each of the X_1 values. This will convert the design into a Draper–Lin design plus two centerpoints. The X_1 values given by the authors do not sum to 0, as they must. Thus the design really isn't a valid design. According to Chapman (personal communication, 2005), the factor-level combinations required by the Draper–Lin design were not feasible, and hence the alterations.

Those alterations created a myriad of problems, however. In addition to the X_1 values not summing to zero, the choice of X_1 values creates a (high) correlation of .707 between X_1 and X_1^2, whereas there is no correlation between linear and pure quadratic terms with a Draper–Lin design. This high correlation probably at least partly explains the existence of the X_1^2 term in the model without the X_1 term. Similarly, the correlation between X_2 and $X_1 X_4$ is $-.641$, which likely helps explain why there is an $X_1 X_4$ term and an X_2^2 term in the model without there being an X_2 term. Finally, when there is a two-factor interaction term in a model ($X_3 X_4$) with neither of the main effects being present, an interaction plot should be constructed, followed by a conditional effects analysis.

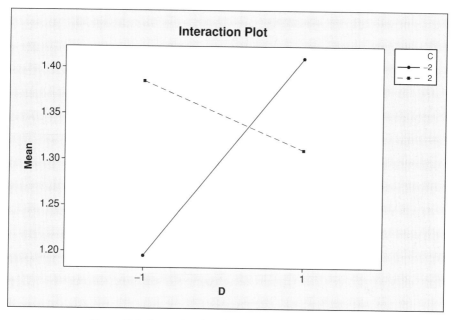

Figure 10.9 Interaction plot using only Plackett–Burman points.

Interaction plots with five levels for each factor do of course create interpretation problems. The advice given later in Section 10.10 is to not use axial points and centerpoints in computing conditional effects. This means that we should also not use such points in constructing interaction plots if the motivation for computing conditional effects is to come from interaction plots. Therefore, the plot in Figure 10.9 is constructed using only the points from the columns of the Plackett–Burman design.

Although each average is computed from only two observations (and thus has a sizable standard error), the strong interaction is apparent, as is the large conditional effect of D at the low (-2) level of C and the large conditional effect of C at the low level of D. Thus, there *are* effects in these two factors; they are simply conditional effects rather than main effects.

Thus, we have shown what resulted in an extremely nonhierarchical model being selected, and how the model is misleading. Analyses of the type given here should *always* be performed whenever a nonhierarchical model is selected.

Example 10.3

We will illustrate the statement made earlier that Draper–Lin designs can have standard errors that differ for terms of the same order. Given below is the Draper–Lin design for four factors constructed using Design-Expert, and using the default value for the number of centerpoints.

A	B	C	D	Y
1.00	-1.00	1.00	1.00	10.1
-1.00	1.00	1.00	1.00	13.1
0.00	0.00	0.00	0.00	12.4
0.00	1.68	0.00	0.00	13.5
1.68	0.00	0.00	0.00	12.7
0.00	0.00	0.00	-1.68	12.6
-1.68	0.00	0.00	0.00	14.2
0.00	-1.68	0.00	0.00	14.3
-1.00	-1.00	-1.00	-1.00	10.8
1.00	-1.00	-1.00	1.00	10.6
0.00	0.00	0.00	0.00	11.9
-1.00	-1.00	1.00	-1.00	12.7
1.00	1.00	1.00	-1.00	13.8
0.00	0.00	0.00	0.00	14.6
0.00	0.00	1.68	0.00	15.2
0.00	0.00	0.00	0.00	16.3
1.00	1.00	-1.00	-1.00	17.1
0.00	0.00	-1.68	0.00	15.5
0.00	0.00	0.00	0.00	16.4
0.00	0.00	0.00	1.68	13.9
-1.00	1.00	-1.00	1.00	11.5

The portion of the output from Design-Expert that shows the standard errors is given below.

Factor	Coefficient Estimate	df	Standard Error
Intercept	14.60	1	0.83
A-A	-0.45	1	0.82
B-B	-0.24	1	0.82
C-C	-0.059	1	0.52
D-D	0.39	1	0.82
AB	1.52	1	1.07
AC	-0.91	1	0.69
AD	-1.65	1	1.07
BC	-0.39	1	0.69
BD	-0.88	1	1.07
CD	0.31	1	0.69
A^2	-0.62	1	0.50
B^2	-0.46	1	0.50
C^2	0.053	1	0.50
D^2	-0.69	1	0.50

Notice that the standard error of the C main effect is less than the standard error of the other three main effects, and the standard errors of the interaction terms also differ. In general, a price has to be paid for small designs of any type and the user must decide if standard errors that differ by this amount are of any concern. (The Draper–Lin design for three factors has equal standard errors for effects of the same type but not the same order as the interaction term coefficients have standard errors that differ from the standard errors of the coefficients of the pure quadratic terms.)

Before leaving this example, a few comments on the computation of quadratic effects seem desirable. The emphasis in this book is on design rather than analysis, as the former cannot be performed by software alone or almost exclusively unless a practitioner has access to expert systems software or software that has a strong expert systems flavor. The computation of quadratic effects is not as simple as the computation of linear effects, partly because there are alternative ways of doing it and there is no computational definition that doesn't have a shortcoming. Therefore, we will not pursue that here. The interested reader is referred to the detailed discussion of Giesbrecht and Gumpertz (2004, pp. 289–298).

10.5.3 Additional Applications

In this section we examine two applications of CCDs and critique the work that was performed.

Example 10.4—Case Study

Wen and Chen (2001) described the application of a CCD after a Plackett–Burman design had been used to identify what seemed to be the important factors. The stated objective was "optimizing EPA production by N. laevis." Of course a screening design is susceptible to the type of problems discussed in Section 10.1.

The CCD that was used by Wen and Chen (2001) was obtained by using a 2^{5-1} design in the factorial part, coupled with six centerpoints and 10 axial points with $(2, -2)$ used as the coordinate values in the star points (i.e., $\alpha = 2$). Note that $\sqrt[4]{16} = 2$, so the condition for rotatability is met with this choice of α. Only the average of the response values for the centerpoints was provided, rather than the individual values, although the individual values were used to estimate σ.

Four response variables were used in the study. The dotplot for the response variable EPA Yield is given in Figure 10.10.

Obviously there is an outlier, which one would reasonably expect to represent an error of some sort, but there was no mention of this observation by Wen and Chen (2001). We will later show a graph that also casts suspicion on the point.

Since the centerpoint observations were not given, it isn't possible to duplicate the analyses by Wen and Chen (2001). It is worth noting, however, that when a full quadratic model in the five factors is fit to the published data, the results in terms of effect significance differ greatly from the results given by Wen and Chen (2001) for each of the response variables. This suggests that the repeated centerpoint values may not be measuring the same thing as the six degrees of freedom that are available for the error term when the average of the values is used. More specifically, the question of whether the response values at the centerpoint are true replicates must be addressed.

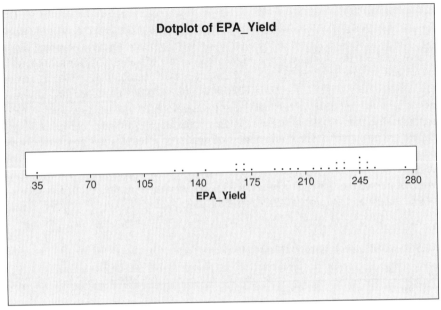

Figure 10.10   Dotplot of EPA yield.

The only factor that shows as being significant for the response EPA Yield when the published data are used is the last factor, temperature, as both the linear and quadratic terms in that factor are significant, the former more so than the latter. The linearity is apparent from Figure 10.11; the quadrature, however, is apparent only if the rightmost point is a valid point. A least squares line fit through the other points in the graph will obviously fall about 100 points above the rightmost point (whose response value is 37.12), suggesting that the first digit should have perhaps been a 1 and the digit is missing. In any event, the point should have been investigated.

As stated, the authors' objective was optimization but we won't pursue our analysis further since some of the data were not published. The moral of this story, however, is that data must be closely examined. "Seek and ye shall find" applies to data analysis as well.

10.6   PROPERTIES OF SPACE-FILLING DESIGNS

Since rotatability is not a relevant issue for a space-filling design because the points are not equidistant from the origin because of the "irregular" configuration of points, and there is also a slight deviation from orthogonality, the designs should be evaluated from other perspectives.

Equileverage designs are discussed extensively in Section 13.6 and have considerable intuitive appeal. An unstated objective in experimental design is that the design points should exert equal, or at least nearly equal, influence on the effect estimates,

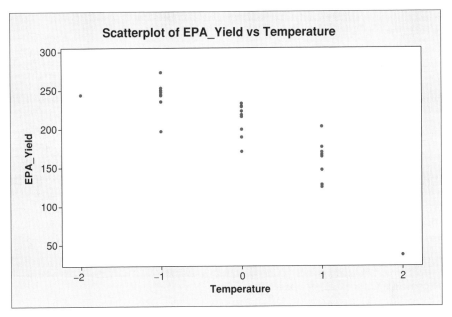

Figure 10.11   Scatterplot of part of Wen and Chen (2001) data.

and equivalently on the regression coefficients in the model representation. That is, the points should have "equal leverage."

Of course, whether or not an observation is influential will depend in part on the Y-coordinate, but we can improve the chances of not having any influential points by using X-values that are equally influential, that is, by constructing the design so that the design points have equal leverages, remembering that the leverages depend upon the model that is fit.

A leverage value reflects the distance that a point is from the "middle" of the group of points. With this semidefinition in mind, it is easy to see why common designs such as two-level full and fractional factorial designs are equileverage designs since the design points are at the vertices of a hyperrectangular region, such as in Figure 4.5, and are thus equidistant from the center, which would have coordinates of zero for all the factors.

A space-filling design has leverages that can differ considerably, especially in low dimensions, due simply to the "space filling." That is, the points in the interior of the design have leverages that are much less, relatively speaking, than the points on the periphery of the design space. For example, for the sphere-packing design shown in Figure 10.6, the leverages for fitting the full second-order model range from .266 to .739—a rather wide range. By comparison, the CCD has leverages that are equal (.625) for fitting the full second-order model, except of course for the centerpoint leverage, which is .20.

Since a space-filling design is not orthogonal and differs considerably from the equileverage state because of the interior points, the motivation for using one of these designs must come from the fact that they are much better at detecting nonlinearities, especially complex nonlinearities, than traditional response surface designs. Since the range of possible values for each coordinate must be specified when constructing a space-filling design, very extreme points cannot be constructed, although, as indicated previously, the leverages can still differ considerably.

One property of space-filling designs that will be undesirable in certain applications is that each factor level is constantly being changed since the number of distinct levels of each factor is equal to the size of the design. This is a *major* problem if at least one factor in the experiment is hard to change. Another problem is that the computer-generated levels are not common decimal fractions so the levels will have to be rounded to the nearest feasible values. The multitude of factor-level combinations also increases the likelihood that levels that cannot be used together will be generated (see the discussion of debarred observations in Section 13.8). If the feasible design space is irregular within the permissible range of levels for each factor, there is also an increased likelihood of having points in such regions when many sections of the presumed design space are given points.

Therefore, although space-filling designs are potentially useful, there will be many scenarios when they cannot be used.

10.7 APPLICATIONS OF UNIFORM DESIGNS

Fang and Lin (2003) state that the development of the UD was motivated by three system engineering projects in 1978, as described in Fang (1980) and Wang and Fang (1981). Thus the design was developed to *meet* a requirement and a specific application.

10.8 BOX–BEHNKEN DESIGNS

Since five levels aren't necessary to fit a second-order model, designs with three levels are an alternative. One such design is due to Box and Behnken (1960). These designs can be described as a combination of the balanced incomplete block idea with a 2^2 factorial. Given in Table 10.2 is the Box–Behnken (BB) design for three factors generated by MINITAB, with the number of centerpoints, 3, being the MINITAB default for three factors.

Notice that the 12 points that are not centerpoints consist of the 2^2 factorial designs for all pairs of factors, with a zero filled in for the column in each design that is not used (i.e., the incomplete block idea). Accordingly, the design is orthogonal since it is composed of orthogonal designs. The design is also rotatable because the designs have equal leverages, and the leverages are equal (except for the leverages at the centerpoints) because the design consists of the three possible pairs of 2^2 designs, which are rotatable, ignoring the other factor. Furthermore, the variance of the fitted response is the same at each of the points that are not centerpoints, and those 12 points are all at the same distance, $\sqrt{2}$, from the center. Thus, we should say that the design is rotatable.

TABLE 10.2 Box–Behnken Design for
Three Factors and Three Centerpoints

A	B	C
−1	−1	0
1	−1	0
−1	1	0
1	1	0
−1	0	−1
1	0	−1
−1	0	1
1	0	1
0	−1	−1
0	1	−1
0	−1	1
0	1	1
0	0	0
0	0	0
0	0	0

There has been moderate disagreement about the rotatability (or not) of BB designs in the literature, however. Specifically, Myers and Montgomery (1995, p. 321) stated that the designs for four and seven factors are rotatable and the designs for the other numbers of factors (i.e., 3, 5, and 6) are near-rotatable. Wu and Hamada (2000, p. 418) differ slightly, stating "The design for $k = 4$ is known to be rotatable and the other designs are nearly rotatable." At the other extreme, the software user's manual for Statistics Toolbox 4.0 states that "Box–Behnken designs are rotatable designs" (see http://ftp.math.hkbu.edu.hk/help/pdf_doc/stats/rn.pdf and http://www.mathworks.com/access/helpdesk/help/toolbox/stats/f56652.html). If we apply the requirement that the mixed fourth moment must be 3 (see Myers and Montgomery, 1995, p. 309), then we conclude that not all BB designs are rotatable. However, the deciding factor should be whether or not the variance of the fitted response is the same for all design points equidistant from the center. If we adopt that criterion, then the designs are rotatable. The reader is asked to address this issue in Exercise 10.27.

Notice also that there is no design point for which all the factors are at either the high level or the low level. This could be advantageous if there is a restricted region of operability such that such factor-level combinations would not be possible. That is, they would be "debarred observations," a term that has been used in the literature and is also used in Section 13.8. Of course the ABC interaction is not estimable because all the points in the ABC column would be zero, but this could not be termed a weakness of the design because three-factor interactions are generally not significant. (Only when a BB design has at least six factors can any three-factor interaction estimate be obtained, but it is undesirable to compute an interaction estimate from only a few design points, especially when it is a small fraction of all the design points.)

Not only is the BB design devoid of these corner points, but it can also have the same type of restricted operability region advantage over the CCD because the axial points in the CCD may also protrude beyond the operating region. Of course, one way that problem could be avoided would be to use a CCI (see Section 10.5.1), but with that design the factorial points might be farther inside the operability region than desired.

In general, if the region of interest is spherical, a BB design is a good design choice.

It would not be particularly meaningful to compute conditional effects for BB designs for a small number of factors when there are large interactions. This can be explained as follows. Notice from Table 10.2 that all the columns for two-factor interactions will have only four nonzero entries. Thus, each interaction estimate is computed from only 4 of the 15 observations. In general, it is not a good idea to compute *any* effect from only about 1/4 of the design points. Even worse, if a conditional effect were computed by one of the levels involved in the linear main effect estimate, the conditional effect would be computed using only two numbers. Thus, interaction effect estimates would be quite shaky with a BB design with three factors and it would be impractical to compute conditional effects. In fact, interaction effect estimates will be quite shaky with *any* BB design, as is explained in the next paragraph.

For four factors the design will of course have 24 points—$\binom{4}{2} \times 4$—plus the number of centerpoints. Since this as well as all BB designs contain sets of 2^2 factorials with the other column entries being zeros, it follows that *any* two-factor interaction in *any* BB design will be estimated using only four points. Each interaction effect estimate will thus be computed as the difference of two averages that are each computed from only two observations, just as would be the case if the *AB* interaction were estimated using a 2^2 design. Interaction effect estimates can thus be expected to be *highly variable*, and this is a serious weakness of these designs. This weakness may not have been pointed out previously.

A two-factor interaction could thus be large simply because the estimator has a large variance, which would necessitate computing conditional effects, but these should not be computed from only two numbers. Furthermore, these designs would not be recommended if an experimenter were interested in seeing response values at the vertices of a cuboidal region, as the vertices are not part of a BB design, so the design region is not cuboidal. (Multifactor second-order designs for cuboidal regions were given by Atkinson, 1973.)

Although BB designs have been viewed as being economical relative to CCDs (see, e.g., the comparison at http://www.itl.nist.gov/div898/handbook/pri/section3/pri3363.htm), this is certainly offset by the deficiencies of the design that have been noted earlier in this section.

10.8.1 Application

Example 10.5

An application of a BB design was given by Palamakula, Nutan, and Khan (2004, references). The objective of the study was to determine a model that would produce an optimized self-nanoemulsified capsule dosage form of a highly lipophilic model compound, Coenzyme Q10. There were three factors used in a BB design, so the design was that given in Table 10.2. There were five response variables but four served as constraints relative to the response variable of interest, which was the cumulative percentage of the drug released after 5 minutes.

The authors focused attention on effects whose p-value was less that .05, and those were the following effects: X_2, X_1^2, X_1X_3, X_2^2, and X_3^2. Note that this is very much a nonhierarchical model. Although it has been argued elsewhere that this is not a bad thing (see Section 4.18), it is disturbing that the quadratic terms in the first and third factors are significant without the linear terms being significant. This coupled with the fact that X_1X_3 is significant without the linear terms being significant mandates further analysis, and the reader is asked to perform that analysis in Exercise 10.26, where the data are given.

10.9 CONDITIONAL EFFECTS?

Conditional effects have been discussed and illustrated extensively in previous chapters, especially Chapter 5. Their degree of usefulness for response surface designs depends upon the design. Obviously conditional effects cannot be computed unless levels are repeated, which means that they cannot be used with UDs. Levels would have to be grouped and averages computed, which would not be particularly meaningful.

Conditional effects could be constructed for a CCD. In general, as was emphasized in Section 4.2.1, conditional effects are most meaningful when the averages are computed from more than a few observations, just as effect averages in general are more meaningful when this condition is met. Only the factorial portion of the CCD should be used in computing conditional effects since the centerpoint values cannot be used since each factor has the same coordinate at the centerpoint, so there is no change in coordinate values for any of the factors. Although the axial points could technically be used, each average would actually be just a single observation, so those conditional effects would have large variances. Therefore, it would be practical to compute conditional effects from only the factorial points, and to proceed analogous to the way that conditional effects were used in Chapters 4 and 5.

As stated in Section 10.4, Sztendur and Diamond (2002) did use a conditional effects approach in arriving at a path of steepest ascent. Specifically, an experiment described by Hsieh and Goodwin (1986) involved nine factors and alluded to the analysis of Bisgaard and Fuller (1994–1995) that identified the significant effects as D, F, and one or more of the confounded interactions BG, CJ, and EH. Sztendur and Diamond (2002) assumed that BG was the significant interaction (a very bold assumption that is apparently without a basis) and consequently used a model with terms in B, D, F, and BG. Their use of the conditional effect of B that is given by $B + BG$ can be viewed as a way around the problem of having a significant interaction term when the method of steepest ascent is to be employed, as that method cannot be used with interactions. Of course the use of the sum $B + BG$ to produce the conditional effect presupposes that the high level of factor G is the best level in terms of maximizing the response. A clue as to whether or not this is likely to be the case can be found simply by comparing the average response at the high level of G with the average response at the low level of G. Another potential problem is that a conditional effect has a different variance from effects computed using all the data. Thus, summing the regression coefficients, as Sztendur and Diamond (2002) did, produces a regression coefficient that has a smaller standard error than the other coefficients in the model. Furthermore, if the design had been unreplicated, a normal probability plot approach could not have been used because that plot requires effect estimates with equal standard errors.

Their particular application is not justifiable because of (1) their arbitrary selection of one of the three interactions, and (2) their forcing a hierarchical model despite the evidence that nonhierarchical models will often be appropriate (Montgomery, Myers, Carter, and Vining, 2005). Nevertheless, this portion of their article is useful because it reminds us that there is a way out of the problem of having a significant interaction in a model when the method of steepest ascent is to be used.

As with factorial designs, statistical analysis that leads to the selection of a non-hierarchical model could signal the need for a conditional effects analysis, and this does happen in one of the chapter exercises as an interaction plot clearly shows a factor main effect, although the factor was not included in the model.

10.10   OTHER RESPONSE SURFACE DESIGNS

There are many other experimental designs that have been proposed as response surface designs. Such designs might be viewed as secondary response surface designs, and we will look at several of these designs in this section and see whether they should receive greater attention by experimenters.

10.10.1   Hybrid Designs

Roquemore (1976) developed a class of designs that are referred to as *hybrid designs* because the designs are related to CCDs but are constructed to satisfy other criteria. There are multiple designs for each number of factors. For example, for three factors the designs were labeled 310, 311A, and 311B.

Myers and Montgomery (1995, p. 362) stated, "It has been our experience that hybrid designs are not used as much in industrial applications as they should" and go on to indicate that the designs can be useful when experimental runs are expensive. They advise against the use of saturated or near-saturated response surface designs, but reason that such designs are going to be used anyway, and when they are used, a hybrid design would be a good choice.

Therefore, we will consider these designs first, for which some concern has been expressed about possible high leverage values. (A large difference in some leverage values would indicate that, if the response values were very close, some design points would exert noticeably more influence on the model coefficients than other specific points. Leverages are explained fully in Section 13.6.)

One of the undesirable features of the hybrid designs is the odd numbers for many of the factor levels, which will require that the values in the plans be rounded when the designs are applied. For example, in one of the designs for four factors the distinct values for the fourth factor are 1.7844, -1.4945, 0.6444, and -0.9075.

The smallest of these designs is for three factors and has 10 design points. That is a saturated design and leverages cannot be computed for a saturated design. One of the two-factor interactions is computed from only four design points, and if large, the conditional effects would have to be computed from only two points. This of course is the same thing that happens with the BB designs so this particular hybrid design shares that weakness.

One of the designs with three factors and 11 points, labeled the 311A by Roquemore (1976), has equal leverages for most of the design points. The design, constructed using Design-Expert, is given below. (Design-Expert cannot be used to construct either design 310 or design 311B.)

A	B	C
0	0	0
$-\sqrt{2}$	0	$-\sqrt{2}/2$
0	0	$-\sqrt{2}$
-1	-1	$\sqrt{2}/2$
1	1	$\sqrt{2}/2$
$\sqrt{2}$	0	$-\sqrt{2}/2$
0	$\sqrt{2}$	$-\sqrt{2}/2$
0	0	$\sqrt{2}$
1	-1	$\sqrt{2}$
0	$-\sqrt{2}$	$-\sqrt{2}/2$
-1	1	$\sqrt{2}/2$

The third and eighth design points have leverages of 0.75; the other points except the centerpoint all have leverages of 0.9375. Thus, this is close to an equileverage design.

The centerpoint has a leverage value of 0 for estimating the model effects, which should be apparent since it is at the center of the design. Of course centerpoints are not used in computing effect estimates, as stated previously, so they have no influence whatsoever on those effect estimates, but leverage values, in general, measure *potential* influence, not actual influence. When leverages for this design are computed using the $\mathbf{X}$ matrix consisting of the first column being 1s followed by the nine columns for the nine effects to be estimated, the leverage for the centerpoint is 1. Thus, it has leverage only in regard to the constant term, which is not of any intrinsic interest. This illustrates why leverages should always be computed from the design matrix, and not from the $\mathbf{X}$ matrix, as we are interested only in the potential influence of points on the effect estimates corresponding to terms in the model. For any orthogonal design (which these designs are not), the estimate of the constant will simply be the average of all the observations, so every observation is weighted equally in that computation. We know that without looking at any leverage values. For nonorthogonal designs, the constant has no simple interpretation other than being the fitted value at each centerpoint.

The other design, 311B, has the same type of leverage structure as the 311A design as eight of the leverages are 0.925 and two are 0.8. The 311B design has the same weakness as the 310 design, however, as most of the design values are odd numbers to four decimal places.

The leverages for the 310 design were given by Jensen (2000); the design is given below.

A	B	C
0	0	1.2906
0	0	-0.1360
-1	-1	0.6386
1	-1	0.6386
-1	1	0.6386
1	1	0.6386
1.1736	0	-0.9273
-1.1736	0	-0.9273
-1	1.1736	-0.9273
1	-1.1736	-0.9273
0	0	0

If, as before, we ignore the fitting of the constant term, the leverage of the second design point is practically zero and the other leverages are practically 1.0, with the centerpoint leverage of course being zero. (If we fit the constant term, the leverages of the 2nd and 11th points are approximately 0.5, but as stated previously, we should compute leverages from the design matrix.)

The three-factor hybrid designs are near-saturated designs as there are 11 observations for estimating the nine effects in the quadratic model plus the constant. The user of Design-Expert receives the following warning message when selecting the design from the menu.

```
Warning! Hybrid designs are minimal point designs. They are
very sensitive to outliers.
```

Of course any saturated or near-saturated design will be sensitive to outliers.

The 16-point hybrid designs for four factors are more useful as half of the two-factor interactions are computed using 8 observations and the other three are obtained using 10 observations.

If hybrid designs are to be used—and they should certainly be considered—it would be a good idea to avoid the designs with fewer than 16 points. Hybrid designs are not available in software such as MINITAB, JMP, or D. o. E. Fusion, however, so their users would have to key in the design. Some of the published hybrid designs can be created with Design-Expert, as indicated previously, although they are a bit hidden as they must be accessed through the menus, selecting Miscellaneous under Response Surface and then selecting Hybrid under Design Type.

10.10.2   Uniform Shell Designs

These designs have been called Doehlert designs, named after their inventor, David Doehlert (1929–1999). Doehlert (1970) and Doehlert and Klee (1972) presented designs that have been termed *uniform shell designs* because the designs for k factors are spread uniformly over a k-dimensional sphere. Despite the word "uniform," these designs should not be confused with UDs, which were discussed in Section 10.3 and which are discussed in more detail in Section 13.9.1.

    Despite not being available in the major statistical software, these designs have been used in many applications, two of which are Dumenil, Mattei, Sergent, Bertrand, Laget, and Phan-Tan-Luu (1988) and De Vansay, Zubrzycki, Sternberg, Raulin, Sergent, and Phan-Tan-Luu (1994).

10.10.3   Koshal Designs

These are designs contributed by Koshal (1933) that are highly simplistic and have the minimum number of design points for estimating effects (i.e., the designs are saturated). The designs are discussed in various places in the literature, including Myers and Montgomery (1995, pp. 357–359). For a second-order design with a single centerpoint, the design is given below.

Row	A	B	C
1	0	0	0
2	1	0	0
3	0	1	0
4	0	0	1
5	2	0	0
6	0	2	0
7	0	0	2
8	1	1	0
9	1	0	1
10	0	1	1

    Rows 2–4 are obviously for estimating the linear effects of a factor (and notice that one factor is being changed at a time), rows 5–7 are for estimating the quadratic effects, and rows 8–10 are for the interaction effects. What may not be obvious when the design is written in the above form is that the design is far from being orthogonal. Specifically, the columns of the design have pairwise correlations of $-.33$. Although supersaturated designs can have pairwise correlations of this magnitude, this is a poor correlation structure for a saturated design. There are better designs.

10.10.4   Hoke Designs

Another class of designs that, like Koshal designs, are also economical were given by Hoke (1974). The properties are almost summarized by the title of the paper and we can add that the designs are also nonorthogonal. The designs generalize those given by Rechtschaffner (1967). Although nonorthogonality will generally be acceptable as long as it is slight, users need to know the extent of the nonorthogonality, preferably by viewing pairwise correlations of the columns of the design matrix. There is no discussion of these correlations in Hoke (1974), however, as the author focused attention on the variances of the estimators of the model parameters. None of the correlations of the Hoke designs are large, although the correlations between the columns of the D2 design for three factors are all $-.184$. This design has only 10 points, which is a small design for fitting a second-order model in three factors. The larger designs have smaller correlations, and the D4 design for five factors has orthogonal columns. The Hoke designs are listed at http://www.math.montana.edu/~jobo/cr/designs.txt.

10.11   BLOCKING RESPONSE SURFACE DESIGNS

It may be necessary to block a response surface design for the same reasons as blocking a factorial design may be necessary, such as not being able to make all the runs in one day. Indeed, this was the case in the experiment that is described in Exercise 14.4. This is especially true because response surface designs generally have more design points than full factorial or fractional factorial two-level designs because of the number of levels used in the designs.

10.11.1   Blocking Central Composite Designs

An obvious way to block a CCD is to use two blocks and, if there were no centerpoints, to have the factorial design runs in one block and the axial points in the other block. Doing so would cause one of the pure quadratic effects to become nonestimable, however, but this can be remedied by adding the appropriate number of centerpoints, which will allow all the quadratic effects to be estimable. As stated previously, however, the quadratic effects are not orthogonal to each other.

When blocking is used, it is desirable to use the number of centerpoints and the value of α that make the blocks orthogonal to columns that represent the factor effects and interactions that are being estimated.

For example, with two factors, it is necessary to have six centerpoints and have $\alpha = \sqrt{2}$ in order to achieve these conditions, with the factorial points plus three centerpoints in one block and the axial points and the other three centerpoints in the other block. With three factors, either two or three blocks could be used. In each case six centerpoints would be used. With two blocks, four of those points would be in the block with the factorial runs and the other two would be in the block with the axial point runs, and the axial point values would all be ± 1.633. If three blocks are used, the six centerpoints are equally divided among the three blocks and the axial point values, which are in one block, are all still ± 1.633. The factorial points occupy the other blocks, and they are split in such a way as to confound the ABC interaction between those two blocks, which of course is the preferred way of running a 2^3 design in two blocks.

We might ask why, for example, the CCD for three factors with 20 points has $\alpha = 1.682$ when there is no blocking and $\alpha = 1.633$ when either two or three blocks are used, whereas a CCD for four factors has $\alpha = 2$, regardless of whether there is blocking or not? This is because there are different conditions that must be met when blocking is used. These conditions are discussed in detail by Myers and Montgomery (2002).

One of the conditions is that the sum of squares of the individual factor coordinates must be the same in all blocks when the block sizes are the same. When the block sizes are not the same, as will generally be the case when the number of factors is not 2 or 4, the sum of squares must be proportional to the block size across the blocks.

To illustrate, consider the following MINITAB-generated CCD for three factors to be run in three blocks. Of course the blocks cannot be equal since there are 20 design points.

Block 1			Block 2			Block 3		
A	B	C	A	B	C	A	B	C
-1	-1	-1	1	-1	-1	-1.633	0	0
1	1	-1	-1	1	-1	1.633	0	0
1	-1	1	-1	-1	1	0	-1.633	0
-1	1	1	1	1	1	0	1.633	0
0	0	0	0	0	0	0	0	-1.633
0	0	0	0	0	0	0	0	1.633
						0	0	0
						0	0	0

Notice that the factorial points are in the first two blocks and the axial points are in the last block, with two centerpoints in each block. In order for the sum of squares condition to be met, the axial value α must be such that $2\alpha^2/8 = 4/6$, with $2\alpha^2$ being the sum of squares of the columns in Block #3, and 4 being the sum of the squares of each column in each of the other two blocks. Solving this equation for α produces $\alpha = 1.633$.

Each of the first two blocks contains one of the 2^{3-1} designs, plus two centerpoints, and we can see that ABC is confounded between the two blocks because the product of the three columns is -1 in the first block (not counting the centerpoints) and $+1$ in the second block.

The value of α necessary for rotatability when a CCD is blocked differs from the value that is needed for orthogonality. This is apparent from the table given by Box and Hunter (1957), which gave blocking schemes for up to seven factors and the value of α for each scheme that is necessary for orthogonality and rotatability. This table was reproduced by Giesbrecht and Gumpertz (2004, p. 410) and by Myers and Montgomery (2002).

10.11.2   Blocking Box–Behnken Designs

A BB design can be blocked as long as at least four factors are used. MINITAB will construct a blocked BB design for four, five, six, or seven factors. Given below is the MINITAB-generated design for four factors, using the default value of 3 for the number of centerpoints and not randomizing the runs (for presentation purposes).

Block	A	B	C	D
1	-1	-1	0	0
1	1	-1	0	0
1	-1	1	0	0
1	1	1	0	0
1	0	0	-1	-1
1	0	0	1	-1
1	0	0	-1	1
1	0	0	1	1
1	0	0	0	0
2	-1	0	-1	0
2	1	0	-1	0
2	-1	0	1	0
2	1	0	1	0
2	0	-1	0	-1
2	0	1	0	-1
2	0	-1	0	1
2	0	1	0	1
2	0	0	0	0
3	-1	0	0	-1
3	1	0	0	-1
3	-1	0	0	1
3	1	0	0	1
3	0	-1	-1	0
3	0	1	-1	0
3	0	-1	1	0
3	0	1	1	0
3	0	0	0	0

The reader is asked in Exercise 10.20 to explain why the design in Table 10.2 could not be run in three blocks.

10.11.3   Blocking Other Response Surface Designs

A Draper–Lin design can be orthogonally blocked as long as there are three, four, or six factors. The designs for 5, 7, 8, 9, and 10 factors cannot be orthogonally blocked because of factorial runs that are not used, the deletion of which causes the design to be nonorthogonal. For these designs, Design-Expert uses the value of α that minimizes the average squared correlation of the block effect with all second-order model coefficients.

For example, given below is the blocked Draper–Lin design for three factors run in two blocks that is generated by Design-Expert when a "small" design is selected from the CCD menu and the orthogonal blocking option is selected and the default value of three centerpoints in each block is used.

Block	A	B	C
1	1	1	-1
1	1	-1	1
1	-1	1	1
1	-1	-1	-1
1	0	0	0
1	0	0	0
1	0	0	0
2	-1.60	0	0
2	1.60	0	0
2	0	-1.60	0
2	0	1.60	0
2	0	0	-1.60
2	0	0	1.60
2	0	0	0
2	0	0	0
2	0	0	0

This design can be orthogonally blocked because the first block contains a 2^2 design (i.e., an orthogonal design) with centerpoints. For five factors, however, there are 11 runs and since that is an odd number, the block that contains them cannot have orthogonal columns, nor can the columns be orthogonal to a column of 1s.

The (seven) options that Design-Expert gives the user regarding the choice of α are worth noting, as they include rotatable, orthogonal blocks, spherical, orthogonal quadratic, practical, face centered, and other. Each of these choices leads to the construction of the indicated design, but "orthogonal quadratic" and "practical" may need some explanation. The former causes the quadratic terms to be orthogonal to the other terms and the latter is the default value for designs with at least six factors. It is defined as the fourth root of the number of factors.

10.12 COMPARISON OF DESIGNS

Some of the well-known response surface designs have been compared in various studies, with the comparison based on one or more efficiency/optimality measures. These measures are discussed in Section 13.7.1. Ozol-Godfrey, Anderson-Cook, and Montgomery (2005) indicate a preference for the CCD and Hoke D6 designs, with the former preferred for the full second-order model and the latter perhaps being superior for the first-order model and the first-order model plus interaction terms. Indeed, the D6 design is the only Hoke design for three factors whose three columns are orthogonal, but the orthogonality does not extend to the interaction effects.

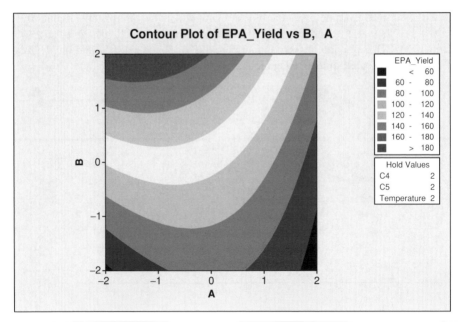

Figure 10.12   Example of a contour plot (MINITAB produced).

10.13   ANALYZING THE FITTED SURFACE

Once a design has been selected, the experiment performed, the data analyzed, and a model fit, the next step is to analyze the fitted surface because optimization is a usual goal of a response surface study. That is, the experimenter is interested in determining the combination of factor settings that will maximize or minimize the expected response, depending upon the objective.

Although it is helpful to see a surface such as Figure 10.1 in which the points are connected, it is generally more useful to look at *contours of constant response*, an example of which is given in Figure 10.12.

This plot is a modification of the plot produced by MINITAB using the RSCON-TOUR command. That plot is different from the form of contour plots typically given in textbooks, as the space between the contours is filled in so as to produce a solid graph, and the values associated with the contours are shown in the legend rather than on the graph. Textbooks do not have graphs with several shades of a color, however, so Figure 10.12 is adapted from the multicolor graph produced by MINITAB.

Figure 10.12 is an example of a "rising ridge"; that is, the response continues to increase as one moves toward the upper left portion of the design space. This suggests that the maximum may be outside the design region. There is nothing that can be done in this situation but use a follow-up experiment that has a design center at the largest response on the contour plot and move from there to hopefully envelop the region of the maximum with the design space for the follow-up experiment.

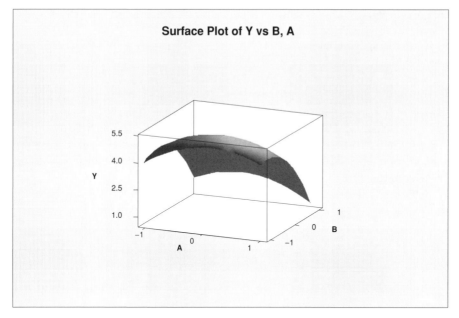

Figure 10.13   Fitted surface that corresponds to Figure 10.1.

The nature of the response surface must be characterized so that the experimenter can determine (1) if the maximum or minimum response appears to lie within the experimental region, and (2) if the maximum or minimum does lie within the experimental region, how much does the response change for small movements in various directions?

Given in Figure 10.13 is the fitted surface for the data from which Figure 10.1 was constructed. Notice that Figure 10.13 differs slightly from Figure 10.1 and of course this is due to the fact that the former is constructed from the following model:

$$\widehat{Y} = 4.847 - 0.017A - 1.550B - 0.918A^2 - 1.318B^2 - 0.325AB \quad (10.1)$$

whereas Figure 10.1 was not constructed from any model. Nevertheless they do not differ by very much and that should certainly be the case since the R^2 value for the fitted model is .998.

The graph suggests that the maximum response lies within the experimental region, and we would guess from the graph that it is in the vicinity of $A = 0$, $B = 0$. In this case it is easy to see from Eq. (10.1) that this is exactly where the maximum occurs because all the coefficients of terms in the model are negative.

In general, it won't be this easy to determine the maximum or minimum in a typical application, so more sophisticated methods must be employed. One problem that must be addressed regardless of the method used is that the coefficients of the squared terms in a second-order model are not independent because the squared terms are not independent (unless they are mean centered, which is not generally done by statistical software).

For example, consider the model

$$Y = 70 + 2A + 3B - A^2 - 2B^2 - AB \qquad (10.2)$$

There is obviously no finite global minimum as setting $A = 0$ and $B \rightarrow -\infty$ results in $Y \rightarrow -\infty$. It appears as though there is a maximum, however, although the maximum is not obvious. What is needed is an analysis of the fitted surface to determine the *stationary point(s)*. With only two factors we can make this determination by using calculus and solving a system of equations. That is,

$$\frac{\partial Y}{\partial A} = 2 - 2A - B = 0$$

$$\frac{\partial Y}{\partial B} = 3 - 4B - A = 0$$

Solving these two equations for A and B produces $A = 5/7 = 0.714$ and $B = 4/7 = 0.571$. This produces $Y = 71.5714$, which we can regard as the "true maximum."

Data were generated using the model in Eq. (10.2) and adding an error term $\epsilon \sim N(0, 0.6)$, and the design values of A and B are those from a CCD with five centerpoints and $\alpha = \sqrt{2}$. The fitted equation is

$$\widehat{Y} = 69.810 + 1.973A + 2.985B - 1.159A^2 - 1.908B^2 - 1.414AB \qquad (10.3)$$

with $R^2 = .992$. Notice that the coefficients of A^2 and AB are estimated with a large percentage error despite the fact that R^2 is quite high. The correlation between the estimators of the quadratic terms is $-.130$; the estimators of the other terms are uncorrelated.

The fitted surface for the fitted equation (10.3) is given in Figure 10.14.

An analysis of this fitted surface produces

$$\frac{\partial \widehat{Y}}{\partial A} = 1.973 - 2.318A - 1.414B = 0$$

$$\frac{\partial \widehat{Y}}{\partial B} = 2.985 - 3.816B - 1.414A = 0$$

Solving these two equations gives $A = 0.483$ and $B = 0.603$. The value for A differs noticeably from the value of A obtained using Eq. (10.2). The value of $\widehat{Y}$ at this point is 71.19, not far from the true maximum of 71.57 from Eq. (10.2).

It is apparent from Figure 10.14 that the maximum occurs when both A and B are between 0 and 1, and if we look closely at the figure, we can see that the maximum occurs when A and B are each close to 0.5. So the point (0.483, 0.603) thus "looks right."

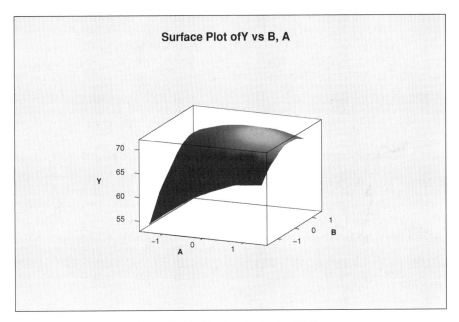

Figure 10.14 Fitted surface for Eq. (10.3).

10.13.1 Characterization of Stationary Points

It is easy to determine whether a stationary point is a maximum or a minimum when there are only two factors since the surface can be easily graphed. This is not the case when there are more than two factors, however. Consequently, there is a need to determine whether a stationary point is a maximum, minimum, or saddle point. (The latter will be illustrated shortly.)

This is done by looking at the eigenvalues of the matrix of coefficients of the second-order terms. We will let W denote that matrix. From Eq. (10.3) we have

$$W = \begin{bmatrix} -1.159 & -1.414 \\ -1.414 & -1.908 \end{bmatrix}$$

(Notice that the coefficient of AB occupies the off-diagonal positions.) Both eigenvalues of this matrix are negative, as can be determined, for example, by using the EIGEN command in MINITAB. Since the eigenvalues are negative, the point must be a maximum. If all the eigenvalues had been positive, the point would have been a minimum, and a mixture of signs means that the point is a saddle point.

We will now illustrate a saddle point. Consider the equation

$$Y = 70 + 2A + 3B - A^2 + 3B^2 - AB \tag{10.4}$$

which is a slight modification of Eq. (10.2), as the coefficient of B^2 is now positive. This causes the signs of the eigenvalues to be mixed, so that using Eq. (10.4) to

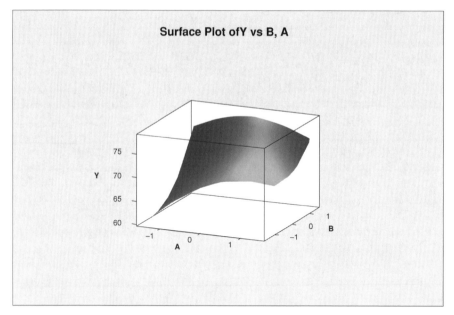

Figure 10.15   Fitted surface for Eq. (10.5).

generate data should result in a fitted surface that resembles a saddle. Generating data with random errors that have the same distribution as before results in the fitted equation

$$\widehat{Y} = 70.1339 + 2.0672A + 2.988B - 0.9301A^2 + 2.9032B^2 - 0.9590AB \qquad (10.5)$$

with the parameter estimates differing only slightly from the corresponding parameters.

The fitted surface for Eq. (10.5) is given in Figure 10.15.

The saddle may not be obvious from Figure 10.15, but should become apparent when the contours of constant response are displayed, analogous to Figure 10.12. This is given in Figure 10.16.

Figure 10.16 isn't a perfect saddle; rather, it looks like a "twisted saddle with one side missing." The point to be made is that the response increases as one starts from the center and moves either up or down, whereas the response decreases if one moves to the left.

10.13.2   Confidence Regions on Stationary Points

In trying to determine optimum factor levels to maximize or minimize the response, it would be helpful to know how well determined the optimum is and how sensitive the response is to slight changes in the factor levels away from the optimum. Just as confidence intervals on parameters are constructed in the application of basic statistical

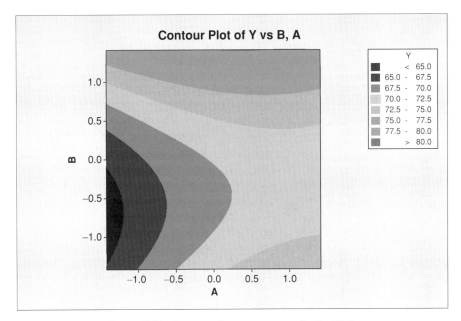

Figure 10.16 Contours of constant response for Eq. (10.5).

procedures, it would be desirable to have a confidence region on the optimum point. (Notice the use of the word "region" instead of "interval" since at least two dimensions will be involved.)

Unfortunately, it is not practical to attempt to compute a confidence region on the optimum point without software. Accordingly, del Castillo, and Cahya (2001) described their software program, accessible from StatLib (http://lib.stat.cmu.edu/TAS) and to be used with MAPLE, and illustrated its use in their article.

10.13.3 Ridge Analysis

Contour plotting is quite useful when there is a small number of factors, but is of very limited use when there is a large number of factors. Ridge analysis, due to Hoerl (1959), is a technique that can be used in determining the optimum combination of factor levels for *any* number of factors. The objective is to determine a path toward the optimum region by using a constrained optimization approach. (Note that this is different from the path of steepest ascent/descent. The latter is determined from a fitted first-order model whereas ridge analysis is generally applied to a second-order model, such as the model given in Eq. (10.1).)

Specifically, the optimum is sought subject to the constraint that $\sum_{i=1}^{k} X_i^2 = R^2$. The method of Lagrangian multipliers is often used to determine the X_i that are the coordinates of the optimum point subject to the constraint, although Peterson, Cahya, and del Castillo (2002) gave a method that does not require the use of Lagrangian multipliers. Their method and the methodology of Gilmour and Draper (2003) are discussed and debated in Peterson, Cahya, and del Castillo (2004) and Gilmour and Draper (2004).

Confidence bands on the path toward the optimum response were introduced by Carter, Chinchilli, Myers, and Campbell (1986), with an improved approach given by Peterson (1993).

Consider the rising ridge shown in Figure 10.12. Since the optimum point appears to lie outside the experimental region, it would be helpful to know the path to take toward the optimum point so that future experimentation could occur along that path. We can make a rough guess of that path from Figure 10.12, but if we had three factors instead of two, we would need multiple contour plots and would have to try to merge the information from those plots.

10.13.3.1 Ridge Analysis with Noise Factors

Peterson and Kuhn (2005) provided an approach for performing a ridge analysis and optimizing a response surface in the presence of noise variables, which extended the work of Peterson (1993). The authors showed how to construct an optimal path for the root mean squared error about a process target. They repeat the message of Hoerl (1985) that contour plots may not be sufficient for understanding high-dimensional response surfaces, especially when there are noise variables. They recommend using an overlaid ridge trace plot to indicate how the process mean differs from the target value as there is movement along the ridge path.

10.13.4 Optimum Conditions and Regions of Operability

It is entirely possible that the methods for determining optimum operating conditions will not be applicable because the region of feasible operating conditions is not hyperrectangular. When this occurs, different methods will have to be employed. Giesbrecht and Gumpertz (2004) discuss this in some detail, especially relative to SAS PROC OPTEX, which permits the selection of a specific subset of points, with the other points spread uniformly over the feasible region.

10.14 RESPONSE SURFACE DESIGNS FOR COMPUTER SIMULATIONS

The use of experimental designs in computer simulation experiments is increasing considerably and response surface designs are among the most frequently used designs for such applications, especially in the field of engineering. For example, Unal, Wu, and Stanley (1997) used a CCD to study a tetrahedral truss for a scientific space platform. Zink, Mavris, Flick, and Love (1999) used a face-centered CCD to improve the wing design of a lightweight fighter jet by assessing a new active aeroelastic wing technology that could only be simulated through physics-based finite element analysis on high-powered computers. See, for example, http://www.statease.com/pubs/rsmsimpexcerpts–chap10.pdf for additional information on the use of response surface designs in computer simulation experiments.

10.15   ANOM WITH RESPONSE SURFACE DESIGNS?

The use of Analysis of Means (ANOM) with various designs has been discussed in earlier chapters. It would not be practical to use ANOM in conjunction with designs like the CCD for various reasons however. First, the levels of a factor are not equally spaced when the CCD is used, and there is only one observation at each of the star points. Therefore, an average computed at a star point wouldn't make any sense. Nor would it make any sense to consider ANOM for use with a UD since the design points generally do not repeat. Although ANOM might be used with a BB design, doing so would present problems since the number of observations at the middle level of each factor differs from the number of observations at each of the other two levels, for each factor. Thus, straightforward use of ANOM would not be possible, although ANOM might be applied by using the data as if there were a single factor with the number of levels equal to three times the number of factors, with an unbalanced ANOM approach then applied.

10.16   FURTHER READING

Response surface methodology, including designs, is a broad field that books have been written about and there is not sufficient space to describe all the methodological advances and interesting applications herein.

   Other applications and methodological papers that may be of interest include the following. Huang et al. (2006), mentioned at the start of this chapter, first used a Plackett–Burman design (see Section 13.4.1) to identify significant factors, then used a full factorial CCD in a biological application. Allen, Yu, and Bernshteyn (2000) eschewed the use of a CCD because sufficient resources were not available to carry it out, and instead used a design that the reader is asked to critique in Exercise 10.28. Vining, Kowalski, and Montgomery (2005) discussed the use of a CCD and a BB design run under a split-plot structure and Kowalski, Borror, and Montgomery (2005) gave a modified path of steepest ascent for split-plot experiments.

   Morris (2000) presented a new method of constructing composite designs and Gilmour (2006) developed a new class of three-level response surface designs for use in biological applications where the run-to-run variation is generally much greater than it is in engineering experiments. Gilmour (2004) extended the work of Edmondson (1991) in developing four-level response surface designs based on irregular two-level fractional factorial designs.

   Draper and John (1998) gave response surface designs for hard-to-change factors, as did Trinca and Gilmour (2001), and response surface designs for both quantitative and qualitative factors were given by Wu and Ding (1998). Akhtar and Prescott (1986) gave designs that were robust to missing observations. Ankenman, Liu, Karr, and Picka (2002) introduced a new class of designs, which they termed *split factorial* designs for response surface applications. Mays and Easter (1997) gave optimal response surface designs in the presence of dispersion effects. Mee (2001) presented noncentral composite designs, which consist of a pair of two-level designs with different centers, as an alternative to the method of steepest ascent. Gilmour and Trinca (2003) gave row–column response surface designs, which are designs that result from blocking. Block and Mee (2001a) presented some new second-order designs and Block and Mee (2001b) presented a table of second-order designs, including those due to Notz (1982).

Box and Draper (1982) gave measures of lack of fit for response surface designs and predictor transformations. Giovannitti-Jensen and Myers (1989) discussed graphical assessment of the prediction capability of response surface designs and Zahran, Anderson-Cook, and Myers (2003) proposed using a fraction of the design space for assessing the prediction capability.

Vining and Myers (1990) introduced the dual response problem, where the focus is on both mean and variance, in the context of RSM, and other papers followed, including more recent papers of Ding, Lin, and Wei (2004) and Jeong, Kim, and Chang (2005).

Myers, Montgomery, Vining, Borror, and Kowalski (2004) gave a response surface literature review.

10.17   THE PRESENT AND FUTURE DIRECTION OF RESPONSE SURFACE DESIGNS

As with statistical methodology in general, the frequency of usage of any "improved" response surface designs will depend upon whether or not these designs are incorporated in software. One relatively recent movement in that direction was the addition of space-filling designs to JMP a few years ago. In his paper "Response surface methodology—current status and future directions," Myers (1999) summarized the then current state of RSM and discussed future directions. One movement in a new direction has been the application of response surface designs and methodology when the appropriate model is a generalized linear model rather than a linear model.

There should be other considerations as well. In particular, designs should be constructed to take into consideration debarred (i.e., impossible) factor-level combinations (Section 13.8) and restricted regions of operability (also Section 13.8). These issues motivate the use of designs other than traditional response surface designs, including optimal designs. Recent work on optimal response surface designs includes Drain, Carlyle, Montgomery, Borror, and Anderson-Cook (2004).

10.18   SOFTWARE

Comments were made in the chapter sections about software availability for the various designs that were covered. In this section we summarize those comments and provide some additional information.

Unfortunately, some of the better designs are not included in major statistical software packages so users of those designs will have to manually construct the designs or find special-purpose software. For example, hardly any software has the capability for Hoke designs and this is true in general for less-known designs.

MINITAB has the capability for CCDs—full, half, and quarter, blocked and un-blocked. Box–Behnken designs are the only other response surface designs available, however, and they also can be either blocked or unblocked. The numerical analysis capabilities are the same type as are available with regression analyses; the graphical analyses include surface and contour plots, examples of which were given in the chapter figures. When there are more than two responses, the contour plots can be overlaid. This is discussed in Section 12.1, although it is stated that contour plots start to lose some value when there are more than two factors.

A 55-page discussion of MINITAB's response surface design capabilities, as well as good practical advice on the use of these designs, is available at http://www.minitab. com/support/docs/rel14/14helpfiles/DOE/ResponseSurfaceDesigns.pdf. One potentially problematic feature of MINITAB, which it shares with other statistical software, is that it will not allow the fitting of nonhierarchical models, such as fitting a quadratic term without fitting the linear term. As discussed by Montgomery et al. (2005), there are conditions under which nonhierarchical models can be justified. Specifically, they state that physical and chemical mechanisms are unlikely to be hierarchical. An example of an (extreme) nonhierarchical model was given in a research paper that is cited in Exercise 10.17. As was shown in Example 10.2, however, in which an *extreme* nonhierarchical model found in the literature was discussed, evidence can often be found that does not support the model. As can be inferred from the end of Section 4.18, a nonhierarchical model should *always* be viewed with suspicion and an investigation performed and additional calculations, including conditional effects, made to determine if the model is justifiable.

Design-Expert has the best capability for response surface designs of the software packages that are discussed here. For example, with Design-Expert a BB design can be created for up to 21 factors, whereas MINITAB and JMP do not go beyond 7 factors. Designs with at most 12 factors can be blocked using Design-Expert. The latter allows the user to specify a "user-defined" design by indicating whether or not each category of points (axial check points, interior points, centroid points, vertices, centers of edges, etc.) is to be included in the design. For example, a design with 21 factors can be constructed with 1024 vertices, 5120 centers of edges, 20 constraint plan centroids, 1024 axial check points, 5140 interior points, and 1 overall centroid, for a total of 12,329 design points! Of course only in something like a computer experiment would such a design ever have a chance of being used . . . and computer experiments aren't free! The response surface design capabilities in Design-Expert 7 can be read in the tutorial at http://www.statease.com/x70ug/DX7-04C-MultifactorRSM-P1.pdf.

D. o. E. Fusion guides the user who has no design in mind, as well as allowing the knowledgeable user to reach a menu from which to select a response surface design. Its capabilities for response surface designs fall between MINITAB and Design-Expert. For example, D. o. E. Fusion allows the user to construct a BB design for up to 10 factors. CCDs of resolution IV and resolution V can be constructed, with these being labeled Central Composite IV and Central Composite V, respectively, on the menu. In addition to CCD and BB designs, "Star" designs can also be constructed. These are designs that consist of runs in which each factor except one is at the middle level, plus centerpoints.

JMP can be used to generate the type of space-filling designs discussed in this chapter, including UDs. It will also generate a standard CCD as well as CCD-Orthogonal, CCD-Orthogonal Blocks, and CCD-Uniform Precision. It will also generate both blocked and unblocked BB designs, but as with MINITAB, it will not generate a BB design for more than seven factors. This is perhaps due to the fact that most sources do not discuss or illustrate BB designs for more than seven factors. An exception is Giesbrecht and Gumpertz (2004), who give BB designs for up to 10 factors. Of course the number of design points becomes quite large with that many factors; the design for 10 factors given by Giesbrecht and Gumpertz (2004) has 170 points.

No response surface design can be constructed for more than eight factors with JMP. For three-level designs, only the L_{27} Taguchi design can be constructed when there are more than four factors, and the factors have to be categorical, not continuous. Thus, it would be difficult to use JMP in the "one-design approach" discussed in Section 10.1 because of the limitation on the number of factors that can be handled.

Statgraphics can be used to create Draper–Lin designs, BB designs, and CCDs, and response surface designs that are robust against time trends can be created with the Gendex DOE toolkit.

10.19 CATALOGS OF DESIGNS

In addition to software, there are also catalogs of response surface designs, just as there are catalogs of 3^{k-p} designs as discussed in Section 6.10. Many second-order designs can be found at http://stat.bus.utk.edu/techrpts/2001/2nd_orderdesigns.htm#hartley2. This includes some response surface designs that were not discussed in this chapter.

10.20 SUMMARY

Classical and modern approaches to RSM have been presented in this chapter. Historically, the standard approach has been to use a three-stage procedure: (1) Use a two-level design as a screening design to identify the important factors, (2) conduct experiments along a path of steepest ascent/descent in an effort to identify the region of optimum response, and (3) once that region appears to have been located, use a response surface design to characterize the nature of the surface (assuming that some second-order model is fit) to try to identify the optimum factor settings.

Such an approach will fail if there are significant interactions as then the wrong set of factors will be identified in the first stage. To guard against this possibility, especially if interactions are suspected, a single 3-level resolution V design might be used both for factor screening and for fitting a second-order model, with the design projected onto the factors that seem to be significant. As pointed out in Section 10.1, however, such a design would require a large number of runs.

It is quite likely that newer designs such as uniform designs for irregular regions will play a greater role in response surface methodology applications in the future, once they become widely available in software.

REFERENCES

Akhtar, M. and P. Prescott (1986). Response surface designs robust to missing observations. *Communications in Statistics: Simulation and Computation*, **15**, 345–363.

Allen, T., L. Yu, and M. Bernshteyn (2000). Low-cost response surface methods applied to the design of plastic fasteners. *Quality Engineering*, **12**(4), 583–591.

Anderson, M. J. and P. J. Whitcomb (2004). *RSM Simplified: Optimizing Processes Using Response Surface Methods for Design of Experiments*. University Park, IL: Productivity Press.

Ankenman, B. E., H. Liu, A. F. Karr, and J. D. Picka (2002). A class of experimental designs for estimating a response surface and variance components. *Technometrics*, **44**(1), 45–54.

Atkinson, A. C. (1973). Multifactor second order designs for cuboidal regions. *Biometrika*, **60**, 15–19.

Bisgaard, S. (1997). Why three-level designs are not so useful for technological applications. *Quality Engineering*, **9**(3), 545–550.

Bisgaard, S. and H. T. Fuller (1994–1995). Analysis of factorial experiments with defects or defectives as the response. *Quality Engineering*, **7**(2), 429–443.

Block, R. M. and R. W. Mee (2001a). Some new second-order designs. Technical Report 2001-3, Department of Operations, Statistics, and Management Science, University of Tennessee. (http://stat.bus.utk.edu/techrpts)

Block, R. M. and R. W. Mee (2001b). Table of second order designs. Technical Report 2001-1, Department of Operations, Statistics, and Management Science, University of Tennessee. (http://stat.bus.utk.edu/techrpts)

Box, G. E. P. (1957). Evolutionary operation: A method for increasing industrial productivity. *Applied Statistics*, **6**(2), 81–101.

Box, G. E. P. (1999–2000). The invention of the composite design. *Quality Engineering*, **12**(1), 119–122.

Box, G. E. P. and D. W. Behnken (1960). Some new three-level designs for the study of quantitative variables. *Technometrics*, **2**, 455–475.

Box, G. E. P. and N. R. Draper (1969). *Evolutionary Operation*. New York: Wiley.

Box, G. E. P. and N. R. Draper (1975). Robust designs. *Biometrika*, **62**, 347–352.

Box, G. E. P. and N. R. Draper (1982). Measures of lack of fit for response surface designs and predictor variable transformations. *Technometrics*, **24**, 1–8.

Box, G. E. P. and N. R. Draper (1987). *Empirical Model Building and Response Surfaces*. New York: Wiley.

Box, G. E. P. and J. S. Hunter (1957). Multi-factor experimental designs for exploring response surfaces. *Annals of Mathematical Statistics*, **28**, 195–241.

Box, G. E. P. and K. B. Wilson (1951). On the experimental attainment of optimum conditions. *Journal of the Royal Statistical Society, Series B*, **13**, 1–45.

Bursztyn, D. and D. M. Steinberg (2001). Rotation designs for experiments in high bias situations. *Journal of Statistical Planning and Inference*, **97**, 399–414.

Carter, W. H., V. M. Chinchilli, R. H. Myers, and E. D. Campbell (1986). Confidence intervals and an improved ridge analysis of response surfaces. *Technometrics*, **28**, 339–346.

Chapman, R. E. and K. Masinda (2003). Response surface designed experiment for door closing effort. *Quality Engineering*, **15**(4), 581–585.

Cheng, S.-W. and C. F. J. Wu (2001). Factor screening and response surface exploration. *Statistica Sinica*, **11**, 553–580; discussion: 581–604.

Cox, D. R. and N. Reid (2000). *The Theory of the Design of Experiments.* Boca Raton, FL: CRC Press.

Croarkin, C. and P. Tobias, eds. (2002). *NIST/SEMATECH e-Handbook of Statistical Methods* (http://www.itl.nist.gov/div898/handbook), joint effort of the National Institute of Standards and Technology and International SEMATECH.

De Vansay, E., S. Zubrzycki, R. Sternberg, F. Raulin, M. Sergent, and R. Phan-Tan-Luu (1994). Gas chromatography of Titan's atmosphere. V. Determination of permanent gases in the presence of hydrocarbons and nitriles with a molecular sieve micropacked column and optimization of the GC parameters using a Doehlert experimental design. *Journal of Chromatography A*, **668**, 161–170.

Del Castillo, E. (1997). Stopping rules for steepest ascent in experimental optimization. *Communications in Statistics: Simulation and Computation*, **26**(4), 1599–1615.

Del Castillo, E. and S. Cahya (2001). A tool for computing confidence regions on the stationary point of a response surface. *The American Statistician*, **55**(4), 358–365.

Ding, R., D. K. J. Lin, and D. Wei (2004). Dual-response surface optimization: A weighted MSE approach. *Quality Engineering*, **16**(3), 377–385.

Doehlert, D. H. (1970). Uniform shell designs. *Applied Statistics*, **19**, 231–239.

Doehlert, D. H. and V. L. Klee (1972). Experimental designs through level reduction of a d-dimensional cuboctahedron. *Discrete Mathematics*, **2**, 309–334.

Drain, D., W. M. Carlyle, D. C. Montgomery, C. Borror, and C. Anderson-Cook (2004). A genetic algorithm hybrid for constructing optimal response surface designs. *Quality and Reliability Engineering International*, **20**, 637–650.

Draper, N. R. (1985). Small composite designs. *Technometrics*, **27**, 173–180.

Draper, N. R. and J. A. John (1998). Response surface designs where levels of some factors are difficult to change. *Australian and New Zealand Journal of Statistics*, **40**, 487–495.

Draper, N. R. and D. K. J. Lin (1990). Small response-surface designs. *Technometrics*, **32**, 187–194.

Dumenil, G., G. Mattei, M. Sergent, J. C. Bertrand, M. Laget, and R. Phan-Tan-Luu (1988). Application of a Doehlert experimental design to the optimization of microbial degradation of crude oil in sea water by continuous culture. *Applied Microbiology and Biotechnology*, **27**, 405–409.

Edmondson, R. N. (1991). Agricultural response surface experiments based on four-level factorial designs. *Biometrics*, **47**, 1435–1448.

Fang, K.-T. (1980). The uniform design: Application of number-theoretic methods in experimental design. *Acta Mathematicae Applicatae Sinica*, **3**, 363–372.

Fang, K.-T. and D. K. J. Lin (2003). Uniform experimental designs and their applications in industry. In *Handbook of Statistics*, Vol. 22, Chap. 4 (R. Khattree and C. R. Rao, eds.). Amsterdam: Elsevier Science B.V.

Fang, K.-T. and R. Mukerjee (2000). A connection between uniformity and aberration in regular fractions of two-level factorials. *Biometrika*, **87**, 193–198.

Ghadge, S. V. and H. Raheman (2006). Process optimization for biodiesel production from mahua (*Madhuca indica*) oil using response surface methodology. *Bioresource Technology*, **97**(3), 379–384.

Gheshlaghi, R., J. M. Scharer, M. Moo-Young, and P. L. Douglas (2005). Medium optimization for hen egg white lysozyme production by recombinant *Aspergillus niger* using statistical methods. *Biotechnology and Bioengineering*, **90**(6), 754–760.

Ghosh, S. and W. S. Al-Sabah (1996). Efficient composite designs with small number of runs. *Journal of Statistical Planning and Inference*, **53**(1), 117–132.

Giesbrecht, F. G. and M. L. Gumpertz (2004). *Planning, Construction, and Statistical Analysis of Comparative Experiments*. Hoboken, NJ: Wiley.

Gilmour, S. G. (2004). Irregular four-level response surface designs. *Journal of Applied Statistics*, **31**(9), 1043–1048.

Gilmour, S. G. (2006). Response surface designs for experiments in bioprocessing. *Biometrics*, **62**, 323–331.

Gilmour, S. G. and N. R. Draper (2003). Confidence intervals around the ridge of optimal response on fitted second-order response surfaces. *Technometrics*, **45**, 333–339.

Gilmour, S. G. and N. R. Draper (2004). Response. *Technometrics*, **46**, 358.

Gilmour, S. G. and L. A. Trinca (2003). Row–column response surface designs. *Journal of Quality Technology*, **35**(2), 184–193.

Giovannitti-Jensen, A. and R. H. Myers (1989). Graphical assessment of the prediction capability of response surface designs. *Technometrics*, **31**, 159–171.

Gorenflo, V. M., J. B. Ritter, D. S. Aeschliman, H. Drouin, B. D. Bowen, and J. M. Piret (2005). Characterization and optimization of acoustic filter performance by experimental design methodology. *Biotechnology and Bioengineering*, **90**(6), 746–753.

Hahn, G. J. (1976). Process improvement through Simplex EVOP. *Chemtech*, **6**, 343–345.

Hardin, R. H. and N. J. A. Sloane (1991). Computer-generated minimal (and larger) response surface designs: (I) The sphere. (available at http://www.research.att.com/~njas/doc/doeh.pdf)

Hartley, H. O. (1959). Smallest composite designs for response surfaces. *Biometrics*, **15**, 611–624.

Hoerl, A. E. (1959). Optimum solution to many variables equations. *Chemical Engineering Progress*, **55**, 69–78.

Hoerl, R. W. (1985). Ridge analysis 25 years later. *The American Statistician*, **39**, 186–192.

Hoke A. T. (1974). Economical second-order designs based on irregular fractions of 3^n factorial. *Technometrics*, **16**, 375–423.

Hsieh, P. and W. Goodwin (1986). Sheet molded compound process improvement. In *Fourth Symposium on Taguchi Methods*, pp. 13–21. Dearborn, MI: American Supplier Institute.

Huang, L. Z., Y. Lu, Y. Yuan, F. Lü, and X. Bie (2006). Optimization of a protective medium for enhancing the viability of freeze-dried *Lactobacillus delbrueckii subsp. bulgaricus* based on response surface methodology. *Journal of Industrial Microbiology and Biotechnology*, **33**(1), 55–61.

Jensen, D. R. (2000). The use of standardized diagnostics in regression. *Metrika*, **52**, 213–223.

Jeong, I.-J., K.-J. Kim, and Y. C. Chang (2005). Optimal weighting of bias and variance in dual response surface optimization. *Journal of Quality Technology*, **37**(3), 236–247.

Khuri, A. I. (1999). Discussion. *Journal of Quality Technology*, **31**(1), 58–60.

Khuri, A. I. (2003). Current modeling and design issues in response surface methodology: GLMs and models with block effects. In *Handbook of Statistics*, Vol. 22, Chap. 6 (R. Khattree and C. R. Rao, eds.). Amsterdam: Elsevier Science B.V.

Khuri, A. I., ed. (2005). *Response Surface Methodology and Related Topics.* Washington, DC: World Scientific.

Khuri, A., and J. Cornell (1996). *Response Surface: Design and Analysis*, 2nd ed. New York: Marcel Dekker.

Kim, B. H. and C. C. Akoh (2005). Modeling of lipase-catalyzed acidolysis of sesame oil and caprylic acid by response surface methodology: Optimization of reaction conditions by considering both acyl incorporation and migration. *Journal of Agricultural and Food Chemistry*, **53**(20), 8033–8037.

Kleijnen, J. P. C., D. den Hertog, and E. Angün (2004). Response surface methodology's steepest ascent and step size revisited. *European Journal of Operational Research*, **159**, 121–131.

Koshal, R. S. (1933). Application of the method of maximum likelihood to the improvement of curves fitted by the method of moments. *Journal of the Royal Statistical Society, Series A*, **96**, 303–313.

Kowalski, S. M., C. M. Borror, and D. C. Montgomery (2005). A modified path of steepest ascent for split-plot experiments. *Journal of Quality Technology*, **37**(1), 75–83.

Lowe, C. W. (1974). Evolutionary operation in action. *Applied Statistics*, **23**(2), 218–226.

Lu, W.-K., T.-Y. Chiu, S.-H. Hung, I.-L. Shih, and Y.-N. Chang (2004). Use of response surface methodology to optimize culture medium for production of poly-γ-glutamic acid by *Bacillus licheniformis*. *International Journal of Applied Science and Engineering*, **2**, 49–58.

Mays, D. P. and S. M. Easter (1997). Optimal response surface designs in the presence of dispersion effects. *Journal of Quality Technology*, **29**, 59–70.

McDaniel, W. R. and B. E. Ankenman (2000). A response surface test bed. *Quality and Reliability Engineering International*, **16**, 363–372.

Mee, R. W. (2001). Noncentral composite designs. *Technometrics*, **43**(1), 34–43.

Mee, R. W. (2004). Optimal three-level designs for response surfaces in spherical experimental regions. Technical Report 2004-3, Department of Statistics, Operations, and Management Science, University of Tennessee (http://stat.bus.utk.edu/techrpts).

Moberg, M., K. E. Markides and D. Bylund (2005). Multi-parameter investigation of tandem mass spectrometry in a linear ion trap using response surface modelling. *Journal of Mass Spectrometry*, **40**(3), 317–324.

Montgomery, D. C., R. H. Myers, W. H. Carter, Jr., and G. G. Vining (2005). The hierarchy principle in designed industrial experiments. *Quality and Reliability Engineering International*, **21**, 197–201.

Morris, M. D. (2000). A class of three-level experimental designs for response surface modeling. *Technometrics*, **42**, 111–121.

Myers (1971). *Response Surface Methodology*. Boston: Allyn and Bacon.

Myers, R. H. (1999). Response surface methodology—current status and future directions. *Journal of Quality Technology*, **31**(1), 30–44; discussion: 45–74.

Myers, R. H. and D. C. Montgomery (1995). *Response Surface Methodology: Process and Product Optimization using Designed Experiments*. New York: Wiley.

Myers, R. H. and D. C. Montgomery (2002). *Response Surface Methodology: Process and Product Optimization Using Designed Experiments*, 2nd ed. New York: Wiley.

Myers, R. H., D. C. Montgomery, G. G. Vining, C. M. Borror, and S. M. Kowalski (2004). Response surface methodology: A retrospective and literature survey. *Journal of Quality Technology*, **36**(1), 53–77.

Nicolai, R. P., R. Dekker, N. Piersma, and G. J. van Oortmarssen (2004). Automated response surface methodology for stochastic optimization models with known variance. In *Proceedings of the 2004 Winter Simulation Conference*, pp. 491–499. (R. G. Ingalls, M. D. Rosetti, J. S. Smith, and B. H. Peters, eds.), The Society for Computer Simulation International, San Diego, CA.

Notz, W. (1982). Minimal point second order designs. *Journal of Statistical Planning and Inference*, **6**, 47–58.

Ozol-Godfrey, A., C. M. Anderson-Cook, and D. C. Montgomery (2005). Fraction of design space plots for examining model robustness. *Journal of Quality Technology*, **37**(3), 223–235.

Palamakula, A., M. T. H. Nutan, and M. A. Khan (2004). Response surface methodology for optimization and characterization of limonene-based Coenzyme Q-10 self-nanoemulsified capsule dosage form. *AAPS PharmSciTech*, **5**(4), Article 66. (This article is available at http://www.aapspharmscitech.org/articles/pt0504/pt050466/pt050466.pdf.)

Peterson, J. J. (1993). A general approach to ridge analysis with confidence intervals. *Technometrics*, **35**, 204–214.

Peterson, J. J., S. Cahya, and E. del Castillo (2002). A general approach to confidence regions for optimal factor levels of response surfaces. *Biometrics*, **58**, 422–431.

Peterson, J. J., S. Cahya, and E. del Castillo (2004). Letter to the editor. *Technometrics*, **46**(3), 355–357.

Peterson, J. J. and A. M. Kuhn (2005). Ridge analysis with noise variables. *Technometrics*, **47**(3), 274–283.

Rechtschaffner, R. (1967). Saturated fractions of 2^n and 3^n factorial designs. *Technometrics*, **9**, 569–575.

Roquemore K. G. (1976). Hybrid designs for quadratic response surfaces. *Technometrics*, **18**, 419–424.

Snee, R. D. (1985). Computer-aided design of experiments—some practical experiences. *Journal of Quality Technology*, **17**(4), 222–236.

Spendley, W., G. R. Hext, and F. R. Himsworth (1962). Sequential applications of simplex designs in optimization and EVOP. *Technometrics*, **4**(4), 441–461.

Steinberg, D. M. and D. Bursztyn (2001). Discussion of "Factor screening and response surface exploration" by S.-W. Cheng and C. F. J. Wu. *Statistica Sinica*, **11**, 596–599.

Sztendur, E. M. and N. T. Diamond (2002). Extensions to confidence region calculations for the path of steepest ascent. *Journal of Quality Technology*, **34**(3), 289–296.

Tang, M., J. Li, L.-Y. Chan, and D. K. J. Lin (2004). Application of uniform design in the formation of cement mixtures. *Quality Engineering*, **16**(3), 461–474.

Tuck, M. G., S. M. Lewis, and J. I. L. Cottrell (1993). Response surface methodology and Taguchi: A quality improvement study from the milling industry. *Journal of the Royal Statistical Society, Series C*, **42**, 671–676.

Trinca, L. A. and S. G. Gilmour (2001). Multistratum response surface designs. *Technometrics*, **43**(1), 25–33.

Unal, R., K. C. Wu and D. O. Stanley (1997). Structural design optimization for a space truss platform using response surface methods. *Quality Engineering*, **9**, 441–447.

Vázquez, M. and A. M. Martin (1998). Optimization of *Phaffia rhodozyma* continuous culture through response surface methodology. *Biotechnology and Bioengineering*, **57**(3), 314–320.

Vining, G. G. and R. H. Myers (1990). Combining Taguchi and response surface philosophies: A dual response approach. *Journal of Quality Technology*, **22**, 38–45.

Vining, G. G., S. M. Kowalski, and D. C. Montgomery (2005). Response surface designs within a split-plot structure. *Journal of Quality Technology*, **37**(2), 115–129.

Wang, Y. and K.-T. Fang (1981). A note on uniform distribution and experimental design. *KeXue TongBao*, **26**, 485–489.

Wen, Z.-Y. and F. Chen (2001). Application of statistically based experimental designs for the optimization of eicosapentaenoic acid production by the diatom *Nitzschia laevis*. *Biotechnology and Bioengineering*, **75**(2), 159–169.

Westlake, W. J. (1965). Composite designs based on irregular fractions of factorials. *Biometrics*, **21**, 324–335.

Wu, C. F. J. and Y. Ding (1998). Construction of response surface designs for qualitative and quantitative factors. *Journal of Statistical Planning and Inference*, **71**, 331–348.

Wu, C. F. J. and M. Hamada (2000). *Experiments: Planning, Analysis, and Parameter Design Optimization*. New York: Wiley.

Xu, H., S. W. Cheng, and C. F. J. Wu (2004). Optimal projective three-level designs for factor screening and interaction detection. *Technometrics*, **46**, 280–292.

Zahran, A. R., C. M. Anderson-Cook, and R. H. Myers (2003). Fraction of design space to assess prediction capability of response surface designs. *Journal of Quality Technology*, **34**, 377–386.

Zink, P. S., D. N. Mavris, P. M. Flick, and M. H. Love (1999). Impact of active aeroelastic wing technology on wing geometry using response surface methodology. Talk given at the *Langley International Forum on Aeroelasticity and Structural Dynamics*, Williamsburg, VA, June 22–25, 1999.

EXERCISES

10.1 Construct and interpret the surface plot for the example given in Section 6.2. Does the plot suggest that something is awry? Explain.

10.2 In their article "Main and interaction effects of acetic acid, furfural, and *p*-hydroxybenzoic acid on growth of ethanol productivity of yeasts," E. Palmqvist, H. Grage, N. Q. Meinander, and B. Hahn-Hägerdal (*Bioengineering and Biotechnology*, **63**(1), 46–55, 1999) stated that they used a modified CCD. The factorial portion was a 2^3 design that was replicated three times, with each replicate run as a block, with the blocks corresponding to days. A reference point, which was a point without any added compounds, was used in each block. The axial points were run in separate blocks, and two blocks were used since two replicates of the axial points were used. A reference point was also used in each axial block, so the total number of reference points was 5, one per block.

The full design, including replicates, consisted of 45 runs. The authors gave the design values in the original units, not in coded units. One of the blocks for the factorial portion (they are all the same, except for the run order) is given below, in juxtaposition to one of the axial blocks.

```
    Factorial Block              Axial Block

0      0        0          0      0        0
2      0.6      0.4        0      1.5      1
8      0.6      0.4        10     1.5      1
2      0.6      1.6        5      1.5      0
8      0.6      1.6        5      1.5      2
2      2.4      0.4        5      0        1
8      2.4      0.4        5      3.0      1
2      2.4      1.6        5      1.5      1
8      2.4      1.6        5      1.5      1
```

(a) Determine the points in the axial block in coded units.
(b) Assess the design. (Remember that the authors stated that this is a "modified central composite design.") Do you consider this to be a useful design?

10.3 Use appropriate software, such as JMP, to construct a uniform design for four factors in 16 runs, using $(-2, 2)$ as the range for each factor. Compare this with the corresponding (rotatable) CCD with 16 runs for four factors. Comment.

10.4 Assume that an experimenter uses a CCD for three factors, using a 2^3 design for the factorial part and using five centerpoints. The latter were used for estimating σ, which produced a result that was much smaller than the estimate obtained from all the available degrees of freedom for obtaining the estimate. Give one possible reason for this discrepancy.

10.5 In their article, Allen, Yu, and Bernshteyn (2000, references) described a scenario in which engineers decided to use response surface methods to select the design parameters, in the absence of "accurate engineering models." A CCD was eschewed, however, since it required 25 runs and management was willing to guarantee only enough resources to perform 12 experimental runs. Instead, a "low cost" design with only 14 runs was used and the design is given below.

Run	A	B	C	D
1	-0.5	-1.0	-0.5	1.0
2	1.0	1.0	-1.0	1.0
3	-1.0	1.0	1.0	1.0
4	1.0	-1.0	-0.5	-0.5
5	0.0	0.0	-1.0	0.0
6	0.0	1.0	0.0	0.0
7	-0.5	-1.0	1.0	-0.5
8	-1.0	0.0	0.0	0.0
9	1.0	1.0	1.0	-1.0
10	-1.0	1.0	-1.0	-1.0
11	0.0	0.0	0.0	-1.0
12	0.5	-0.5	0.5	0.5
13	0.5	-0.5	0.5	0.5
14	0.5	-0.5	0.5	0.5

(a) Is this a sufficient number of design points to fit a second-order model in four factors? Explain.

(b) What are the properties of the design? Specifically, are there any effects that can be estimated orthogonally? Are there any effects for which the correlation between the estimated effects is undesirable? Explain.

(c) Would you recommend that this design be used? Explain.

10.6 In the article cited in the previous exercise, the authors also gave the results of a study with only 12 experimental runs, with the design and data given below.

Row	A	B	C	D	Y
1	1.25	1.7	12.5	10.00	55.95
2	2.00	2.1	10.0	10.00	101.76
3	1.00	2.1	20.0	10.00	101.23
4	2.00	1.7	12.5	6.25	52.93
5	1.50	1.9	10.0	7.50	59.93
6	1.50	2.1	15.0	7.50	80.54
7	1.25	1.7	20.0	6.25	60.87
8	1.00	1.9	15.0	7.50	72.02
9	2.00	2.1	20.0	5.00	102.70
10	1.00	2.1	10.0	5.00	51.36
11	1.50	1.9	15.0	5.00	59.42
12	1.75	1.8	17.5	8.75	81.94

(a) Notice that the design values were given in raw units rather than coded units. Convert these to coded units.

(b) Can the full second-order model be fit to these data? If not, how might one proceed to determine the effects to estimate?

(c) The authors fit a model with 10 terms. How would you determine significant effects? Would you use ANOVA or some other approach?

(d) Two of the terms in the model fit by the authors were A and A^2. Compute the correlation between the terms for the settings used for A. Now subtract the mean of A from each value of A, square those values, and then compute the correlation between the mean-centered values, and the square of those values. Comment. What would you recommend?

(e) Would you recommend that a different design be used that would be essentially as economical as this one? Explain.

10.7 Response surface methodology has long been used in the food industry. An example is the article "A response surface methodology approach to the optimization of whipping properties of an ultrafiltered soy product" by Carol L. Lah, Munir Cheryan, and Richard E. DeVor (*Journal of Food Science*, **45**, 1720–1726). The objective was to determine optimum conditions for whipping a full-fat soy protein product produced by ultrafiltration. The initial design and the data are given below.

Row	A	B	C	D	E	F	G	Y_1	Y_2
1	-1	-1	-1	-1	-1	-1	-1	97.0	60
2	1	-1	-1	-1	1	1	-1	47.0	100
3	-1	1	-1	-1	1	1	1	76.5	93
4	1	1	-1	-1	-1	-1	1	50.2	0
5	-1	-1	1	-1	1	-1	1	195.0	57
6	1	-1	1	-1	-1	1	1	60.4	93
7	-1	1	1	-1	-1	1	-1	68.3	90
8	1	1	1	-1	1	-1	-1	64.5	100
9	-1	-1	-1	1	-1	1	1	-6.1	100
10	1	-1	-1	1	1	-1	1	16.0	55
11	-1	1	-1	1	1	-1	-1	83.0	100
12	1	1	-1	1	-1	1	-1	9.3	100
13	-1	-1	1	1	1	1	-1	40.3	68
14	1	-1	1	1	-1	-1	-1	24.9	98
15	-1	1	1	1	-1	-1	1	101.0	98
16	1	1	1	1	1	1	1	-10.4	100

The two response variables are overruns (Y_1) and stability (Y_2) and the factors are variables like time, speed, and temperature.

(a) Notice that this is not a CCD. What type of design is it? Be specific. What is the likely purpose of using such a design in view of the title of the article?

(b) Fifteen additional runs were made, using only four of the factors, so that a CCD in those four factors is formed. Do you agree that three of the factors should be dropped from the model for each of the two response variables? If so, which three do you believe must have been dropped?

(c) Would you have proceeded differently? If so, how?

10.8   List the design points for an inscribed CCD (CCI) for four factors. Is the design rotatable? Explain. If not, could the design be made rotatable? Explain.

10.9   What could be the motivation for using a CCI design instead of a standard CCD?

10.10   Explain why a surface plot constructed from a CCD with four factors almost certainly does not represent the true surface.

10.11   Assume that $k = 3$ and a confidence region is constructed on a point that is apparently optimum. Explain why this is termed a "region" instead of an "interval." What will influence the size of the region?

10.12   We consider again the data that were partially analyzed in Exercise 5.39 in Chapter 5 and continue the analysis from the standpoint of response surface methodology. In the experiments described in that exercise, two 2^{5-1} designs were used, with centerpoints used with the first design. The method of steepest ascent was employed after the first experiment, with the second 2^{5-1} design centered at the best combination of steepest ascent design points. Would you have used a different type of design for the second experiment? In particular, do you believe that the experimenters appropriately characterized the nature of the response surface with the results of the second experiment in view of the fact that there were significant interaction effects when the first experiment was performed? (They recognized the significance of one interaction.) Could the method of steepest ascent be used with interactions? Read the article and comment on what they did. What would you have done differently, if anything. Explain.

10.13   In a recent article, Gheshlaghi, Scharer, Moo-Young, and Douglas (2005, references) used a conventional approach to optimization using RSM in a three-stage operation. The first stage consisted of the use of a 2_V^{5-1} design with five centerpoints to try to identify significant factors and interactions; the second stage consisted of experimental runs made along what was believed to be the path of steepest ascent; and the final stage consisted of the use of a CCD in three of the five original factors in the region that was believed to contain the optimum combination of factor levels, with the objective being to maximize lysozyme concentration.

   Of interest is the following quote from their paper: "The traditional method of optimization involves varying one factor at a time, while keeping the others constant" (p. 754). Whether or not such an inefficient approach is still traditional may be debatable, but perhaps it is still traditional in the authors' field(s).

(a) When only significant terms were included in the model, the model was $Y = 88.25 + 5.5X_1 + 15.38X_2 + 7.13X_3$. Seven design points were used along what was hoped was the path of steepest ascent. Two points are given below but only one coordinate is given for each point. Fill in the missing coordinates.

Point number	X_1	X_2	X_3
22	0.50	—	—
23	—	2.8	—

(b) The final model equation chosen by the experimenters was

$$Y = 208.8 + 7.49X_1 + 3.49X_2 - 7.65X_1^2$$
$$-7.65X_2^2 - 8.71X_3^2 + 4.63X_1X_2 - 4.62X_2X_3$$

Notice that the equation contains the quadratic term in X_3 as well as an interaction term in that variable, without the linear term also being included. Without seeing the data, what would you be inclined to investigate about X_3?

(c) The response values at the high level of factor C in the factorial portion of the design were 182, 187, 170, and 192, and the response values at the low level of C were 171, 181, 176, and 206. The sum of the first pair of four numbers is 731 and the sum of the second pair is 734, the closeness of the two sums suggesting that there is no C effect. The treatment combinations for these eight numbers were, in order, c, ac, bc, abc, (1), a, b, and ab. If you do an appropriate analysis, using only these numbers and ignoring the two axial points on C and the centerpoints, would you conclude that there is no C effect? Explain.

(d) Notice that the stated coefficients for the two interaction terms differ by .01. Using the data in part (b) and thinking about which data points in the CCD are used in computing interaction effects, what should the two coefficients be, recalling the relationship between effect estimates and model coefficients for factorial designs that was given in Appendix B to Chapter 4?

(e) After the CCD was run, 48 experimental runs had been made, yet only the data from the CCD runs were used in computing the model equation. Thinking about the projective properties of fractional factorial designs, as discussed in Section 5.11, would it be possible, or at least practical, to combine those runs with the runs from the CCD? Explain. There were seven points made along the path of steepest ascent, the first two of which were the focus of part (b). Could the response values for those points be used in any way to obtain model coefficients? Explain.

10.14 Gorenflo, Ritter, Aeschliman, Drouin, Bowen, and Piret (2005, references) stated "According to the principle of hierarchy, nonsignificant terms were kept in the model if they were contained in other interaction terms that were found to be significant..."

(a) Do you agree with this policy? Could this decision pose any special problems in RSM? Explain. (Note from the equation in part (b) of Exercise 10.13 that such a policy is not followed by all experimenters.)

(b) The following table is from Gorenflo et al. (2005), which lists the p-values of significant effects and one nonsignificant effect for a full quadratic model that is fit using a CCD for five factors that was a 2^{5-1} design in the factorial part. Notice that the table illustrates the above quote.

Parameter	Value	SE	P-value
Intercept	14.56	0.09	<.0001
BX	0.54	0.05	<.0001
PR	1.20	0.03	<.0001
PI	0.44	0.02	<.0001
ST	0.004	0.025	0.86
RR	0.15	0.02	<.0001
PR2	0.40	0.03	<.0001
BX*PR	0.23	0.03	<.0001
BX*PI	0.11	0.02	<.0001
BX*RR	0.16	0.03	<.0001
PR*PI	0.27	0.02	<.0001
PR*ST	0.14	0.03	<.0001
PR*RR	0.13	0.03	0.0001
PI*ST	0.09	0.02	<.0001
ST*RR	0.11	0.03	0.0005

Note: ST main effect kept in model due to principle of hierarchy (significant higher order terms contain respective main effect, e.g., PR ST, ST RR, PI ST); BX—Bioreactor cell concentration; PR—Perfusion rate; PI—Power input; ST—Stop time; RR—Recirculation ratio.

There would be 10 two-factor interactions if all of them were included in the model; notice that 8 of them have very small p-values. Similarly, four of the five main effects are highly significant. The raw data were not given by Gorenflo et al., so the data would have to be obtained to do further analyses. If you had the data, what would you look for/suspect? If after careful analyses, you were in agreement with the numbers given in the table above, would you include ST in the model and would you be concerned about the magnitude of the interactions? Explain.

10.15 The case study given in Section 10.5.3 utilized one of the response variables given by Wen and Chen (2001, references). Another response variable was cell dry weight (DW). The values of that response variable and the corresponding treatment combinations are given below for the factorial part of the design, with the factors simply labeled as A, B, C, D, and E. The axial points were in the order -2, 2 and ordered on the factors A–E. The corresponding response values for the 10 axial points and the average of the six centerpoints were 8.00, 7.70, 7.48, 8.06, 7.51, 7.49, 7.97, 6.15, 1.60, and 7.69 for the average.

```
Treatment comb.:    e     d     c    cde    b    bde   bce    a    ade
Response value:   4.90 7.74 7.23 6.13 6.84 7.03 5.57 6.55 7.05
Treatment comb.:   ace   acd   abe   ab   abd   abc  abcde
Response value:   5.80 4.32 7.40 4.25 6.40 5.11 5.75
```

(a) Recognizing the limitations caused by the fact that only the average of the centerpoint observations is given, analyze the data and arrive at a fitted model.

(b) Are there any outliers for this response variable as was the case for the response variable EPA Yield, as was shown in Figure 10.10? If so, remove the outlier(s), making the assumption that this is simply bad data, and reanalyze the data. If you conclude that there are no outliers, give appropriate graphical and/or other support for your conclusion.

10.16 Consider part (c) of Exercise 10.13 and construct the BC interaction plot. What would you tell the experimenters based on what you see in that plot?

10.17 Lu, Chiu, Hung, Shih, and Chang (2004, references) used RSM to study the effects of L-Glutamic acid, citric acid, glycerol, and NH_4Cl on the production of poly-γ-glutamic acid under certain conditions. The strategy that they employed was to first use a 2^4 factorial design for the purpose of determining the path of steepest ascent. Experiments were conducted along that path as long as the first-order model was adequate. A CCD was then used for those four factors with ± 2 as the axial point values, and four centerpoints were used. In this study, the experimenters knew what factors they wanted to study; hence, there was no need for a screening design.

(a) The axial point values combined with the centerpoints means that each factor was measured at five equispaced levels. Accordingly, could the design be the equivalent of some 5^{4-k} design? Why, or why not?

(b) The fitted equation given by the authors was $\widehat{Y} = 35.34 - 3.82X_1^2 - 4.34X_2^2 - 3.83X_3^2 - 7.20X_4^2$, with X_1 through X_4 denoting the four factors and only significant terms used in the model. Notice that there are no linear (or interaction) terms, so the model is (strongly) nonhierarchical, and the coefficient of each term is negative. Without necessarily knowing anything about the subject matter, would you be inclined to question this equation just on general grounds? Why, or why not?

(c) The data and the design as they appeared in the paper are given below, except that here the factors are arbitrarily labeled as A–D. Note that although the first column is labeled "Run," there is no indication of the actual run order that was used.

Run	A	B	C	D	Y
1	31.0	18.0	140.0	6.0	22.17
2	31.0	18.0	140.0	16.0	23.16
3	31.0	18.0	152.0	6.0	23.67
4	31.0	18.0	152.0	16.0	23.03
5	31.0	34.0	140.0	6.0	23.52
6	31.0	34.0	140.0	16.0	28.97
7	31.0	34.0	152.0	6.0	27.11
8	31.0	34.0	152.0	16.0	22.78
9	37.0	18.0	140.0	6.0	36.66
10	37.0	18.0	140.0	16.0	28.67
11	37.0	18.0	152.0	6.0	18.87
12	37.0	18.0	152.0	16.0	23.16
13	37.0	34.0	140.0	6.0	28.14
14	37.0	34.0	140.0	16.0	21.09
15	37.0	34.0	152.0	6.0	27.23
16	37.0	34.0	152.0	16.0	26.77
17	28.0	26.0	146.0	11.0	27.67
18	40.0	26.0	146.0	11.0	29.41
19	34.0	10.0	146.0	11.0	23.94
20	34.0	42.0	146.0	11.0	31.07
21	34.0	26.0	134.0	11.0	24.98
22	34.0	26.0	158.0	11.0	32.07
23	34.0	26.0	146.0	1.0	22.26
24	34.0	26.0	146.0	21.0	21.32
25	34.0	26.0	146.0	11.0	36.86
26	34.0	26.0	146.0	11.0	34.82
27	34.0	26.0	146.0	11.0	34.86
28	34.0	26.0	146.0	11.0	34.80

(d) Do you agree with the model that was selected by the authors? They stated that their model had an R^2 value of .707. Is that correct?

(e) Use whatever supplementary graphs and/or numerical analyses that are necessary to explain why the coefficients of the quadratic terms are negative. After looking at these results, what questions, if any, would you ask of the experimenters/authors?

10.18 If you were going to determine the significant factors from a group of candidate factors and then try to determine the optimum combination of levels of those factors so as to maximize the response, would you do so by starting with a three-level design or with a two-level design? Explain.

10.19 Explain why the method of steepest ascent cannot be applied to a fitted model that contains an interaction term.

10.20 (Harder problem) Explain why the Box–Behnken design in Table 10.2 cannot be run in three blocks with one centerpoint in each block.

10.21 (Harder problem) There is no Box–Behnken design for two factors. Explain why such a design could not be constructed.

10.22 (Harder problem) Explain why it is not possible to construct a Box–Behnken design with only two nonzero numbers in each row of the design.

10.23 As discussed in Section 10.5.2.1, specific columns of a Plackett–Burman design are used in constructing Draper–Lin designs. Determine the difference in the D-values for the 16-run designs for four factors constructed using columns 1, 2, 3, and 6 of the 8-run Plackett–Burman design and columns 1, 2, 3, and 4 of that design.

10.24 Assume that a Draper–Lin design for four factors is used and the 16 response values are 10.2, 10.7, 11.5, 12.6, 13.4, 12.2, 15.1, 10.9, 11.9, 11.3, 12.2, 14.1, 15.8, 13.6, 13.9, and 15.0. Fit the full second-order model and compare the standard errors of the estimates. Is the difference in precision of any concern? Explain.

10.25 Consider a full CCD for five factors with 48 points: 32 factorial points, 10 axial points, and 6 centerpoints. Compute the correlation between the quadratic effect columns, then do the same for a half CCD for the same number of factors with 32 points: 16 factorial points, 10 axial points, and 6 centerpoints. Compare the two sets of correlation coefficients and comment.

10.26 The results of an experiment in which a Box–Behnken design for three factors was used was described in Example 10.5. The response values and the factors in raw units are given below.

Limonene (mg)	Cremophor (mg)	Capmul GMO-50 (mg)	%dissolved in 5 minutes
81	57.6	7.2	44.4
81	7.2	7.2	6.0
18	57.6	7.2	3.75
18	7.2	7.2	1.82
81	32.4	12.6	18.2
81	32.4	1.8	57.8
18	32.4	12.6	68.4
18	32.4	1.8	3.95
49.5	57.6	12.6	58.4
49.5	57.6	1.8	24.8
49.5	7.2	12.6	1.60
49.5	7.2	1.8	12.1
49.5	32.4	7.2	81.2
49.5	32.4	7.2	72.1
49.5	32.4	7.2	82.06

(a) Perform an analysis of the data, using conditional effects and any supplementary computations if necessary, and compare your conclusions with
those mentioned in Example 10.5 and with the conclusions reached in the
article, which is available at http://www.aapspharmscitech.org/articles/
pt0504/pt050466/pt050466.pdf.
(b) Consider Figure 3 in Palamakula, Nutan, and Khan (2004), which the
authors claim shows the "quadratic effect of interactions." Do you agree
with that conclusion? Why, or why not?

10.27 Consider the Box–Behnken design for three factors that was given in Table
10.2. Simulate a set of response values and determine the variance at each of
the design points that is not a centerpoint. What do you observe and what do
you conclude about the rotatability (or not) of the design?

10.28 Allen et al. (2000, references) were faced with an experimental situation for
which there were four factors to be investigated but there were sufficient
resources for only 12 experimental runs. The design that they used is given
below.

A	B	C	D
-0.5	-1.0	-0.5	1.0
1.0	1.0	-1.0	1.0
-1.0	1.0	1.0	1.0
1.0	-1.0	-0.5	-0.5
0.0	0.0	-1.0	0.0
0.0	1.0	0.0	0.0
-0.5	-1.0	1.0	-0.5
-1.0	0.0	0.0	0.0
1.0	1.0	1.0	-1.0
-1.0	1.0	-1.0	-1.0
0.0	0.0	0.0	-1.0
0.5	-0.5	0.5	0.5

(a) Would you recommend that this design be used for fitting a full quadratic
model if there were *no* cost constraints? Why, or why not?
(b) Is this an orthogonal design? If not, is the degree of nonorthogonality
likely to be of any consequence? Explain.
(c) How would one determine what effects in a full quadratic model are
significant with this design? (*Hint*: The discussion in Section 13.7 may
be helpful.)

CHAPTER 11

Repeated Measures Designs

Repeated measures designs are designs for which, as the name suggests, repeated measurements are made on the experimental units, which are typically people as the design is often used in clinical work and in education. Certainly, repeated measures on the same subject are a necessary part of any design for studying rates of learning as a function of treatment effects. In clinical work, subjects may be measured over time after receiving some type of treatment.

At the outset, we should note that not much attention is given to repeated measures designs in most books on experimental design; exceptions are Hinkelmann and Kempthorne (2005), Giesbrecht and Gumpertz (2004), Kuehl (2000), and Winer, Brown, and Michels (1991). In particular, the latter has one chapter on single-factor designs with repeated measures and a chapter on multifactor designs with repeated measures. The best book that is devoted almost exclusively to repeated measures designs is probably Davis (2002). Other books on the subject include Lindsey (1999), Crowder and Hand (1990), Vonesh and Chinchilli (1997), Girden (1992), and Davidian and Giltinan (1995). Although somewhat outdated, Fleiss (1986) has two chapters on repeated measures designs. Review papers on the subject include Keselman, Algina, and Kowalchuk (2001), which is on the analysis of data from repeated measures designs, as is Everitt (1995), who provides illustrative examples.

There are both advantages and disadvantages associated with these designs. One advantage is that information on the time trend of the response variable is available under different treatment conditions. Furthermore, each subject serves as his or her own control, which permits the use of a smaller sample size. More specifically, subject-to-subject variation is not problematic when each subject receives all the treatments instead of each subject receiving only one treatment. This is especially important when the experimental unit/subject is a person because it can be difficult to select people who are homogeneous on characteristics that should be essentially constant for a given study.

11.1 ONE FACTOR

The simplest example of a repeated measures experiment is when a group of people are measured on a certain characteristic both before and after some type of treatment, such as an industrial training program or a medication program. Of course this is a classic example of when a paired t-test would be used as the method of analysis.

A paired t-test cuts in half the number of observations used in the analysis, relative to a pooled t-test, since differences are formed. There are four components of the total variation when ANOVA is used, with the initial breakdown being between people and within people, with the latter broken down into treatments and residual.

Example 11.1

An experiment is conducted using nine experimental subjects in two breathing chambers, with each subject spending time in each chamber. One chamber had a high concentration of carbon monoxide, while the other did not. Our scientific interest is in establishing that exposure to carbon monoxide significantly increases respiration rate.

Subject	With CO	Without CO
1	30	30
2	45	40
3	26	25
4	24	23
5	34	30
6	51	49
7	46	41
8	32	35
9	30	28

Of course an F-test in ANOVA is nondirectional and the alternative hypothesis for this experiment is obviously directional, but we will use this as an example of the computations with each approach. (Of course with any experimentation the objective is generally to determine that some treatment is the best.)

The simplest *possible* model for the repeated measures design for this application is

$$Y_{ij} = \mu + A_i + B_j + \epsilon_{ij} \quad i = 1, 2, \ldots, 9 \quad j = 1, 2 \qquad (11.1)$$

with μ denoting the mean, A_i the effect of the ith subject, and B_j the effect of the jth treatment. Of course the subject effect is considered to be a random effect. Notice that this is simply the model for two-factor ANOVA without interaction.

The model given by Eq. (11.1) may not be appropriate, however, and frequently won't be appropriate. In particular, observations made on the same experimental unit will frequently be dependent. This dependency might manifest itself in one or more ways, such as carryover effects or correlation because the subjects respond differently to treatments, so that there will be correlation between treatments within subjects. Carryover effects and designs for carryover effects are discussed in Section 11.4. If much time elapses between the repeat measurements (which would be one way to try to avoid carryover effects), there might also be a period effect, perhaps caused by changing environmental conditions that serve as extraneous factors. Of course if there is a period effect, it would be confounded with the treatment effect. If experimental subjects were very slow to receive each treatment, there might be period effects within each treatment.

The bottom line is that careful thought must be given to the model in a repeated measures experiment and appropriate assumptions may not be easily determined.

The ANOVA table for the repeated measures data in this example is as follows.

Source	df	SS	MS	F
Between subjects	8	1289.78		
Within subjects	9	42.50		
Treatments	1	16.06	16.06	4.86
Residual	8	26.44	3.305	
Total	17	1332.28		

The between subjects and within subjects sum of squares are computed in the same way as in one-factor ANOVA if we regard subjects as the factor. Since there are multiple treatments for each subject, a treatment sum of squares can thus be broken out of the within subjects sum of squares. Another way to view this is to recognize that differences within subjects must be due at least in part to differences between treatments, with the remainder unaccounted for by the treatments.

Since we can analyze these data using a paired t-test, it stands to reason that the results must be numerically equivalent. That is, the numerical value of the t-statistic must be equal to $\sqrt{4.86}$. It can be shown that the value of the t-statistic is $1.89/0.857 = 2.205$ which, within rounding, does equal $\sqrt{4.86}$.

The analysis using a paired t-test is much simpler than the ANOVA approach since there are fewer computations to perform, but of course a paired t-test cannot be used if there are multiple factors or more than two levels of a single factor. (There will usually be more than two levels of a single factor.) Nevertheless, the paired t-test is certainly more intuitive than the ANOVA approach and should thus be used whenever possible.

The analysis of repeated measures data for one factor with more than two levels is straightforward, with the analysis being of the same general form as the analysis when there are only two levels. (Regarding the analysis of data from repeated measures designs, it is worth noting that this capability is absent in some of the leading software, such as MINITAB and JMP, with the latter providing capability only through a mixed model approach (i.e., not for a single factor or for multiple factors with other models).)

11.1.1   The Example in Section 2.1.2

In the example in Section 2.1.2, nine observations on each subject were averaged to produce a single number, which permitted the data to be analyzed as either a pooled t-test or its ANOVA equivalent. Although such condensing of data frequently occurs in practice, it would be better to use all the data. There might, for example, be a period effect and it would be useful to be able to test for this. We will see how a possible period effect can be isolated in subsequent sections in this chapter.

11.2   MORE THAN ONE FACTOR

The modeling and analysis that accompanies repeated measures designs is much more involved when there are multiple factors.

Example 11.2

Milliken (2004) gave an example with two factors in a semiconductor experiment, which was performed to investigate factors that influence the thickness of the oxide layer that develops on the surface of silicon wafers. The two factors that were used in the experiment were furnace temperature (900, 1000, and 1100° F) and position in the furnace (top, middle, and bottom). There is a factorial arrangement of levels because all nine combinations of furnace and position are used, with nine wafers from one lot used in each replicate of the experiment, with three wafers going into each of three furnaces. This process was then repeated four times with wafers from a different lot used in each replication. The three temperatures were randomly assigned to a furnace run but position obviously could not be randomly assigned to the part of the furnace, which was the experimental unit for the temperature factor, so the design is thus a repeated measures design. The model for the experiment is given by

$$Y_{ijk} = \mu + A_i + B_j + C_{ij} + D_k + AD_{ik} + E_{ijk}$$

with μ denoting the mean, A_i the temperature effect, B_j the lot effect, C_{ij} the furnace effect, D_k the furnace position effect, AD_{ik} the interaction between temperature and furnace position, and E_{ijk} the effect of positions within furnaces. The latter is the error term. The E_{ijk} are not independent because the positions are (obviously) not randomly assigned to the parts of the furnace. Consequently, the E_{ijk} are correlated on the third subscript, which of course corresponds to the position effect.

The covariance structure must be determined. Milliken (2004) suggests testing a small number of reasonable covariance structures and selecting the one that seems most appropriate, pointing out that PROC Mixed in SAS provides several information criteria that can be used in determining which covariance structure best fits the data. (Crowder (2001) considered the analysis of data from repeated measures designs when the covariance structure has been misspecified.)

Milliken (2004) provided a thorough analysis of the data, analyzing the data for each of five covariance structures. The position × temperature (AD) interaction effect is significant at the .05 level for four of those structures. Of those, the consequence of this is that the temperature and position main effects should not be computed and interpreted in the usual way. This especially applies to the analysis using the Toeplitz covariance structure since the p-value for temperature is .09.

Senn, Stevens, and Chaturvedi (2000) discussed when it can be advantageous to analyze repeated measures data using summary measures. Street, Eccleston, and Wilson (1990) gave some tables of optimal, small repeated measures designs and Kushner (1998) similarly studied optimal and efficient repeated measures designs under the assumption of uncorrelated observations. Carriére (1999) gave methods for analyzing data from repeated measures designs when there are missing observations. Kowalchuk, Keselman, Algina, and Wolfinger (2004) discussed the analysis of repeated measures data using mixed model adjusted F-tests.

11.3 CROSSOVER DESIGNS

The general idea is that there is a change (i.e., a "crossover") in the sequence in which treatments are applied to the subjects. An obvious question to ask is "Why the reversal?" Recall that in a randomized block design, the units are assumed to be homogeneous within blocks and each treatment is applied to only one experimental unit. Therefore, it shouldn't make any difference whether the ordering of treatments within a block is, say, $A\ B\ C$ or $C\ A\ B$. If there is expected to be a difference in the order in which the treatments are applied, however, such as a practice effect since each experimental subject receives more than one treatment, then obviously attention must be given to the sequencing.

The simplest crossover design is a design with two treatments and two periods, with half of the subjects receiving treatment A first followed by treatment B, and the other half receiving treatment B first followed by treatment A. For this design to work, there must be a washout period between adjacent treatment periods. Senn (1996) described the proper use of this design but did not advocate its usage. When there is no carryover effect (to be discussed shortly), the data can be analyzed using a two-sample t-test after first forming the differences $B_{1i} - A_{1i}, i = 1, 2, \ldots, n_1$, and $A_{2i} - B_{2i}, i = 1, 2, \ldots, n_2$, assuming n_1 subjects in the first group and n_2 subjects in the second group, with B_{1i} denoting the measurement obtained on the ith subject, the first period, and the "B" treatment, and similarly for the other three.

The primary motivation for using crossover designs is that they are economical relative to parallel group designs in which the subjects are randomly allocated among the treatments. Crossover designs eliminate most of the between-subject variation and have the same power to detect treatment differences as parallel group designs with far more subjects. This is important when there is a limited number of subjects, as will often be the case when human or animal subjects are used.

Crossover designs are used in various types of applications, especially sensory testing (see, e.g., Wakeling and MacFie, 1995; Kunert, 1998; Périnel and Pagès, 2004 and Kunert and Sailer, 2006), telecommunications (Lewis and Russell, 1998), bioenvironmental and public health studies (Tudor, Koch, and Catellier, 2000), medical applications (Matthews, 1994), and dentistry (Claydon, Addy, Newcombe, and Moran, 2005). Early applications were in a variety of areas including weather modification experiments (Mielke, 1974), psychological experiments (Keppel, 1973), clinical trials (Grizzle, 1965), bioassay (Finney, 1965), tea-tasting experiments, crop experiments, cow-lactating experiments, and other applications. Methods of selecting crossover designs are discussed by Loughin, Johnson, Ives, and Nagaraja (2002).

Crossover designs have been used extensively in medical applications, especially in clinical trials for studying chronic ailments, and in other areas (see, e.g., Max, 2003; Senn, 2003; and Carriére, 1994). These designs have also been the primary designs used for bioavailability and bioequivalence studies. Unfortunately, Garcia, Benet, Arnau, and Cobo (2004) observed that in the 40 published crossover studies that they studied for the period 2000–2003, 18 of them did not give effect estimates and associated standard errors, thus preventing their use in any type of meta-analysis. This omission is in violation of the CONSORT (Consolidated Standards of Reporting Trials) recommendations. In comparing crossover designs and parallel designs, Garcia et al. (2004) found that parallel designs "... need, in order to achieve the same power, between 4 and 10 times more subjects than the corresponding cross-over design. . . ."

The literature contains some slightly conflicting information about what a crossover design actually is. For example, Cochran and Cox (1957, p. 127) state that a crossover design "closely resembles" a Latin square design, whereas Montgomery (1996, p. 204) presents a particular crossover design as a set of Latin square designs, with the corresponding analysis. There has also been some confusion in the literature regarding the correct analysis of data from crossover designs. This is explained at the end of the Section 11.4.

It is clear, however, that a crossover design, which is also called a *changeover design*, is a type of repeated measures design since the subjects receive more than one treatment. Various researchers have extolled the benefits of crossover designs versus parallel (group) designs (i.e., completely randomized designs), for which each subject receives only one treatment. The main problem with parallel designs is generally the inhomogeneity of subjects that creates substantial between-subject variation and, consequently, can make it difficult to detect significant effects. This variation is either removed or largely reduced when treatment comparisons are made entirely or almost entirely within subjects. More specifically, crossover designs can have more power than parallel designs even when the latter have as many as 10 times the number of subjects (Louis, Lavori, Bailar, and Polansky, 1984).

Crossover designs are not without shortcomings, however, as there can be carryover effects from one treatment to the next since subjects are receiving more than one treatment. Furthermore, if treatments are spread out very far in time so as to eliminate or at least minimize carryover effects, experiments may require a long time to run. We continue our discussion of carryover effects after first looking at an example in which a crossover design was purportedly used.

Specifically, we consider the following example from the literature.

Example 11.3

Lawson (1988) described the use of what he termed a crossover design in an industrial application. The design layout was as follows:

```
C  B  D  A  D  A  A  B  B  C  B  C  B  D  C  D
D  D  A  B  A  B  D  D  C  B  D  D  A  B  B  A
B  C  B  D  C  C  C  C  D  D  C  A  C  A  A  C
A  A  C  C  B  D  B  A  A  A  A  B  D  C  D  B
```

with the rows of the design representing the four furnaces, the columns designating replicates, and the letters representing four different mixtures of three reducing agents that were randomly assigned to the four furnaces *during each time period*. Of course this assignment of the mixtures to the furnaces causes the design to be unbalanced in that the different mixtures do not occur the same number of times in each furnace. In particular, mixture *B* is used only twice in the third furnace. This problem could have been avoided by using four 4 × 4 Latin squares.

Although this design layout was termed a crossover design by Lawson (1988), it actually has the basic characteristic of a randomized block design at each period, with the design becoming highly unbalanced when the "blocks" are formed by stringing the periods together. Of course in a randomized block design the experimental units within each block are considered to be homogeneous. This design should not be used if a furnace effect is expected so that furnace should be blocked on.

The raw data were not given by Lawson (1988), so a reanalysis of the data cannot be given here. Any such analysis would be flawed if there were a furnace effect, however. Indeed there was a furnace effect for the two dependent variables (1) average furnace efficiency in percent, and (2) percent of product lost in the slag, the *p*-value being less than .01 for each dependent variable.

The problem with a crossover design, as has been pointed out by various authors, is that it breaks down when there is carryover; that is, when there is a residual effect of a treatment such as the measurement on a subject in the second period influenced by the measurement on the same subject from the other treatment in the first period. As is illustrated by Giesbrecht and Gumpertz (2004, p. 261), any carryover effect is confounded with other effects. Another possible problem with crossover designs (and in general when the experimental unit is a person) is that subjects might drop out of the study. Low, Lewis, and Prescott (1999) assessed the robustness of crossover designs to subjects dropping out and Low, Lewis, and Prescott (2002) also addressed the dropout problem. Boon and Roes (1999) considered various design and analysis issues when crossover designs are used in phase I clinical trials. Federer and Kerschner (1998) compared different classes of crossover designs. Kempton, Ferris, and David (2001) considered optimal crossover designs under the assumption that carryover effects are proportional to direct effects, with the latter being the treatment effects. This paper motivated the work of Bailey and Kunert (2005). Stufken (1996) reviewed the optimality properties of crossover designs, and other work on optimal crossover designs includes Kunert and Stufken (2002), Bose and Dey (2003), and Bose and Mukherjee (2003). Part A of Bose (2002), which is available on the Internet (see Bose, 2002), is based on Bose and Mukerjee (2003) and Part B is reproduced by Bose and Dey (2003).

Most of the optimality results in the literature are based on the model of Cheng and Wu (1980), which assumed that (1) carryover effects stop after the first period, (2) there is no interaction between the treatments applied in successive periods to the same subject, (3) the subject effects are fixed effects, and (4) the errors are independent with mean zero and constant variance. Unfortunately, as discussed by Bose (2002), one or more of these assumptions is likely to be violated in the type of applications of the design that have been used in recent years. In particular, as stated previously, subjects are generally assumed to be a random factor and problems with the second assumption will be discussed shortly. The independent errors assumption might also be shaky in certain applications. Donev (1998) considered crossover designs with correlated observations, and Matthews (1987) considered crossover designs with both carryover effects and autocorrelated observations.

As discussed by Bose (2002), optimal crossover designs are very sensitive to the model assumptions. Bose and Mukherjee (2000, 2003) did obtain optimal designs when some or all of these assumptions are relaxed, however. In particular, the second assumption has been criticized as being a weakness of the model, as discussed by Kunert (1987), for example. Sen and Mukerjee (1987) developed optimal designs for direct and carryover effects designs when the second assumption is relaxed, and assumed that each treatment has a different carryover effect for each treatment in the next period. Unfortunately, this assumption generally required too many parameters for the model to be useful. Hedayat and Afsarinejad (2002) offered a compromise model that assumed two carryover effects: one when a treatment follows itself and the other when the next treatment is any other treatment. These have been termed *self* and *mixed* carryover effects, respectively, in the literature, as in Hedayat and Afsarinejad (2002), and Kunert and Stufken (2002, 2005).

Hedayat and Stufken (2003) appropriately pointed out that a design that is efficient under various plausible models is preferable to one that is optimal under one model but performs poorly under certain other models. Their primary focus was performance under two models. Similarly, John, Russell, and Whitaker (2004) described an algorithm for constructing crossover designs that are efficient under a range of models.

A moderate amount of work has been performed to develop useful crossover designs. For example, Russell and Dean (1998) gave crossover designs when only a small number of subjects is available. See also Bate and Jones (2006) for the construction of crossover designs.

11.4 DESIGNS FOR CARRYOVER EFFECTS

When carryover is believed to exist, a crossover design should be balanced for carryover effects. A crossover design is balanced for (first-order) carryover effects if each treatment follows every other treatment equally often. The simplest designs of this type are the ones introduced by Williams (1949), with Patterson (1952) presenting more general crossover designs for carryover effects. Carriére and Reinsel (1993) proved that the designs of Patterson (1952) are optimal for estimating direct treatment effects when there are two periods. John et al. (2004) gave an algorithm for constructing crossover designs that are efficient under various models for explaining carryover effects. One such model, a mixed effects model, was given by Putt and Chinchilli (1999), who assumed that carryover extended over only one period.

The designs given by Williams are, not surprisingly, called Williams squares and they are also referred to as column-complete Latin squares. One such Williams square is

A	B	C	D
B	D	A	C
C	A	D	B
D	C	B	A

with the letters representing the treatments, the columns representing the subjects, and the rows representing the treatment periods. Notice that each treatment comes first in one treatment period and in the other three periods, it is preceded by each of the other treatments. Thus, if there is a carryover effect from one treatment period to the next period, this design essentially adjusts for that carryover. For example, if there is a strong carryover from treatment A, the measurement using treatment B in the first treatment period will be affected, but the measurements using treatment C in the second time period and treatment D in the third time period should be similarly affected.

The design given above is not unique as Cochran and Cox (1957, p. 134) gave the following design, rearranged slightly here with the columns representing the treatment periods, which also has the required property.

A	B	D	C
B	C	A	D
C	D	B	A
D	A	C	B

Actually, more than one square should be used for the same reason that multiple Latin square designs should be used, as was discussed in Section 3.3.5. The designs presented by Williams (1949) were for an even number of treatments. Much more recently, Newcombe (1995, 1996) gave designs for an odd number of factors. For example, the design given for five factors and 15 subjects is

Subject	Sequence
1	ABDCE
2	BCEDA
3	CDAEB
4	DEBAC
5	EACBD
6	ACDBE

7	BDECA
8	CEADB
9	DABEC
10	EBCAD
11	ABEDC
12	BCAED
13	CDBAE
14	DECBA
15	EADCB

Although the designs given by Williams (1949) are Latin squares, the corresponding analysis methods that were provided break out the sum of squares for residual effects as well as a sum of squares for "direct effects." Those methods require that at least two squares be used, but more should actually be used because, for example, there will be only four degrees of freedom for the error term when there are three treatments and two squares. Of course there is the obvious question of whether or not this is necessary since the design adjusts for residual effects.

Williams designs can be analyzed by using the "Williams" option in the CROSSDES package in *R* (see Section 11.7), and Hinkelmann and Kempthorne (2005, p. 700) illustrate the analysis of data from a Williams design using SAS PROC GLM.

One question that arises is whether or not a preliminary test for carryover effects should be performed, with the outcome of the test influencing the form of the analysis, similar to letting interactions that aren't significant become part of the error term so as to increase the degrees of freedom for the error term. The general recommendation is that there should be no preliminary test for carryover effects.

Whether or not there is any carryover effect depends on what is being measured. Cochran and Cox (1957) gave an example of a crossover design to compare the computing speeds of two machines. One person used each of two machines to compute a sum of squares of 27 observations. There were 10 such datasets, which comprised the replications of the design. Since it was felt that the second computation of a sum of squares might be faster than the first computation, the order in which the machines were used was reversed in half of the replications.

Thus, the design was of the following form:

```
A  B  A  B  B  A  B  A  A  B
B  A  B  A  A  B  A  B  B  A
```

with the columns representing the replications.

The analysis of such a design is similar to the analysis of a Latin square design in that row and column effects are isolated and separated from error. Sums of squares for rows and columns are computed in the usual way and treatments are tested against error.

A variation of this design (as described by Cochran and Cox, 1957) is to break up the above configuration into five 2 × 2 Latin square designs. The advantage of this is that row-to-row variation can be broken up into that variation within squares. This is an advantage if the replicates are such that row-to-row variation within the replicates can be expected to differ and thus differ from the overall row-to-row variation. Of course the column-to-column variation becomes that variation within squares when the Latin squares are used.

Sometimes carryover effects can result from factors not being reset. (Recall the discussion of factors not being reset in Section 4.20.) Giesbrecht and Gumpertz (2004, p. 257) gave an example of an actual experiment that involved fuel additives and the engineers could never be sure that all of the fuel additive from one test had been removed from the metal surfaces in the carburetor by the time the next test was performed. Consequently, a design that was balanced for carryover effects was used.

A safe strategy would be to proceed as if carryover effects are present and to analyze the data accordingly. Weerahandi (2004, p. 268) stated that at least the treatment effects, carryover effects, and period effects should be modeled when a crossover design is used. Period effects may not be estimable, however, as is illustrated in the following clinical trials example.

Example 11.4

Weerahandi (2004) gave an example of a clinical trial experiment that was originally given by Senn (2002). The objective of the experiment was to compare two treatments, a single inhaled dose of 200 µg Salbutamol (S) and a single inhaled dose of 12 µg Formoterol (F). The response variable was the peak expiratory flow, a measure of lung function There were 13 children involved in the experiment, each of whom had moderate or severe asthma. There were two periods involved, with a washout period of one day between those two periods. Seven children received the sequence FS over the two periods, and the other six children received the sequence SF.

Our objective, of course, is to determine if there is a difference in the two means, μ_F and μ_S, and if so, to estimate that difference. The first question we might ask is "Is this a good design for accomplishing that objective?" Since there are two groups and two periods, we can use a 2×2 table to represent the parameters that are involved in each cell. If we let $A = F$ and $B = S$, so as to use the conventional A and B notation, we then have the following tabular representation with the rows representing the groups and the columns representing the periods.

$\mu_A + \gamma$	$\mu_B + \gamma + \pi + \lambda_A$
$\mu_B - \gamma$	$\mu_A - \gamma + \pi + \lambda_B$

with μ_A and μ_B denoting the two means of interest, γ the sequence effect (if it exists), λ_A and λ_B the carryover effects for each treatment from the first to the second period, and π the period effect.

We see immediately that we cannot form a linear combination of the four cells to estimate μ_A and μ_B individually, and also we cannot estimate their difference unless we assume $\lambda_A = \lambda_B = \lambda$. With this assumption, we can obviously estimate $\delta = \mu_A - \mu_B$ as $\widehat{\delta} = \frac{1}{2}(\overline{Y}_{11} + \overline{Y}_{22}) - \frac{1}{2}(\overline{Y}_{12} + \overline{Y}_{21})$, with $\overline{Y}_{ij}$ denoting the average observation in the ith row and jth column of the table.

Which other parameters are estimable? In order to estimate γ, we need to use a linear combination for which the coefficients will be $+1$ for the first row and -1 for the second row so that the other parameters will drop out, still assuming of course that $\lambda_A = \lambda_B = \lambda$. Doing so will produce a coefficient of 4 for γ. Thus, γ is estimated as $\widehat{\gamma} = \frac{1}{4}(\overline{Y}_{11} + \overline{Y}_{12} - \overline{Y}_{21} - \overline{Y}_{22})$.

Are any other parameters estimable? We observe that we cannot estimate either π or λ individually, although their sum could be estimated as $\frac{1}{2}(\overline{Y}_{12} + \overline{Y}_{22} - \overline{Y}_{11} - \overline{Y}_{21})$. Thus, the two effects are confounded.

Since we are interested in comparing the two means, we would expect that we would be able to construct a t-test. This would not be a routine t-test, however, as it couldn't be an independent sample t-test because the data aren't independent. It also could not be a paired t-test because there are two sequences, not one. There is also not a single variance since there are two periods and two groups of patients.

There is, however, an analogy with a paired t-test because with that test the variance of the difference $Y_1 - Y_2$ is equivalent to $\sigma_1^2 + \sigma_2^2 - 2\sigma_{12}$, with σ_{12} denoting the covariance between Y_1 and Y_2 and the first two terms denoting the variance of each. Here the covariance is between the periods and since there are two groups, the variances and covariance must be assumed to be the same for the two groups. Otherwise, a standard analysis approach could not be performed. With this assumption, the analysis is performed by merging the data from the two groups.

Let $\sigma^2 = \sigma_1^2 + \sigma_2^2 - 2\sigma_{12}$. With n_1 observations in the first group and n_2 observations in the second group, σ^2 is thus estimated by pooling the data from the two groups, completely analogous to what is done with a pooled t-test. Since $\text{Var}(\overline{Y}_{22} - \overline{Y}_{21}) = \sigma^2/n_2$ and $\text{Var}(\overline{Y}_{11} - \overline{Y}_{12}) = \sigma^2/n_1$ and the two differences are independent, it follows that $\text{Var}(\widehat{\delta}) = \text{Var}\{1/2((\overline{Y}_{22} - \overline{Y}_{21}) - (\overline{Y}_{11} - \overline{Y}_{12}))\} = \frac{\sigma^2}{4}\left(\frac{1}{n_1} + \frac{1}{n_2}\right)$. Therefore, with δ_0 denoting the hypothesized value of δ, the t-statistic is

$$t = \frac{\widehat{\delta} - \delta_0}{\frac{s}{2}\sqrt{\frac{1}{n_1} + \frac{1}{n_2}}}$$

which has $n_1 + n_2 - 2$ degrees of freedom.

Remember that the analysis is based on the assumption that $\widehat{\delta}$ is approximately normally distributed and $\lambda_A = \lambda_B = \lambda$. The second assumption cannot be tested with the design that is used. Note in particular that $\lambda_A - \lambda_B = 4\gamma$ (from the 2×2 table), so the effects cannot be separated. In general, analyses should not be based on assumptions that cannot be tested. In this instance, if λ_A and λ_B differ considerably, the t-test result will be impacted.

Could the design have been improved? In particular, could the problem of having to assume $\lambda_A = \lambda_B$ and not being able to test the assumption have been avoided? This can be accomplished by using four groups instead of two and using AB, BA, AA, and BB as the four sequences in the four groups, respectively. This design is due to

Balaam (1968). As shown by Weerahandi (2004, p. 278), expressions for $\widehat{\lambda}_A$ and $\widehat{\lambda}_B$ can be easily obtained. These estimators are functions of both $\widehat{\mu}_A$ and $\widehat{\mu}_B$, however. A simpler expression is obtainable if we look at $\widehat{\lambda}_A - \widehat{\lambda}_B$, which can be written as $\frac{1}{2}(\overline{Y}_{12} + \overline{Y}_{32} - \overline{Y}_{22} - \overline{Y}_{42})$, with $\overline{Y}_{ij}$ as defined previously except that this time there are four rows, corresponding to the four groups. Although $\widehat{\mu}_A$ and $\widehat{\mu}_B$ are unbiased, different weights are assigned to the cell means in the 4×2 table in the process of producing the unbiasedness.

It is possible in certain types of experiments, such as crop experiments, for carryover effects to extend over more than one or two periods. Patterson (1968) introduced serial designs for such applications.

Motivated by the use of crossover designs in the telecommunications industry, Lewis and Russell (1998) presented crossover designs when there are carryover effects from two factors. They gave methods for constructing crossover designs that are based on the Latin squares of Williams (1949) and of Russell (1991). The reader is also referred to Senn (2002) for additional reading on crossover designs and extensive discussion of carryover effects, to Raghavarao (1990) for the use of crossover designs in industry, and to Chapter 19 in Hinkelmann and Kempthorne (2005). Other recommended sources on crossover designs include Patterson (1951, 1973), Afsarinejad (1983), Fletcher (1987), Russell (1991), Ratkowsky, Evans, and Alldredge (1993), Wakeling and MacFie (1995), Vonesh and Chinchilli (1997), Goad and Johnson (2000), and Weerahandi (2004). In particular, Jones and Kenward (2003) gave a historical perspective as they traced the early history of crossover trials through crop-rotation experiments, feeding trials, and bioassay, in addition to giving a comprehensive catalog of designs and downloadable SAS programs. Dallal (2001) is also recommended for computer analysis of crossover designs and for other information. In particular, he points out that care must be exercised in reading the literature on crossover designs, noting for example that an error in Grizzle (1965) that was later corrected in Grizzle (1974) is apparently still misleading people, as noted by Grieve (1982) and as is evident from some later publications. Possible time effects have also been considered by various authors, and one proposed solution is a switchback design, which requires an extra period so as to allow for a switchback to the treatment used in the first time period. Accordingly, a simple switchback design would be of the form $A \rightarrow B \rightarrow A$ and $B \rightarrow A \rightarrow B$ (see, e.g., Oman and Seiden, 1988).

11.5 HOW MANY REPEATED MEASURES?

An obvious question to address is, "How many repeated measurements are to be made on each experimental unit?" This issue has been considered by various authors, including Vickers (2003), who addressed the issue for medical applications, such as patients being evaluated every three months after thoracic surgery. Of course power is increased as the number of repeated measurements is increased, but there is a point of diminishing returns.

11.6   FURTHER READING

The optimality and efficiency of crossover designs has been studied over a period of four decades, with the history traced by Hedayat and Yang (2004), who refer readers to Stufken (1996) for additional references. See also, Hedayat and Yang (2003, 2005) and Hedayat, Stufken, and Yang (2005) for optimal crossover designs. Practitioners who are not particularly interested in the optimality results of crossover designs may wish to read Stufken (1996), which is an expository article. Although the analysis of data from a crossover design is usually numerical, Pontius and Schantz (1994) applied the sliding-square plot idea of Rosenbaum (1989) for the graphical analysis of data from a two-period crossover design.

Missing data in small repeated measures designs can be problematic and can cause a considerable loss of power for detecting significant effects. Huang and Carriére (2006) developed a strategy for multiple imputations and compared analysis using multiple imputations with non-imputation incomplete data methods, such as that given by Carriére (1999). Unfortunately, they found that the non-imputation methods were superior to the use of multiple imputations.

Other papers that may be of interest include Hedayat and Afsarinejad (1975, 1978), Marshall and Jackson (1993), Senn and Lambrou (1998), Dean, Lewis, and Chang (1999), and Prescott (1999).

11.7   SOFTWARE

One problem that faces the user of crossover designs and other repeated measurement designs is that software that has this capability seems, at first glance, not to be as plentiful as software for more common designs, such as factorial designs. For example, Design-Expert 7.0 does not list this capability, nor does Release 14 of MINITAB. Both software packages can be used to analyze repeated measures data, however. (For MINITAB, see http://www.minitab.com/support/answers/answer.aspx?ID=413 for an explanation.)

The help file material in JMP indicates that the latter can be used to analyze repeated measures data, using either a univariate, mixed model approach or a multivariate approach.

The most commonly used statistical procedures for repeated measures data undoubtedly include SAS PROC GLM and SPSS GLM. SAS PROC MIXED is necessary for modeling the covariance structure, however, rather than simply assuming a certain structure. Clearly, the former will generally be the preferred approach. JMP can also be used to analyze repeated measures data. The construction of balanced carryover designs, including those of Williams, Patterson, and Afsarinejad, using R, the freeware version of S-Plus, is available using the CROSSDES package (see http://www. ibiblio.org/pub/languages/R/CRAN/doc/packages/crossdes.pdf). Cross Over 1.0, discussed in Jones and Kenward (2003), is another software package for crossover designs. Useful information on the analysis of crossover design data with various statistical software packages is given by Dallal (2001), who points out that such data can be analyzed by any software that has the capability for repeated measures analysis of variance.

It should be noted that these software procedures are for the analysis of repeated measurement data; the user of these designs must first select an appropriate design.

11.8 SUMMARY

Repeated measurement designs are used in various applications, especially experiments involving people. This is one topic that is generally discussed more extensively in research articles than in books, with Davis (2002) and Winer et al. (1991) being notable exceptions. With some experimental designs, such as screening designs, it isn't necessary to formulate a tentative model. With other designs, only a good idea of the types of effects that are likely to be significant is needed. Repeated measurement designs present some unique problems, however, as the possibility of carryover effects and possible or probable lack of independence should affect decisions regarding the design and the subsequent analysis.

An extensive, although somewhat outdated, review of repeated measurement designs was given by Afsarinejad (1990).

REFERENCES

Afsarinejad, K. (1983). Balanced repeated measurements designs. *Biometrika*, **70**, 199–204.

Afsarinejad, K. (1990). Repeated measurements designs — a review. *Communications in Statistics — Theory and Methods*, **19**(11), 3985–4028.

Bailey, R. A. and J. Kunert (2006). On optimal cross-over designs when carry-over effects are proportional to direct effects. *Biometrika*, **93**(3). (available at http://www.designtheory.org/library/preprints/rabjkpre.pdf)

Balaam, L. N. (1968). A two-period design with t^2 experimental units. *Biometrics*, **24**, 61–73.

Bate, S. T. and B. Jones (2006). The construction of nearly balanced and nearly strongly balanced uniform cross-over designs. *Journal of Statistical Planning and Inference*, **136**(9), 3248–3267.

Boon, P. C. and K. C. B. Roes (1999). Design and analysis issues for crossover designs in phase I clinical trials. *Journal of Biopharmaceutical Statistics*, **9**, 109–128.

Bose, M. (2002). Crossover designs: Analysis and optimality using the calculus for factorial arrangements. In *Design Workshop Lecture Notes*, International Statistical Institute, Kolkata, India, November 25–29, pp. 183–192. (available at http://www.isid.ac.in/~ashish/workshop/mausumiw3.pdf.)

Bose, M. and A. Dey (2003). Some small and efficient cross-over designs under an additive model. *Utilitas Mathematica*, **63**, 173–182.

Bose, M. and B. Mukherjee (2000). Cross-over designs in the presence of higher order carryovers. *Australian and New Zealand Journal of Statistics*, **42**, 235–244.

Bose, M. and B. Mukherjee (2003). Optimal crossover designs under a general model. *Statistics and Probability Letters*, **62**(4), 413–418.

Carriére, K. C. (1994). Crossover designs for clinical trials. *Statistics in Medicine*, **13**, 1063–1069.

Carriére, K. C. (1999). Methods of repeated measures data analysis with missing values. *Journal of Statistical Planning and Inference*, **77**, 221–236.

Carriére, K. C. and G. C. Reinsel (1993). Optimal two-period repeated measures designs with two or more treatments. *Biometrika*, **80**, 924–929.

Cheng, C.-S. and C.-F. Wu (1980). Balanced repeated measures designs. *Annals of Statistics*, **8**, 1272–1283.

Claydon, N. C., M. Addy, R. Newcombe, and J. Moran (2005). The prevention of plaque re-growth by toothpastes and solutions containing block copolymers with and without polypeptide. *Journal of Clinical Periodontology*, **32**(6), 545–548.

Cochran, W. G. and G. M. Cox (1957). *Experimental Designs*, 2nd ed. New York: Wiley.

Crowder, M. (2001). On repeated measures analysis with misspecified covariance structure. *Journal of the Royal Statistical Society, Series B*, **63**, 55–62.

Crowder, M. J. and D. J. Hand (1990). *Analysis of Repeated Measures*. London: Chapman and Hall.

Dallal, G. E. (2001). The computer-aided analysis of crossover studies. (available at http://www.tufts.edu/~gdallal/crossovr.htm)

Davidian, M. and D. M. Giltinan (1995). *Nonlinear Models for Repeated Measurement Data*. Boca Raton, FL: CRC Press.

Davis, C. S. (2002). *Statistical Methods for the Analysis of Repeated Measurements*. New York: Springer.

Dean, A. M., S. M. Lewis, and J. Y. Chang (1999). Nested changeover designs. *Journal of Statistical Planning and Inference*, **77**, 337–351.

Donev, A. N. (1998). Crossover designs with correlated observations. *Journal of Biopharmaceutical Statistics*, **8**, 249–262.

Everitt, B. S. (1995). The analysis of repeated measures: A practical review with examples. *The Statistician*, **44**, 113–135.

Federer, W. T. and R. P. Kerschner (1998). Comparison of classes of changeover designs. *Journal of Combinatorics, Information, and System Sciences*, **23**, 379–391.

Finney, D. J. H. (1965). Crossover designs in bioassay. *Proceedings of the Royal Society*, **145B**, 42–61.

Fleiss, J. L. (1986). *The Design and Analysis of Clinical Experiments*. New York: Wiley.

Fletcher, D. J. (1987). A new class of change-over designs for factorial experiments. *Biometrika*, **74**, 649–654.

Garcia, R., M. Benet, C. Arnau, and E. Cobo (2004). Efficiency of the cross-over design: An empirical investigation. *Statistics in Medicine*, **23**(24), 3773–3780.

Giesbrecht, F. G. and M. L. Gumpertz (2004). *Planning, Construction, and Statistical Analysis of Comparative Experiments*. Hoboken, NJ: Wiley.

Girden, E. R. (1992). *ANOVA: Repeated Measures*. London, UK: SAGE Publications.

Goad, C. L. and D. E. Johnson (2000). Crossover tests: A comparison of ANOVA tests and alternative analyses. *Journal of Agricultural, Biological, and Environmental Statistics*, **5**, 69–87.

Grieve, A. P. (1982). Correspondence: The two-period changeover design in clinical trials. *Biometrics*, **38**, 517.

Grizzle, J. E. (1965). The two-period change-over design and its use in clinical trials. *Biometrics*, **21**, 467–480.

Grizzle, J. E. (1974). Correction. *Biometrics*, **30**, 727.

Hedayat, A. S. and K. Afsarinejad (1975). Repeated measurements designs, I. In *A Survey of Statistical Design and Linear Models*, pp. 229–242. (J. N. Srivastava, ed.). Amsterdam: North-Holland.

Hedayat, A. S. and K. Afsarinejad (1978). Repeated measurements designs, II, *The Annals of Statistics*, **6**, 619–628.

Hedayat, A. S. and K. Afsarinejad (2002). Repeated measurement designs for a model with self and mixed carryover effects. *Journal of Statistical Planning and Inference*, **106**, 449–459.

Hedayat, A. S. and J. S. Stufken (2003). Optimal and efficient crossover designs under different assumptions about the carryover effects. *Journal of Biopharmaceutical Statistics*, **13**, 519–528.

Hedayat, A. S. and M. Yang (2003). Universal optimality of balanced crossover designs. *The Annals of Statistics*, **31**, 978–983. (see http://statistics.unl.edu/faculty/yang/crossover.pdf)

Hedayat, A. S. and M. Yang (2004). Universal optimality for selected crossover designs. *Journal of the American Statistical Association*, **99**, 461–466.

Hedayat, A. S. and M. Yang (2006). Optimal and efficient crossover designs for comparing test treatments with a control treatment. *The Annals of Statistics*, **33**, 915–943.

Hedayat, A. S., J. Stufken, and M. Yang (2006). Optimal and efficient crossover designs when subject effects are random. *Journal of the American Statistical Association*, **101**, 1031–1038.

Hinkelmann, K. and O. Kempthorne (2005). *Design and Analysis of Experiments, Volume 2: Advanced Experimental Design*. Hoboken, NJ: Wiley.

Huang, R. and Carriére, K. C. (2006). Comparison of methods for incomplete repeated measures data analysis in small samples. *Journal of Statistical Planning and Inference*, **136**(1), 235–247.

John, J. A., K. G. Russell, and D. Whitaker (2004). CrossOver: An algorithm for the construction of efficient crossover designs. *Statistics in Medicine*, **23**(17), 2645–2658.

Jones, B. and M. G. Kenward (2003). *Design and Analysis of Cross-over Trials*, 2nd ed. New York: Chapman and Hall.

Kempton, R. A., S. J. Ferris, and O. David (2001). Optimal change-over designs when carryover effects are proportional to direct effects of treatments. *Biometrika*, **88**, 391–399.

Keppel, G. (1973). *Design and Analysis: A Researcher's Handbook*. Englewood Cliffs, NJ: Prentice-Hall.

Keselman, H. J., J. Algina, and R. K. Kowalchuk (2001). The analysis of repeated measures designs: A review. *British Journal of Mathematical and Statistical Psychology*, **54**, 1–20.

Kowalchuk, R. K., H. J. Keselman, J. Algina, and R. D. Wolfinger (2004). The analysis of repeated measurements with mixed-model adjusted F tests. *Educational and Psychological Measurement*, **64**(2), 224–242.

Kuehl, R. O. (2000). *Design of Experiments: Statistical Principles of Research Design and Analysis*, 2nd ed. New York: Duxbury.

Kunert, J. (1987). An example of universal optimality in a full-rank model. *Metrika*, **34**, 217–223.

Kunert, J. (1998). Sensory experiments as crossover studies. *Food Quality and Preference*, **9**, 243–253.

Kunert, J. and J. Stufken (2002). Optimal crossover designs in a model with self and mixed carryover effects. *Journal of the American Statistical Association*, **97**, 898–906.

Kunert, J. and J. Stufken (2005). Optimal crossover designs for two treatments in the presence of mixed and self carryover effects. Technical Report No. 2005-11, Department of Statistics, University of Georgia.

Kunert, J. and O. Sailer (2006). On nearly balanced trials for sensory trials. *Food Quality and Preference*, **17**(3–4), 219–227.

Kushner, H. B. (1998). Optimal and efficient repeated-measurements designs for uncorrelated observations. *Journal of the American Statistical Association*, **93**, 1176–1187.

Lawson, J. S. (1988). A case study of effective use of statistical experimental design in a smoke stack industry. *Journal of Quality Technology*, **20**(1), 51–62.

Lewis, S. M. and K. G. Russell (1998). Crossover designs in the presence of carryover effects from two factors. *Journal of the Royal Statistical Society, Series C (Applied Statistics)*, **47**, 379–391.

Lindsey, J. K. (1999). *Models for Repeated Measurements*, Vol. 19. Oxford, UK: Oxford University Press.

Loughin, T. M., D. E. Johnson, S. E. Ives, and T. G. Nagaraja (2002). Methods for selecting crossover designs with applications to an experiment with two factors in a split plot. *Journal of Agricultural, Biological, and Environmental Statistics*, **7**, 143–156.

Louis, T. A., P. W. Lavori, J. C. Bailar, III, and M. Polansky (1984). Crossover and self-controlled designs in clinical research. *New England Journal of Medicine*, **310**(1), 24–31.

Low, J. L., S. M. Lewis, and P. Prescott (1999). Assessing robustness of crossover designs to subjects dropping out. *Statistics and Computing*, **9**, 219–227.

Low, J. L., S. M. Lewis, and P. Prescott (2002). An application of Pólya Theory to crossover designs with dropout. *Utilitas Mathematica*, **63**, 129–142.

Marshall, R. J. and R. T. Jackson (1993). Analysis of case-crossover design. *Statistics in Medicine*, **12**, 2333–2341.

Matthews, J. N. S. (1987). Optimal crossover designs for the comparison of two treatments in the presence of carryover effects and autocorrelated errors. *Biometrika*, **74**, 311–320.

Matthews, J. N. S. (1994). Modeling and optimality in the design of crossover studies for medical applications. *Journal of Statistical Planning and Inference*, **42**, 89–108.

Max, M. B. (2003). The design of clinical trials for treatment of pain. In *Interactive Textbook on Clinical Symptom Research*, Chap. 1 (M. B. Max and J. Lynn, eds.) (Available at http://symptomresearch.nih.gov/tablecontents.htm.)

Mielke, P. W., Jr. (1974). Square rank test appropriate to weather modification cross-over design. *Technometrics*, **16**, 13–16.

Milliken, G. A. (2004). Mixed models and repeated measures: Some illustrative industrial examples. In *Handbook of Statistics, Vol. 22: Statistics in Industry*, Chap. 5 (R. Khattree and C. R. Rao, eds.). Amsterdam, The Netherlands: Elsevier Science B.V.

Montgomery, D. C. (1996). *Design and Analysis of Experiments*, 4th ed. New York: Wiley.

Newcombe, R. G. (1995). Residual effect of chlorhexidine gluconate in 4-day plaque regrowth crossover trials, and its implications for study design. *Journal of Periodontal Research*, **30**, 319–324.

Newcombe, R. G. (1996). Sequentially balanced three-squares cross-over designs. *Statistics in Medicine*, **15**(20), 2143–2147.

Oman, S. D. and E. Seiden (1988). Switch-back designs. *Biometrika*, **75**(1), 81–89.

Patterson, H. D. (1951). Change-over trials. *Journal of the Royal Statistical Society, Series B*, **13**, 256–271.

Patterson, H. D. (1952). The construction of balanced designs for experiments involving sequences of treatments. *Biometrika*, **39**, 32–48.

Patterson, H. D. (1968). Serial factorial design. *Biometrika*, **55**(1), 67–81.

Patterson, H. D. (1973). Quenouille's changeover designs. *Biometrika*, **60**, 33–45.

Périnel, E. and J. Pagès (2004). Optimal nested crossover designs in sensory analysis. *Food Quality and Preference*, **15**, 439–446.

Pignatiello, J. (1984). SPC questions, queries, and quandaries: Two-stage nested designs. *ASQ Statistics Division Newsletter*, **6**(1).

Pontius, J. S. and R. M. Schantz (1994). Graphical analyses of a two-period crossover design. *The American Statistician*, **48**, 249–253.

Prescott, P. (1999). Construction of uniform-balanced cross-over designs for any odd number of treatments. *Statistics in Medicine*, **18**, 265–272.

Putt, M. and V. N. Chinchilli (1999). A mixed effects model for the analysis of repeated measures cross-over studies. *Statistics in Medicine*, **18**, 3037–3058.

Raghavarao, D. (1990). Crossover designs in industry. In *Statistics, Textbooks and Monographs*, Vol. 109. New York: Marcel Dekker, Chap. 18, pp. 517–530.

Ratkowsky, D. A., M. A. Evans, and J. R. Alldredge (1993). *Cross-over Experiments: Design, Analysis and Applications*. New York: Marcel Dekker.

Rosenbaum, P. R. (1989). Exploratory plots for paired data. *The American Statistician*, **43**, 108–109.

Russell, K. G. (1991). The construction of good change-over designs when there are fewer subjects than treatments. *Biometrika*, **78**, 305–313.

Russell, K. G. and A. M. Dean (1998). Factorial cross-over designs with few subjects. *Journal of Combinatorics, Information and System Sciences*, **23**, 209–235.

Sen, M. and R. Mukerjee (1987). Optimal repeated measures designs under interaction. *Journal of Statistical Planning and Inference*, **17**, 81–91.

Senn, S. (1996). The AB/BA crossover: How to perform the two-stage analysis if you can't be persuaded that you shouldn't. In *Liber Amicorum Roel van Strik* (B. Hansen and M. De Ridder, eds.) Rotterdam: Erasmus University, pp. 93–100. (available online at http://www.senns.demon.co.uk/ROEL.pdf)

Senn, S. (2002). *Cross-over Trials in Clinical Research*, 2nd ed. New York: Wiley.

Senn, S. (2003). Within-patient studies: Cross-over trials and n-of-1 studies. In *Interactive Textbook on Clinical Symptom Research*, Chap. 6 (M. B. Max and J. Lynn, eds.). (Available at http://symptomresearch.nih.gov/tablecontents.htm.)

Senn, S and D. Lambrou (1998). Robust and realistic approaches to carry-over. *Statistics in Medicine*, **17**, 2849–2864.

Senn, S., L. Stevens, and N. Chaturvedi (2000). Repeated measures in clinical trials: Simple strategies for analysis using summary measures. *Statistics in Medicine*, **19**, 861–877.

Street, D. J., J. A. Eccleston, W. H. Wilson (1990). Tables of small optimal repeated measurements designs. *Australian Journal of Statistics*, **32**, 345–359.

Stufken, J. (1996). Optimal crossover designs. In *Handbook of Statistics*, Vol. 13: *Design and Analysis of Experiments*, pp. 63–90 (S. Ghosh and C. R. Rao, eds.). Amsterdam: North Holland.

Thaler, J. S. (1999). Induced resistance in agricultural crops: Effects of jasmonic acid on herbivory and yield in tomato plants. *Environmental Entomology*, **28**(1), 30–31.

Tudor, G. E., G. G. Koch, and D. Catellier (2000). Statistical methods for crossover designs in bioenvironmental and public health studies. In *Handbook of Statistics*, Vol. 18: *Bioenvironmental and Public Health Statistics*, pp. 571–614 (P. K. Sen and C. R. Rao, eds.) Amsterdam: North Holland.

Vickers, A. J. (2003). How many repeated measures in repeated measures designs? Statistical issues for comparative trials. *BMC Medical Research Methodology*, **3**, 1–22.

Vonesh, E. F. and V. M. Chinchilli (1997). *Linear and Nonlinear Models for the Analysis of Repeated Measurements*. New York: Marcel Dekker.

Wakeling, I. N. and H. J. H. MacFie (1995). Designing consumer trials balanced for first and higher orders of carryover effect when only a subset of k samples from t may be tested. *Food Quality and Preference*, **6**, 299–308.

Weerahandi, S. (2004). *Generalized Inference in Repeated Measures: Exact Methods in MANOVA and Mixed Models*. Hoboken, NJ: Wiley.

Williams, E. J. (1949). Experimental designs balanced for the estimation of residual effects of treatments. *Australian Journal of Scientific Research*, **2**, 149–168.

Winer, B. J., D. R. Brown, and K. M. Michels (1991). *Statistical Principles in Experimental Design*, 3rd ed. New York: McGraw-Hill.

EXERCISES

11.1 There were two 4×4 Latin square designs given in Section 11.4 that would each be suitable for use when carryover effects from use of a crossover design are expected. Is the following Latin square design also suitable for that purpose? Why, or why not?

$$
\begin{array}{cccc}
A & B & C & D \\
B & C & D & A \\
C & D & A & B \\
D & A & B & C \\
\end{array}
$$

11.2 (Harder problem) Show algebraically that the analysis of repeated measures data with two levels of one factor has the same corresponding relationship as one-factor ANOVA and an independent sample t-test for independent data. That is, the square of the t-statistic in a paired t-test is equal to the F-statistic for the treatments effect, as was illustrated in Section 10.1.

11.3   VOICE.MTW is one of the sample data files that comes with the MINITAB software. A professor in a theater department wanted to determine if a voice training class improved a performer's voice quality. There were 10 students who took the class and six judges rated each student on a scale of 1–6, with 6 being the best score. The judges scores were as follows:

	Student #								
1		2		3		4		5	
Before	After	Before	After	Before	After	Before	After	Before	After
5	3	2	2	5	3	4	3	5	4
4	5	3	3	5	4	5	4	6	5
4	5	3	1	6	3	5	5	6	6
5	6	3	2	4	5	4	3	6	5
4	5	2	2	2	3	3	3	4	3
5	5	3	3	5	5	4	4	5	5

6		7		8		9		10	
Before	After	Before	After	Before	After	Before	After	Before	After
3	4	2	2	2	2	5	2	3	3
3	3	2	2	4	4	2	3	3	3
4	3	1	1	5	2	2	2	3	2
4	3	4	3	5	4	5	3	4	4
3	3	2	2	4	3	3	2	3	4
4	4	2	3	5	4	4	4	4	4

Analyze the data and determine whether or not the voice training appears to have been worthwhile.

11.4   Repeated measures can occur with virtually any type of design, so the term "repeated measures design" should perhaps be viewed as a misnomer. Thaler (1999, references) used repeated measures in agricultural experimentation. This paper is available online at www.botany.utoronto.ca/ResearchLabs/thalerLab/Environmental%20Entomology.pdf. Read the article and determine on what the repeated measures occur. You will notice mention of other experimentation without repeated measures. Comment on the author's experimental design approach. In particular, would you suggest that anything different be done in future experimentation of this type?

11.5   (Harder problem) Pignatiello (1984, see references) described the following scenario that was presented to him by an engineer working for a military contractor.

Five parts were randomly sampled from each of two lots and the value of a quality characteristic was recorded. This process was repeated so that each of the 10 parts was measured twice, the rationale for doing so being that individual readings vary due to measurement error. Assume that the two lots are the only lots that are of interest. The data are given below:

		Lot 1						Lot 2		
Parts	1	2	3	4	5	1	2	3	4	5
Measurements	10.2	8.4	9.2	10.5	11.3	10.9	14.1	11.8	12.8	13.3
	9.1	10.4	9.1	10.1	10.4	12.6	11.5	11.4	10.7	12.5

These data were analyzed as having come from a nested design with the error sum of squares, with 10 df, computed as one would compute an error sum of squares in the presence of replication. Would you have treated the error term differently? Specifically, is there a repeated measures aspect to the data, and if so, does the repeated measures aspect dictate that the data be analyzed differently from the analysis of replicated design points without repeated measurements? Explain. Analyze the data and draw appropriate conclusions.

11.6 Assume that an experiment is performed to determine if there are differences in two types of athletic shoes, A and B, relative to running speeds, with the distance being 50 yards. Twelve individuals are recruited for the study and the times are as follows:

Subject	Times	
1	B (6.3)	A (6.1)
2	B (5.9)	A (6.0)
3	A (6.0)	B (5.8)
4	B (6.5)	A (6.5)
5	A (5.7)	B (5.6)
6	A (5.9)	B (6.0)
7	A (6.1)	B (6.2)
8	B (6.0)	B (5.9)
9	A (6.3)	B (6.1)
10	B (6.2)	A (6.1)
11	A (6.4)	B (6.2)
12	B (6.1)	A (6.0)

(a) Analyze the data under the assumption that there are no carryover effects.
(b) Considering the nature of the experiment, do you believe that carryover effects should exist? Why, or why not? If you believe that they should exist, what would you recommend?

11.7 Consider Exercise 3.31 in Chapter 3. What complications do repeated measures present relative to the study in which a 3×3 Latin square design was used (which were apparently not addressed in that study)?

CHAPTER 12

Multiple Responses

Only a single response variable was assumed in the previous chapters. In practice, however, there are usually multiple responses, such as multiple product measurements that are of interest and that ideally should be maintained at a target value while, say, maximizing product yield. There are many different types of scenarios that occur in practice. For example, if there are three response variables, the objective might be to maximize two of them and minimize the other one. In this chapter we will look at several examples with various objectives.

When multiple responses are analyzed, a model is first fit for each response. Then some give-and-take must occur in the optimization since a combination of factor levels that would optimize one response almost certainly would not optimize the other responses. Assume for the sake of illustration that there are two response variables, which are of equal importance and each is to be maximized, and each variable is a function of the same two factors. One way to picture the necessary "give-and-take" would be to construct overlaid contour plots. Almost certainly, the indicated maximum on each contour plot would not point to the same combination of levels of the two factors, so some compromise would be necessary. The user could make this compromise based on the visual information. Most practical problems will have either more than two response variables or more than two factors, however, so such an approach has limited usefulness. Nevertheless, this approach is illustrated in Section 12.1.

Certain simplifying but unrealistic assumptions were made in some of the early research on multiple response optimization, such as assuming that all response variables are functions of the same independent variables and, even worse, that the model for each response variable has the same general form (see, e.g., Khuri and Conlon, 1981). While such assumptions simplify the analyses considerably, simplified analyses based on unrealistic assumptions are of little value.

Even if such assumptions were tenable, there is still the problem that response variables are often correlated. When this is the case, it won't be possible to try to change factor levels so as to move in the direction of a desirable value for one response

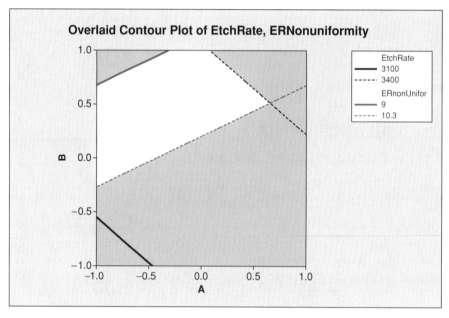

Figure 12.1 Overlaid contour plot for two responses.

variable without moving the expected value of another response variable that may already be very close to its most desirable value.

In the next section we look at a numerical optimization approach for finding the optimal solution, although there is no guarantee that the global optimum has been achieved as there might be multiple local optima. It is also desirable to look at graphs to see how sensitive the optimum solution in terms of the fitted values is to changes in the predictors.

Before multiple response optimization was implemented in statistical software, it was common to try to picture optimum and near-optimum solutions by using overlaid contour plots. Figure 10.12 is an example of a contour plot for a single response variable. If there were two responses, the contour plot for the second response could be overlaid on the contour plot shown in Figure 10.12. Such an approach will work fine when there are two response variables and two predictors, but dimensionality becomes a problem when there are more than two predictors.

12.1 OVERLAYING CONTOUR PLOTS

Figure 12.1 is an overlaid contour plot with two of the response variables in a dataset that is analyzed in Section 12.5. Here, for the sake of illustration, we assume that there are only two responses and two factors. A model for each response must be fit, and we fit a model with only main effects for each response.

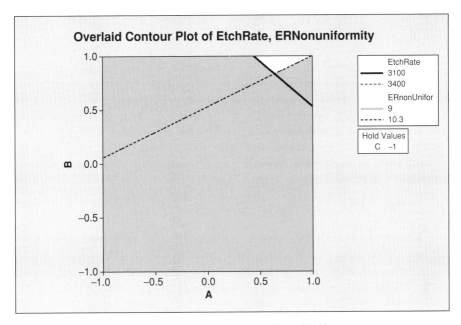

Figure 12.2 Overlaid contour plot with factor C held constant.

The plot was constructed by specifying an acceptable range for Etch Rate, which is to be maximized, of 3100 to 3400, and an acceptable range of Etch Rate Nonuniformity, which of course would be minimized, of 9 to 10.3. The region in white is the feasible region in terms of the two factors, A and B. The experimenter would be quite pleased if the Etch Rate were 3400 or above, but below 3100 would be totally unacceptable, with the range between them being "just okay." Similarly, Etch Rate Nonuniformity of 9 or below would be most desirable and above 10.3 totally unacceptable. Notice that the combination of 9 and 3400 cannot be achieved, at least not within the range of factor levels used in the experiment. It might be achieved with a higher level of B than was used in the experiment, but such a level might not be feasible. If it were not feasible, then some compromise, as mentioned at the beginning of the chapter, would be necessary.

Figure 12.2 is a contour plot of the data on the same two response variables, but this time the model for each response variable consists of the main effects of all three factors that were in the dataset. Since there are three factors, one of them must be set at a fixed level, and in Figure 12.2 factor C was set at the low level.

Notice that the feasible region has shrunk considerably and there is no contour for the most desirable value of Etch Rate Nonuniformity. If factor C had been fixed at the high level, a different, and undesirable, contour plot would have resulted, as the reader is asked to show in Exercise 12.8. Thus, contour plots start to lose some value when there are more than two factors.

12.2   SEEKING MULTIPLE RESPONSE OPTIMIZATION WITH DESIRABILITY FUNCTIONS

There are various methods for achieving optimization with multiple responses, with the best known method based on *desirability functions*, as explained in, for example, Derringer and Suich (1980), who modified the desirability function approach given by Harrington (1965), which is discussed in Section 12.6 (see also Derringer, 1994). The general idea is to compute a desirability value, d_i, for each response that is a measure of how close the fitted value with the optimal settings of the factors is to the desired value, and use these values to form the *composite desirability* for k response variables given by

$$D = (d_1 * d_2 * \cdots * d_k)^{1/k} \tag{12.1}$$

Derringer and Suich (1980) sought to maximize D by using a pattern research method similar to the one given by Hooke and Jeeves (1961). Unfortunately, that method is somewhat unreliable. Since that is a "hill-climbing" technique, it is susceptible to producing local solutions rather than global solutions. For this reason, it has been recommended that the procedure be used twice, using the supposed optimum solution as the starting point when the procedure is used the second time. A better approach, in general, when there is a fear of being derailed by local optima is to use many random starting points. Of course this option is available to users only when the algorithm is available for direct use, not when it is part of a statistical software package that controls how the algorithm is used. Algorithms are discussed further in Section 12.7; see http://www.mit.jyu.fi/palvelut/sovellusprojektit/ovi/methods.html#hookejeeves for a comparison of different types of algorithms.

For example, assume that there is only one response variable and the objective is to minimize impurity, with any value of at most 5 considered to be equally desirable, with 5.5 declared to be the largest acceptable value. A fitted value of at most 5 would then have a desirability of 1 because the goal was met, whereas a fitted value of 5.25 would have a desirability, d, of

$$\begin{aligned} d &= \frac{\text{Upper} - \widehat{y}}{\text{Upper} - \text{Lower}} \\ &= \frac{5.5 - 5.25}{5.50 - 5.00} \\ &= 0.50 \end{aligned} \tag{12.2}$$

Clearly the desirability value must be between 0 and 1, inclusive, and this is also true for the composite desirability, which will be 1 only if all the individual desirability values are 1, and will be 0 if at least one individual desirability is 0.

The desirability for a response to be maximized would be similarly computed. For example, if a fitted value of at least 15 is desired with 13.5 being the smallest acceptable value, a fitted value of 14.3 would have a d-value of $d = (\widehat{y} - \text{lower})/(\text{upper} - \text{lower}) = (14.3 - 13.5)/(15 - 13.5) = 0.533$. (Derringer and Suich (1980) presented only the maximization case, pointing out that the minimization of $\widehat{Y}$ is equivalent to the maximization of $-\widehat{Y}$. Loosely speaking, if we replace $\widehat{y}$ by $-\widehat{y}$ in the expression for d in the maximization case, then of course "lower" would be replaced by "upper" in the numerator of the expression because of the change of sign.)

When there is a target value for the response (termed the "most desirable" value by Derringer and Suich, 1980), the d-value is computed as

$$d = \frac{\widehat{y} - \text{Lower}}{\text{Target} - \text{Lower}} \qquad \text{Lower} \le \widehat{y} \le \text{Target} \qquad (12.3)$$

$$= \frac{\widehat{y} - \text{Upper}}{\text{Target} - \text{Upper}} \qquad \text{Target} \le \widehat{y} \le \text{Upper}$$

Thus, in each case $d = 1$ if $\widehat{y} = \text{Target}$.

Although the Derringer and Suich (1980) desirability function approach is undoubtedly the best known approach and is almost certainly the most frequently used approach since it is implemented in the leading statistical software packages, it is not the only transformation approach that has been proposed. Various other methods, including the global criterion method, are described and compared by Tabucanon (1988), which unfortunately is out of print. The global criterion method, which unfortunately does not work for target values, is described in detail in Kros and Mastrangelo (2004).

12.2.1 Weight and Importance

The presentation of desirability to this point has focused on a single response variable. Assume that there are four response variables. The importance of hitting a target value or achieving a maximum or minimum value may vary considerably over the different response variables. If so, different values may be used for Importance, and different values may also be used for Weight.

We will consider Weight first, which determines the shape of the desirability function for each response. Assume that response i is to be maximized and that a value of $\widehat{Y}_i$ more than slightly less than an acceptable maximum value quickly becomes almost unacceptable with movement away from the maximum. More specifically, the desirability is not linear for values less than the acceptable maximum. For such a scenario, d_i^r should be used, with $r > 1$. Since d_i is of course less than 1.0 when the maximum is not achieved, a value of r much greater than 1 could drive d_i^r close to zero, depending on the value of d_i. Similarly, if a response is to be minimized and thus a $\widehat{Y}_i$ value very close to the acceptable minimum is essential, r much greater than 1 would drive d_i^r close to zero. For a target value, two constants could be used for the two equations in Eqs. (12.3), which might reflect that missing the target on the high side is worse than missing it on the low side.

The set of desirabilities with each raised to the appropriate power would then be used in place of the d_i in Eq. (12.1).

The user who does not specify weights for the individual desirability functions is implicitly assuming that each function is a linearly increasing function within the range of specified lower and upper response values if the response is to be maximized, and a linearly decreasing function between the particular specified values if the response is to be minimized. If there is a target value, then linearity is assumed on each side of the target value. The shape of each desirability function is automatically shown when Design-Expert is used, for example.

Whereas Weight is used with each individual response, Importance should be viewed relative to all the responses. For example, if there are four responses ($R1$, $R2$, $R3$, and $R4$), with $R1$ and $R2$ to be maximized and $R3$ and $R4$ to be minimized, the maximization of $R1$ may be more important than the maximization of $R2$, and the minimization of $R3$ may be more important than the minimization of $R4$. It would then be appropriate to assign a greater "Importance" value to $R1$ than $R2$, and $R3$ should be assigned a greater importance value than $R4$. In MINITAB, the Importance values can range from 0.1 to 10, with 1.0 being the default value. In Design-Expert, the possible Importance values range from "$+$" to "$+++++$," with "$+++$" being the default.

12.3   DUAL RESPONSE OPTIMIZATION

One of the most visible applications of multiple response optimization during the past 20 years has been *dual response optimization*. The word "dual" in this context does not mean that there are two responses; rather there are two functions of the same response. The first function is that presented in Section 12.1: to maximize or minimize the response, or hit a target value. The second function involves the variance of that response. Specifically, the three dual problems given by Taguchi (see, e.g., Taguchi, 1986), which are approached by modeling the mean and variance separately, were (1) minimize the variance while hitting a target value for the response, (2) maximize the response while keeping the variance constant, and (3) minimize the response while keeping the variance constant.

Significant work in this area includes Vining and Myers (1990), Lin and Tu (1995), who pointed out some deficiencies in the method given by Vining and Myers (1990); Copeland and Nelson (1996), who proposed solving dual response problems with the use of direct function minimization; Tang and Xu (2002), who proposed a goal programming approach; Koksoy and Doganaksoy (2003), who let the mean and variance be on equal footing and used standard nonlinear multiobjective programming techniques; del Castillo and Montgomery (1993), who used a nonlinear programming approach; and Ding, Lin, and Wei (2004), who used a weighted mean squared error approach.

12.4   DESIGNS USED WITH MULTIPLE RESPONSES

It should be noted that multiple responses cause complexities primarily in analysis rather than in design. Multiple responses do force some design considerations, how-ever. In particular, if we relax the assumption that the same model can be used for each response (and later this will be illustrated), we then have to think about the factors that might be related to *at least one* of the response variables. Of course in the extreme case where each response is a function of different variables, multiple response optimization would then reduce to a set of single response optimizations. Such a scenario would undoubtedly be extremely rare, however.

Essentially any design could be used with multiple response optimization, just as any design could be used when there is a single response. The literature contains some articles on specific designs used for which there are multiple responses, such as the analysis of split-plot designs with multiple responses, as discussed by Ellekjaer, Fuller, and Ladstein (1997–1998). Chen, Hedayat, and Suen (1998) presented optimal designs for experiments with multiple responses.

It is important that enough design points be used so that parameter estimates are obtained with good precision. If not, the variance of a predicted response could be large, with the consequence that predicted response values may be obtained with somewhat low precision, which would mean that an optimum solution may also have low precision. We should keep in mind that optimal solutions are functions of data and specifically functions of parameter estimates, rather than obtained from theoretical models. Therefore, "optimal solutions" have a variance, just as do the statistics that produce them.

This suggests that a parsimonious model be fit for each response. Another consideration, however, is what to do when an analysis suggests a nonhierarchical model. Those who advocate hierarchical models would state that nonsignificant effects be included to make the model hierarchical. A nonhierarchical model may have been caused by large interactions. The principle of parsimony should be weighed against the consequences in multiple response optimization if a "full model" of some sort is not used.

12.5 APPLICATIONS

Example 12.1

We consider a dataset given and analyzed extensively by Czitrom, Sniegowski, and Haugh (1998) that was analyzed briefly in Ryan (2000), but which will be analyzed in more detail here, using an approach that is different from that used by Czitrom et al. (1998).

An experiment was performed involving integrated circuits to determine the effects of three manufacturing factors (bulk gas flow (A), CF_4 flow (B), and power (C)) on three response variables: selectivity, etch rate, and etch rate nonuniformity. The objective was to determine factor settings that would maximize selectivity, which is the ratio of the rate at which oxide is etched to the rate at which polysilicon is etched, maximize etch rate (which would also maximize manufacturing throughput), and of course minimize etch rate nonuniformity.

The design that was used was a 2^3 with two centerpoints. We will analyze these data with an eye especially toward determining whether or not the same model form seems appropriate for each of the three response variables. In particular, we will compare R^2 values and also assess the absolute magnitude of those values.

The data that are given in Table 12.1 are in the general (uncoded) form in which the data were given in Czitrom et al. (1998).

Of course analyses cannot be performed with raw-form data, however, as was illustrated in Section 4.1.

TABLE 12.1 Data from a Silicon Wafer Experiment

Run	Bulk Gas Flow (sccm)	CF$_4$ Flow (sccm)	Power (watts)	Selectivity	Etch Rate (Å/min)	Etch Rate Nonuniformity
1	60	5	550	10.93	2710	11.7
2	180	5	550	19.61	2903	13.0
3	60	15	550	7.17	3021	9.0
4	180	15	550	12.46	3029	10.3
5	60	5	700	10.19	3233	10.8
6	180	5	700	17.5	3679	12.2
7	60	15	700	6.94	3638	8.1
8	180	15	700	11.77	3814	9.3
9	120	10	625	11.61	3378	10.3
10	120	10	625	11.17	3295	11.1

For the response variable Selectivity, the main effects of the first two factors are the only significant effects (at $\alpha = .05$) when a model with all estimable effects is fit. The R^2 value is .9445. This alone is not the best way to select a model, however. Subsequent analysis shows that the AB interaction has a p-value of .037 when used in the model with A and B, so we might use the model with A, B, and AB as the three terms. This gives an R^2 value of .9747. The AB effect estimate is -1.467. Since this is not large relative to the A and B main effect estimates, which are 6.527 and -4.927, respectively, those main effect estimates are not degraded. Since the interaction is somewhat borderline significant and the p-value is well above .01, we will not use the interaction term in the model, also for reasons of simplicity.

For the Etch Rate response variable, the main effects of all three factors are the only significant effects and the R^2 value is .9451.

For the Etch Rate Nonuniformity response variable, the main effects of the three factors are again the significant effects and $R^2 = .9819$.

The fact that the last two response variables would be fit with the same model is not surprising since the two variables are closely related.

It is obvious that all three models fit the data very well, but if we include the main effect of the third factor in the model for the first response so as to have the same model for all three factors, we are placing a term in the model for the first response variable that has a p-value of .236. This is not all that bad, however, because it means that the optimum value of that variable will be relatively insensitive to the value of the third factor in the process of using multiresponse optimization to determine the optimum levels of the three factors for all three response variables, as factors that are not significant have comparatively small coefficients. (Here the coefficient of the third factor in the model for the first response is -1.32, compared to 9.13 and -6.95 for the coefficients of the first and second factors, respectively.) We will see that the (estimated) optimum level of the third factor differs very little over the analyses that follow.

The original settings of the factors were, in uncoded units, $A = 90$, $B = 5$, and $C = 625$, which correspond to -0.5, -1, and 0, respectively, in coded units. These would produce fitted values of Selectivity = 12.43, Etch Rate = 3068, and Etch Rate Nonuniformity = 11.6 from the model that Czitrom et al. (1998) used for each response. Those models are different for Selectivity and Etch Rate from the models selected in this example, however. Specifically, their model for Etch Rate included the AB and AC interactions in addition to the three main effect terms, and their model for Selectivity included the three main effect terms plus the AB interaction.

This led, through their use of contour plots, to the selection of $A = 180$, $B = 15$, and $C = 550$ in the original units, which correspond to 1, 1, and -1, respectively, in the coded units. These settings produce fitted values of Selectivity = 12.45, Etch Rate = 3048 (less than the value for the original settings), and Etch Rate Nonuniformity = 10.3. Note that their optimum solution was one of the 10 design points in the experiment.

Ryan (2000) did not use contour plots but rather used statistical software and arrived at factor settings of $A = 169.07$, $B = 14.09$, and $C = 561.04$. This solution was based on the assumption that maximizing the first two response variables and minimizing the third variable were of equal importance. The fitted values using these settings are Selectivity = 12.48, Etch Rate = 3091, and Etch Rate Nonuniformity = 10.37. This combination of fitted values was superior to that given by Czitrom et al. (1998), so the factor settings would be preferable.

The "optimal" solution to a given problem might be expected to improve over time simply because algorithms and desirability functions have improved over time. Multiple response optimization through the use of desirability functions does have various pitfalls however, as discussed at the beginning of this chapter, and the best solution is to this day not easily obtained. In particular, some well-known statistical software have failed to find the optimal solution for certain problems. For this reason, it is prudent to use different software and compare the results. If the solution obtained with one software package has a higher composite desirability than the solution obtained with another software package, then the first solution is obviously superior, although it might not be the optimal solution. If a user has any doubts about the solution obtained with particular statistical software being the optimal solution, it would be best to obtain, if possible, information about the algorithm that is used. It would also be a good idea to start with the supposed optimal solution and make small changes in the factor levels and see how this affects the composite desirability, provided that this can be performed dynamically, as can be done with JMP, for example.

Given in Figure 12.3 is the JMP output for this example, after specifying minimum, middle, and maximum values of 12.45, 13.725, and 15, respectively, for Selectivity; 3068, 3184, and 3300 for Etch Rate; and 10, 10.15, and 10.30 for Etch Rate Nonuniformity.

(*Note*: Using different models for the responses in JMP creates some problems, which require special treatment, which is why the first row label is of a different form than the other row labels.)

Before discussing these results, it should be noted that running the optimization in JMP using this solution as the starting values will produce a slightly different solution. Therefore, we should always think of any multiple optimization "solution" as just an

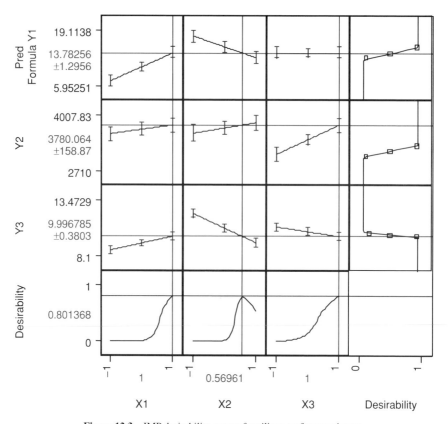

Figure 12.3 JMP desirability output for silicon wafer experiment.

estimate of the optimum solution, and it might be a poor estimate, depending on the nature of the problem and the type of algorithm that the selected software uses.

The individual graph panels in the first three columns of panels show the sensitivity of the composite desirability and the fitted values of each response to changes in each of the three factors. (The top panel in the third column has a horizontal line across the different levels of factor C because that factor was not used in the model for the first response variable.) We can see that the composite desirability is highly sensitive to changes in the second factor, as is indicated by the desirability graph for that factor in the last row. This can also be seen dynamically in JMP by positioning one's cursor at the vertical line, which corresponds to the value of the second factor, and sliding the level one way or the other. This suggests that algorithms that differ slightly could give slightly different solutions for the second factor, and we will see this later for this example.

The panels also suggest that higher levels of the first and third factors may be desirable and the fact that the desirability increases sharply past the midpoint of the range of each factor explains why the optimal solution has each factor set at the highest level used in the experiment.

The 95% confidence intervals on the mean response for each response variable are given at the beginning of each row of panels. The vertical bars in the first three rows and first three columns are 95% prediction intervals for each of the three response variables, for the low, middle, and high values of each of the factors. The coded levels of the factors would of course have to be converted to the actual units.

It is of interest to compare the JMP results with the results obtained using Design-Expert, which are given below.

```
Constraints
                 Lower     Upper     Lower     Upper
Name    Goal     Limit     Limit     Weight    Weight     Importance
A     is in range          -1          1         1            3
B     is in range          -1          1         1            3
C     is in range          -1          1         1            3
R1    maximize    12.45    15           1         1            3
R2    maximize    3068     3300         1         1            3
R3    minimize    10       10.3         1         1            3

Solutions
Number    A       B       C       R1         R2        R3        Desirability
1        1.00    0.55    1.00    13.9791    3760.47   10.0001      0.843
2        1.00    0.55    0.99    13.9712    3757.67   10           0.842
3        1.00    0.55    0.98    13.96      3753.6    10           0.840
4        0.97    0.53    1.00    13.9247    3756.32   10           0.833
```

We observe that the solutions produced by Design-Expert and JMP are almost identical, differing only slightly on the level of the second factor. The composite desirabilities differ more than slightly, however, and this is because a different desirability function form is used by each software, as will be explained shortly.

This can be seen as follows. For the Design-Expert solution, the individual desirabilities are computed as follows. For Selectivity, which was to be maximized,

$$d = \frac{\hat{y} - \text{Lower}}{\text{Upper} - \text{Lower}}$$

$$d = \frac{13.9791 - 12.45}{15 - 12.45}$$

$$= 0.599647$$

The desirability is 1.0 for Etch Rate since the fitted value of 3760.47 is at least equal to and is actually much greater than the value, 3300, that is considered to be completely acceptable. For Etch Rate Nonuniformity, which was to be minimized, the desirability is virtually 1.0 since the fitted value is practically equal to the largest value that is considered to be completely acceptable. Therefore, the composite desirability is simply $(0.599647 * 1 * 1)^{1/3} = 0.843$, which of course was the value given in the Design-Expert output.

We can force the same solution using JMP by fixing the value of the second factor at 0.55. Doing so gives a composite desirability of 0.787, which again we cannot compare with the Design-Expert composite desirability since different functions are used. The fitted values also differed noticeably, part of which was probably due to rounding the level of the second factor.

Regarding the difference in the composite desirabilities, Design-Expert uses the desirability function form used by Derringer and Suich (1980), as we have just verified, whereas JMP does not. The Derringer and Suich (1980) approach does not give smooth functions, which limits the possible shapes of the desirability functions. In contrast, JMP defines the desirability functions based on the lower, middle, and upper values specified by the user, resulting in piecewise smooth functions. For example, moving the middle values for the responses to be maximized close to their maximum values and moving the middle value for the response very close to the minimum value cause a considerable drop in the composite desirability with very small changes in the factor levels.

Regardless of which solution we go by, each of the discussed solutions *appears* to be far superior to the solutions given by Czitrom et al. (1998), and is also superior to the solution given by Ryan (2000), which was obtained using different software from what has been used here. There is one constraint that should be used, however, and we will see how that affects the solution. The solution given by Czitrom et al. (1998) had the second factor set at its highest level, which was 15 in the original units. The manufacturing engineer wanted to increase the CF_4 Flow from 5 to 15 because the manufacturing equipment that was to be used for the etching process is under better control when the CF_4 Flow is at least 15. The value of CF_4 Flow in the solution given by Ryan (2000) was 14.09, as stated previously, which may or may not have been good enough for the engineer.

We know from Figure 12.3 that increasing the value of the second factor is going to cause a sharp decrease in the desirability. A fixed factor level cannot be specified using JMP, other than locking in the "current" value in an analysis, as changes are made by sliding a vertical line in the graph using a mouse, although of course this is an inexact method. Accordingly, if we use a minimum acceptable value, in design units, of 0.999999 for the second factor and a maximum value of 1.0, we might expect to have a difficult time reaching a fitted value of at least 12.45 for the first response because the coefficient of the second factor in the fitted equation is -2.486. Thus, a considerable increase in the value of the second factor will result in a considerable decrease in the fitted value for the first response. The fitted value is 12.71, however, and the composite desirability given by JMP is only 0.537. The other fitted values were 3832.62 and 9.40.

Using Design-Expert, if we try to set the level of CF_4 Flow at 15 (i.e., the coded level set to 1), we obtain the results shown in Table 12.2.

As can be seen from the output, Design-Expert used nine starting points in seeking the optimum solution. Factors A and B are each set at the high level in each of the five solutions, with the level of factor C then obtained in such a way as to produce composite desirability values that happen to be equal. As expected, the value of factor

TABLE 12.2 Solutions Given by Design-Expert

Constraints

Name	Goal	Lower Limit	Upper Limit	Lower Weight	Upper Weight	Importance
A	is in range	−1	1	1	1	3
B	is equal to 1.00		1	1	1	3
C	is in range	−1	1	1	1	3
R1	maximize 12.45	15	1	1	1	3
R2	maximize 3068	3300	1	1	1	3
R3	minimize 10	10.3	1	1	1	3

Solutions

Number	A	B	C	R1	R2	R3	Desirability
1	1.00	1.00	1.00	12.8488	3816	9.375	0.539
2	1.00	1.00	−0.01	12.8487	3473.34	9.83171	0.539
3	1.00	1.00	0.19	12.8487	3541.83	9.74042	0.539
4	1.00	1.00	0.55	12.8487	3662.81	9.57918	0.539
5	1.00	1.00	0.02	12.8487	3486.32	9.81441	0.539

5 Solutions found
Number of Starting Points: 9

A	B	C
−1.000	1.000	1.000
0.000	0.000	0.000
−1.000	−1.000	−1.000
−1.000	−1.000	1.000
1.000	1.000	1.000
1.000	−1.000	1.000
1.000	1.000	−1.000
1.000	−1.000	−1.000
−1.000	1.000	−1.000

C varies directly with $R2$ and inversely with $R3$ since the coefficient of factor C in the model for $R2$ is positive and it is negative in the model for $R3$.

Upon seeing this output, it would be prudent for the user to simply increase the maximum $R2$ value, since $R2$ is to be maximized and all the solutions exceeded the upper bound, and run it again. Raising the maximum to 4000 predictably reduces the number of solutions to one and that solution is the first solution, with a composite desirability of 0.501.

Czitrom et al. (1998) set the first factor at its largest design level and this level is also part of the "optimal" solutions given here, so it is not possible to better their solution for the first response variable without fitting a different model—which they did. They used a significance level of .10 in their analysis, which is a liberal value. A parsimonious model is desirable because using terms that are not significant will inflate the variance of the predicted response, as shown by Walls and Weeks (1969).

It is important to keep in mind that regardless of the desirability form that is used, the composite desirability is a statistic that has sampling variability just as do the parameter estimates in the fitted models and the predicted values. If the predicted responses have a large variance for a given combination of factor levels, then we would expect those variances to translate, in an unknown and undoubtedly complicated way (which has apparently not been investigated in the literature), into a variance of the composite desirability that may be undesirable. (As shown in Figure 12.3, the predicted responses using JMP are given with a 95% prediction interval, as stated previously, so the user can see whether or not the interval is too large. Theoretical results, including the distribution of desirability indices, are given by Govaerts and Le Bailly de Tilleghem, 2005).

In this example the overall desirability for each of the given solutions was not particularly good because the predicted response for selectivity was well short of the goal of 15 (chosen arbitrarily for this example). Reducing the goal value to, say, 12.5 increases the composite desirability considerably.

It is worth noting that the predicted response values are based on some rather strong assumptions, namely, that each model is the true model so that the model holds for points that were not used in the experiment. Furthermore, the variance of the predicted responses will not be small when there are only 10 design points.

There are two sets of values that can be specified when Design-Expert is used: Weight and Importance for each response. A value of 0.1 to 10 for Weight can be assigned to each response in Design-Expert, with the set of these values indicating the relative importance of the response variables. The default value is 1.0. A Weight factor cannot be specified in JMP, but an Importance value can be specified. The weights specified for each response determine the shape of the desirability function when the objective is to hit a target value and when the response is to be maximized or minimized, but they can have very little effect on the solution when the objective is to maximize or minimize the response. For example, using the last variation of Example 12.1, weights of 0.1, 4, and 10 for the three response variables do not cause the optimal solution to change, but the use of different weights will strongly affect the desirability values.

The set of Importance values could have a considerable effect on the solution, however. These values are used to specify the relative importance of optimizing each response. In Design-Expert the importance values can be 1, 2, 3, 4, or 5, as stated in Section 12.2, whereas JMP will apparently accept virtually any positive number.

To illustrate, let's assume that it is far more important to have a small value for Etch Rate Nonuniformity than a large value for the other two variables. In Design-Expert it would be logical to assign an Importance value of 1 to each of the first two response values and 5 to Etch Rate Nonuniformity. This does not affect the solution, however, as the solution given is the first solution listed in Table 12.2. Similarly, the JMP solution is also unaffected when a high Importance value is given to $R3$ and a low value given to the other two responses.

In general, the values assigned for Weight and Importance may not have any effect on the optimal solution, with the effect seen only in the composite desirability. The upper and lower limits specified as well as any factor constraints are likely to exert more influence on the solution than the Weight and Importance values. For example, using Weight values of 1, 0.1, and 10 in Design-Expert for the three responses, respectively, has no effect on the output other than to give a "second-best solution" that has a slightly lower desirability than the optimal solution, which is not affected by the considerable disparity in the weights. Similarly, using various combinations of Weight and Importance does not lead to a different solution. (Again, this assumes that the second factor is set equal to 15 and the upper and lower limits on the fitted response values are unchanged.)

The effect that these have on the composite desirability in JMP can be seen by changing the relative weights in the Prediction Profiler graphical display and observing how the composite desirability changes dynamically. (There is more discussion of software for multiple response optimization in Section 12.8.)

Example 12.2

As a second example, we describe and analyze partially a case study given by Myers (1985). A 3^2 design with three replicates was used and there were four response variables. We will use this as an example primarily for the purpose of determining whether or not optimization *should* be performed, based on the model that is selected for each response.

The experiment involved the drug gentamicin. The manufacturing of two reagents, a particle reagent and an antibody, was not well controlled when new process technology was introduced. Experimentation was needed to determine the volumes of the reagents to be used in an analytical test pack in order to maintain uniform pack performance properties. There were four response variables, Final Blank Absorbance and three rates for calibrators. The data, using the uncoded factor levels, are given below.

Particle Reagent	Antibody	Final Blank Absorbance	Calibrator Rates Blank	4μg/ml	16μg/ml
20	10	.24250	79	8	-1
35	10	.38315	316	-1	32
50	10	.44030	309	326	146
20	25	.37970	85	12	5
35	25	.61950	485	319	48
50	25	.73550	418	490	220
20	40	.51080	83	79	9
35	40	.81405	539	306	78
50	40	.95405	432	542	240
20	10	.23960	70	9	2
35	10	.36490	297	246	29
50	10	.43890	307	329	141
20	25	.38320	84	15	5
35	25	.62520	492	281	59
50	25	.73140	435	428	192
20	40	.50990	77	-2	8

35	40	.81370	544	315	72
50	40	.94520	450	529	236
20	10	.25640	53	8	2
35	10	.37920	310	237	28
50	10	.43530	312	321	129
20	25	.38455	87	14	-2
35	25	.61195	483	337	53
50	25	.73495	435	355	80
20	40	.51400	84	18	9
35	40	.79975	541	35	70
50	40	.95600	425	523	253

Part of the output from Design-Expert is given below.

```
                   ANOVA for selected factorial model

Analysis of variance table [Classical sum of squares - Type II]
           Sum of                  Mean               F       p-value
Source     Squares      df         Square            Value    Prob > F
Model       1.29         8         0.16             4277.82   < 0.0001
  A-A       0.50         2         0.25             6672.88   < 0.0001
  B-B       0.74         2         0.37             9783.36   < 0.0001
  AB        0.049        4         0.012             327.53   < 0.0001
Pure Error 6.791E-004  18         3.773E-005
Cor Total   1.29        26
```

The "Pred R-Squared" of 0.9988 is in reasonable agreement with the "Adj R-Squared" of 0.9992.

We immediately notice that $R^2 = .999$, even though the design is replicated and there are 8 degrees of freedom for the model terms and 26 degrees of freedom overall. Although such an extremely high R^2 value is not uncommon in a few areas of application, such as calibration, this high value should cause one to question the data. Myers (1985) performed an analysis using the median response values and found that of the eight estimable effects (linear and quadratic for each of the two factors), and four interaction components obtained from all possible combinations of linear and quadratic for each factor, the numbers of effects that were significant at the .05 level of significance were 7, 8, 7, and 5 for the four response variables, respectively. Thus, there were 27 significant effects out of 32 for the four response variables combined. Certainly there should not be such a very high percentage of estimable effects that are significant.

The following output for the second response variable, again from Design-Expert, shows that when the actual response values are used, every estimable effect is significant since each parameter estimate's 95% confidence interval does not include zero, and does not even come close to including zero. Thus, this more or less agrees with the conclusions of Myers (1985). (Note that the "1" and "2" refer to the linear and quadratic effects, respectively.)

Term		Coefficient Estimate	df	Standard Error	95% CI Low	95% CI High
Intercept		304.89	1	1.55	301.63	308.14
A[1]		-226.89	1	2.19	-231.49	-222.29
A[2]		140.33	1	2.19	135.73	144.94
B[1]		-76.78	1	2.19	-81.38	-72.18
B[2]		28.89	1	2.19	24.29	33.49
A[1]	B[1]	66.11	1	3.10	59.60	72.62
A[2]	B[1]	-60.78	1	3.10	-67.29	-54.27
A[1]	B[2]	-21.56	1	3.10	-28.06	-15.05
A[2]	B[2]	12.56	1	3.10	6.05	19.06

According to Myers (1985) "The median response values were routinely used in the analysis, to make the fit less sensitive to blunders." Bad data points can cause terms to be significant that should not have been significant, but this won't necessarily inflate R^2. In Exercise 12.2 the reader is asked to determine if there are any outliers and to perform the optimization, if it is practical to do so considering what has been pointed out in this example.

12.6 MULTIPLE RESPONSE OPTIMIZATION VARIATIONS

As stated in Section 12.2, the original desirability function approach was given by Harrington (1965). For maximization he used $d = \exp[-\exp(a - bY)]$; for minimization, $d = 1 - \exp[-\exp(a - bY)]$, and for a target value $d = \exp|\frac{Y-T}{b}|^n$, with $a \leq Y \leq b$ being the acceptable range for the response variable Y, with n a constant.

Various purported improvements have been presented in the literature since the original work of Harrington (1965) and Derringer and Suich (1980). These were compared by Wurl and Albin (1999) and used in a case study. The work has focused on improved desirability functions. For example, Del Castillo, Montgomery, and McCarville (1996) pointed out that gradient-based optimality algorithms cannot be employed when the individual desirability functions illustrated in Section 12.1 and employed by Derringer and Suich (1980) are used. This is because gradient-based optimization methods, which are superior to hill-climbing techniques such as the Hooke–Jeeves method, as discussed in Section 12.1, require that the objective function have continuous first derivatives, but this requirement is not met by Eq. (12.2).

To remedy this problem, Del Castillo et al. (1996) proposed a piecewise continuous desirability function that corrects for the nondifferentiable points in the desirability functions used by Derringer and Suich (1980) by using a local polynomial approximation. The composite desirability function given in Eq. (12.1) is still used, and is computed using the piecewise desirability function for each response. The authors used the generalized reduced gradient (GRG) algorithm of Lasdon, Waren, Jain, and Ratner (1978), which has been used in Microsoft Excel, in optimizing the function. In the authors' example, however, the GRG algorithm produced practically the same solution as the Hooke–Jeeves algorithm. Undoubtedly, this will not always happen, however, as the solution obtained using the GRG algorithm could differ considerably from the solution obtained with the Hooke–Jeeves algorithm.

Réthy, Koczor, and Erdélyi (2004) proposed a modified desirability function approach that utilized a loss function approach that was in the spirit of Taguchi (loss functions) and Six Sigma, as did Ribardo and Allen (2003). See also the methods of Shah, Montgomery, and Carlyle (2004), Tong and Su (1997) and Ames, Mattucci, Macdonald, Szonyi, and Hawkins (1997). Tabucanon (1988) may also be of interest.

Ortiz, Simpson, Pignatiello, and Heredia-Langner (2004) proposed a genetic algorithm (GA) approach and compared its performance against the GRG algorithm used in conjunction with the composite desirability. The GA algorithm outperformed the GRG algorithm in their study.

Peterson (2004) proposed an entirely different approach, one that takes into consideration correlation between the response variables, which the desirability function approach does not do, and also the precision with which the parameters are estimated. Similarly, Vining (1998) proposed a mean squared error method that also utilized the variance–covariance structure of the multiple responses. Methods that utilize the correlation structure are important because multiple responses will often, if not usually, be correlated, as stated previously, just as measures such as height and weight are correlated in people.

Kim and Lin (2000) suggested an approach using an exponential desirability function that takes into account the differences in predictability of the models for the different responses as well as priorities for the response variables. A good physical interpretation of the objective function value is one of the touted advantages of the proposed approach; the authors do note some possible disadvantages, however.

Various methods have been compared by Kros and Mastrangelo (2001, 2004), with Kros and Mastrangelo (2004) concluding that the combination of (all of) "maximize," "minimize," and "hit the target" for response variables in a problem may dictate how likely the target is to be achieved.

Unpublished work on new desirability functions includes Gibb, Carter, and Myers (2001), who proposed using $d = [1 + \exp(-\frac{Y-a}{b})]^{-1}$, $d = [1 - \exp(-\frac{Y-a}{b})]^{-1}$, and $d = \exp[-\frac{1}{2}(\frac{Y-T}{b})^2]$, for maximization, minimization, and the target value cases, respectively.

Unfortunately, practitioners who would like to use any of these alternative approaches may be stymied by the fact that they are not presently available in statistical software. Regarding software, Ramsey, Stephens, and Gaudard (2005) showed how to perform multiple optimization using JMP.

Example 12.3

Returning to a more conventional example, we consider the example with two response variables given by Vining (1998). Engineers performed an experiment involving reaction time (A), reaction temperature (B), and the amount of catalyst (C) in an effort to determine the optimum levels of these factors to maximize the conversion of a polymer (Y_1) and achieve a target value of 57.5 for the thermal activity (Y_2). A central composite design was used with $\alpha = 1.682$ and six centerpoints were used. Vining (1998) used this example to illustrate the compromise approach he advocated, which was mentioned earlier in this section. The data are given below.

A	B	C	Y_1	Y_2
-1	-1	-1	74	53.2
1	-1	-1	51	62.9
-1	1	-1	88	53.4
1	1	-1	70	62.6
-1	-1	1	71	57.3
1	-1	1	90	67.9
-1	1	1	66	59.8
1	1	1	97	67.8
-1.682	0	0	76	59.1
1.682	0	0	79	65.9
0	-1.682	0	85	60.0
0	1.682	0	97	60.7
0	0	-1.682	55	57.4
0	0	1.682	81	63.2
0	0	0	81	59.2
0	0	0	75	60.4
0	0	0	76	59.1
0	0	0	83	60.6
0	0	0	80	60.8
0	0	0	91	58.9

We can use this example to illustrate the different solutions provided by different software, along the lines of what was shown in Example 12.1 in Section 12.5.

Vining (1998) stated that the engineers could justify a second-order model with nine terms: the three linear and the three pure quadratic effects, plus all three interaction terms. A preliminary analysis, shown below, suggests that the data do not support such a model, however, as the effects that have p-values less than .05 are B, C, B^2, C^2, AC, and BC. (A separate analysis showed that the ABC interaction was not significant.)

```
Response Surface Regression: Y1 versus A, B, C

Estimated Regression Coefficients for Y1

Term         Coef   SE Coef       T        P
Constant   81.091     1.924  42.153    0.000
A           1.028     1.276   0.806    0.439
B           4.040     1.276   3.166    0.010
C           6.204     1.276   4.861    0.001
A*A        -1.834     1.242  -1.476    0.171
B*B         2.938     1.242   2.365    0.040
C*C        -5.191     1.242  -4.179    0.002
```

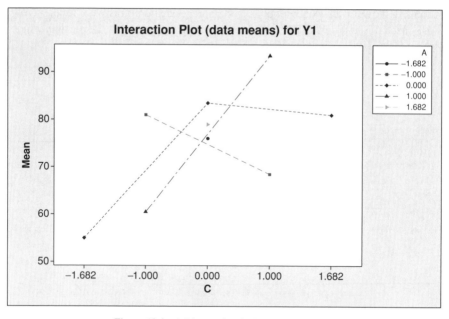

Figure 12.4 *AC* interaction for Example 12.3 data.

```
A*B              2.125     1.668    1.274    0.231
A*C             11.375     1.668    6.821    0.000
B*C             -3.875     1.668   -2.324    0.043

S = 4.717       R-Sq = 92.0%       R-Sq(adj) = 84.8%
```

We can see that the AC interaction effect is by far the largest effect, being almost twice as large as the next largest effect. This is obviously the explanation for the nonhierarchical model. We can see the problem if we construct the AC interaction plot (Fig. 12.4).

We need only consider the factorial points since the non-factorial points are not used in the computation of the interaction effects. The graph shows that there is a large A effect at the high level of C and a large A effect at the low level of C. There is not much difference in the response values at the two axial point values for A, and the response values at $A = 0$ do not affect the regression coefficient of the linear effect of A if a regression approach is used to determine the significant effects. Hence, the p-values are unaffected. This figure shows that the engineers were justified in assuming that there was an A effect, but the true nature of the effect cannot be seen without looking at this graph and/or computing conditional effects.

Conditional effects were discussed in a mostly general way for response surface designs in Section 10.9. Here the question arises as to whether or not a very large interaction, such as the AC interaction in this example, could mask a real quadratic effect in either A or C. This is an important question since Vining (1998) stated that "the engineers could justify" the model that included the quadratic term in factor

A but the analysis shows the quadratic term to not be significant. For designs with more than two levels, there is not a simple way to decompose interaction effects into the difference of conditional main effects, as was shown for two-level designs in Chapter 4, Appendix A. For example, for designs with three levels, the middle level is used in computing the quadratic effect of a factor, under one set of definitions, but is not used in computing the linear × linear interaction component(s). We should remember, however, as mentioned in Section 6.1.2 for three-level designs, that there is not just a single way to compute effect estimates when designs have more than two levels, as is also discussed for three-level designs by Giesbrecht and Gumpertz (2004, p. 291).

The situation is more complicated for response surface designs, some of which have five levels. If we use a linear model approach for the analysis of data from such designs, there is not a simple way to express a quadratic effect estimate since quadratic effects are neither orthogonal to each other nor to a column of ones.

Therefore, there is not a clean way of looking at quadratic effect estimates separately, let alone developing a relationship between the magnitude of quadratic effects and the magnitude of interaction effects. Consequently, we will not pursue this further and for this scenario it would have perhaps been prudent to simply rely on the engineers' judgment and include the A^2 term.

Although it may be annoying to use a model with coefficients for main effect terms that cannot be justified with a conditional effects analysis, the relevant question is whether or not the fitted values move in the proper directions as the factor levels are changed.

Continuing with the example, Vining (1998) stated that the engineers were comfortable with a model for Y_2 that contained only the A and C main effects. Although there are some interactions that are large relative to the B main effect, they are small relative to the A and C effects, so a model with only those two terms seems appropriate.

The smallest acceptable value for Y_1 was considered to be 80 and 100 or above was most desirable. The acceptable range for Y_2 was considered to be 55–60, with a target value of 57.5, as stated previously.

Design-Expert gives the following solution(s).

Name	Goal	Lower Limit	Upper Limit	Lower Weight	Upper Weight	Importance
A	is in range	-1.682	1.682	1	1	3
B	is in range	-1.682	1.682	1	1	3
C	is in range	-1.682	1.682	1	1	3
Y1	maximize	80	100	1	1	1
Y2	is target = 57.5	55	60	1	1	1

Solutions

Number	A	B	C	Y1	Y2	Desirability
1	-0.49	1.68	-0.56	95.1754	57.4999	0.871 Selected
2	-0.49	1.68	-0.56	95.1742	57.5001	0.871
3	-0.48	1.68	-0.58	95.1738	57.4999	0.871
4	-0.54	1.68	-0.48	95.0832	57.5001	0.868
5	-0.91	-1.68	0.11	83.6089	57.5001	0.425
6	-0.91	-1.68	0.12	83.6081	57.5001	0.425

The best solution(s) does not differ noticeably from the solutions given by Vining (1998) for the Derringer and Suich (1980) approach, which is what is being used here. This is especially true in regard to the fitted value of Y_1, which Vining gave as 95.21, almost the same as given in the Design-Expert output, and there is agreement with the fitted value of 57.5 for Y_2. Although Vining (1998) did not state that one of the response variable objectives was more important than the other one, changing the relative Importance values in Design-Expert has very little effect on the solution(s). Specifically, giving an Importance value of 1 for Y_1 and 5 for Y_2 gives a fitted value of 97.1753 for Y_1, compared to 97.1754 when they have equal Importance values. The only noticeable difference in the factor levels is that $C = -0.57$ with this unequal weighting, rather than -0.56 with the equal weighting. Reversing the Importance values gives the same solution as when using equal Importance values. Of course the composite desirability values differ greatly as the composite desirability is naturally quite high (.955) when the Importance values are 1 for Y_1 and 5 for Y_2 since the optimal solution hits the target value for Y_2. Reversing the Importance values gives a composite desirability of .794.

When JMP is used, convergence to an optimal solution is not achieved, but the last solution computed is given below.

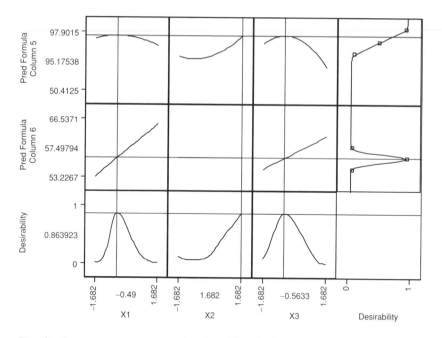

Despite the nonconvergence, notice that this solution is essentially the same as the solution given by Design-Expert. This is interesting because JMP and Design-Expert use different desirability function approaches, as was explained in Section 12.5.

One way to observe the "non-uniqueness" of the JMP solution is to run "Maximize Desirability" from this solution and notice the small change in the solution that results each time this is done, along with the accompanying message of nonconvergence.

Interest should focus, however, on the more important issue of trying to minimize the variance of the predicted responses by using parsimonious models. Unless a poor optimization algorithm is being used, and that does not appear to be the case with JMP, a solution that is far from the true optimal solution should very rarely occur. Sampling variability due especially to a small design and/or too many terms in the model should always be considered and minimized to the degree possible.

The approach recommended by Vining (1998) minimizes a multivariate general squared error loss function, under the assumption that the variance–covariance matrix of the response variables is known. The loss function contains a matrix **C** that can be used to represent costs or simply a differential weighting of the relative importance of the various response variables. The two solutions given by Vining (1998), the second of which assigns a very high weight to the second response variable, are both inferior to the best solution(s) given by Design-Expert, as in particular, neither solution is particularly close to the target value for Y_2.

12.7 THE IMPORTANCE OF ANALYSIS

As stated in other parts of the book, the emphasis has been on design and not analysis, although in some places there has been considerable analysis. In multiple response optimization, however, the emphasis is on analysis, which begins with an appropriate model fit to each response variable. A parsimonious model is desirable so as to minimize the variance of the fitted responses, remembering that the "optimal solution" is conditioned on the dataset, and the solution may not even be the optimal solution for the given data.

The variance of the predicted responses should be minimized as doing so will minimize the sampling variance of the optimal solutions, which is quite important, regardless of whether the computing algorithm is effective in identifying the optimal solution for a given dataset.

The datasets should of course also be carefully scrutinized, especially checking to see if there are any bad data points. Accordingly, the reader is asked in Exercise 12.9 to check and see if there are any suspicious data points relative to the second response variable and its fitted model, and if so, to determine the effect that deleting the point has on the optimal solution, especially since the design becomes unbalanced if a deleted point is not a centerpoint.

For additional information, see Chiao and Hamada (2001), who proposed a method for analyzing correlated multiple response data from designed experiments.

12.8 SOFTWARE

One of the unique features of Design-Expert is that up to 30 combinations of factor levels can be produced graphically that all have a desirability value of 1.0, although of course the levels cannot be read exactly from the graph. Another unique feature, and this was pointed out by Reece (2003, p. 375), is that the terms to use in the model for each response can be selected from the half-normal probability plot. What is somewhat odd about this is that the initial plot doesn't show any effects as being significant. That is, neither Lenth's method (see Section 4.10) nor any other method is used to identify significant effects, so this has to be done by the user. (It would be better if the half-normal probability plot showed the significant effects.) When an effect is selected, a line is fit to the remaining points so the line changes dynamically, but there is still no identification of significant effects from the ones that have not been hand selected. Although opinions will likely differ on the value of this setup, Reece (2003, p. 375) stated, "The use of the half-normal probability plot in this manner is a most useful approach to analyzing relatively simple designs without resorting to formal regression analysis".

Although data analysis diagnostics are not being used in this book since the emphasis is on design construction and identification of useful designs, it is worth noting that Design-Expert has excellent (regression) diagnostic capabilities that can be used very easily with multiple response optimization or with any of the other design features. This combined with the other features and the guidance (in blue) that appears with the output when the model for each response is fit essentially put Design-Expert at or at least near the top of the list of statistical software for multiple response optimization.

MINITAB can handle multiple responses through its multiple response optimizer. As a graphical aid, up to 10 contour plots can be overlaid; the use of this capability for analyzing two responses was illustrated in Figures 12.1 and 12.2. Although different models can be fit when the multiple response optimizer is used in menu mode, those models would have to be determined before using that capability because the default is to fit the model with all estimable terms for each response. Once the models are fit in menu mode, they cannot be modified in that mode because the procedure remembers the models that were fit and there is no facility that would permit any changes. Therefore, the models should be selected using regression methods before the optimizer is used. This is true regardless of whether menu mode or command mode is used. If a user does not remember to do this before the optimizer is used in menu mode, then it will have to be done afterward and the models specified in command mode, which is not as convenient. It is not terribly inconvenient, however, as a "trick" can be used of cutting and pasting the main command and subcommands that are displayed when menu mode is used after appropriate modification of the first subcommand.

In discussing MINITAB, Reece (2003, p. 435) stated, "The system does not provide convenient mechanisms for simplifying models to remove terms that are not significant. Therefore, the author simply used the full regression models for each response." Although this was in regard to the general fitting of regression models to multiple responses rather than directly regarding the multiple response optimizer, this may be a typical response and shows that the author used menu mode. Of course models with only significant terms can be fit very easily in command mode just by doing it in two stages.

JMP also has multiple response capability but the same model must be used for each response with the "Fit Model" capability. A workaround is described in the *JMP Statistics and Graphics Guide* when different models are to be fit, but this requires storing the "prediction formula" (i.e., prediction equation) for each response and using the Profiler option in the Graph menu. This is easily done, however, but Design-Expert is slightly easier to use and also has somewhat superior capabilities.

There are other statistical software packages that have multiple response optimization capabilities (see Reece, 2003, p. 334). The software packages that received the highest rating in the category of "Optimizations" were Design-Expert, Echip, Cornerstone (which is not widely known and is actually referred to as a design module), and JMP. (Although the company that produces the Echip software is listed as being out of business, the software is still supported.)

When using software for which the initial fitted model, such as a full quadratic model, is automatically fit for each response, the user must intervene and identify the significant terms, and then fit an appropriate subset model.

It is worth noting that the optimal solution is not guaranteed with any of these software packages. A convergence criterion (or multiple criteria) must be used and it is important that software use a good optimization routine to minimize the risk of identifying local optima as the global optimum. Cornerstone requires a convergence criterion as part of its input, whereas, at the other extreme, such input is not permitted with MINITAB. It is also worth noting that the optimal solution is, at best, optimal only in regard to maximizing the composite desirability. The combination of factor levels that is required for the optimal solution might not be feasible, as the combination may not be in a feasible region or the particular combination may be debarred for some reason. (See Section 13.9 for a discussion of restricted regions of operability and debarred observations.) Myers and Montgomery (1995, p. 252) make a similar suggestion in stating ". . . the researcher should do confirmatory runs at that condition to be sure that he or she is satisfied that all responses are in a satisfactory region." Although constraints can be specified with Design-Expert, debarred observations cannot be specified in the multiple response optimization routine (or in statistical software in general). Therefore, the user should check to see if the combination of factor levels produced by the optimization routine is feasible.

12.9 SUMMARY

The use of multiple response optimization methods will undoubtedly increase over time as their implementation in statistical software becomes better known to practitioners. No longer is it necessary to use the Solver routine in Excel to solve multiple response optimization problems, as was done in the late 1990s by at least one researcher whose work has been cited in this chapter.

Undoubtedly algorithmic approaches based on the Derringer and Suich (1980) approach are the most frequently used, although alternative approaches were proposed many years ago.

As new and improved methods are developed, it will be interesting to see how fast software developers move to implement them. This will undoubtedly depend on their perception of frequency of use of the routines in their software. Certainly the frequency of use pales in comparison to the use of ANOVA and regression, but in practice experimental design applications usually have multiple response variables, so there is a great need for the best multiple response optimization methods to be implemented in the leading statistical software packages.

The selection of "maximize," "minimize," or "hit the target" is obviously dictated by the problem, but there is some evidence that the combination of these in a particular problem will dictate how likely the target is to be achieved (Kros and Mastrangelo, 2004).

A desirability function approach can also be used when there is only a single response, which of course is a much simpler problem. This is discussed in Section 14.1.

REFERENCES

Ames, A. E., N. Mattucci, S. Macdonald, G. Szonyi, and D. M. Hawkins (1997). Quality loss functions for optimization across multiple response surfaces. *Journal of Quality Technology*, **29**(3), 339–346.

Chen, H., A. S. Hedayat, and C.-Y. Suen (1998). A family of optimal designs for experiments with multiple responses. *Journal of Combinatorics, Information and Systems Sciences*, **23**, 259–269.

Chiao, C.-H. and M. Hamada (2001). Analyzing experiments with correlated multiple responses. *Journal of Quality Technology*, **33**(4), 451–465.

Copeland, K. and P. Nelson (1996). Dual response optimization via direct function minimization. *Journal of Quality Technology*, **28**, 331–336.

Czitrom, V., J. Sniegowski, and L. D. Haugh (1998). Improved integrated circuit manufacture using a designed experiment. In *Statistical Case Studies: A Collaboration Between Academe and Industry* (R. Peck, L. D. Haugh, and A. Goodman, eds.). Philadelphia: Society of Industrial and Applied Mathematics; Alexandria, VA: American Statistical Association.

Del Castillo, E. and D. C. Montgomery (1993). A nonlinear programming solution to the dual response problem. *Journal of Quality Technology*, **25**, 199–204.

Del Castillo, E., D. C. Montgomery, and D. R. McCarville (1996). Modified desirability functions for multiple response optimization. *Journal of Quality Technology*, **28**(3), 337–345.

Derringer, G. (1994). A balancing act: Optimizing a product's properties. *Quality Progress*, **27**(6), 51–58. (http://www.statease.com/pubs/derringer.pdf)

Derringer, G. and R. Suich (1980). Simultaneous optimization of several response variables. *Journal of Quality Technology*, **25**, 199–204.

Ding, R., D. K. J. Lin, and D. Wei (2004). Dual response surface optimization: A weighted MSE approach. *Quality Engineering*, **16**(3), 377–385.

Ellekjaer, M. R., H. T. Fuller, and K. Ladstein (1997–1998). Analysis of unreplicated split-plot experiments with multiple responses. *Quality Engineering*, **10**(1), 25–36.

Gibb, R. D., W. H. Carter, Jr., and R. H. Myers (2001). Incorporating experimental variability in the determination of desirable factor levels. Unpublished manuscript.

Giesbrecht, F. G. and M. L. Gumpertz (2004). *Planning, Construction, and Statistical Analysis of Comparative Experiments*. Hoboken, NJ: Wiley.

Govaerts, B. and C. Le Bailly de Tilleghem (2005). Distribution of desirability in multi-criteria optimization using desirability functions based on the cumulative distribution function of the standard normal. (manuscript, see http://www.stat.ucl.ac.be/ISpub/dp/2005/dp0531.pdf)

Harrington, E. C., Jr. (1965). The desirability function. *Industrial Quality Control*, **21**(10), 494–498.

Hooke, R. and T. A. Jeeves (1961). Direct search solution of numerical and statistical problems. *Journal of the Association of Computing Machinery*, **8**(2), 212–229.

Khuri, A. I. and M. Conlon (1981). Simultaneous optimization of multiple responses represented by polynomial regression functions. *Technometrics*, **23**(4), 363–375.

Kim, K. and D. K. J. Lin (2000). Simultaneous optimization of mechanical properties of steel by maximizing exponential desirability functions. *Journal of the Royal Statistical Society, Series C*, **48**, 311–325.

Koksoy, O. and N. Doganaksoy (2003). Joint optimization of mean and standard deviation using response surface methods. *Journal of Quality Technology*, **35**(3), 239–252.

Kros, J. F. and C. M. Mastrangelo (2001). Comparing methods for the multiple-response design problem. *Quality and Reliability Engineering International*, **17**, 323–331.

Kros, J. F. and C. M. Mastrangelo (2004). Comparing multi-response design methods with mixed responses. *Quality and Reliability Engineering International*, **20**, 527–539.

Lasdon, L. S., A. D. Waren, A. Jain, and M. Ratner (1978). Design and testing of a generalized reduced gradient code for nonlinear programming. *ACM Transactions on Mathematical Software*, **4**(1), 34–50.

Lin, D. and W. Tu (1995). Dual response surface optimization. *Journal of Quality Technology*, **27**, 34–39.

Myers, G. C. (1985). Use of response surface methodology in clinical chemistry. In *Experiments in Industry: Design, Analysis and Interpretation of Results* (R. D. Snee, L. B. Hare, and J. R. Trout, eds.). Milwaukee, WI: American Society for Quality Control.

Myers, R. H. and D. C. Montgomery (1995). *Response Surface Methodology: Process and Product Optimization Using Designed Experiments*. New York: Wiley. (2nd ed. in 2002)

Ortiz, F., Jr., J. R. Simpson, J. J. Pignatiello, and A. Heredia-Langner (2004). A genetic algorithm approach to multiple-response optimization. *Journal of Quality Technology*, **36**(4), 432–450.

Palamakula, A., M. T. H. Nutan, and M. A. Khan (2004). Response surface methodology for optimization of and characterization of limonene-based coenzyme Q10 self-nanoemulsified capsule dosage form. *AAPS PharmsciTech*, **5**(4), 1–8.

Peterson, J. J. (2004). A posterior predictive approach to multiple response surface optimization. *Journal of Quality Technology*, **36**(2), 139–153.

Ramsey, P. J., M. L. Stephens, and M. Gaudard (2005). Multiple optimization using the JMP® statistical software. Presentation made at *the Kodak Research Conference*, May 9, 2005. (available at www.northhavengroup.com/pdfs/MultipleOptimization.pdf)

Reece, J. E. (2003). Software to support manufacturing systems. In *Handbook of Statistics*, Vol. 22, Chap. 9 (R. Khattree and C. R. Rao, eds.). Amsterdam: Elsevier Science B.V.

Réthy, Z., Z. Koczor, and J. Erdélyi (2004). Handling contradicting requirements using desirability functions. *Acta Polytechnica Hungarica*, **1**(2), 5–12.

Ribardo, C. and T. T. Allen (2003). An alternative desirability function for achieving 'Six Sigma' quality. *Quality and Reliability Engineering International*, **19**, 227–240.

Ryan, T. P. (2000). *Statistical Methods for Quality Improvement*, 2nd ed. New York: Wiley.

Schmidt, R. H., B. L. Illingworth, J. C. Deng, and J. A. Cornell (1979). Multiple regression and response surface analysis of the effects of calcium chloride and cysteine on heat-induced whey protein gelation. *Journal of Agricultural and Food Chemistry*, **27**, 529–532.

Shah, H. K., D. C. Montgomery, and W. M. Carlyle (2004). Response surface modeling and optimization in multiresponse experiments using seemingly unrelated regressions. *Quality Engineering*, **16**(3), 387–397.

Tabucanon, M. T. (1988). *Multiple Criteria Decision Making in Industry*. New York: Elsevier.

Taguchi, G. (1986). *Introduction to Quality Engineering: Designing Quality into Products and Processes*. White Plains, NY: Kraus International Publications.

Tang, L. C. and K. Xu (2002). A unified approach for dual response optimization. *Journal of Quality Technology*, **34**(4), 437–447.

Tong, L.-I. and C.-T. Su (1997). Optimizing multiple response problems in the Taguchi method by fuzzy multiple attribute decision making. *Quality and Reliability Engineering International*, **13**, 25–34.

Vining, G. G. (1998). A compromise approach to multiresponse optimization. *Journal of Quality Technology*, **30**(4), 309–313.

Vining, G. G. and R. H. Myers (1990). Combining Taguchi and response surface methodologies: A dual response approach. *Journal of Quality Technology*, **22**, 38–45.

Walls, R. C. and D. L. Weeks (1969). A note on the variance of the predicted response in regression. *The American Statistician*, **23**, 24–26.

Wisnowski, J. W., G. C. Runger, and D. C. Montgomery (1999–2000). Analyzing data from designed experiments: A regression tree approach. *Quality Engineering*, **12**(2), 185–197.

Wurl, R. C. and S. L. Albin (1999). A comparison of multiresponse optimization: Sensitivity to parameter selection. *Quality Engineering*, **11**(3), 405–415.

EXERCISES

12.1 A desirability function analysis was used in the article "Stencil printer optimization study" by J. V. Stephenson and D. Drabenstadt (*Surface Mount Technology*, November 1999), which is available on the Internet at http://smt.pennnet.com/Articles/Article_Display.cfm?Section=Archives&Subsection=Display&ARTICLE_ID=111949&KEYWORD=printing. Is this the same type of desirability function that was presented in Section 12.2. If not, explain what is actually being done.

12.2 Consider Example 12.2 and the motivation for using medians that was given by Myers (1985). Recall that an unrealistically high fraction of estimable effects was significant, with all the estimable effects for the second response being significant. Does there appear to be any outliers on any of the responses that could have caused this result? If not, would you suggest that optimization be performed? Explain. If optimization seems reasonable, perform the optimization and, if possible, compare your results with those given in Figure 5 of Myers (1985).

12.3 One of the sample datafiles that comes with MINITAB is FACTOPT.MTW, which contains data for a 2^3 design with two replications, for a chemical reaction experiment. The three factors were reaction time, reaction temperature, and type of catalyst, and the two responses were yield and cost. The former is of course to be maximized and the latter is to be minimized. Use lower limit $= 35$ and upper limit $= 45$ for yield and lower limit $= 28$ and upper limit $= 35$ for cost and determine the optimal levels of the three factors. The data are as follows.

Row	Time	Temp	Catalyst	Yield	Cost
1	50	200	A	48.4665	31.7457
2	20	200	A	45.1931	31.0513
3	50	200	B	49.2040	36.8941
4	50	150	B	45.5991	32.6394
5	20	150	A	42.7636	27.5306
6	50	150	A	44.7592	29.3841
7	20	200	B	44.7077	34.6241
8	20	150	B	43.3937	30.5424
9	50	200	A	49.0645	32.3437
10	50	150	B	45.1531	33.0854
11	50	200	B	48.6720	37.4261
12	20	200	B	45.3297	35.2461
13	50	150	A	45.3932	28.7501
14	20	150	B	43.0617	30.2104
15	20	150	A	43.2976	28.0646
16	20	200	A	44.8891	30.7473

(a) Determine the optimal factor settings, assuming that yield and cost are equally important from a desirability function perspective.

(b) Now assume that the importance of maximizing yield is given a weight of 0.7 and of minimizing cost is given a weight of 0.3 and determine the optimal factor settings.

12.4 The MINITAB sample datafile RSOPT.MTW contains data on four factors: hot bar temperature, dwell time, hot bar pressure, and material temperature. The experiment focused on the seal strength of a bag, which was one of the response variables, with the variability in the seal strength being the other response variable. The objective was to determine the settings of the four factors so as to hit a target of 26 pounds for the seal strength, with 24 as the minimum acceptable value and 28 as the upper limit. The variance of the seal strength is to be between 0 and 1. Assume that these objectives are of equal importance. The data are as follows.

Row	HotBarT	DwelTime	HotBarP	MatTemp	Strength	VarStrength
1	200	1.00	50	110	12.4470	5.28200
2	225	0.75	100	90	24.7047	1.63296
3	125	0.75	100	90	20.6865	1.34636
4	175	0.75	100	90	29.1000	0.95000
5	175	0.75	100	50	27.4284	1.69595
6	200	0.50	150	110	30.3010	3.45200
7	150	1.00	150	110	23.2990	2.21400
8	175	0.75	100	90	28.3000	0.83000
9	175	0.75	0	90	25.9942	0.83801
10	200	0.50	50	110	27.6490	1.07600
11	175	0.75	100	90	28.2000	0.91000
12	175	0.75	100	90	28.7000	0.92000
13	175	0.75	100	90	27.4000	0.97000
14	150	0.50	150	70	12.0010	2.99600
15	200	1.00	150	110	15.6990	6.49000
16	200	1.00	50	70	8.2510	6.74600
17	150	0.50	50	110	12.2010	4.14400
18	200	0.50	50	70	26.7490	0.80000
19	175	1.25	100	90	21.3752	2.92266
20	150	0.50	50	70	10.5010	4.46000
21	150	1.00	50	70	15.6990	0.99400
22	175	0.75	100	90	28.9000	0.86000
23	200	1.00	150	70	13.7030	4.49400
24	175	0.25	100	90	25.5021	1.49727
25	150	0.50	150	110	15.7010	3.54000
26	175	0.75	200	90	30.0581	1.14190
27	175	0.75	100	130	30.0516	2.58797
28	200	0.50	150	70	28.4010	0.84000
29	150	1.00	150	70	21.5990	2.13800
30	175	0.75	100	90	28.5000	0.94000
31	150	1.00	50	110	19.7990	2.69000

It can be observed, although perhaps not too easily, that the design used is a central composite design with axial points of $\alpha = \pm 2$ in coded form and seven centerpoints. Fit a model to each response and perform the optimization, using the information on the objectives given at the start of the problem.

12.5 Derringer and Suich (1980, references) gave an example of an experiment with four response variables and three factors, and a central composite design with six centerpoints was used. The experiment involved the development of a tire tread compound, with the four response variables being PICO Abrasion Index (Y_1), 200 percent Modulus (Y_2), Elongation at Break (Y_3), and Hardness (Y_4). The first two were to be maximized, with 120 being the smallest acceptable value for Y_1 and a value of at least 170 being perfectly acceptable. For Y_2, the minimum acceptable value was 1000 and 1300 or more was completely acceptable. The value of Y_3 was to be between 400 and 600 and the target value was assumed to be 500. Similarly, Y_4 was to be between 60 and 75, and 67.5 was taken as the target value.

The three ingredients (i.e., factors) used in the experiment were hydrated silica level (A), silane coupling agent level (B), and sulfur level (C). The data are given below.

A	B	C	Y_1	Y_2	Y_3	Y_4
-1	-1	-1	103	490	640	62.5
1	-1	-1	120	860	410	65.0
-1	1	-1	117	800	570	77.5
1	1	-1	139	1090	380	70.0
-1	-1	1	102	900	470	67.5
1	-1	1	132	1289	270	67.0
-1	1	1	132	1270	410	78.0
1	1	1	198	2294	240	74.5
-1.633	0	0	102	770	590	76.0
1.633	0	0	154	1690	260	70.0
0	-1.633	0	96	700	520	63.0
0	1.633	0	163	1540	380	75.0
0	0	-1.633	116	2184	520	65.0
0	0	1.633	153	1784	290	71.0
0	0	0	133	1300	380	70.0
0	0	0	133	1300	380	68.5
0	0	0	140	1145	430	68.0
0	0	0	142	1090	430	68.0
0	0	0	145	1260	390	69.0
0	0	0	142	1344	390	70.0

The authors fit the same full quadratic model for each response and obtained a composite desirability of 0.583. This resulted from the following fitted values: $\widehat{Y}_1 = 129.5$, $\widehat{Y}_2 = 1300$, $\widehat{Y}_3 = 465.7$, and $\widehat{Y}_4 = 68$. These values resulted from using $X_1 = -0.05$, $X_2 = 0.145$, and $X_3 = -0.868$. Do you agree with their analysis? In particular, is the full quadratic model appropriate for each response variable? If not, give your solution and compare it with the authors' solution. If your solution is superior, what would you recommend to the authors?

12.6 Palamakula, Nutan, and Khan (2004, references) used a Box–Behnken design with three factors with the objective of optimizing self-nanoemulsified capsule dosage form of a highly lipophilic model compound, coenzyme Q10 (CoQ). The factors studied were R-(+)-limonene (A), surfactant (B), and cosurfactant (C). There were five response variables, although the authors studied only the first two. The two variables that are to be maximized are the percentage of the drug dissolved after the first 5 minutes (Y_1), and the percentage that is dissolved in 15 minutes (Y_2). The turbidity (Y_3) and particle size (Y_4) values should be under 150. This did happen in every run of the experiment except run #4, for which particle size exceeded 1000. The zeta potential (Y_5) should be under 25. The data are given below, with the factor values given in raw form rather than coded form.

A	B	C	Y_1	Y_2	Y_3	Y_4	Y_5
81	57.6	7.2	44.4	99.6	63.2	10.9	12.1
81	7.2	7.2	6.0	1.34	140	49.8	62.3
18	57.6	7.2	3.75	13.1	9.73	14.9	70.4
18	7.2	7.2	1.82	1.44	12.9	>1000	58
81	32.4	12.6	18.2	36.1	16.8	32.6	35.2
81	32.4	1.8	57.8	72.9	9.37	39.3	27
18	32.4	12.6	68.4	89.9	5.13	38.6	16.8
18	32.4	1.8	3.95	76.08	5.13	24.7	12.9
49.5	57.6	12.6	58.4	87.9	14.3	16.0	9.09
49.5	57.6	1.8	24.8	39.7	4.53	31.3	16.8
49.5	7.2	12.6	1.6	2.97	108	26.0	74.9
49.5	7.2	1.8	12.1	26.5	41.7	123	49.7
49.5	32.4	7.2	81.2	94.6	7.67	13.3	5.02
49.5	32.4	7.2	72.1	88.2	7.70	29.5	4.11
49.5	32.4	7.2	82.06	95.4	8.23	14.0	6.45

Maximize Y_1 and Y_2, assuming that they are of equal importance, while checking to make sure that the constraints on Y_3, Y_4, and Y_5 are met. Note the following. The authors' solution had $\widehat{Y}_1 = 81.6$ and $\widehat{Y}_2 = 95.9$, resulting from, in coded form, $A = 0.0344$, $B = 0.216$, and $C = 0.240$. Although the authors initially used a full quadratic model for each response variable, the article does not clearly indicate whether or not this model was also used in the optimization. (a) Do you agree with the authors' results? If not, support your solution and compare it to the authors' solution. (b) Are there large interactions that are creating problems relative to the main effect coefficients? Explain. If so, what you suggest be done about the problem?

12.7 Del Castillo, Montgomery, and McCarville (1996, references) gave an example of a wire bonding process optimization problem from the semiconductor industry to illustrate the improved optimization approach that they proposed. A Box–Behnken design was used for three factors: flow rate (A), flow temperature (B), and block temperature (C). Wires were bonded between two positions and the maximum, beginning, and finish bond temperature recorded for each position, for a total of six response variables. The data are given below with the factor levels given in the raw units and the first three response variables are "maximum," "beginning," and "finish," respectively, for the first position, followed by these variables in the same order for the second position.

A	B	C	Y_1	Y_2	Y_3	Y_4	Y_5	Y_6
40	200	250	139	103	110	110	113	126
120	200	250	140	125	126	117	114	131
40	450	250	184	151	133	147	140	147
120	450	250	210	176	169	199	169	171
40	325	150	182	130	122	134	118	115
120	325	150	170	130	122	134	118	115
40	325	350	175	151	153	143	146	164
120	325	350	180	152	154	152	150	171
80	200	150	132	108	103	111	101	101
80	450	150	206	143	138	176	141	135
80	200	350	183	141	157	131	139	160
80	450	350	181	180	184	192	175	190
80	325	250	172	135	133	155	138	145
80	325	250	190	149	145	161	141	149
80	325	250	180	141	139	158	140	148

There is a target value for each response variable, as well as a minimum and maximum for each. These are given below.

Response variable	Minimum	Target	Maximum
Y_1	185	190	195
Y_2	170	185	195
Y_3	170	185	195
Y_4	185	190	195
Y_5	170	185	195
Y_6	170	185	195

Thus, the maximum bond temperature at each of the two positions has the same target value, and the other four response variables also have the same target value. The responses were considered to be equally important. The authors fit the same model to the last three response variables and the same model to the second and third variables. The model for the first response variable differed from these two models so there were three different models altogether. The authors obtained the following solution: $A = -0.0074$, $B = 1.0$, and $C = 0.7417$, with fitted values for the response variables of 186.9, 173.0, 170.1, 190.0, 170.9, and 182.4. Thus, the target value for the fourth response variable was hit exactly and the fitted value for the last response variable was close to the target value. Fit appropriate models for each response variable and perform the optimization. How does your solution compare to the authors' solution, especially in regard to the target values?

12.8 Consider the example in Section 12.1 with the data given in Table 12.1. Fit the model with only main effect terms in factors A, B, and C for the response variables Etch Rate and Etch Rate Nonuniformity. Then construct the contour plot, analogous to Figure 12.2, but this time fix factor C at the high level.

Compare your graph to Figure 12.2 and comment. In particular, does your graph show a feasible region?

12.9 Consider Example 12.3. Determine if there are any unusual data points (i.e., outliers) when the main effects of factors A and C are the only terms in the model. If there are any such points, what action would you suggest be taken regarding the point(s)? In particular, the design becomes unbalanced if deleted points are not centerpoints. Does this create a major problem? Note that Design-Expert still produces a fitted model and an optimal solution when there is a missing factorial point or axial point (or both). What does this suggest about how the data are used in obtaining the fitted model? If you find one or more suspicious data points and believe that they should be deleted, redo the optimization with the deleted point(s). If you deleted any points, does the deletion have much effect on the optimal solution?

12.10 Consider Example 6.3 in Section 6.4. There were three response variables and 30 observations per treatment combination used in a 2×3 design. (See the journal article for the data.) When multiple optimization is performed, it is worthwhile to consider that not only is the optimal solution a bit "random" even for a fixed set of data, as indicated in Section 12.5, relative to JMP solutions, but also of course the solution depends upon the response values, which are naturally random. In view of this but considering the number of observations per treatment combination, how well determined would the optimization likely be? If possible, read the article that is cited in Section 6.4 and try to determine an "optimal" solution.

12.11 Derringer (1994, references) gave some illustrative examples, one of which was a discrete case application in which there were four product formulations and three formulation properties: tensile strength, hardness, and elongation. There was a target value of 2000 pounds per square inch for tensile strength, with a small value of hardness desired, as well as a large value of elongation. Hardness was considered to be twice as important as tensile strength and elongation was considered to be four times as important as tensile strength. The four formulations with the individual desirabilities are given below.

Candidate formulation	Tensile strength	Desirability	Hardness	Desirability	Elongation	Desirability
1	1750	0.40	45	0.10	550	0.15
2	2000	1.00	30	0.97	500	0.00
3	1600	0.07	35	0.87	525	0.05
4	1600	0.07	30	0.97	585	0.47

(a) Determine the best candidate formulation, using the information that is given.
(b) How sensitive is the selection of the best formulation to the stated weights?

12.12 Wisnowski, Runger, and Montgomery (1999–2000, references) gave an example with four response variables and we will use the first three. The latter are the percentages of three bacterial populations killed in the production of a soap product. The design, given below along with the response values, was a 2_V^{5-1} design with four centerpoints, except that the centerpoints were not such on factor B since only two levels for that factor were used.

A	B	C	D	E	Bacteria 1	Bacteria 2	Bacteria 3
0	-1	0	0	0	25.53	18.60	88.38
-1	-1	-1	1	-1	54.90	12.80	99.95
1	-1	-1	-1	-1	36.53	32.76	99.96
0	-1	1	1	1	23.47	27.47	94.31
1	1	1	1	1	20.92	52.05	97.47
0	1	0	0	0	61.70	36.63	99.90
1	-1	-1	1	1	41.48	28.34	64.62
-1	1	-1	-1	-1	80.23	62.57	99.99
-1	1	1	1	-1	61.70	51.07	99.99
0	1	0	0	0	46.13	36.36	99.17
1	-1	1	1	-1	74.44	67.44	99.52
-1	-1	-1	-1	1	31.89	20.49	87.86
1	1	1	-1	-1	56.77	92.23	99.99
-1	1	-1	1	1	3.33	29.51	99.82
-1	1	1	-1	1	20.00	38.90	98.50
1	1	-1	1	-1	77.62	98.76	99.99
-1	-1	1	1	1	9.46	-8.28	93.90
-1	-1	1	-1	-1	80.26	47.77	99.98
1	-1	1	-1	1	35.28	24.84	95.58
1	1	-1	-1	1	38.55	13.69	99.99

(a) Does there appear to be any bad data points. If so, determine what correction action to take.

(b) Recognizing that each of the response variables should be maximized, determine the best combination of each factor to use to accomplish this after fitting a model for each response.

12.13 Consider again Example 12.1. The AB interaction was not used in the model for the first response variable for two stated reasons. Use this term in the model and redo the optimization, using appropriate software. Does this make much difference? Explain.

12.14 Khuri and Conlon (1981, references) gave two examples, one of which was from Schmidt, Illingworth, Deng, and Cornell (1979, references). The latter investigated the effects of cysteine (A) and calcium chloride (B) combinations on the textural and water-holding characteristics of dialyzed whey protein concentrate gel systems. There were four texture characteristics: hardness (Y_1), cohesiveness (Y_2), springiness (Y_3), and "combustible water" (Y_4). Each response was to be maximized. The data from the use of a central composite design were as follows.

A	B	Y_1 (kg)	Y_2	Y_3 (mm)	Y_4 (g)
-1	-1	2.48	0.55	1.95	0.22
1	-1	0.91	0.52	1.37	0.67
-1	1	0.71	0.67	1.74	0.57
1	1	0.41	0.36	1.20	0.69
-1.414	0	2.28	0.59	1.75	0.33
1.414	0	0.35	0.31	1.13	0.67
0	-1.414	2.14	0.54	1.68	0.42
0	1.414	0.78	0.51	1.51	0.57
0	0	1.50	0.66	1.80	0.44
0	0	1.48	0.66	1.79	0.50
0	0	1.41	0.66	1.77	0.43
0	0	1.58	0.66	1.73	0.47

Khuri and Conlon fit the same (quadratic) model to each response variable. Would you do the same? If not, fit appropriate models to each response and perform the optimization.

CHAPTER 13

Miscellaneous Design Topics

The field of experimental design is extremely broad and there is not sufficient space in any design book to cover every important topic in detail. This chapter covers various topics that might be considered special topics because they aren't covered in any depth, if even mentioned, in most design books, but that doesn't lessen their importance. Some of these topics are covered herein in some detail; other topics are only briefly discussed. Many topics are covered and there is a large number of references at the end of the chapter to point readers in appropriate directions for further reading.

13.1 ONE-FACTOR-AT-A-TIME DESIGNS

For decades practitioners have been strongly advised to use factorial and other multi-factor designs rather than varying one factor at a time while holding the other factors at fixed levels, as they have done for decades and still continue to do. An example of this is a case study given by Chokshi (2000), who stated "While this is a departure from traditional approaches such as "vary-one-factor-at-a-time" and "trial-and-error" approaches. . . ." Czitrom (1999) stated that many scientists and engineers will continue to use one-factor-at-a-time (OFAT) experiments ". . . until they understand the advantages of designed experiments over OFAT experiments, and until they learn to recognize OFAT experiments so they can avoid them." (This paper is available on the Internet at http://www.amstat.org/publications/tas/czitrom.pdf.) Many additional sources could be quoted to show that varying one factor at a time continues to be used in a large fraction of all designed experiments. Voelkel (2005) considered the efficiency of fractional factorial designs relative to OFAT experiments, under the somewhat restrictive condition that there are no interactions, and found that it is desirable to scale back the levels of fractional designs from the levels that would have been used if an OFAT design had been employed. McDaniel and Ankenman (2000) stated that "some one-factor-at-a-time experimentation may hold promise in the small factor change problem because they can detect significant main effects and

---

Modern Experimental Design   By Thomas P. Ryan
Copyright © 2007 John Wiley & Sons, Inc.

may allow for fewer factor changes." (The small factor change problem is that where an experimenter seeks a certain amount of improvement in the response variable, such as moving it closer to a target value, while changing the factor levels as little as possible.) A point to be made is that opinions differ regarding the efficacy and potential of OFAT experimentation. Another important point is that there is a difference between OFAT experimentation and OFAT *designs*. It is obvious that most writers are referring to the former, whereas the latter should not automatically be dismissed.

We thus need to make a distinction between OFAT experimentation used by practitioners and the OFAT designs that have appeared in the literature, especially in recent years. To illustrate, consider the example given in Table 3 of Czitrom (1999). There were two factors: time (in seconds) and temperature (in degrees Celsius) with 8, 9, and 10 being the values used for time and 980, 1000, and 1020 the values used for temperature. Six runs were given in the following order: (10, 980), (10, 1000), (10, 1020), (8, 1000), (9, 1000), and (10, 1000). (Here we might assume that the combination of time and temperature in use was (10, 1000).)

Notice that only three more runs would be required to investigate all combinations of time and temperature, and there are only five distinct runs in the OFAT design since the (10, 1000) combination is repeated. This is the type of OFAT design/experimentation that practitioners are likely to use, especially practitioners who are unaware of the fact that OFAT designs with better structure are given in the literature. Better OFAT designs are discussed in the remainder of this section.

One of the main arguments against an OFAT approach is that interactions cannot be detected. This has been discussed and illustrated in many books, including Montgomery and Runger (2003, p. 509). Arguments against OFAT experimentation have been based on the assumption that factor changes for all factors can be easily made. This is very often not the case, however, and within the past 10 years this has been emphasized in the literature, as in Ganju and Lucas (1997) and Wang and Jan (1995), and is discussed in Section 13.2.

Consider the simplest case of two factors of interest but one is almost impossible to vary. This then reduces to a one-factor experiment, for which a *t*-test could be performed, assuming multiple observations per level of the factor that can be varied. Similarly, a three-factor experiment reduces to a two-factor experiment, and so on.

When we recognize that hard-to-change factors occur very frequently in practice, we must either (a) consider the use of factorial designs with the minimum number of factor changes for hard-to-change factors, or (b) consider designs other than factorial designs.

We should also recognize that interactions *can* be estimated with OFAT designs. The OFAT designs date from at least Daniel (1973) and have recently been "revived" by Qu and Wu (2005). Daniel (1973) considered five types of OFAT designs but not every type will be considered here (see also, Daniel, 1994).

In a standard OFAT design with two-level factors, there will be $(n^2 + n + 6)/2$ runs for n factors. Specifically, there will be one run with all of the factors at the high level, subsequent runs in which each factor in turn is set at the low level while the other factors remain at the high level, one run with all factors set at the high level, and $(n - 1)(n - 2)/2$ runs with ith and jth factors at the high level and the other factors at the low level $1 \leq i < j \leq n - 1$. (Note that the latter is *not* the same as having all pairs of factors at the high level with the other factors at the low level since $j \leq n - 1$.)

All of the points in these designs will not be used to estimate a given effect, however, as the name of the designs would suggest. Indeed, the standard OFAT design has 13 points for four factors, and we cannot have factorial designs for two-level factors that have an odd number of points since the number of design points at the high and low levels for each factor would be unequal. Thus, we have to view these designs differently from the way we view factorial designs.

Instead, the user of such designs would make comparisons as they occur, and there is some discussion of this in Daniel (1973).

Consider the discussion of conditional effects in Section 4.2.1 and consider the fact that the lack of effect sparsity for conditional effects would cause a problem with normal probability plots of conditional effects *if* such a plot could be constructed, since Lenth's method and all other methods break down when there is not effect sparsity. What happens when there are many conditional effects and we use a OFAT design?

With either OFAT approach, conditional effects could be obtained only by conditioning on all of the other factors being at their high level (or their low level with the alternative approach). Although this might sometimes be of interest, this would be stronger conditioning than conditioning on a level of a factor and might be called treatment combination conditioning rather than factor-level conditioning. In essence, the conditional effects that are obtained would be the same as the effect estimates. Obviously, there would be essentially no power to detect effects when only two numbers are compared, so replication would be almost essential unless the resultant numbers were to differ greatly (and both be valid values).

In a strict OFAT design, one-factor level is changed at each run (unlike a standard OFAT design), with the treatment combinations progressing in $n + 1$ runs from all factors being at the high level to all factors being at the low level. Thus, this type of OFAT design has fewer run-to-run factor-level changes than does a standard OFAT design. As with the standard OFAT design, conditional effects would have to be obtained from two runs, the difference being that with the strict OFAT design, the other factors would not always be at their high (low) level, but rather the treatment combinations in the pairs of runs that would be used in computing the conditional effects would differ over the pairs.

Obviously, only main effects can be estimated with these designs of $n + 1$ runs for n two-level factors, but it is also possible to construct OFAT designs of resolution IV and resolution V. Specifically, a resolution IV design would be produced by folding over an OFAT design. (As described in Section 5.9, a foldover design results from combining a design with its mirror image.)

The construction of strict OFAT designs of resolution V is discussed by Qu and Wu (2005), with such designs having originally been given by Daniel (1973). In particular, Daniel (1973) gave a strict OFAT design for five factors and 16 runs, with the treatment combinations used in the following order: (1), a, ab, abc, $abcd$, $abcde$, $bcde$, cde, de, e, ae, ade, $acde$, acd, ac, and ace. Notice that literally one factor level is changed between successive runs. Also notice that the structure of this design differs considerably from the structure of the design that was given at the beginning of the section. The design given by Daniel (1973) would be an ideal design if all of the factors were hard to change, although such a state of affairs would certainly be extremely rare.

How might such a design be used? The main effect of factor A could be estimated after the second run, although that estimate would be confounded with all interactions involving A. The next opportunity to estimate that main effect comes after the seventh run, as the sixth and seventh runs could be paired with the first two, and the estimate from those four runs would not be confounded with any interactions. Similarly, the other four main effects could be estimated, in addition to 3 of the two-factor interactions, with the other 7 two-factor interactions confounded in groups of 2 and 3.

What has been gained, if anything, relative to a 2_V^{5-1} design? The obvious gain is a considerable reduction in the number of factor-level changes, although the amount of the reduction, of course, depends on the number of factor changes that would be dictated by the randomization of runs when the fractional factorial design is used. In other ways, this design is strongly inferior to the fractional factorial design as the effect estimates from the latter have a smaller variance since all 16 points are used to estimate each effect and all of the two-factor interactions are estimable.

One of the touted advantages of OFAT designs is the possibility of early termination of the experiment, thus saving resources. Assume that for the strict OFAT that was just considered, the difference between the response values at (1) and at a differ by an order of magnitude. If no errors were made, this would suggest a possibly large A effect, although this effect is confounded with many interaction effects. If the experimenter proceeds and estimates the main effect after the seventh run and the estimate is still very large, this would seem to constitute clear evidence of a large A effect, with no further experimentation involving factor A apparently necessary, although further experimentation with A would be necessary to investigate interaction effects involving factor A.

Daniel (1973) indicated the effect and linear combinations of effects that were estimable after each run, and it should be realized that one cannot use additional runs in estimating a given effect. For example, if we look at the entire set of runs, factor B is at the high level for 5 of the runs and at the low level for the other 11 runs. Thus, we will have this type of imbalance if we proceed beyond run 10. Daniel (1973) used a different approach, however, and indicated that factor B was estimable after run 8, by which point 3 runs were at the high level and 5 runs were at the low level. Daniel (1973) is not using all eight data points in the estimation of the B effect, however.

Qu and Wu (2005) gave an example of a six-factor strict OFAT design of resolution V with 22 runs and, as in the example of Daniel (1973), indicated the effects that are estimated after each run. The runs in order were $(1), a, ab, abc, abcd, abcde, abcdef$, $bcdef, cdef, def, ef, f, af, aef, adef, acdef, acde, acd, ac, acf, acef$, and ace. The first four main effects, A–D, are estimated after runs 8–11, respectively, E and F are both estimated after run 12, and either one or two interactions are estimated after each of the succeeding runs. Qu and Wu (2005) indirectly gave a measure of the nonorthogonality of OFAT designs by giving the D-efficiency for strict and standard OFAT designs with the number of factors between 4 and 10 and compared these designs against some other small, economical designs (see Section 13.7.1 for a discussion of D-efficiency). D-efficiency values indirectly measure nonorthogonality since only orthogonal designs have a D-efficiency value of 1.0.

The advantages of OFAT designs under certain conditions must be weighed against the fact that there is no randomization of runs since the runs must be made in a prescribed order and there is not a fixed number of runs as in factorial designs. Consequently, changing environmental conditions could undermine the results, which is why it is even more important that statistical process control checks (such as the check runs discussed in Section 4.14) be used with OFAT designs than with factorial designs, although such checks should certainly be used with both types of designs, and preferably with virtually all types of designs.

All things considered, OFAT designs should be considered when experimental runs are extremely expensive and when there are hard-to-change factors. The designs should be used with caution, however, because any small design has shortcomings and OFAT designs are nonorthogonal designs.

13.2 COTTER DESIGNS

Cotter designs (Cotter, 1979) are similar to OFAT plans and are intended for use when there are many factors, few resources, and it is believed that interactions may exist. The designs have $2k + 2$ runs for k factors. The second "2" is composed of one run with all of the factors at the high level and one run with all of the factors at the low level. The $2k$ runs result from each of the factors in turn set at the high level while the other factors are set at the low level, and vice versa.

This design should be considered only for $k \geq 4$ because for $k = 3$ it produces a 2^3 design (as might be expected because $2k + 2 = 2^k$ when $k = 3$), and for $k = 2$ it would produce a 2^2 design with two points replicated. The design for $k = 4$ is given below:

A	B	C	D
-1	-1	-1	-1
1	-1	-1	-1
-1	1	-1	-1
-1	-1	1	-1
-1	-1	-1	1
-1	1	1	1
1	-1	1	1
1	1	-1	1
1	1	1	-1
1	1	1	1

Designs that have an "ad hoc" look to them and a small number of design points relative to the number of factors will generally be nonorthogonal, and these designs are nonorthogonal. In general, although the designs are in JMP, for example, and the design given above was created using JMP, they are of very questionable value and the default in JMP has them suppressed from the list of screening designs.

They should not be used where there are many factors, although this is when they were intended to be used. This can be explained as follows. The correlation between any pair of factors, and thus between the main effects estimates is

$$\frac{(2k + 2) - 2(4)}{2k + 2} = \frac{2k - 6}{2k + 2}$$

for $k \geq 4$. This result can be explained as follows: For any pair of columns, the number of rows for which the product of the corresponding column elements is not 1 (i.e., is -1) is 4. The number of dot products that are 1 is thus reduced by that number, and the dot product is reduced further by the fact that the sum of the dot products for the pairs with unlike signs is 4. (Hence, 2(4).) This explains the numerator of the fraction, which also explains the correlation since the denominator is fixed.

Notice that the correlation approaches 1 as $k \to \infty$, with the correlations being $1/3$ when $k = 5$. Even this correlation is too large and the criticisms of the deleterious effects of moderate correlations that have been made about some supersaturated designs (see Section 13.4.2) can also be applied to Cotter designs.

Another problem with these designs is that although they are meant to be used when interactions are suspected, only a small percentage of interactions can be estimated when k is large because $\binom{k}{2}$ grows at a much faster rate than $2k + 2$. For example, $\binom{k}{2} = 45$ when $k = 10$, but $2k + 2 = 22$ and the 10 main effect estimates must be obtained from the 21 degrees of freedom.

13.3 ROTATION DESIGNS

A relatively new class of designs was proposed by Bursztyn and Steinberg (2001, 2002). These designs are obtained by rotating (hence, the name) a two-level factorial design. Specifically, let D denote an $n \times k$ design matrix for a two-level fractional factorial design and let R be any $k \times k$ orthogonal matrix. Then the design $D^* = (1/c)DR$ is a rotation design, with c a scaling constant that scales the design to the unit cube. Some choices of R may be more useful than others, and Steinberg and Bursztyn (2001) state that a useful choice for R is $R = HSH'$ when $k = 2^t$, with H denoting the standard Hadamard matrix for estimating the effects with the 2^t design and S is a block diagonal matrix. (A Hadamard matrix is a square, orthogonal matrix with entries of $+1$ and -1.)

A primary feature of these rotation designs is that they are intended to be used for factor screening and fitting a model higher than a first-order model in the same stage, thus obviating the usual two-stage procedure.

13.4 SCREENING DESIGNS

The term "screening design" is used often in the literature. As the name implies, the objective with a screening design is to "screen out" seemingly unimportant factors so that the apparently important factors can be identified and used in subsequent experiments as part of a sequential experimentation strategy. An example of such a strategy using a screening design is given in Chokshi (2000).

The emphasis is thus on main effects when these designs, which generally have two levels, are used, and the successful use of these designs requires that main effects dominate interaction effects.

Screening designs have a small number of design points relative to the number of factors. A design with $(k + 1)$ runs for k factors is often called an *orthogonal main effects plan* and it can also be called a *saturated design*, the latter being a design that just has enough degrees of freedom to estimate the effects that one wishes to estimate. There are also designs with fewer than $(k + 1)$ runs and these are called *supersaturated designs*, which are covered in Section 13.4.2.

One analysis problem that must be addressed when designs with a small number of runs relative to the number of factors are used is the determination of significant effects. It is entirely possible that a majority of the main effects will be significant in a particular application, which will create a problem if Lenth's method or any one of the alternative methods is used. With a supersaturated design, a technique such as forward selection must be used to identify the main effects that are to be estimated since not all of them can be estimated. Although such an approach is not mandatory when a saturated design is used, it would not be a bad idea to use both a variable selection approach and a normal probability plot approach and compare the results, especially when there is a small number of runs.

Assume that there are seven candidate factors/main effects and eight runs are to be used, with each factor at two levels. This is a 2^{7-4} design, a $(1/16)$th fraction of a 2^7 design. Since only main effects are estimable, these are thus aliased with the interaction effects—all of them. Specifically, each main effect is aliased with 3 two-factor interactions, plus higher-order interactions. Now let's add a factor so that we have eight factors. Can we construct a saturated design with nine experimental runs? Obviously not if we still want to use two levels because 9 is not a multiple of 2. Notice that 9 is also not a power of 2, but we can construct designs with a number of runs not a power of 2, provided that the number of runs is a multiple of 4. Such designs are presented in the next section.

13.4.1 Plackett–Burman Designs

We can construct two-level (orthogonal) designs with 12, 20, 24, 28, and larger numbers of runs, up to 100. These are called *Plackett–Burman* (PB) *designs* and were given by Plackett and Burman (1946). The projective properties of these designs were given by Lin and Draper (1992), Tyssedal (1993), and Box and Tyssedal (1996a).

These designs were actually given for studying up to k factors in $N = (k + 1)$ runs, but they are equivalent to 2^{k-p} designs when N is a power of 2, with $2^{k-p} = k + 1$. Examples include the 2^{7-4} and 2^{15-11} designs. Statistical software is readily available for constructing these designs, as Design-Expert will construct the design for (up to) 11, 19, 23, 27, and 31 factors, with the number of runs of course being one more than each of these numbers. MINITAB will construct PB designs for 12, 20, 24, 28, 32, 36, 40, 44, and 48 runs, and of course up to one less for the number of factors for each design with the indicated number of runs. Thus, PB designs can be constructed with MINITAB for 2–47 factors, using the appropriate run size. Of course the run size must be an even number, as the number of low and high levels of each factor would not otherwise be equal.

The value of these designs is that they provide (possibly) economical alternatives to 2^{k-p} designs when runs are quite expensive. That is, with the 2^{k-p} series we have 8, 16, 32, 48, ... runs, whereas the PB designs give us three possibilities between 16 and 32 runs, and one design between 8 and 16 runs.

These designs are somewhat controversial. In particular, Anderson and Whitcomb (2000; http://www.statease.com/pubs/aqc2004.pdf) recommend that they be avoided and Montgomery, Borror, and Stanley (1997) criticize the design as having a complex alias structure (which is true). Snee (1985b) offers an opposing view, however, and Hamada and Wu (1992) and Wu and Hamada (2000, p. 356) provide an analysis strategy that permits the estimation of some interactions. We will later show how all of the two-factor interactions involving factors that are declared to be significant can be estimated for a particular scenario.

The designs have an easily discernible pattern. To illustrate, for $n = 8$ the first row of the design has the following signs: $+ + + - + - -$. The next six rows are obtained by successively moving the sign on the far right to the far left, with the last row being all minus signs. Thus, the design for n = 8, generated using MINITAB, is as follows:

Run	A	B	C	D	E	F	G
1	1	-1	-1	1	-1	1	1
2	1	1	-1	-1	1	-1	1
3	1	1	1	-1	-1	1	-1
4	-1	1	1	1	-1	-1	1
5	1	-1	1	1	1	-1	-1
6	-1	1	-1	1	1	1	-1
7	-1	-1	1	-1	1	1	1
8	-1	-1	-1	-1	-1	-1	-1

As stated earlier in this section, this design must be a 2^{k-p} design since the number of runs is a power of 2. This is not obvious, however, because of the way the runs are listed by MINITAB. If we look at the first three columns, however, we can see that this does constitute a 2^3 design in those factors, which becomes apparent if we put the runs in standard order, as follows:

Run	A	B	C	D	E	F	G
8	-1	-1	-1	-1	-1	-1	-1
1	1	-1	-1	1	-1	1	1
6	-1	1	-1	1	1	1	-1
2	1	1	-1	-1	1	-1	1
7	-1	-1	1	-1	1	1	1
5	1	-1	1	1	1	-1	-1
4	-1	1	1	1	-1	-1	1
3	1	1	1	-1	-1	1	-1

Now it is clear that the first three columns comprise a 2^3 design. (Actually, any column triplet constitutes a 2^3 design.) It may also be apparent that the generators for the other four factors are $D = -AB$, $E = -BC$, $F = -AE$, and $G = -AC$. Thus, the defining relation is $I = -ABD = -BCE = -AEF = -ACG$ plus all generalized interactions of these four effects. Of course this isn't the way we would normally construct a 2^{7-4} design, but we can see that the PB design with eight runs is in fact one of these designs.

Designs with larger values of n are constructed in the same manner, as can be seen for the following PB design for $n = 12$, which was also generated using MINITAB but which cannot be a 2^{k-p} design because 12 is not a power of 2.

Run	A	B	C	D	E	F	G	H	J	K	L
1	+	−	+	−	−	−	+	+	+	−	+
2	+	+	−	+	−	−	−	+	+	+	−
3	−	+	+	−	+	−	−	−	+	+	+
4	+	−	+	+	−	+	−	−	−	+	+
5	+	+	−	+	+	−	+	−	−	−	+
6	+	+	+	−	+	+	−	+	−	−	−
7	−	+	+	+	−	+	+	−	+	−	−
8	−	−	+	+	+	−	+	+	−	+	−
9	−	−	−	+	+	+	−	+	+	−	+
10	+	−	−	−	+	+	+	−	+	+	−
11	−	+	−	−	−	+	+	+	−	+	+
12	−	−	−	−	−	−	−	−	−	−	−

Assume that the design for $n = 8$ has been used for the seven indicated factors and only the main effects for factors A, B, and C were found to be significant using Lenth's method (see Section 5.1). This tells us all we need to know about the 3 two-factor interactions involving those factors because, as stated, $D = -AB$, $E = -BC$, and $G = -AC$. That is, the interactions are not significant because the main effects with which they are confounded are not significant.

When all seven factors are used, the fact that D, E, and G are not significant is important information because the interpretation of the main effects of A, B, and C is relatively straightforward in the absence of two-factor interaction effects involving those factors because the conditional effects will not differ greatly when interaction effects are small (see Chapter 4, Appendix A). Of course if the main effects for D, E, and G were judged significant, there would then be a dilemma because of the confounding. The *absence* of significant effects for these factors is helpful in this case because declaring the main effect of factor E to be not significant is the same as declaring to be not significant the two-factor interaction (involving factors B and C) that is confounded with factor E.

We have thus seen how a PB design can be used advantageously relative to interactions, even though screening designs such as this type of design are generally viewed as being for main effect determination almost exclusively.

In general, users of experimental designs should always remember that it will rarely be necessary to estimate all of the effects that can be estimated with a design of at least moderate size, so columns in a design matrix can often be used for something other than the original purpose.

The alias structure for a PB design with eight runs is relatively straightforward, although somewhat involved, since this is actually a fractional factorial design. The alias structure is much more complicated when the run size is 12 (and in general when the run size is not a power of 2), as in the 12-run design each main effect is partially aliased with several two-factor interactions. (The term *partially aliased* is used here to designate that an effect is neither orthogonal to nor confounded with another effect, but rather can be expressed as a linear combination of effects, possibly a long string of effects. For example, the full alias structure for the PB design with 12 runs and 11 factors is extremely long, with the *first part* of the (partial) alias structure for factor E given by Design-Expert as $[E] = E - 0.333 * AB - 0.333 * AC + 0.333 * AD - 0.333 * AF - 0.333 * AG - 0.333 * AH - 0.333 * AJ + 0.333 * AK + 0.333 * AL - 0.333 * BC + 0.333 * BD \ldots$ (The entire partial alias structure up through three-factor interactions is not given here as it would occupy about half a page.)

Partial aliasing is actually worse than complete confounding relative to what was illustrated for the 8-run design because complete confounding would allow us to make the kinds of substitutions that were made in the 8-run design illustration. For example, the AB interaction could not be estimated using one of the columns in the matrix for the 12-run design because the column that results from multiplying together the A and B columns in the design is not given by any of the columns in the matrix.

This does not mean that the AB interaction could not be estimated, however, because if A and B are orthogonal (as of course they are since the PB design is orthogonal), then AB is orthogonal to A and B. This is easy to prove: Let a_i denote the ith observation in the column of the design matrix for factor A and similarly let b_i denote the ith observation in the column of the design matrix for factor B. Since the columns of the PB design matrix are orthogonal, $\sum_{i=1}^{n} a_i b_i = 0$. Since $a_i b_i$ is $(AB)_i$, it follows that AB is orthogonal to A if $\sum_{i=1}^{n} a_i b_i a_i = 0$. Of course we may write this as $\sum_{i=1}^{n} a_i^2 b_i = (1) \sum_{i=1}^{n} b_i$, since each $a_i^2 = 1$. It follows that $(1) \sum_{i=1}^{n} b_i = 0$, since $\sum_{i=1}^{n} b_i = 0$. Thus, AB is orthogonal to A and the companion proof that AB is orthogonal to B should be obvious. Of course by definition, AB cannot be orthogonal to every column, however, and in fact $AB = -F$.

We can estimate as many effects as we have degrees of freedom for estimating effects; an effect that we estimate doesn't necessarily have to be an effect that we intended to accommodate with our choice of design. More specifically, it is worth noting (although it may be quite obvious) that, for example, AB is not part of the design matrix, but it can be part of the model matrix, and in general any term can be part of the model matrix as long as the factors that comprise it are included in the design matrix, provided the degrees of freedom are sufficient to accommodate it.

There is a problem, however, if, for example, the use of a 12-run PB design for 11 factors results in factors A, B, C, and D identified as significant, with the AB and CD interactions believed to be important and added to the model. The problem is that doing so creates correlations between the following effect estimates: CD and A, CD and B, AB and C, and AB and D. Thus, orthogonality is lost when we go beyond having one 2-factor interaction and the main effects of the two factors that comprise the interaction estimated.

The absolute value of these correlations is only $1/3$, however. Hamada and Wu (1992) gave an analysis strategy for use with PB designs that they recommended for use when correlations are small to moderate, with the strategy involving the use of forward selection to identify significant effects. They used some examples of datasets to illustrate their approach.

PB designs for $n = 20$, 24, and 28 are given by Wu and Hamada (2000, pp. 331–332) Since a PB design is a special class of an orthogonal array design, it can be found among the large collection of orthogonal array designs at http://www.research.att.com/~njas/oadir. Specifically, the design is given at http://www.research.att.com/~njas/oadir/oa.12.11.2.2.txt. (*Note*: In order to see that the PB designs given at the latter URL are equivalent to the design produced by MINITAB, it is necessary to use the correspondence $0 = +$ and $1 = -$, with the levels at the Web site given by 0 and 1 and the MINITAB levels of course given by $+$ and $-$. Of course to construct these designs we only need to know the first row of the design, as explained previously.

Nelson (1982) recommended the use of a foldover PB design to deal with the complex alias structure of the PB design and to produce a resolution IV design. Ahuja, Ferreira, and Moreira (2004) used this idea in an experiment that they performed and termed the design the reverse PB design, a name that was not used by Nelson (1982). This is discussed further in the next section.

It is also possible to use half fractions of PB designs, as discussed by Lin (1993a). See Samset and Tyssedal (1998) for an extensive catalog of PB designs of various sizes and their foldovers and half fractions. This includes a nice summarization of the correlation structure for each design. This technical report can be accessed through the Internet. See the URL given in the References section.

13.4.1.1 Projection Properties of Plackett–Burman Designs

PB designs have very good projection properties, as is extolled by Box, Hunter, and Hunter (2005, pp. 283–284). Lin and Draper (1992) and Box and Bisgaard (1993) showed that some of the saturated PB designs whose number of runs is not a power of 2 have better projection properties than do the saturated fractional factorial designs.

Properties of PB designs are discussed extensively by Samset and Tyssedal (1998). The smaller designs have projectivity of 3, meaning that the designs contain at least a full 2^3 design. The larger PB designs, such as for 68, 72, 80, and 84 runs, have projectivity 4, meaning that they contain at least a full 2^4 design. As pointed out by Samset and Tyssedal (1998), a half fraction of a PB design of projectivity 4 has projectivity 3. Of course if we take a half fraction of a PB design that is almost saturated, then we obtain a design that is supersaturated (see Section 13.4.2), which requires the use of special methods for the data analysis, such as stepwise regression.

13.4.1.2 Applications
PB designs are used extensively in industry, but they are also used extensively in biotechnology, as is apparent from the literature, such as Balusu, Paduru, Seenayya, and Reddy (2004).

Another example is the use of a PB design for investigating shipworm bacterium that was given by Ahuja et al. (2004), including a folded over PB design, which the authors called a "reverse Plackett–Burman design," following the recommendation of Nelson (1982) of "reversing" the signs in the PB design in order to de-alias effects. (Ahuja et al. (2004) seem to have invented the name and it seems more appropriate to use conventional terminology, especially since the term "foldover" is well established.") Other peculiar terminology included "... 80 runs (20 runs in the original design and their 20 replicates, and 20 runs in the fold-over design with 20 replicates)." If we read this literally, the design would have 800 runs, not 80. The design is actually a replicated PB design with its foldover also replicated.

The experiment involved 18 factors, so a design with 20 runs was chosen to permit the estimation of all of the main effects. The replication does not, of course, permit the estimation of any more effects than are possible without the replication, as the replication simply creates a pure error term in the ANOVA table with 40 degrees of freedom. The foldover, however, does convert the design to a resolution IV design, so that main effects are clear of two-factor interactions.

There were some errors made in the analysis of the data and the discussion of the design. In particular, the authors don't seem to understand that the foldover of a PB design creates a resolution IV design, even though they recognize that the main effects and two-factor interactions are not confounded, even partially, with the foldover design that they used. In particular, they stated that an alternative design would have been a resolution IV design with 96 runs (a design with 48 original runs that are replicated). The type of design to which the authors are referring is not clear.

Another error, which resulted from hand computation instead of using software, is that the authors gave the expression for the mean square of a factor as equal to (effect estimate)^2/20, whereas the correct expression is 20*(effect estimate)^2. This error resulted in some very small mean squares being computed with the consequence that no F-statistic exceeded 2.78 and some of the statistics were practically zero. The authors concluded that only three of the factors were significant, but that was also an error because, with the error term having 61 degrees of freedom and each factor of course having one degree of freedom since there are two levels for each factor, $F_{1,61,.05} = 4.0$. Thus, technically, the authors should have concluded that nothing was significant.

Finding that only 3 of the 18 factors were significant led the authors to state "Thus the analysis of Plackett–Burman design alone may lead to erroneous conclusions." The problem was actually the authors' analysis. As long as the design is carried out properly and the experimenters have done a good job of brainstorming and selecting a set of factors to study, the analysis of main effects using a resolution IV design should not be misleading. A correct analysis cannot be performed by the reader because the 80 data values were not given, but such an analysis would presumably show more than three significant factors. The lesson to be learned here is that it is best to rely on software, especially when something other than a very simple design is being used.

There are many other examples in the literature of the use of a PB design in biotechnology, including an experiment described by Wen and Chen (2001). They used a PB design to identify significant factors and then used a central composite design for the factors that emerged as being significant in the analysis of the PB design data (see Exercise 13.7).

Case Study

Another example of the use of a PB design, this time in the training of engineers, was given by Anthony (2002). I used catapult experiments to train industrial personnel in experimental design, as have other short course instructors and consultants. The general idea is to use a small catapult (which catapults a ping-pong ball) to identify and illustrate effects that influence the distance the ping-pong ball is catapulted. In this particular training exercise, the team of engineers identified seven factors that were deemed worthy of study. These factors were ball type/color, rubber band type, stop position, peg height, hook position, cup position, and release angle. The design that was selected was a PB design with 12 runs. The design layout for these seven factors, denoted by $A–G$ respectively, is given below, in addition to the values of the response variable, Y.

A	B	C	D	E	F	G	Y
Y	BR	4	1	1	5	170	119
Y	BL	1	4	1	5	Full	161
P	BL	4	1	4	5	Full	253
Y	BR	4	4	1	6	Full	249
Y	BR	1	4	4	5	170	114
Y	BL	4	1	4	6	Full	227
P	BL	4	4	1	6	170	214
P	BR	4	4	4	5	170	327
P	BR	1	4	4	6	Full	304
Y	BR	1	1	4	6	170	60
P	BL	1	1	1	6	170	18
P	BR	1	1	1	5	Full	78

This is how the design was presented in Anthony (2002), but there is a not-so-obvious error in the design layout, which becomes apparent when we try to analyze the data. The design should be orthogonal but it isn't, which is the message that one receives when trying to analyze the data using, say, MINITAB. For factor B, one of the BRs must be a BL so that there are six of each. (Note that this is an example of what was discussed in Section 4.10 regarding bad data.)

But which one is incorrect? If this were the only error in the layout (which it is), we could determine the error using a process of elimination since for each pair of factors, each pair of levels must occur three times. Starting with factor A, we conclude that the error must be in row 1, 4, 5, or 10; looking at factor C tells us that it must be 5, 9, 10, or 12; from examining factor D, we can tell that it must be 4, 5, 8, or 9. At this point, we have determined that if there is only one error, it has to be that the BR in row 5 must be a BL because that is the only number that is common to all three sets of numbers. Of course there could be more than one error, however, so we need to continue with the other factors. Doing so results in the sets (5, 8, 9, 10), (1, 5, 8, 12), and (1, 5, 8, 10) for factors E, F, and G, respectively. Since "5" also appears in each of these sets (and is the only number to appear in all six sets), we know that there is only one error, which we have identified. We could also identify the error by generating the design with software and matching that up with the design in the article. This also leads to the identification of the single error.

Without knowledge to the contrary, we will assume that this was just a printing error that occurred when the article was printed and that the experiment was properly conducted. In his analysis of these data, Anthony (2002) presented a Pareto chart of the standardized effect estimates, with $\alpha = .10$ as the cutoff for judging significant effects. Five of the seven main effects were judged significant, with the model containing only main effects. (The same result is obtained when a normal probability plot is used since both methods in MINITAB are based on Lenth's pseudostandard error.)

Although this result would seem to suggest that the team of engineers was very successful in identifying which factors to study, in practice we would not expect to see such a high proportion of effects be significant.

Therefore, we need to see why this happened. There are 11 degrees of freedom available for estimating effects, but only seven of them were used; the others being used to estimate the error term. If we try to use the other four degrees of freedom to estimate four interaction terms, we have a problem because the interactions are not orthogonal to (all of) the main effects. Furthermore, the correlations can be moderate in size, which can create problems with interpretation. (E.g., adding the AE interaction to the model creates correlations of .33 in absolute value between that interaction and most of the main effects). The AE interaction plot is given in Figure 13.1.

Despite the difficulty of interpreting this interaction because of its partial aliasing with several main effects, one thing is inarguable: the three greatest distances all occurred with the pink ball and the high setting on the hook position—something that is unlikely to occur due only to chance. It is also worth noting that there is (apparently)

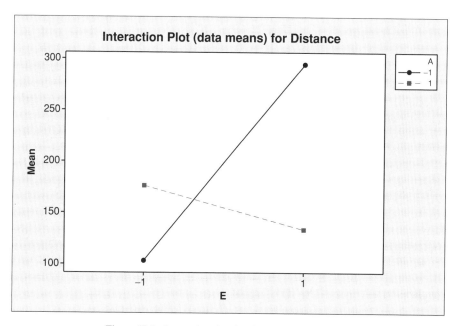

Figure 13.1 Interaction plot showing extreme interaction.

a very large hook position (E) effect with the pink ball. Since the decision was made to run a second experiment using only the pink ball and varying the other four factors in a factorial design, we would expect to see a sizable hook position effect in that experiment.

If we follow Hamada and Wu (1992) and Wu and Hamada (2000), we can employ stepwise regression with a combination of main effects and interactions in an effort to gain insight into possible interaction effects. By using a combination of their three-step approach and stepwise regression applied to all main effects and two-factor interactions, we arrive at a model with the following effects: AE, EG, F, AB, CE, and AC, BF, and AF, with the terms entering the model in this order. This model, which has nine terms, has an R^2 of .9998, which is considerably higher than the R^2 of .9657 for the model with the seven main effects (plus the constant term). Thus, the model with almost all interaction terms explains essentially all of the variation in the distances, but we have to realize that estimating eight effects with 11 degrees of freedom is spreading the data very thin. Of course a model with almost all interaction terms would be highly suspect as an algorithmic selection of such a model would beg for a conditional effects analysis.

Nevertheless, these results suggest that a subsequent experiment (and there was one) should permit the estimation of these seven interactions so as to determine if these interactions are real. This was not done, however, because only the other four factors from the initial experiment (C, D, E, and G) were used in a 2^4 with two replicates follow-up experiment, with factor F set at the high level and, as mentioned, the pink ping-pong ball used.

An oddity occurs with the data from the PB design because if simple linear regressions are run using F, AE, and EG in turn as the predictor variables, the R^2 values are .0003, .374, and .069, respectively, which adds to .4433. When all three terms are used in a regression model, $R^2 = .929$, which is obviously far in excess of .4443. This is an example of "the whole is greater than the sum of the parts," as discussed by Ryan (1997, p. 177) and is similar to the example given therein. This peculiarity can occur only when effect estimates are correlated, as they are here.

So does this mean that there really is an F effect, even though there appeared to be no chance of that when only main effects were considered? The "partial alias" structure for a 12-run design with seven factors is given by Wu and Hamada (2000, p. 351). (This is also automatically produced by the Design-Expert software.) Assume, for the sake of illustration, that the only significant effects were F, AE, and EG, which is not far from the actual situation since they account for 93% of the variation in the response. Then, from Wu and Hamada (2000), $E(\widehat{F}) = F - \frac{1}{3}AE - \frac{1}{3}EG$. Since both signs are negative, these interactions could create the false impression that F is not significant if the true F effect is positive and of about the same order as the sum of $-\frac{1}{3}AE$ and $-\frac{1}{3}EG$. Of course we don't know the true effect nor can we estimate it unambiguously here, but this is a plausible explanation.

In the follow-up experiment (i.e., the 2^4 with two replications), the effects that were found to be significant were, in terms of the original letter designations: D, G, E, C, DE and CG, in the order of effect size. It is worth noting that Anthony (2002) used a Pareto effect analysis of these data, just as was done with the first dataset, although that isn't necessary here since the design is replicated. In fact, an analysis using ANOVA also shows a significant CDE interaction, although the p-value at .044 is borderline.

Since the objective with such an experiment is to determine the factor-level combinations that maximize distance, the use of the 2^4 design is questionable, as is the decision to fix the levels of A and F for that experiment, as interactions involving factor A of the magnitude of the one shown in Figure 13.1 could be missed with that strategy. Since the A effect was strongly significant in the first experiment and the non-significance of the F effect was apparently due to its partial aliasing with certain large two-factor interactions, it might have been better to include at least one of these two factors in a $1/2$ fraction so as to, in particular, determine the F effect untangled from interactions and to see if any two-factor interactions involving factor A were significant in the follow-up experiment. A 2_V^{6-1} design would have allowed the estimation of all two-factor interactions, and if unreplicated would have required the same number of runs as the replicated 2^4 design.

13.4.2 Supersaturated Designs

These are designs for which the number of design points is less than the number of terms in a model. This of course means that not all effects can be estimated, and it also means that the design is not orthogonal.

Such a design might seem to have little value, but these designs, with the idea first proposed by Satterthwaite (1959) for use in random balance designs and studied systematically by Booth and Cox (1962), can be used advantageously when a small number of significant effects is expected. Since this early work, Allen and Bernshteyn (2003) studied the then-existing supersaturated designs and that motivated them to develop an improved class of designs that maximized the probability of identifying active factors. Liu, Ruan, and Dean (2005) subsequently constructed a class of supersaturated designs by augmenting the k-circulant designs of Liu and Dean (2004). The new class of designs contains the designs of Eskridge, Gilmour, Mead, Butler, and Travnicek (2004) as a special case.

In recent years there has been research interest in the construction of three-level and mixed-level designs that are supersaturated. The former are discussed in Yamada and Lin (1999), Yamada, Ikebe, Hashiguchi, and Niki (1999), and in Section 15.5 of Hinkelmann and Kempthorne (2005). Mixed-level supersaturated designs are discussed in Fang, Lin, and Ma (1999), Fang, Lin, and Liu (2003), and Xu and Wu (2005).

Some care must be exercised in the selection of a design, however, as a supersaturated design that departs considerably from an orthogonal design could produce misleading results. This can especially occur if the departure from orthogonality is more than slight, as found by Abraham, Chipman, and Vijayan (1999). In particular, stepwise regression is one method that has been used for identifying the effects that should be estimated, but stepwise regression can easily fail to make the appropriate determination when the correlations between the columns in the design are not quite small. Accordingly, Abraham et al. (1999) recommended that all possible regressions be used instead of stepwise regression (as had been recommended by some researchers) in trying to identify significant effects. (See Ryan (1997) for information on these regression methods.)

One analysis method that is presently advocated (see Lin, 2003) is to use a nonconvex penalized least squares approach, as described in Li and Lin (2002, 2003), with the second of the last two papers available at http://www.sinica.edu.tw/~jds/JDS-134.pdf. It should be noted that this method, which requires special software, is really an estimation method rather than a variable selection method. Nevertheless, it performed well in the numerical study performed by Li and Lin (2003). More recently, Lu and Wu (2004) proposed a method of successive dimensionality reduction that they claim is easier to understand and implement than other recently proposed methods.

13.4.2.1 *Applications*

Two applications of supersaturated designs were described by Bandurek (1999). We will discuss the first of these and critique the analysis, which was rather peculiar.

Booth Dispensers Ltd. manufactures soft drink coolers and cooler carbonators for use in vending machines. One major customer for one of the carbonators needed to be convinced that the product would work in their vending machines. A team of people from both the supplier and the customer identified 20 factors that could affect performance. An orthogonal array with eight runs was selected for the experiment. These are too few runs for a statistical analysis, such as by using stepwise regression as the correlations would undermine such an approach. The experimenters seemed to realize this, so they ordered the responses from best to worst, with the level of each factor recorded on the sheet of paper along with the response value. The experimenters wanted to see if any factors would have level 1 at the top of the pile and level 2 at the bottom, which would suggest to them that such factors were significant. They "shuffled the deck" before looking at the results, then tried to combine engineering judgment with the results of the experiment. Although the results one year later apparently suggested that the experiment was a success, the analysis was quite crude and is not recommended. It is somewhat reminiscent of Taguchi's "pick-the-winner strategy" and Shainin's variables search approach, both of which have low power for detecting significant effects. (See Ledolter and Swersey (1997) for information on the latter.)

13.4.3   Lesser Known Screening Designs

Lin (2003) presents some screening designs that are not well known, one of which is *p-efficient* designs. These are saturated designs introduced by Lin (1993a,b) that are reasonably efficient for estimating the parameters of projective submodels, from whence the design derives its name. The designs are near-orthogonal. Box and Tyssedal (1996b) considered the projective properties of two-level 16-run orthogonal arrays that are due to Hall (1961). See also the simplex designs of Mee (2002).

In general, as has been stressed in this book beginning with the start of Chapter 5, properties of the design that corresponds to the initial model are not what's important since a model with fewer terms than the number of effects that can be investigated with the design used is almost certainly going to be used due to effect sparsity. The properties of the design for the model that is actually used is obviously what is important.

13.5   DESIGN OF EXPERIMENTS FOR ANALYTIC STUDIES

Investigations involving experimental designs are generally enumerative studies; that is, the objective is to assess the current state of affairs regarding the relationship between a response variable and one or more factors. With an analytic study the objective is to predict the future, with the terms *analytic studies* and *enumerative studies* due to Deming (1953, 1975). See also Hahn and Meeker (1993).

In an enumerative study, blocks are generally assumed to be selected at random, and block × treatment interactions are assumed not to exist. In analytic studies, however, blocks should be selected to cover extreme conditions and to identify block × treatment interactions. That is, if, say, process yields are to be predicted, we would want to know how important certain process variables are when the operator on duty is the company's best, versus the importance of those variables when the worst operator is on duty. Thus, it would be much better to deliberately select the particular operators that will serve as blocks rather than simply selecting them at random. For more discussion of blocking in analytic studies, see León and Mee (2000).

13.6 EQUILEVERAGE DESIGNS

An unstated objective in experimental design is that the design points should exert equal, or at least nearly equal, influence on the effect estimates, and equivalently on the regression coefficients in the model representation.

Of course whether or not an observation is influential will depend in part on the Y-coordinate, but we can improve the chances of not having any influential points by using design points that are equally influential; that is, by constructing the design so that the design points have equal leverages.

A leverage value reflects the distance that a point is from the "middle" of the group of points. With this semidefinition in mind (a formal definition will soon be given), it is easy to see why common designs such as two-level full and fractional factorial designs are equileverage designs since the design points are at the vertices of a hyperrectangular region, such as in Figure 4.5, and are thus equidistant from the center, which would have coordinates of zero for all the factors.

It is important to keep in mind that leverage values are determined by the model that is fit, however, not just by the design. Therefore, we cannot speak solely of leverages for particular designs. As we have seen in previous chapters, designs allow various models to be fit and it is the model in conjunction with the design that determines the leverage values.

To illustrate this point, assume that a 2^3 design is employed and only main effects are used in the model. Let the vector x_i denote a column whose first value is 1 and remaining values are the ith value for each of the terms in the model, and let X denote the model matrix, whose first column is a column of 1s and other columns contain the design values for the terms in the model. The n leverages for the n observations are then given by $x_i^T (X^T X)^{-1} x_i$. For example, if only main effects are fit using the data from the 2^3 design, the leverages are all 0.5, whereas if the two-factor interactions are added to the model, the leverages become 0.875. In an equileverage design the leverages are p/n, with p denoting the number of parameters in the model. When only the main effects are fit, that number is 4, so the leverages are $4/8 = 0.5$. When the three 2-factor interactions are added, the leverages are then $7/8 = 0.875$. For a saturated design, the leverages of course are 1.0, although leverages for a saturated design are not given by certain software packages (e.g., MINITAB).

If we construct a design whose leverages are equal or at least almost equal, we reduce the likelihood of having any influential data points, as a point could then only be influential through its Y-value. Despite the fact that this is obviously desirable, researchers have almost completely ignored equileverage designs as there are apparently only two research papers in which the term has been used, and only one paper devoted to the construction of such designs.

Equileverage designs are not without weaknesses, however, as they cannot be used to detect nonlinearity since the points are on the faces of a sphere and no points lie in the interior of the sphere. Consequently, an equileverage design would have to be modified slightly to provide nonlinear detection capability.

It is important that equileverage designs be considered and this can be motivated by the fact that various diagnostics break down when leverages are 1.00, as occurs with some very common designs such as some designs given by Box and Wilson (1951), Box and Behnken (1960), and Roquemore (1976), as discussed by Jensen (2000).

13.6.1 One Factor, Two Levels

The simplest and most intuitive way to introduce and understand equileverage designs is to consider a design for a single factor with two levels. Thus, starting from the model for one factor (see Section 2.1.1):

$$Y_{ij} = \mu + A_i + \epsilon_{ij}$$

We may equivalently write the model as a simple linear regression model:

$$Y_{ij} = \beta_0 + \beta_1 X_i + \epsilon_{ij}$$

In simple linear regression, it is well known that the leverages may be written as

$$h_{ii} = \frac{1}{n} + \frac{(X_i - \overline{X})^2}{\sum_{j=1}^{n}(X_j - \overline{X})^2}$$

Obviously, the leverages will be equal only if all of the X_i are equidistant from $\overline{X}$, and that will happen only if n is even and half of the design points are at one value and the other half at the other value.

It can also be seen that $\text{Var}(\widehat{\beta_1})$ is minimized with $n/2$ design points at each of the two X_i values, as becomes apparent when we recall that $\text{Var}(\widehat{\beta_1}) = \sigma^2 / \sum(X - \overline{X})^2$. It could also be shown that $\text{Var}(\widehat{\beta_0})$ is independent of the choice of the X_i. (As an aside, if a no-intercept model had been fit, the leverages would still be equal but would of course be different from the leverages when the intercept is in the model.)

Unfortunately, however, this design would not permit the detection of nonlinearity, and we generally want to use designs that allow us to detect model misspecification since all models are wrong. (Model-robust designs are discussed briefly in Section 13.20.) If, however, we start from an equileverage design, and then make suitable modifications to facilitate the detection of possible nonlinearity, we will be able to avoid high leverage points.

13.6.2 Are Commonly Used Designs Equileverage?

Since high leverages are outliers in the design space, we might not suspect that there are designs with high leverage values. It is true that many of the commonly used designs are equileverage designs. For example, 2^k and 2^{k-p} designs are equileverage designs for any model that can be fit with one of these designs provided that no centerpoints are used, and the same can be said of Box–Behnken designs without centerpoints. Similarly, central composite designs, discussed in Section 10.5, are equileverage designs if no centerpoints are used, but of course centerpoints are used with central composite designs.

In general, the equileverage property is lost for factorial designs with more than two levels (but not always), as well as for other types of designs with more than a few levels. For example, the leverages of a 5^2 design for fitting only the linear terms are .04, .06, .08, .12, .14, and .20. Thus, the leverages differ considerably, although .20 would not be considered a high leverage value using any reasonable rule of thumb.

Given below is a table of designs and the corresponding leverages, with the number after the leverage value indicating how many times the indicated leverage value occurs, with the design used to fit only linear terms with no interactions.

Design	Leverages
2×3	$(.5833)4; (.3333)2$
$2 \times 2 \times 3$	$(.25)4; (.375)8$
$2 \times 2 \times 2 \times 3$	$(.1667)8; (.2292)16$

Notice as we move down the table that the design moves closer to being equileverage as the factor that spoils the equileverage property (i.e., the last one) becomes a progressively smaller part of the design.

Each of the last two designs is an equileverage design for fitting the main effects and all two-factor interactions as the leverages are all 0.8333 for the first design and 0.625 for the second design. (There are not enough degrees of freedom to fit the interaction term in the first model as doing so would use all of the five degrees of freedom.)

13.7 OPTIMAL DESIGNS

Although not as popular as two-level full and fractional factorial designs, optimal designs have been used in many applications and are especially useful where there are restricted regions of operability (see Section 13.8). Various statistical software packages have the capability of producing designs that are optimal for a given optimality criterion. (These are discussed in the next section.)

Optimal designs are not without their critics, however, and an optimal design won't necessarily be the best design to use in a given situation. For example, consider a single covariate and a response variable, and a simple linear regression model is posited as being a good model that represents their relationship. The design that minimizes the variance of the slope estimate has half of the design points at one value of the covariate and the other half at the other value, as indicated previously. A regression equation (fitted line) would not be very practical as, in particular, it would not allow for nonlinearity between the two points to be tested. A similar point is made by Fedorov and Leonov (2005) regarding logistic regression models. Matthews and James (2005) make the important point that in general applications of optimal designs, the number of distinct points will usually be limited, so there may not be enough distinct data points to permit model verification.

In addition to designs that are optimal in terms of a statistical criterion, designs can also be constructed that are optimal in terms of cost. These are considered in Section 13.12.

13.7.1 Alphabetic Optimality

Some of the properties that a good design should possess were listed in Section 1.2. There has been much discussion in the literature about construction of designs that meet a certain optimality criterion, and the collection of such criteria has been informally referred to as alphabetic optimality. For example, letting $\mathbf{X}$ denote the design matrix, for a collection of $m > n$ candidate design points, the set of n points that maximizes $|\mathbf{X}'\mathbf{X}|$ is the D-optimal design, *for that set of candidate points.*

Since the design matrix is part of the criterion, this means that a design that meets the criterion is D-optimal only if the model that is assumed is correct, but of course the model is never correct.

So there are two problems: a design can be D-optimal only for the candidate points that the experimenter has in mind, but the guarantee of D-optimality for those points hinges on the false assumption that the model is correct. (See Myers and Montgomery (1995, p. 384) for a somewhat similar but more extensive discussion than that given here.) It is important to note, however, that the criticism of optimality strictly applying only for the postulated model has motivated research on model-robust optimal designs, so not all optimal designs have this model-dependency weakness. In particular, DuMouchel and Jones (1994) addressed the model-dependency problem for D-optimal designs.

Another problem with D-optimal designs, as pointed out by Montgomery, Loredo, Jearkpaporn, and Testik (2002), is that if we start with a standard design such as a 2^3, which is in fact the D-optimal design for 3 two-level factors, as can be verified with experimental design software such as Design-Expert, and then request that one point be added at a time to satisfy the D-optimal criterion, the points that are added won't necessarily be the points that an experimenter would want to add. Specifically, if a few points were to be added to a 2^3 design, conventional thinking is that the points should be added to the center so that lack of fit can be tested and the experimental error variance can be estimated. This does not happen, however, as when one point is added, the coordinates are $(-1, -1, -1)$ which of course is a corner point. If another point were added, it would have the same coordinates, as would a third added point and a fourth added point. Thus, the corner point continues to be replicated. This does produce a pure error estimate with four degrees of freedom for $n = 12$, but we might question having that estimate obtained from a point that is well removed from some of the other points rather than being at the center of the design space. Certainly, the latter would generally be preferable.

For these reasons, a D-optimal design is certainly subject to criticism. One alternative is to use a Bayesian D-optimal design, which can be used to check for adequacy of the fitted model. Let k denote the number of primary terms that are to be used in the model, and let q denote the number of potential terms. A Bayesian D-optimal design maximizes the determinant of $(X^T X + K)$, with K a $(p + q) \times (p + q)$ diagonal matrix whose first p diagonal elements are equal to zero and last q diagonal elements equal to 1. The latter is how the Bayesian D-optimal design uses the potential terms to force in runs that permit possible model inadequacy of the model containing only the primary terms to be checked.

This design can be generated using JMP and is illustrated in the help file for "Bayesian D-optimal Designs" for the simple case of a 2^2 design with the potential terms being the two pure quadratic terms. When five points are requested instead of four and the estimation of the pure quadratic terms specified as "if possible," JMP adds a centerpoint. If six points are requested, an additional point with coordinates $(1, 0)$ is added.

There are other optimality criteria (in particular, G-optimality, A-optimality, E-optimality, I-optimality, and L-optimality), which are discussed in detail in books on optimal design such as Silvey (1980) and Pukelsheim (1993)), but alphabetic optimality has received criticism over the years (and understandably so), such as from Box (1982) and Myers (1999). The latter, in discussing response surface methodology, stated ". . . in many cases alphabetic optimality is thus not appropriate." Box et al. (2005, p. 460) stated, "The attempt to compress the information about a design into a single number seems mistaken."

As explained by Atkinson and Bailey (2001), the basic optimum design ideas were set forth by Kiefer (1959), whose paper read before the Royal Statistical Society apparently had a rough reception, especially in the written comments.

This is not to suggest that optimum designs should be abandoned, as optimal mixture designs, in particular, have proven useful (Piepel, 2004). In general, optimal designs are valuable when there are constraints on factor-level combinations, as always occurs with mixture designs since the component factors must sum to 100 percent, in addition to the type of constraints discussed in the next section that can occur with nonmixture designs. Under such circumstances, the commonly used designs are no longer applicable, at least not without modification, so automated generation of designs can be quite useful.

Thus, optimal designs are no longer deserving of all the criticism that these designs have received. We should keep in mind, however, that sequential use of designs is usually preferred, whenever possible, and when we emphasize the sequential nature of experimentation, we logically reduce attention on the need for any one design to be optimum in any sense.

Optimal designs must be constructed by computer and one of the best ways of doing so is by using GOSSET, an extremely comprehensive experimental design program developed by Neil Sloane and Ron Hardin at AT&T. Originally proprietary software, it is now freeware. It does not run on all major platforms, however, as it runs only on the Unix, Linux, and Mac operating systems. It can be used to generate designs that are I-, A-, D-, or E-optimal. Other notable software programs that will generate optimal designs include Design-Expert, which will generate D-optimal designs using the CONVERT algorithm (Piepel, 1988) to find the vertices of the design. For example, the design given in Table 13.1 for five factors at two levels each was generated using Design-Expert, assuming a model with all main effects and two-factor interactions. Notice that 21 design points were generated from $2^5 = 32$ candidate points, the number

TABLE 13.1 D-optimal Design for five Factors Generated Using
Design-Expert

Row	A	B	C	D	E
1	1	−1	1	−1	1
2	1	1	−1	−1	1
3	−1	−1	−1	−1	−1
4	−1	−1	1	−1	−1
5	−1	1	−1	−1	−1
6	1	−1	−1	1	1
7	1	1	−1	1	−1
8	−1	−1	−1	1	−1
9	1	1	1	1	−1
10	1	1	−1	−1	−1
11	1	1	1	1	1
12	1	−1	−1	−1	1
13	−1	1	−1	1	1
14	−1	1	1	1	1
15	−1	−1	−1	−1	1
16	−1	1	1	1	−1
17	1	−1	1	1	−1
18	1	1	1	−1	−1
19	−1	−1	1	1	1
20	−1	1	1	−1	1
21	1	−1	−1	−1	−1

of points being determined by the algorithm and not by the user. (This is not true for all software packages that generate optimal designs, however.)

Of course the design is not orthogonal, as it obviously could not be for an odd number of design points and an even number of levels, but the departure from an orthogonal design is not great. (This can be demonstrated by using the five columns for the design and computing the "correlation matrix." Specifically, all of the pairwise correlations are .028 and −.028.)

A method for generating large D-efficient designs was given by Kuhfeld and Tobias (2005). (D-efficiency is explained at the end of this paragraph.) The authors presented ". . . experimental design tools that originated in industrial statistics but have been significantly extended to handle problems in marketing research." The approach they gave is claimed to generate 115,208 orthogonal arrays. The Appendix of their paper contains a large number of the orthogonal array parent designs, ranging from 4 runs to 513. (Note that these designs were termed D-efficient rather than D-optimal. It is commonplace to describe a "near-optimal" design, using a measure of the efficiency of the design relative to a, perhaps hypothetical, orthogonal design of the same size. Of course for a very large number of factors and a huge number of design points, we should not expect to be able to construct an optimal design with any algorithm. The authors refer to one instance in which their macro ran for 2 hours and produced a design with a D-efficiency of 80.521%. A formal definition of D-efficiency is given by Waterhouse (2005, p. 9). Simply stated, it is the determinant of a given design matrix divided by the determinant of the D-optimal design, for a given size, with that fraction raised to the $1/p$ power, with p denoting the number of parameters to be estimated.

Heredia-Langner, Carlyle, Montgomery, Borror, and Runger (2003) reviewed the proposed methods for generating optimal designs and presented a genetic algorithm, used in computer science, which has certain advances over other methods. In particular, the candidate points need not be enumerated. Berger and Wong (2005) is a relatively small edited work that contains contributed chapters on applications of optimal designs in various fields. In Chapter 10, "The Optimal Design of Blocked Experiments in Industry," the authors, P. Goos, L. Tack, and M. Vandebroek, contend that it is important to incorporate cost considerations in optimal design theory, while admitting "... the literature on that topic is very scarce." They do provide an example in which they illustrate how an optimal design can be constructed that uses cost information when cost information on hard-to-change factors is available, at least in ratio form relative to the costs of factors that are not hard to change.

It should be noted that there are many potential applications of optimal designs when not all of the relevant factors can be controlled. This issue is addressed by Lopez-Fidalgo and Garcet-Rodríguez (2004) and in the papers that they cited. Applications of optimal designs in various industries are discussed briefly in Section 13.7.2.

13.7.2 Applications of Optimal Designs

Optimal designs seem to be used somewhat infrequently when experimental designs are employed, at least in some fields of application. For example, Dette, Melas, and Strigul (2005) state that "... at present, there are only a few real applications of optimal designs in microbiological practice." An exception is the use of D-optimal designs for identifying parameters of the nitrification model for activated sludge batch experiments (Ossenbruggen, Spanjers, and Klapwik, 1996). Other examples of applications of optimal designs for microbiological models, with references, are given in Table 6.2 of Dette et al. (2005). Matthews and James (2005) made several cautionary comments, similar to those in Section 13.7.1, in discussing the application of optimal designs in the measurement of cerebral blood flow using the Kety–Schmidt technique. In particular, they made the point that the model used in their application was relatively new, so they would prefer to have more design points than the limited number of points that generally occurs with optimal designs. When there is not much, if any, evidence to support a particular model, a design such as a uniform design (see Section 13.9.1) might be a better choice.

There are various other fields in which optimal designs have been used, as indicated by Berger and Wong (2005). These include toxicology, pharmacokinetic studies, bioavailability studies for compartmental models, rhythmometry, cancer research, and various other fields. Although this may sound impressive, we should keep in mind that there is a huge number of statistically designed experiments and undoubtedly experiments in which optimal designs are used are a very small fraction of the total number of experiments.

13.8   DESIGNS FOR RESTRICTED REGIONS OF OPERABILITY

Implicit in the use of factorial and fractional factorial designs is the tacit assumption that the design points correspond to feasible operating conditions. Frequently, this will not be the case, however, as certain combinations of factor levels may not be feasible, and might even cause hazardous conditions, as in chemical reactions. Kennard and Stone (1969) were apparently the first to discuss irregular experimental regions in the literature; Snee (1985a) gave 10 examples of experiments with irregular regions and illustrated those regions graphically; Czitrom (2003, pp. 20–24) described three experiments for which there was a restricted factor space; and Bingham (1997) described an application to fastener quality in which certain factor levels could not be used together. Similarly, Cheng and Li (1993) described an experiment conducted in Taiwan in 1990 in which no measurements could be made when all of the factors were at their lowest levels because a bar-code verifier couldn't detect a concentration. Montgomery et al. (2002) discussed experimental designs for restricted regions in terms of design efficiencies. Giesbrecht and Gumpertz (2004, p. 413) state "In practice, it is more common to find severe constraints that limit the investigation to irregular regions." Those constraints will usually be in the form of linear inequality constraints, but Reece (2003, p. 386) reports encountering complex constraint relationships between factors that require second-order expressions to describe.

Example 13.1

Reece and Shenasa (1997) gave an example of an experiment for which there was a restricted region of operability that unfortunately was not discovered until the design for the experiment had been constructed. Specifically, a 2_{IV}^{7-3} design with five centerpoint replicates was constructed. (It is worth noting, especially relative to the message in Section 1.7, that prior analysis of data from the relevant process showed that it was stable and suitable for experimentation.)

Before the experiment was run, the pumping capacity of the furnace system that was to be used in the experiment was checked. It was found that the system could not provide 150 MTorr units of pressure when the total flow of gases was 400 SCCM. Consequently, the four treatment combinations that had this combination of levels of these two factors had to be either removed or modified. The team of experimenters decided to modify those four treatment combinations in such a way that the total flow of gasses was set at 200 whenever the pressure was 150. Six addition runs were added, using the RS/1 software, so as to create a D-optimal design. Only five of these runs were used, however, due to time constraints.

Example 13.2

Another example of a restricted factor space (or *debarred observations*, using terminology, although perhaps not the best, which has been used in the literature) was given by Buncick, Ralston, Denton, and Bisgaard (2000). There were five factors, and a 2_V^{5-1} design with $I = ABCDE$ was originally selected for use. Four of the 16 treatment combinations had to be eliminated, however, for the following reason. Two of the factors were heat (none, 160) and substrate holder rotation (off, on). When there was no rotation, it was not possible to accurately position the substrate over the sputter source after the substrate was heated. Of course this creates an unbalanced design so the decision was made to use only the runs at the high level of the sputter deposition power setting. The number of runs used would thus be eight (half of 16) minus the number of debarred observations that were at the high level of that factor, which was 2. Thus, the number of runs was cut to 6. The runs were analyzed as two 2_{III}^{3-1} designs with two overlapping runs. Thus, the debarred observations presented a complication that required a method of dealing with it.

Since restricted factor spaces frequently occur, methods of constructing designs under such conditions should be carefully considered. Kennard and Stone (1969) used their CADEX algorithm to select design points to uniformly cover the experimental region. This does not mean, however, that we can automatically obtain a design with good properties by using such algorithms, as a good design may not be feasible. This is illustrated later in the section.

Zahran, Anderson-Cook, Myers, and Smith (2003) considered the modification of a 2^2 factorial design to accommodate restricted design spaces. Specifically, they considered the case where the design point $(-1, 1)$ is not feasible and must be replaced. Replacing that point but retaining the other three points will result in a nonorthogonal design, so the experimenter will have to choose between the nonorthogonality that results from replacing only the impossible point, or changing three of the four points but retaining orthogonality. The latter is achieved by moving just outside the region of nonoperability, then changing two of the other three points so as to create a rectangle.

For example, assume that the impossible point is $(1, 1)$, as in Zahran et al. (2003). In order to create an orthogonal design with minimal disturbance of the coordinates of two of the other three points, we would need to know more than just the fact that $(1, 1)$ is an impossible point. For example, is $(.9, .9)$ also an impossible point?

Assume that the region of inoperability is given by the square that is enclosed by the points $(.95, .95)$, $(.95, 1)$, $(1, .95)$, and $(1, 1)$. A very simple way to construct an orthogonal design would be to use, for example, the points $(.94, .94)$, $(-.94, .94)$, $(.94, -.94)$, and $(-.94, -.94)$, which would then of course be rescaled so that the $(-1, +1)$ notation could be used. This would be accomplished by using a scale factor of 0.94. Of course with the rescaling, the points would correspond to the factor values in the original units to which the $(-.94, +.94)$ notation corresponds in the coded units.

To illustrate, assume that the original units were temperature (°F), with values of 300 and 400, and pressure (in psi) with values of 200 and 300. Then .94 for temperature would correspond to $.94(50) + 350 = 397$ and $-.94$ would correspond to $-.94(50) + 350 = 303$. The pressure values would be similarly obtained and can be shown to be 203 and 297, as should be intuitively apparent.

A possible objection to this design from an experimenter would be that all of the design points are odd numbers, and another possible objection is that each coordinate of the other three points is changed, albeit slightly.

Zahran et al. (2003) did not consider this approach to altering the original design points. Instead, they stated that an orthogonal design could be constructed using the design points as $(-1, -1)$, $(-1, a)$, $(a, -1)$, and (a, a). But such a design, *without recentering*, is not orthogonal unless $a = 1$, which would be the original design. The recentering would have to convert the center of the design from the original $(0, 0)$ to $[(a - 1)/2, (a - 1)/2]$. Then the scaling factor would have to be $(a + 1)/2$, with the latter being half the distance between the coordinates for each factor. Then by using the coordinate values for each factor (i.e., a and -1), we can easily see that we obtain the usual $+1$ and -1 for each coordinate value with the appropriate scaling.

Specifically, $[(a - (a - 1)/2]/[(a + 1)/2] = 1$, and $[(-1 - (a - 1)/2]/[(a + 1)/2] = -1$. Thus, the design points given by Zahran et al. (2003) really do constitute an orthogonal design, but only if we view these points as not the coded values but rather view them as coordinate points which when properly coded will produce the usual $+1$ and -1 values. One way to view the design points given by Zahran et al. (2003) is to view them as uncoded values (since the center is not $(0, 0)$), so that the same coding convention would be applied as is applied when the starting point is original units of factors. Another way of looking at it is the following: if the four points are the corner points of a rectangle, then an orthogonal design can be constructed with appropriate centering and scaling. Think of it this way, if we have two variables (factors) whose pairs of values form a rectangle, then there is complete absence of a linear relationship between the variables. Then the correlation between the two variables would be zero, which would mean that the variables could be made orthogonal with appropriate center and scaling.

One thing to keep in mind is that if centerpoints are to be used, they will not be the original centerpoints because those points will not be in the center of the rescaled design. For example, using the temperature and pressure values, the original center would have been 350 for temperature and 250 for pressure. If, as in the example above, we use (.94, .94) as the coded values, then those coordinates would correspond to (397, 297). Similarly, the other points would have original-unit coordinates of (300, 200), (300, 297), and (397, 200). The center of $[(a - 1)/2, (a - 1)/2]$ is thus $(-.003, -.003)$. This must of course correspond to (348.5, 248.5) in the original units, and we can see that the latter is the center for this set of four points.

We can avoid these entanglements if we simply work with the raw units. Zahran and Anderson-Cook (2003), who present a more general equation for determining the coordinates of the point that replaces the ineligible point, gave an example with two factors that we will use for illustration. The design region was stated as 250–300 for the first factor and 50–75 for the second factor, but values of the first factor above 270 and values of the second factor greater than 70 were believed to cause problems. This simply means that a 2^2 design would be constructed using 250 and 270 for the first factor and 50 and 70 for the second factor.

The JMP-generated 2^2 design for this scenario is given below:

Run	X1	X2
1	250	50
2	270	50
3	250	70
4	270	70

This design can be generated by starting with the desired factor levels and then entering the appropriate constraints, or more directly, simply starting with (250, 270) and (50, 70), respectively, as the levels of the two factors.

Zahran and Anderson-Cook (2003) gave the restricted region in terms of coordinates corresponding to coordinates of the endpoints of the unrestricted region being (0, 0), (0, 1), (1, 0), and (1, 1). The corresponding coordinates of the restricted region were shown to be (0, 0), (0, 0.8), (0.7, 0), and (0.7, 0.8). (Similarly, the coordinates could be written using the more customary $(-1, 1)$ range, which would be $(-1, -1)$, $(-1, 0.6)$, $(0.4, -1)$, and $(0.4, 0.6)$. Of course these numbers also look a bit unorthodox, and coding would have to be performed to produce the more customary set of four coordinates, with each coordinate being either 1 or -1.

Regardless of how we write the coordinates, we have a 2^2 design—and a rectangular region—so the design is orthogonal. If the coordinates are written as in the last set, however, the values would have to be converted to the raw units before the experiment could be run. Since the penultimate set of coordinates would be obtained as $X1^* = (X1 - 250)/(200/7)$ and $X2^* = (X2 - 50)/(25)$, the values of $X1$ would thus be obtained as $(200/7)X1^* + 250$, and the values of $X2$ would be obtained as $25X2^* + 50$. Again, we can avoid the unnecessary algebra just by working with the raw units.

Zahran et al. (2003) also considered ways to construct a design with four or five points to accommodate the restricted region that satisfy each of a few other optimality criteria. Specifically, they focused attention on GH-optimality and Q-optimality, in addition to D-optimality. The Gendex DOE toolkit can be used to construct designs for restricted regions.

This issue of a restricted factor space was also addressed by Cheng and Li (1993), who considered the construction of orthogonal fractional factorial designs under such conditions. Consider the case of a single combination of factor levels that is infeasible. If this combination involved all of the factors, the solution to the problem would be trivial: simply do not use the fraction that contains the infeasible combination.

The problem can be complex, however, if not all of the factors are involved in the combination. For example, assume that a 2_V^{5-1} design is to be used for factors $A–E$ and that the high levels of factors A and C cannot be used together, this being the only restriction. Of course, "high" is arbitrary, so the first question to ask is "can the high levels be reduced slightly and the pair of high levels still be used together?" If a drastic reduction would be necessary, then it would be best to determine if this combination can be avoided by using one of the two fractions. Of course the defining relation would be $I = ABCDE$. There are eight combinations with A and C at their high levels, so the question then is whether or not they would all be in the same fraction.

There are at least two ways to look at this. Since the two fractions are defined by the treatment combinations that have an odd number of letters in common with the defining relation and an even number of letters in common with it, respectively, we can quickly see that not all of the eight treatment combinations will be in one fraction (e.g., ac is even but abc is odd). Thus, the defining relation $I = ABCDE$ won't work, nor would any other reasonable choice for the defining relation. Using Proposition 1 of Cheng and Li (1993), we arrive at the conclusion that the defining relation would have to be $I = -AC$, which of course would be totally unacceptable as this would confound main effects. Thus, we cannot construct a 2_V^{5-1} design.

Of course a 2_{IV}^{5-2} design is a less-desirable design, but some experimentation (or trial and error) would show that for each possible defining relation that produces a set of four fractions, the requirement of not having A and C at their high levels cannot be met. Stated differently, none of the four fractions, which of course each contains eight treatment combinations, contains all eight of the ineligible combinations. Thus, there is not a practical fractional factorial that would accommodate this restriction, as we would have to drop back to an entirely unsatisfactory defining relation such as $I = A = C = AC$ in order to be able to construct a fraction that contains the eight treatment combinations. (Here the eight treatment combinations with A and C at their high levels would be in one fraction, the odd–odd fraction, so any of the other three fractions would work, although being totally impractical since we would not be able to estimate the main effects of A and C.)

Is the problem of constructing fractional factorials to accommodate unacceptable combinations harder when there are more factors involved in the ineligible combinations? Not necessarily. After much work, Cheng and Li (1993) showed that there is one 2^{5-1} design that will accommodate two ineligible combinations, one involving three of the five factors and the other involving four of the factors.

Clearly, an algorithmic approach is needed for determining designs that will accommodate ineligible combinations, and Cheng and Li (1993) described how the Franklin and Bailey (1977) algorithm would have to be modified to accomplish that. Determining feasible designs in the presence of ineligible combinations is a difficult task, especially in the relative absence of software that will accomplish this.

JMP, however, permits the specification of a restricted design space and this can be used to specify ineligible combinations. As a simple example, consider a 2^2 design, which also happens to be D-optimal on the interval $(-1, 1)$ on each factor. If the "Custom Design" feature is invoked and "D-optimal" is selected, and if the constraint $X1 + X2 \geq -1$ is entered, then the design point $(-1, -1)$ is replaced by the point $(0, -1)$. Now consider a 2^4 design that must be modified as the scientists say that both $X1$ and $X2$ cannot be at the high level because of physical considerations. Again the Custom Design feature must be selected and when the constraint $X1 + X2 \leq 1$ is entered, the design that is returned is the 2^4 design except that three values of $X2$ are changed from 1 to 0 when $X1 = 1$. If a D-optimal design is specified with this constraint and JMP is used, the design that is produced is the one that is given below:

A	B	C	D
-1	1	-1	-1
-1	1	-1	1
-1	1	1	-1
-1	1	1	1
1	-1	1	1
-1	-1	-1	-1
-1	-1	-1	-1
1	-1	-1	-1
-1	-1	1	-1
-1	-1	-1	1
-1	1	-1	1
-1	1	1	-1
-1	-1	1	1
1	-1	1	-1
-1	-1	1	1
1	-1	-1	1

Notice that the sum of the levels of the first two factors is indeed less than two and that there is great imbalance in the levels of the first two factors, especially the first factor, for which the high level occurs only four times in the 16 runs. This is the type of price that is paid when a D-optimal design is used with constraints. Notice also that the correlation between the columns for the first two factors is $-.447$, with the correlation induced by the constraint. (This is the only nonzero correlation, however.)

As indicated previously, just having a suitable algorithm won't necessarily be sufficient, as a suitable orthogonal design may not exist. Thus, nonorthogonal designs will undoubtedly have to be considered in many applications. Supersaturated designs might also have to be considered because if there are too many debarred combinations, the number of feasible combinations might be less than the number of parameters in a tentative model. Of course the magnitude of the correlations between the parameter estimates will have to be assessed and carefully considered if such a design is to be used, and effect estimates will be sacrificed since effect estimates don't have any meaning for nonorthogonal designs.

One potential danger when infeasible treatment combinations exist and a suitable design does not exist is that if we back off too far from a region of infeasibility because no practical design exists with the ineligible combinations, we may fail to detect an interaction or two that exists. As a simple example, assume that treatment combination ab is infeasible with a 2^2 design and that an interaction plot would show two lines that would be a considerable distance apart in the vicinity of the ab combination (but not at the other end) if we backed off only very slightly from the raw unit values to which ab corresponds and then constructed a 2^2 design for a slightly smaller factor space. Now assume that we back off considerably from that point so that the factor space is slightly more than half of the original space. Now the lines may not have nearly the vertical separation in the vicinity of the "new" ab combination as was the case previously.

For example, consider Figure 13.2. The moderate interaction is due to the ab combination. If we think about shrinking the factor space so that the factor-level

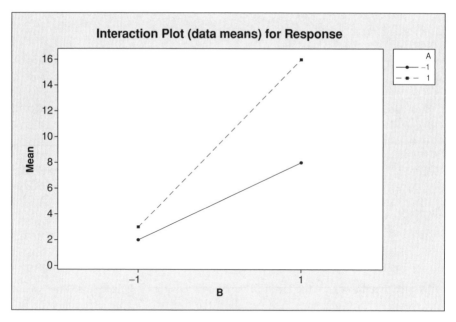

Figure 13.2 2^2 design showing moderate interaction.

combinations $(1, 1)$ and $(1, -1)$ are replaced by $(0, 1)$ and $(0, -1)$, respectively, before rescaling, then there would be only a weak interaction signal.

Obviously, there is a need for research on this topic so that the best design for a given combination of debarred combinations can be identified. In particular, for fractional factorial designs it would be of interest to identify the best fractional factorial design that can be constructed for a given set of debarred combinations, and compare the properties of that design against the properties of the design that results from excluding the debarred combinations from the 2^k runs.

13.9 SPACE-FILLING DESIGNS

Space-filling designs are useful when there is little or no information about the effects of factors on the response but the true model is suspected to be highly nonlinear. These designs are useful for modeling systems that are deterministic or near-deterministic. For such modeling situations, variability is of little or no concern because variability is believed to be practically nonexistent.

A *uniform design* is one type of space-filling design. When there is no prior information, the points might as well be regularly spaced over the design region, which is essentially what is accomplished with these designs. These designs, when used in conjunction with nonparametric regression and possibly semiparametric regression, are potentially beneficial in many applications and industries, including the pharmaceutical, biotechnological, chemical, and process industries. They are also useful in computer experiments when the postulated model is quite complex.

13.9.1 Uniform Designs

This is one type of space-filling design that was originally proposed by K.-T. Fang and Y. Wang in 1980 and presented in Fang (1980) and Wang and Fang (1981). As stated by Fang, Lin, Winker, and Zhang (2000), the design can be considered to be a "nonparametric experimental design." The design was originally motivated by three system engineering projects in 1978 (Fang and Lin, 2003). According to Professor Fang, uniform designs have been used in various fields such as chemistry and chemical engineering, pharmaceutics, quality engineering, system engineering, survey design, computer sciences, and natural sciences. Fang et al. (2000) gave a case study of "launching a dynamic system for a one-off ship-based detector."

The concept of the uniform design is quite simple. To illustrate, assume that we have a single factor and n design points are to be used, with the points uniformly spaced. Assume that the design space on the single factor is (0, 1). The points are to be $(1/n)$ units apart, so that the distance between the first point and the nth point is thus $(n-1)/n$. Since the set of n points should obviously be centered in the interval (0, 1), it follows that with a denoting the first point, $a + (n-1)/n + b = 1$, with b denoting the distance between the nth point and 1. Similarly, $a - b = 0$. Solving these two equations gives $a = 1/2n$. Thus, the design points are $\frac{1}{2n}, \frac{3}{2n}, \frac{5}{2n}, \ldots, \frac{2n-1}{2n}$, which can be written as $\frac{2i-1}{2n}, i = 1, 2, \ldots, n$.

If we wanted to use the usual $(-1, 1)$ as the design space for the factor, then the distance between the points should be $2/n$ (double the previous distance since the design space has been doubled), with the first equation now being $a + 2(n-1)/n + b = 1$ and the second equation to be solved is $a - b = -1$. Solving these two equations thus gives $a = -\frac{(n-1)}{n}$. The design points are thus $-\frac{(n-1)}{n}, -\frac{n-3}{n}, \ldots, \frac{n-1}{n}$, which can be written in the form $-\frac{(n-(2i-1))}{n}, i = 1, 2, \ldots n$.

Note that the selection of the number of levels of each factor to use determines the number of runs since each level is used only once.

Many uniform designs are also orthogonal designs; many others are near-orthogonal. Although the uniform design is a simple and easy-to-understand concept in one dimension, it is neither in more than one dimension. In fact, as pointed out by Fang and Lin (2003), the definition of a uniform design that they give in their Eq. (5.1) is not uniquely satisfied when there are at least two factors.

Their definition of a uniform design is as follows. Let s denote the number of factors and define the experimental region as the unit cube $C^s = [0, 1]^s$. Let $\mathcal{P} = \{x_1, x_2, \ldots, x_n\} \subset C^s$ denote the set of n experimental points, with x_i denoting a vector of values that gives the ith value of each factor. Let M be a measure of uniformity of $\mathcal{P}$ with a small value of M desirable since such values represent better uniformity than larger values. Let $Z(n, s)$ denote the collection of sets of n points on C^s. A set $\mathcal{P}^* \in Z(n, s)$ is termed a uniform design if it has the minimum M-value over $Z(n,s)$.

As Fang and Lin (2003) explain, many different measures of uniformity have been defined (see pp. 142–143 of their chapter).

The construction of uniform designs for at least two factors requires either software, a catalog, or an applet. One of the latter is given at http://www.math.hkbu.edu.hk/UniformDesign. The output from the latter is in the form that orthogonal arrays are usually given, with 1 being the lowest level for each factor, and the highest level is n. This is the same form that is used by Fang and Lin (2003), and the designs are referred to as U-type (uniform) designs.

If the user prefers to see the design given such that each level is contained in the interval $(-1, 1)$, then it would be better to use appropriate software. For example, JMP allows the user to specify the interval when space-filling designs are constructed, and the following JMP output gives a uniform design for four points and two factors, with the design space for each factor specified as $(-1, 1)$. (JMP has only somewhat recently added uniform design capability.)

```
Space Filling Design

Factors

X1    Continuous

X2    Continuous

2 Factors
Space Filling Uniform Design

Factor Settings

Run        X1         X2
1      -0.73363   -0.24587
2       0.73363    0.24587
3      -0.24587    0.73363
4       0.24587   -0.73363
```

If the interval $(1, 4)$ is specified, JMP constructs the design given below:

```
Factor Settings

Run        X1         X2
1       2.86881    1.39956
2       1.39956    2.13119
3       3.60044    2.86881
4       2.13119    3.60044
```

This is different from the U-type uniform design mentioned previously, thus illustrating the nonuniqueness of uniform designs for at least two factors.

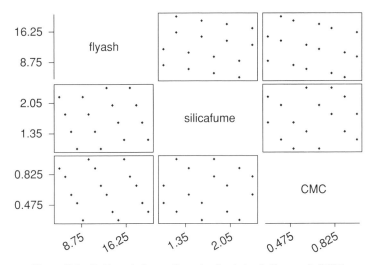

Figure 13.3 Uniform design configuration for design in Tang et al. (2004).

Case Study

Tang, Li, Chan, and Lin (2004) described a successful application of a uniform design in product formation in the cement manufacturing industry. The objective in the experiment they described was to determine an optimal composition of materials so that the cement grouting material that was formed had the desired properties. There were two response variables, one a measure of compressive strength and the other a coefficient of bleeding, and there were four factors: the percentages of fly ash, silica fume, carboxyl methyl cellulose, and cement. Since these are percentages, this is really a mixture experiment with only three of the percentages free to vary. Therefore, the design consisted of the first three factors.

A uniform design was used because 16 levels were used for the first factor, 8 levels for the second factor, and 8 factors for the third factor. The large number of levels for each factor rules out traditional mixture designs, such as those discussed in Section 13.11. Instead, a uniform design $U_{16}(16 \times 8^2)$ with 16 runs was used, with each factor constrained as follows:

$$5 \leq x_1 \leq 20 \qquad 1 \leq x_2 \leq 2.4 \qquad 0.3 \leq x_3 \leq 1.0$$

The uniformity of the design in two dimensions can be seen in the matrix scatterplot as given in Figure 13.3.

The run numbers given in Table 1 of Tang et al. (2004) show that the runs were not randomized as the percentage for the first factor, fly ash, ran consecutively from 5 to 20. Factors having differing numbers of levels makes this a somewhat unorthodox uniform design that would have to be constructed using special software or tables. JMP, for example, which has the capability for uniform design construction, cannot construct uniform designs with differing numbers of factor levels. This can be done in

Design-Expert, but technically the latter does not construct uniform designs. Rather, it creates "distance-based" designs that attempt to spread the design points evenly over the region but it does so by maximizing the distance between points. The designs that are generated have coordinates that are quite different from the coordinates in the designs produced by JMP. The Design-Expert documentation advises against the use of these designs, however. The differing numbers of levels of course means that some levels of the factors with eight levels will be repeated.

We will not analyze the data from the experiment in this section since this was a mixture experiment, but will note that the apparent lack of randomization regarding the runs does, strictly speaking, invalidate the model hypothesis tests that were performed for each of the two response variables, although that is not of any great concern since the p-values were of the order of 10^{-10} and smaller and these tests were not a major component of the article.

To what extent are uniform designs used in practice? This may be debatable but the following quote given in Freeman (2004, p. 38), which is available online at http://cipd.mit.edu/education/thesis_files/ion_freeman_thesis_2004.pdf with this quote resulting from one of the phone interviews of practitioners, is worth pondering: "Orthogonal array can require too many experiments, and not give the best coverage. I don't need orthogonality; I need the best information. Uniform design techniques are far more popular in the practice." The last statement is probably not true but may be true 20 years from now.

13.9.1.1   *From Raw Form to Coded Form*
There is an easy and obvious correspondence between the levels of a factor in raw units and in coded units when a U-type uniform design is used. For example, let's assume that an experimenter wishes to use temperature levels of 340, 370, 400, and 430°F. The coded values would be simply 1, 2, 3, and 4. If 420 were to be used instead of 430, a uniform design would not be possible because of the unequal spacing that would result from the switch.

We have a different situation if we use the values in the JMP output as those values are not equally spaced on each factor since this is not a U-type uniform design. Therefore, the temperature levels of 340, 370, 400, and 430 cannot be transformed to those coded values since the temperature values are equally spaced. There is such a small difference, however, that these temperature values could be used anyway.

Uniform space-filling designs essentially mimic the continuous uniform distribution for a given factor, so we can think of points for two factors as essentially mimicking a bivariate continuous uniform distribution.

13.9.2   Sphere-Packing Designs

Uniform designs can be contrasted with "sphere-packing" designs, which maximize the minimum Euclidean distance between the points. See, for example, the discussion given at http://www.warwick.ac.uk/statsdept/staff/JEHS/jehpack.htm. For example, a sphere-packing design for two factors and four points is simply the 2^2 design; for three factors and eight points, it is the 2^3 design; and so on. Of course, this is because the design points for these factorial designs are at the vertices of the spherical region and are thus the maximum distance apart. The answer is not so obvious if the number of design points is not of the form 2^k, however.

For example, the following is the sphere-packing design for six runs and two factors given by JMP:

```
X1 Continuous
X2 Continuous

2 Factors
Space Filling Sphere Packing

Factor Settings

Run        X1           X2
1      -0.33333      1.00000
2       1.00000      1.00000
3       1.00000     -1.00000
4      -0.33333     -1.00000
5      -1.00000      0.00000
6       0.33333     -0.00000
```

13.9.3 Latin Hypercube Design

The Latin hypercube design (LHD) is the other commonly used space-filling design, and it is essentially a compromise between the uniform design and the sphere-packing design. A design of this type is constructed by dividing the desired range for each factor into n intervals and then randomly selecting a point within each interval. A randomly selected point for the first factor is then matched with a randomly selected point for the second factor and subsequent factors, and then the process is repeated $n - 1$ times until the coordinates for all n points have been obtained, with the number of levels of each factor equal to the number of runs. The design is due to McKay, Beckman, and Conover (1979).

The design for four points and two factors is given below:

```
X1 Continuous
X2 Continuous

2 Factors
Space Filling Latin Hypercube

Factor Settings

Run        X1           X2
1      -0.33333     -1.00000
2      -1.00000      0.33333
3       0.33333      1.00000
4       1.00000     -0.33333
```

Notice that the levels of each factor are equally spaced, which is one of the characteristics of these designs. Notice also that the number of levels of each factor is equal to the number of runs, just as is the case with the uniform design. Although it differs in this respect from the sphere-packing design, it shares with that design the maximizing of the Euclidean distance between design points, but subject to the constraint of having even spacing between the factor levels.

The design with four points is orthogonal because the points form a parallelogram, which will generally not be the case if we use more design points for the two factors, especially if there is an odd number of design points, and similarly when more factors are used.

Case Study

Bailey and Jones (2004) gave a case study that illustrated the use of an LHD, and which could also be viewed as an example that illustrates (many) debarred combinations. The case study, which the authors stated that they intentionally simplified to illustrate space-filling designs and which was perhaps also simplified for a general audience, involved the use of a polymerase chain reaction (PCR). The main purpose of this reaction is to make many copies of native DNA, with the response variable being the amount of the product DNA fragment that is formed. There were eight factors to be studied: primers A, B, and C, annealing temperature, $MgCl_2$, dNTP, polymerase, and cycles. As stated by the authors, the reaction involves repeatedly heating and cooling the reaction mixture, inducing the DNA to melt, and separating the double strands.

Although the scientists could be guided by experience in choosing reasonable ranges for the factors, few combinations of factor levels work for a particular PCR. Since they don't know which combinations will work, they need to sample a large number of levels of each factor.

Accordingly, the design that was used was an LHD with 48 chosen as the number of runs to be made. Of course the objective is to uniformly spread the points over the eight-dimensional space. Although the eight-dimensional coverage is obviously not completely captured by a scatterplot matrix, the matrix that was given in Bailey and Jones (2004) and which is given in Figure 13.4 does provide two-dimensional glimpses into the uniformity.

It is worth noting that some of the factors, including annealing temperature, are hard to change, and Bailey and Jones (2004) pointed out that the actual runs may have differed from those in the design. Because of the way the design points are constructed, replicated points and/or replicated coordinates are rare because the coordinates are generated at random. Consequently, it is not possible to minimize the number of factor-level changes and still adhere to the design. If factors are very difficult or impractical to change, an experimenter might be expected to simply deviate from the design, with the LHD not being an appropriate design in the first place when there are hard-to-change factors.

Latin hypercube designs and related designs are compared in terms of prediction error by Morris, Mitchell, and Ylvisaker (1993). Although LHDs can be generated using some statistical software, such as JMP and MathWorks using the model-based calibration toolbox (see http://www.mathworks.com/products/mbc/

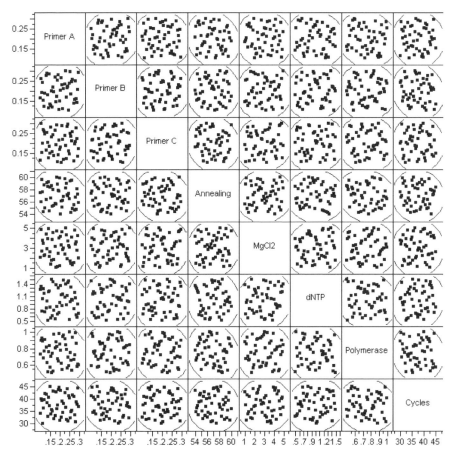

Figure 13.4   Scattering of points in Latin hypercube design.

description1.html and page 67 of http://www.mathworks.com/access/helpdesk/help/pdf_doc/mbc/mbcmodel.pdf), tables of LHDs for up to 22 factors can be found in Cioppa (2002).

13.10 TREND-FREE DESIGNS

Recall the discussion in Section 1.7 about the importance of processes being in a state of statistical control, as the lack of this could undermine a designed experiment in which complete randomization was used.

Consider the following example. Assume that a 2^2 design has been used, without replication, and the randomized run order was (1), b, a, ab. Thus, the combinations at the high level of factor A were run last. A trend, for whatever reason, which causes the response values to increase without a change in the experimental conditions could produce a false signal that factor A was significant.

Assume that control cannot be achieved, and quite frankly, achieving control such that there is an unchanging mean and variance would be extremely difficult, in general. One way to try to overcome the problem would be to use a trend-free design. With such designs, treatments are assigned to experimental units in a systematic manner rather than by using randomization.

There has been comparatively little work on such designs, which essentially began with the work of Cox (1951). A significant early work was Daniel and Wilcoxon (1966), who gave sequences of treatment combinations for full and fractional factorial two-level designs for use in the presence of time trends. Their work was not extended for many years and Daniel (1982, personal communication) wondered if it would ever be extended. Finally, John (1990) introduced a foldover method developed from their plans and discussed the trend-free properties of systematic run orders based on their method. Trend-free block designs were introduced by Bradley and Yeh (1980) and in 1983, they showed that such designs do not always exist. Bradley, Yeh, and Notz (1985) introduced nearly trend-free block designs. Other work in the area of trend-free designs includes Hill (1960).

Trend-free designs can be constructed with the Gendex DOE toolkit.

13.11 COST-MINIMIZING DESIGNS

The explicit consideration of developing designs to minimize costs has received very little attention in the literature, just as cost minimization relative to statistical methodology in general has received only scant attention. Significant early work on cost-minimizing experimental designs was performed by Draper and Stoneman (1968) and Dickinson (1974). Subsequent work followed, most of which is summarized by Kim (1997), who also considered trend-free designs. This dissertation is available at http://scholar.lib.vt.edu/theses/available/etd-7497-15232/unrestricted/etd.PDF. Bisgaard (2000) provides the framework for cost comparisons between split-plot designs and fully randomized designs and gives some illustrative examples. Tack and Vandebroek (2002) considered cost-efficient block designs; Park, Montgomery, Fowler, and Borror (2006) considered cost-constrained G-efficient response surface designs, and Allen, Yu, and Bernshteyn (2000) described a response surface design application when there were cost constraints.

As mentioned in Section 13.8.1, Goos, Tack, and Vandebroek (2005) state that it is important to incorporate cost considerations in optimal design theory but admit that there is not much literature on the subject.

Hard-to-change factors were discussed in Section 4.19. Certainly the prevalence of such factors should motivate the development of cost-efficient designs for hard-to-change factors.

13.12 MIXTURE DESIGNS

Mixture designs are used, as the name practically implies, to aid in the determination of an optimal combination of components for which the total amount is fixed, with

100 percent being the total if the components are expressed as percentages, as is the usual practice.

Since there is a constraint on the total mixture and there will also generally be constraints on the individual components, the experimental space is not rectangular or hyperrectangular. Thus, the types of designs that were given in previous chapters, such as Chapters 4 and 5, are not applicable.

Cornell (2002) is an excellent source of information on mixture designs and the reader is referred to that source for comprehensive information on these designs. Software that is available for the design and analysis of mixture experiments includes MIXSOFT (see, e.g., http://members.aol.com/mixsoft).

13.12.1 Optimal Mixture Designs or Not?

As stated in Section 13.7, optimal designs have been very controversial but have been used advantageously in mixture experiments (Piepel, 2004, personal communication). Mixture designs are used extensively in pharmaceutical applications and there are many documented applications of the use of optimal designs, including Bodea and Leucuta (1997). Some critical comments about optimal designs as mixture designs have been voiced by Myers and Montgomery (1995, p. 595).

13.12.2 ANOM

Just as ANOM can be used for various common designs, as illustrated in Chapters 2 and 4 and discussed in Chapter 3, it can also be used for assessing the significance of components in an axial mixture design, as illustrated by Nelson (1993) and in Chapter 7 of Nelson, Wludyka, and Copeland (2005).

13.13 DESIGN OF MEASUREMENT CAPABILITY STUDIES

An important use of experimental designs is in assessing measurement capability. A typical study involves parts and operators, with parts being random and operators usually random, but fixed if the operators used in the experiment are the same ones who work on the process. Burdick, Borror, and Montgomery (2003) discuss the design of a measurement system study in their review paper, with such a study referred to more specifically as a gage repeatability and reproducibility study.

13.14 DESIGN OF COMPUTER EXPERIMENTS

Engineers frequently encounter complex systems with computer experiments required to optimize such systems. Classical designs, such as factorial designs, are useful when there is a large sample size. Space-filling designs are often an attractive alternative, however, because models used in computer experiments do tend to be rather complex. Chen, Tsui, Barton, and Allen (2003) devote considerable space to the discussion of such designs in computer experiments, and their review paper would be a good starting point for the study of experimental designs in computer experiments, as would the book by Santner, Williams, and Notz (2003). Simpson, Lin, and Chen (2001) compared space-filling designs and discussed sampling strategies for computer experiments. The latter simply stated that designs for computer experiments should "provide information about all portions of the experimental region," and "allow one to fit a variety of models." More recently, Jin, Chen, and Sudjianto (2005) presented an algorithm that was purported to save considerably on computing time for constructing optimal designs for computer experiments.

A useful user's guide (literally) for designing computer experiments, written at a basic level, is Kleijnen, Sanchez, Lucas, and Cioppa (2005), which is available online and which has a five-page online supplement that contains design matrices for the designs used in the authors' study (see the references for both). See also the comparison study in Bursztyn and Steinberg (2006). For additional reading on this topic, see Fang, Li, and Sudjianto (2006), which includes material on space-filling designs and other topics. See also the case study of Liu and Fang (2006).

13.15   DESIGN OF EXPERIMENTS FOR CATEGORICAL RESPONSE VARIABLES

Although measurement data provide more information than categorical data, the latter (such as go, no-go data) are often used due to the expense of making accurate measurements, faulty measurement systems, and so on. One problem that can occur when data are dichotomized is that the resultant data may not be informative. That is, the data may be highly asymmetric, with the data being almost all of one type. It would obviously be difficult to model data that are, say, 99 percent "go" and 1 percent "no-go," as a design with a sizable number of runs would be required to virtually ensure that some no-go data were obtained. Otherwise, the experiment would be a waste of time. Joseph and Wu (2004) gave an approach, incorporating experimental design, that is intended to overcome this problem and the interested reader is referred to their paper for details, as it won't be discussed here since it is outside the scope of this book. Another article that may be of interest is Bisgaard and Weiss (2002–2003), who took an experiment performed by Taguchi with the response variable being function/malfunction and performed an alternative analysis that provided greater insight.

13.16   WEIGHING DESIGNS AND CALIBRATION DESIGNS

These designs are probably most often associated with the National Institute of Standards and Technology (NIST) and with K. S. Banerjee. In particular, Banerjee (1975) is a well-known book on weighing designs, as is Jones and Schoonover (2002), written by two veteran NIST workers. These books are recommended, as are the seminal articles of Banerjee (1950, 1951), in addition to the many articles of Bronislaw Ceranka, in particular. There is considerable literature on weighing designs and these designs and calibration designs are covered in the next two sections.

13.16.1 Calibration Designs

The general idea with these designs is to determine the weight of one or more objects. The designs are referred to as *calibration designs* by NIST because their objective is to assign the correct weight to a test item by comparing it to a reference standard, whose weight is assumed to be known. There is an extensive catalog of these designs in the NIST *e-Handbook of Statistical Methods*. There are designs for different purposes and for different applications; see Section 2.3.4 of the *e-Handbook* at http://www.itl.nist.gov/div898/handbook/mpc/section3/mpc34.htm.

Just as statistical process control methods are used in conjunction with the designs discussed in previous chapters, so too are they used with calibration designs, as is discussed in Section 2.3.5 of the *e-Handbook*.

The general form of the solution that provides the estimates of the test weights for a calibration design is a restrained least squares method described in Zelen (1962), with the restraint (or "constraint") being the known weight of a reference standard if only one is used, or the sum of the weights of the reference standards if more than one standard is used. See also Lawson and Hanson (1995). The details are also given in Cameron, Croarkin, and Raybold (1977) and are as follows.

Assume that what is known is a linear combination of weights of reference standards of the form

$$r_1\beta_1 + r_2\beta_2 + \cdots + r_k\beta_k = m$$

with the β_i being the weights that are to be estimated (individually) and the r_i are constants. When least squares without constraints is used, the sum of the squared deviations between the observed and fitted values is minimized. When constrained least squares is used, the method of Lagrangian multipliers is used so that the quantity to be minimized is, say, W, with

$$W = \text{sum of squared deviations} + \lambda(r_1\beta_1 + r_2\beta_2 + \cdots + r_k\beta_k - m)$$

The constrained least squares normal equations are then those that result from minimizing W.

Example 13.3

We will use the simplest example of a calibration design, a (1, 1, 1), to illustrate the design and the mechanics involved. The "(1, 1, 1)" refers to one reference standard that is known, plus a check standard and a test standard with unknown values. The design matrix and observed values, Y_1, Y_2, and Y_3, are as follows:

Reference standard	Check Standard	Test item	Observations
1	-1	0	Y_1
1	0	-1	Y_2
0	1	-1	Y_3

Here the runs are the rows, just as is the case with other statistical designs. Notice that the layout is essentially a balanced incomplete block (BIB) design as each pair of treatments occurs once in each "block" when we view the rows as blocks. A restraint on the solution is, of course, not imposed when a BIB design is used, however.

The solution, which consists of estimates of the unknown values of the check standard and test item, cannot be obtained using standard software, however, because the reference standard is assumed known and is thus not something to be estimated. This is somewhat analogous to estimating main effects and interactions for an experiment with a factorial design when one of the main effects is assumed known.

The solution might seem to be simply the solution to a system of linear equations subject to a restraint that represents the known reference standard, i.e., just by using straight algebra. That isn't the case, however, unless the system of equations is the set of constrained least squares equations. Indeed, as indicated earlier in this section, that is what is done. Furthermore, it should be apparent that the equations that would result from using the design and the data without least squares will generally be inconsistent.

The mechanics for this example won't be given here because they are somewhat involved and also because they are available on the Web (see http://www.itl.nist.gov/div898/handbook/mpc/section3/mpc3321.htm).

Using what is provided therein, it can be shown that the estimate for the test item is

```
Test item estimate = ⅓(-Y₁-2Y₂-Y₃) + Reference standard
```

Readers are also referred to http://www.itl.nist.gov/div898/handbook/mpc/section3/mpc332.htm for general information on solutions to calibration designs.

13.16.2 Weighing Designs

We may also use these type of designs without reference standards. When this is done, it is more appropriate to refer to the designs as weighing designs, as Banerjee (1975) did in the title of his book. That is, there are k objects to be weighed and a design must be constructed to permit their weights to be estimated.

In general, the criteria for good weighing designs depend partly on the intended use for the resulting values, as explained by Cameron et al. (1977). For example, if the weights are to be used independently of each other, it would be desirable to use a design that gives the smallest variance of each of the effect estimators, whereas if the weights are to be used in combination, it would then be appropriate to use a design that minimizes the variance of a certain linear combination or gives small variances for linear combinations that are likely to be used.

Cameron et al. (1977) gave the following categories of weighing designs: (1) designs for nominally equal groups, (2) designs for the 2, 2, ..., 1, 1, 1, ... series, (3) the 5, 3, 2, 1 and 5, 2, 2, 1 series, (4) binary and miscellaneous series, and (5) designs for direct reading with constant load. (The numbers, such as 5, 3, 2, 1, refer to the relationships between the weights in a reference set. For example, the weights might be 5 grams, 3 grams, 2 grams, and 1 gram, or 500, 300, 200, and 100.)

Regarding the first group, designs for nominally equal groups, the (1, 1, 1) calibration design was given in Section 13.16.1. The notation is used differently to designate a weighing design. For the latter, the notation refers to the reference set consisting of three equal weights. The first restraint, as given by Cameron et al. (1977), is that the sum of the first two weights is known and the second restraint is that the third weight is known.

Since there are three weights and all pairwise comparisons are made, there must be three observations since $\binom{3}{2} = 3$. Similarly there are $\binom{4}{2} = 6$ observations for four equal weights, $\binom{5}{2} = 10$ observations for five equal weights, and so on.

Designs for trend elimination were given by Cameron et al. (1977) in juxtaposition to the designs with equal weights. These designs are due to Cameron and Hailes (1974). These designs are used when responses are time dependent due to temperature and atmospheric changes.

The second group of designs, the 2, 2, ..., 1, 1, 1, ... series does not possess the simplicity and symmetry of the designs for equal weights. This should be apparent from the 2, 1, 1, 1 design that is given below:

Observations	2	1	1	1
Y_1	+	−		−
Y_2	+		−	−
Y_3	+	−	−	
Y_4		+		−
Y_5		+	−	
Y_6			+	−

The most commonly used designs have been those in the third group, the 5, 3, 2, 1 and 5, 2, 2, 1 designs. The 5, 3, 2, 1, 1 design, which is due to Hayford (1893), is the following:

Observations	5	3	2	1	1
Y_1	+	−	−	+	−
Y_2	+	−	−	−	+
Y_3	+	−	−		
Y_4	+	−		−	−
Y_5		+	−	−	
Y_6		+	−		−
Y_7			+	−	−
Y_8			+	−	

Various examples of weighing designs can be found in Banerjee (1951), in particular.

There has been a considerable amount of research on weighing designs since Banerjee (1975). In particular, Craigen (1996) provided a table of weighing design matrices that has since had an addition or two. (A missing entry in that table was provided by Arasu and Torban, 1999.)

13.17 DESIGNS FOR ASSESSING THE CAPABILITY OF A SYSTEM

Hubele, Beaumariage, Baweja, and Hong (1994) discussed the use of experimental designs to assess the capability of a system. The authors described a designed experiment that was used to help determine the capability of a vision system at AlliedSignal. A design with blocking was used, but unlike most block designs, the blocks were not random effects, but rather corresponded to systematic effects. The result was a better understanding of system performance.

13.18 DESIGNS FOR NONLINEAR MODELS

All the designs presented to this point in the book presume the use of a linear model (i.e., a model that is linear in the parameters). The true, unknown model is almost certainly nonlinear, but a linear model is often a suitable proxy for the nonlinear model. In linear model applications of experimental design, the objectives are to determine what effects are important and to estimate those effects. In nonlinear applications, however, the objectives are to construct designs that facilitate model discrimination and parameter estimation. The work on designs for nonlinear models starts from Box and Lucas (1959) and there is a vast amount of literature on the subject, especially in regard to the construction of optimum designs. More recent references on designs for nonlinear models include Dette, Melas and Wong (2005), who constructed efficient designs for the Michaelis–Menten enzyme kinetic model, Hackl, Muller, and Atkinson (2001), Fedorov and Hackl (1997), and Atkinson and Haines (1996).

13.19 MODEL-ROBUST DESIGNS

The term "model-robust designs" refers to designs that are robust to model misspecification, as in Li and Nachtsheim (2000). One of the disadvantages of the designs that they proposed, however, is that the designs are frequently not orthogonal designs, although the departure from orthogonality is generally very small. They found that their designs were "surprisingly robust" to model misspecification, whereas fractional factorial designs were generally found to not be robust. Some useful information relative to this work can be found at http://webpages.csom.umn.edu/oms/wli/Optimal/html/robust.htm and a catalog of the designs can be found at http://home.hccnet.nl/kees.duineveld/fd.en.ffd2/factorial.html. Other recent work in this topic area includes Goos, Kobilinsky, O'Brien, and Vandebroek (2005), Heredia-Langner, Montgomery, Carlyle, and Borror (2004), and Cooney and Verseput (2000).

13.20 DESIGNS AND ANALYSES FOR NON-NORMAL RESPONSES

There are many response variables that are not normally distributed, such as percentages. A non-normal response variable can often be transformed to be approximately normal. For example, count data are not normally distributed, although data with large counts should be approximately normally distributed. A standard transformation for count data is

$$y = \frac{\sqrt{c} + \sqrt{c+1}}{2}$$

with c denoting a count. This approach was used by Bisgaard and Fuller (1994–1995) in analyzing data given by Hsieh and Goodwin (1986), for which the response variable was the number of defects. Similarly, an arcsine transformation could be used for a binomial response variable.

If transformation to near-normality is not possible, then the question of how to analyze data from a designed experiment with a non-normal response variable must be addressed. Lewis, Montgomery, and Myers (2001a) gave several examples of designed experiments with non-normal responses and showed that using a generalized linear model (GLM) approach is an excellent alternative to transforming the data. Furthermore, Lewis et al. (2001b) stated that a GLM approach is often superior to a transformation approach for a confidence interval on the mean response and for a prediction interval. They also showed, however, that the results are highly sensitive to the proper selection of the link function. So a GLM approach is not devoid of potential problems. For information on generalized linear models, readers are referred to McCullagh and Nelder (1989) and Dobson (2001).

13.21 DESIGN OF MICROARRAY EXPERIMENTS

The design of microarray experiments is a new area that is growing rapidly. A microarray has been defined in many different ways but the bottom line is that it is used to study genes. One particular definition, given at http://www.medterms.com/script/main/art.asp?articlekey=30712, is "A tool used to sift through and analyze the information contained within a genome. A microarray consists of different nucleic acid probes that are chemically attached to a substrate, which can be a microchip, a glass slide or a microsphere-sized bead." Microarray experiments are thus biological experiments, with important implications also for the agricultural and pharmaceutical sciences. Experimental designs are necessary for assessing the many sources of uncertainty in microarray experiments.

Classical experimental designs would not be helpful in microarray experiments because biologists are not interested in main effects and interaction effects. Therefore, new experimental approaches had to be developed to answer questions such as the following that are of interest to biologists: (1) "Which mRNA samples should be competitively hybridized on the same slide?," and (2) "How many times should each slide by replicated?"

Since this is a new field, research papers on the subject date from the early years of this century. Pioneering, primary work on designs for microarray experiments was performed by Kerr and Churchill (2001) and Glonek and Solomon (2004), who used concepts from optimal design theory to suggest efficient designs for some of the common cDNA microarray experiments. More specifically, Glonek and Solomon (2004) proposed optimal designs for factorial and time course experiments, which arise frequently in microarray experiments. The approach enabled efficient designs to be constructed subject to the information available on effects of most interest to biologists, as well as the number of arrays available for the experiment and other practical constraints. They showed that their designs are superior to the popular reference designs, which they claim to be inefficient.

Most biologists performing simple gene expression comparisons, such as between treated and control cells, carry out replicate experiments on different slides. Speed and Yang (2002) referred to these as *technical replicates*, using the term to refer to the situation where the target mRNA in each hybridization was from the same RNA extraction, which is labeled independently for each hybridization. Since estimates of differential gene expression based on technical replicates tend to be correlated, the correlation structure must be taken into account and Speed and Yang (2002) did so.

The popularity of microarray experiments has spawned various books that contain material on methods for designing such experiments. These include Simon, Korn, McShane, Radmacher, Wright, and Zhao (2004) and Allison, Page, Beasley, and Edwards (2006). Significant articles include Wit, Nobile, and Khanin (2005), Scholtens, Miron, Merchant, Miller, Miron, Iglehart, and Gentleman (2004), and Emptage, Hudson-Curtis, and Sen (2003).

13.22 MULTI-VARI PLOT

A *multi-vari plot* is a technique that would undoubtedly be used more extensively with experimental design data if it were better known. As the name essentially implies, the idea is to identify sources of variation, but the method is not well known outside the group of people who are knowledgeable in statistical quality improvement techniques. Thus, the thrust is different from the well-known techniques that seek to identify factors that affect location.

The plot is similar to an ANOM plot in that the extreme values within cells of a design are connected, thus showing the variation within each cell of, say, a factorial design, and of course that variation is assumed to be equal across the cells. An example of a multi-vari plot for a one-factor design is given by Ryan (2000, p. 330), and examples are also given in Delott and Gupta (1990).

Czitrom, de Amezua Ayala, and Sanchez (1998) used a multi-vari plot in a some-what different manner, however, as all of the 120 observations for a replicated 2×3 factorial design (in a split-plot arrangement) are shown as circles. As with plotting techniques, the best way to plot the data is the way that provides the most insightful information.

13.23   EVOLUTIONARY OPERATION

Although sequential experimentation often produces satisfactory results, the dynam-ics may be continually changing so that what is optimal today may not be optimal in three months, and/or the response surface may be sufficiently complicated that determining the best combination of factor levels may be extremely difficult. Conse-quently, frequent, almost continuous, experimentation may be needed. Evolutionary Operation (EVOP) can be used in such situations.

Designed experiments are performed off-line, so as not to interfere with production processes. EVOP is for online experimentation, however. Since the experimentation is done online, only small changes in factor levels are made from current levels. His-torically, only a small number of two-level factors have generally been used in EVOP programs, with apparently only full factorials employed, perhaps with blocking. There are various forms of EVOP that have been proposed, with the most frequently used ap-proach undoubtedly being Box-EVOP, so named for its inventor, George Box (1957). This entails the use of a 2^k design with a decision made after a specified number of cycles to change or not change the factor levels, with the decision determined by whether or not the 2σ confidence intervals for the main effect estimates include zero.

Significant interactions create problems for EVOP, just as they do when 2^k designs are used, in general. When interactions cannot be "transformed away," Ryan (2000, p. 481) suggests looking at the best combination of factor levels and making a decision based on the average for that combination. Although this isn't the same as using conditional effects, it is in the same spirit of breaking the data down to a finer level than is used when ANOVA is performed.

Although there is not very much in the literature on EVOP, Hoerl (1998) listed EVOP as a component of a typical Six Sigma black-belt training program. Despite this, it seems safe to assume that EVOP is still underutilized in industry, just as Hahn and Dershowitz (1974) concluded that it was underutilized over 30 years ago. Another recommended source is Box and Draper (1969).

13.24   SOFTWARE

Whereas most of the experimental design methods presented in previous chapters are readily available in the leading statistical software, this is not the case for some of the methods presented in this chapter. For example, in the comparison study of Reece (2003), which was mentioned in previous chapters, the author rated the software in regard to "Designs with Factor Constraints," with each software package receiving a rating of 1 to 5. Four of the packages did not receive a rating because they did not have that capability at the time of the survey; six received a rating of 4, with only one, RS/Discover, receiving a rating of 5. JMP and Design-Expert were among the well-known packages that received a rating of 4.

As we might expect, there was a high correlation between these ratings and the ratings for "Algorithmic Designs" since, for example, an algorithmic design is typically constructed when there are factor constraints, as was discussed in Section 13.8. Of course opinions vary regarding important experimental design capabilities for software. In discussing MINITAB, Reece (2003) stated, "Not supported are the algorithmic designs needed for some forms of sequential experimentation or for dealing with constrained process spaces. The absence of these designs is a significant contributing factor to rating the software 'Good' as opposed to a higher evaluation." (These comments were in regard to Version 13.31 of MINITAB.) Release 14 and Release 15 of MINITAB *do* have capability for D-optimal designs, however, which is for use with response surface designs and mixture designs and can be generated in either menu mode or command mode. For the latter, the command is OPTDESIGN and the input is a set of candidate points, with the user specifying the number of points for the D-optimal design. The output from that command is the set of row numbers from the set of candidate points that comprise the D-optimal design. Although computing D-optimal and other optimal designs can be computer intensive, the OPTDESIGN command is, for example, executed almost instantly when used by this writer to obtain a design with 16 points from 100 candidate points, with the coded levels of each of the three factors being -1, 0, and 1. Various information is provided in addition to the design, including leverage information. It is useful to see if there are any large leverages and to see whether or not the leverages are about the same.

D-optimal designs can also be constructed with Design-Expert 7.0, for up to 30 factors. Of course specifying a D-optimal design for a large number of factors can cause the program to run for more than a short time, however. Design-Expert 7.0 also allows multiple linear constraints to be specified that can be used to construct a restricted region of operability, to which a D-optimal design could be applied. This was illustrated at http://www.statease.com/x7pup/answers/Prob%207-2.pdf.

13.25 SUMMARY

There were some designs presented in this chapter that are not well known, and consequently they are seldom used. Other designs, however, have been used extensively. Of course the extent to which any particular type of design is used is determined primarily by software capability. Plackett–Burman designs are used extensively and this is undoubtedly due to the fact that the designs can be constructed using MINITAB, Design-Expert, and JMP, plus other software. At the other extreme, uniform designs, Doehlert designs, and Cotter designs have for the most part not found their way into statistical software, although it was stated in Section 13.9.1 that JMP fairly recently added uniform design capability.

REFERENCES

Abraham, B., H. Chipman, and K. Vijayan (1999). Some risks in the construction and analysis of supersaturated designs. *Technometrics*, **41**(2), 135–141.

Ahuja, S. K., G. M. Ferreira, and A. R. Moreira (2004). Application of Plackett–Burman design and response surface methodology to achieve exponential growth for aggregated shipworm bacterium. *Biotechnology and Bioengineering*, **85** (6), 666–675.

Allen, T. T. and M. Bernshteyn (2003). Supersaturated designs that maximize the probability of identifying active factors. *Technometrics*, **45**, 90–97.

Allen, T., L. Yu, and M. Bernshteyn (2000). Low-cost response surface methods applied to the design of plastic fasteners. *Quality Engineering*, **12**(4), 583–591.

Allison, D. B., G. R. Page, T. M. Beasley, and J. W. Edwards, eds. (2006). *DNA Microarrays and Related Genomics Techniques: Design, Analysis, and Interpretation of Experiments*. New York: Chapman and Hall/CRC.

Anderson, M. and P. Whitcomb (2000). *DOE Simplified: Practical Tools for Effective Experimentation*. Portland, OR: Productivity, Inc.

Anthony, J. (2002). Training for design of experiments using a catapult. *Quality and Reliability Engineering International*, **18**, 29–35.

Arasu, K. T. and D. Torban (1999). New weighing matrices of weight 25. *Journal of Combinatorial Designs*, **7**, 11–15.

Atkinson, A. C. and R. A. Bailey (2001). One hundred years of the design of experiments on and off the pages of *Biometrika*. *Biometrika*, **88**, 53–97.

Atkinson, A. C. and L. M. Haines (1996). Designs for nonlinear and generalized linear models. In *Handbook of Statistics*, Vol. 13: *Design and Analysis of Experiments*, Chap. 14, pp. 437–475 (S. Ghosh and C. R. Rao, eds.). Amsterdam: Elsevier Science B.V.

Bailey, M. and B. Jones (2004). Space filling designs for pre-screen experiments. *JMPer Cable*, Issue 13, Winter, 8–10. SAS Institute Inc., Cary, NC.

Balusu, R., R. M. R. Paduru, G. Seenayya, and G. Reddy (2004). Production of ethanol from *Clostridium thermocellum* SS19 in submerged fermentation: Screening of nutrients using Plackett–Burman design. *Applied Biochemistry and Biotechnology*, **117**(3), 133–142.

Bandurek, G. (1999). Practical application of supersaturated designs. *Quality and Reliability Engineering International*, **15**, 123–133.

Banerjee, K. S. (1950). How balanced incomplete block designs may be made to furnish orthogonal estimates in weighing designs. *Biometrika*, **37**, 50–58.

Banerjee, K. S. (1951). Some observations on the practical aspects of weighing designs. *Biometrika*, **38**, 248–251.

Banerjee, K. S. (1975). *Weighing Designs for Chemistry, Medicine, Economics, Operations research, Statistics*. New York: Dekker.

Berger, M. P. F. and W. K. Wong, eds. (2005). *Applied Optimal Designs*. Hoboken, NJ: Wiley.

Bingham, T. C. (1997). An approach to developing multi-level fractional factorial designs. *Journal of Quality Technology*, **29**(4), 370–380.

Bisgaard, S. (2000). The design and analysis of $2^{k-p} \times 2^{q-r}$ split-plot experiments. *Journal of Quality Technology*, **32**, 39–56.

Bisgaard, S. and H. T. Fuller (1994–1995). Analysis of factorial experiments with defects or defectives as the response. *Quality Engineering*, **7**(2), 429–443.

Bisgaard, S. and P. Weiss (2002–2003). Experimental design applied to identify defective parts in assembled products. *Quality Engineering*, **15**(2), 347–350.

Bodea, A. and S. E. Leucuta (1997). Optimization of hydrophilic matrix tablets using a D-optimal design. *International Journal of Pharmacy*, **153**, 247–255.

Booth, K. H. V. and D. R. Cox (1962). Some systematic supersaturated designs. *Technometrics*, **4**, 489–495.

Box, G. E. P. (1957). Evolutionary operation: A method for increasing industrial productivity. *Applied Statistics*, **6**, 81–101.

Box, G. E. P. (1982). Seminar talk. Department of Statistics, University of Wisconsin-Madison.

Box, G. E. P. and D. Behnken (1960). Some new three level designs for the study of quantitative variables. *Technometrics*, **2**(4), 455–475.

Box, G. E. P. and S. Bisgaard (1993). What can you find out from 12 experimental runs? *Quality Engineering*, **5**, 663–668.

Box, G. E. P. and N. R. Draper (1969). *Evolutionary Operation*. New York: Wiley.

Box, G. E. P. and H. L. Lucas (1959). Design of experiments in nonlinear situations. *Biometrika*, **46**, 77–90.

Box, G. E. P. and K. P. Wilson (1951). On the experimental attainment of optimum conditions. *Journal of the Royal Statistical Society, Series B*, **13**(1), 1–45.

Box, G. E. P. and J. Tyssedal (1996a). Projective properties of certain orthogonal arrays. *Biometrika*, **83**(4), 950–955. (This article is also available as Report No. 116, Center for Quality and Productivity Improvement, University of Wisconsin-Madison and can be downloaded at http://www.engr.wisc.edu/centers/cqpi/reports/pdfs/r116.pdf.)

Box, G. E. P. and J. Tyssedal (1996b). Sixteen run designs under high projectivity for factor screening. *Communications in Statistics: Simulation and Computation*, **30**(2), 217–228. (This article is also available as "Projective properties of the sixteen run two-level orthogonal arrays", Report No. 135, Center for Quality and Productivity Improvement, University of Wisconsin-Madison and can be downloaded at http://www.engr.wisc.edu/centers/cqpi/reports/pdfs/r135.pdf.)

Box, G. E. P., J. S. Hunter, and W. G. Hunter (2005). *Statistics for Experimenters: Design, Innovation, and Discovery*, 2nd ed. Hoboken, NJ: Wiley.

Bradley, R. A. and C. M. Yeh (1980). Trend-free block designs: Theory. *The Annals of Statistics*, **8**, 883–893.

Bradley, R. A. and C. M. Yeh (1983). Trend-free block designs: Existence and construction results. *Communications in Statistics, Theory and Methods*, **12**, 1–24.

Bradley, R. A., W. I. Notz, and C. M. Yeh (1985). Nearly trend-free block designs. *Journal of the American Statistical Association*, **80**, 985–992.

Buncick, M. C., A. R. K. Ralston, D. D. Denton, and S. Bisgaard (2000). Characterization of a plasma sputter deposition process by fractional factorial design. *Quality Engineering*, **12**, 371–385.

Bundick, R. K., C. M. Borror, and D. C. Montgomery (2003). A review of methods for measurement systems capability. *Journal of Quality Technology*, **35**, 342–354.

Bursztyn, D. and D. M. Steinberg (2001). Rotation designs for experiments in high-bias situations. *Journal of Statistical Planning and Inference*, **97**(2), 399–414.

Bursztyn, D. and D. M. Steinberg (2002). Rotation designs: Orthogonal first-order designs with higher order projectivity. *Applied Stochastic Models in Business and Industry*, **18**(3), 197–206.

Bursztyn, D. and D. M. Steinberg (2006). Comparison of designs for computer experiments. *Journal of Statistical Planning and Inference*, **136**, 1103–1119.

Cameron, J. M. and G. E. Hailes (1974). Designs for the calibration of small groups of standards in the presence of drift. NBS Technical Note 844, National Bureau of Standards, United States Department of Commerce.

Cameron, J. M., M. C. Croarkin, and R. C. Raybold (1977). Designs for the calibration of standards of mass. NBS Technical Note 952, National Bureau of Standards, United States Department of Commerce.

Chen, V. C. P., K.-L. Tsui, R. R. Barton, and J. K. Allen (2003). A review of design and modeling in computer experiments. In *Handbook of Statistics*, Vol. 22, Chap. 7 (R. Khattree and C. R. Rao, eds.). Amsterdam: Elsevier Science B. V.

Cheng, C.-S. and C.-C. Li (1993). Constructing orthogonal fractional factorial designs when some factor-level combinations are debarred. *Technometrics*, **35**(3), 277–283.

Chokshi, D. (2000). Design of experiments in rocket engine fabrication. In *Proceedings of the 54th Annual Quality Congress*, American Society for Quality, Milwaukee, WI. (http://www.asq.org/members/news/aqc/54_2000/14037.pdf)

Cioppa, T. M. (2002). *Efficient Nearly Orthogonal and Space-Filling Experimental Designs for High-Dimensional Complex Models*. (Ph.D. Dissertation). Monterey, CA: Operations Research Department, Naval Postgraduate School. (available online at http://library.nps.navy.mil/uhtbin/Hyperion-image/02sep_Cioppa_PhD.pdf)

Cooney, G. A., Jr. and R. P. Verseput (2000). The design of model-robust experiments. *Quality and Reliability Engineering International*, **16**, 373–389.

Cornell, J. A. (2002). *Experiments with Mixtures: Designs, Models, and the Analysis of Mixture Data*, 3rd ed. New York: Wiley.

Cotter, S. (1979). A screening design for factorial experiments with interactions. *Biometrika*, **66**(2), 317–320.

Cox, D. R. (1951). Some systematic experimental designs. *Biometrika*, **38**, 312–323.

Craigen, R. (1996). Weighing matrices and conference matrices. In *CRC Handbook of Combinatorial Designs*, pp. 496–504. (C. J. Colbourn and J. H. Dinitz, eds.). Boca Raton, FL: CRC Press.

Czitrom, V. (1999). One-factor-at-a-time versus designed experiments. The *American Statistician*, **53**(2), 126–131.

Czitrom, V., J. C. de Amezua Ayala, and A. R. Sanchez (1998). Matching implant doses during design of experiments. *Quality and Reliability Engineering International*, **14**, 211–217.

Czitrom, V. (2003). Guidelines for selecting factors and factor levels for an industrial designed experiment. In *Statistics in Industry, Handbook of Statistics*, Vol. 22 (R. Khattree and C. R. Rao, eds.). Amsterdam: Elsevier Science B. V.

Daniel, C. (1973). One-at-a-time plans. *Journal of the American Statistical Association*, **68**, 353–360 (June).

Daniel, C. (1994). Factorial one factor at a time experiments, dedicated to the memory of W. Edwards Deming. *The American Statistician*, **48**, 132–135.

Daniel, C. and F. Wilcoxon (1966). Factorial 2^{n-p} plans robust against linear and quadratic trends. *Technometrics*, **8**, 259–278.

Delott, C. and P. Gupta (1990). Characterization of copperplating process for ceramic substrates. *Quality Engineering*, **2**(3), 269–284.

Deming, W. E. (1953). On the distinction between enumerative and analytic surveys. *Journal of the American Statistical Association*, **48**, 244–245.

Deming, W. E. (1975). On probability as a basis for action. *The American Statistician*, **29**, 146–152.

Dette, H., V. B. Melas, and N. Strigul (2005). Design of experiments for microbiological models. In *Applied Optimal Designs*, Chap. 6 (M. P. F. Berger and W. K. Wong, eds.). Hoboken, NJ: Wiley.

Dette, H., V. B. Melas, and W. K. Wong (2005). Optimal design for goodness-of-fit of the Michaelis-Menten enzyme kinetic function. *Journal of the American Statistical Association*, **100**(472), 1370–1381.

Dickinson, A. W. (1974). Some run orders requiring a minimum number of factor level changes for the 2^4 and 2^5 factorial designs. *Technometrics*, **16**, 31–37.

Dobson, A. J. (2001). *An Introduction to Generalized Linear Models*, 2nd ed. London: Chapman Hall/CRC.

Draper, N. R. and D. M. Stoneman (1968). Factor changes and linear trends in eight-run two-level factorial designs. *Technometrics*, **10**, 301–311.

DuMouchel, W. and B. Jones (1994). A simple Bayesian modification of D-optimal designs to reduce the dependence on an assumed model. *Technometrics*, **36**(1), 37–47.

Emptage, M. R., B. Hudson-Curtis, and K. Sen (2003). Treatment of microarray experiments as split-plot designs. *Journal of Biopharmaceutical Statistics*, **13**, 159–178.

Eskridge, K. M., S. G. Gilmour, R. Mead, N. A. Butler, and D. A. Travnicek (2004). Large supersaturated designs. *Journal of Statistical Computation and Simulation*, **74**, 525–542.

Fang, K.-T. (1980). The uniform design: Application of number-theoretic methods in experimental design. *Acta Mathematicae Applacatae Sinica*, **3**, 363–372.

Fang, K.-T. and D. K. J. Lin (2003). Uniform experimental designs and their applications in industry. In *Handbook of Statistics*, Vol. 22, Chap. 4 (R. Khattree and C. R. Rao, eds.). Amsterdam: Elsevier Science B. V.

Fang, K.-T., R. Li, and A. Sudjianto (2006). *Design and Modeling for Computer Experiments*. New York: Chapman and Hall.

Fang, K.-T., D. K. J. Lin, and C. X. Ma (1999). On the construction of multi-level supersaturated designs. *Journal of Statistical Planning and Inference*, **86**, 239–252.

Fang, K.-T., D. K. J. Lin, and M. Q. Liu (2003). Optimal mixed-level supersaturated design. *Metrika*, **58**, 279–291.

Fang, K.-T., D. K. J. Lin, P. Winker, and Y. Zhang (2000). Uniform design: Theory and application. *Technometrics*, **42**(3), 237–248.

Fedorov, V. V. and P. Hackl (1997). *Model-Oriented Design of Experiments*. New York: Springer-Verlag.

Fedorov, V. V. and S. L. Leonov (2005). Response driven designs in drug development. In *Applied Optimal Designs*, Chap. 5 (M. P. F. Berger and W. K. Wong, eds.). Hoboken, NJ: Wiley.

Franklin, M. F. and R. A. Bailey (1977). Selection of defining contrasts and confounded effects in two-level experiments. *Applied Statistics*, **25**, 64–67.

Freeman, I. C. (2004). *Objective Comparison of Design of Experiments Strategies in Design and Observations in Practice*. (Master of Science in Engineering and Management Thesis). Cambridge, MA: Engineering Systems Division, Massachusetts Institute of Technology.

Ganju, J. and J. M. Lucas (1997). Bias in test statistics when restrictions in randomization are caused by factors. *Communications in Statistics – Theory and Methods*, **26**(1), 47–63.

Giesbrecht, F. G. and M. L. Gumpertz (2004). *Planning, Construction, and Statistical Analysis of Comparative Experiments*. Hoboken, NJ: Wiley.

Glonek, G. F. V. and P. J. Solomon (2004). Factorial and time course designs for cDNA microarray experiments. *Biostatistics*, **5**, 89–111.

Goos, P., L. Tack, and M. Vandebroek (2005). The optimal design of blocked experiments in industry. In *Applied Optimal Designs*, Chap. 10 (M. P. F. Berger and W. K. Wong, eds.). Hoboken, NJ: Wiley.

Goos, P., A. Kobilinsky, T. O'Brien, and M. Vandebroek (2005). Model-robust and model-sensitive designs. *Computational Statistics and Data Analysis*, **49**(1), 201–216.

Hackl, P., W. G. Muller, and A. C. Atkinson, eds. (2001): *MODA 6: Advances in Model-Oriented Design and Analysis: Proceedings of the 6th International Workshop on Model-Oriented Design and Analysis*. Santa Clara, CA: Telos, The Electronic Library of Science. (Telos is an imprint of Springer-Verlag, New York, NY.)

Hahn, G. J. and A. F. Dershowitz (1974). Evolutionary operation today—some survey results and observations. *Applied Statistics*, **23**(2), 214–218.

Hahn, G. J. and W. Q. Meeker (1993). Assumptions for statistical inference. *The American Statistician*, **47**, 1–11.

Hall, M. J. (1961). Hadamard matrices of order 16. *Jet Propulsion Laboratory*, Research Summary 1, 21–26.

Hamada, M. and C. F. J. Wu (1992). Analysis of designed experiments with complex aliasing. *Journal of Quality Technology*, **24**(3), 130–137.

Hayford, J. A. (1893). On the least squares adjustment of weighings. U. S. Coast & Geodetic Survey Report for 1892, Appendix 10, USCGS.

Heredia-Langner, A., W. M. Carlyle, D. C. Montgomery, C. M. Borror, and G. C. Runger (2003). Genetic algorithms for the construction of D-optimal designs. *Journal of Quality Technology*, **35**(1), 28–46.

Heredia-Langner, A., D. C. Montgomery, W. M. Carlyle, and C. M. Borror (2004). Model-robust optimal designs: A genetic algorithm approach. *Journal of Quality Technology*, **36**(3), 263–279.

Hill, H. M. (1960). Experimental designs to adjust for time trend. *Technometrics*, **2**, 67–82.

Hinkelmann, K. and O. Kempthorne (2005). *Design and Analysis of Experiments, Volume 2, Advanced Experimental Design*. Hoboken, NJ: Wiley.

Hoerl, R. W. (1998). Six Sigma and the future of the quality profession. *Quality Progress*, **31**(6), 35–42.

Hsieh, P. I. and D. E. Goodwin (1986). Sheet molded compound process improvements. In *Fourth Symposium on Taguchi Methods*, Dearborn, MI: American Supplier Institute, pp. 13–21.

Hubele, N. F., T. Beaumariage, G. Baweja, and S.-C. Hong (1994). Using experimental design to assess the capability of a system. *Journal of Quality Technology*, **26**(1), 1–11.

Jensen, D. (2000). The use of studentized diagnostics in regression. *Metrika*, **52**, 213–223.

Jin, R., W. Chen, and A. Sudjianto (2005). An efficient algorithm for constructing optimal design of computer experiments. *Journal of Statistical Planning and Inference*, **34**(1), 268–287.

John, P. W. M. (1990). Time trend and factorial experiments. *Technometrics*, **32**, 275–282.

Jones, F. E. E. and R. M. Schoonover (2002). *Handbook of Mass Measurement*. Boca Raton, FL: CRC Press.

Joseph, V. R. and C. F. J. Wu (2004). Failure amplification method: An information maximization approach to categorical response optimization. *Technometrics*, **46**(1), 1–12; discussion: 12–31.

Kennard, R. W. and L. Stone (1969). Computer aided design of experiments. *Technometrics*, **11**, 137–148.

Kerr, M. K. and G. A. Churchill (2001). Experimental design for gene expression microarrays. *Biostatistics*, **2**, 183–201.

Kiefer, J. (1959). Optimum experimental designs (with discussion). *Journal of the Royal Statistical Society, Series B*, **21**, 272–319.

Kim, K. (1997). *Construction and Analysis of Trend-Free Factorial Designs under a General Cost Structure*. (Ph.D. dissertation). Blacksburg, VA: Department of Statistics, Virginia Polytechnic Institute and State University.

Kleijnen, J. P. C., S. M. Sanchez, T. W. Lucas, and T. M. Cioppa (2005). A user's guide to the brave new world of designing simulation experiments. *INFORMS Journal on Computing*, **17**(3), 263–289. (available online at http://www.nps.navy.mil/orfacpag/resumePages/sanchsLinkedPages/UserGuideSimExpts.pdf and the 5-page online supplement is available at http://joc.pubs.informs.org/Supplements/Kleijnen.pdf.)

Kuhfeld, W. F. and R. D. Tobias (2005). Large factorial designs for product engineering and marketing research applications. *Technometrics*, **47**, 132–141.

Lawson, C. L. and R. J. Hanson (1995). *Solving Least Squares Problems* (*Classics in Applied Mathematics Series, Vol. 15*). Philadelphia: Society for Industrial and Applied Mathematics.

Ledolter, J. and A. Swersey (1997). Dorian Shainin's variables search procedure: A critical assessment. *Journal of Quality Technology*, **29**, 237–247.

León, R. V. and R. W. Mee (2000). Blocking multiple sources of error in small analytic studies. *Quality Engineering*, **12**(4), 497–502.

Lewis, S. L., D. C. Montgomery, and R. H. Myers (2001a). Examples of designed experiments with nonnormal responses. *Journal of Quality Technology*, **33**(3), 265–278.

Lewis, S. L., D. C. Montgomery, and R. H. Myers (2001b). Confidence interval coverage for designed experiments analyzed using generalized linear models. *Journal of Quality Technology*, **33**(3), 279–292.

Li, R. Z. and D. K. J. Lin (2002). Data analysis of supersaturated design. *Statistics and Probability Letters*, **59**, 135–144.

Li, R. and D. K. J. Lin (2003). Analysis methods for supersaturated design: Some comparisons. *Journal of Data Analysis*, **3**, 103–121.

Li, W. and C. J. Nachtsheim (2000). Model-robust factorial designs. *Technometrics*, **42**, 345–352.

Lin, D. K. J. (1993a). A new class of supersaturated designs. *Technometrics*, **35**, 28–31.

Lin, D. K. J. (1993b). Another look at first-order saturated designs: The p-efficient designs. *Technometrics*, **35**, 284–292.

Lin, D. K. J. (2003). Industrial experimentation for screening. In *Handbook of Statistics*, Vol. 22, Chap. 2 (R. Khattree and C. R. Rao, eds.). Amsterdam: Elsevier Science B. V.

Lin, D. K. J. and N. R. Draper (1992). Projective properties of Plackett and Burman designs. *Technometrics*, **34**, 423–428.

Liu, M. Q. and K. T. Fang (2006). A case study in the application of supersaturated designs to computer experiments. *Acta Scientiarum Mathematicarum* **26**(B4), to appear. (A preprint is available at http://www.math.nankai.edu.cn/keyan/pre/preprint06/06-06.pdf.)

Liu, Y. and A. M. Dean (2004). k-circulant supersaturated designs. *Technometrics*, **46**, 32–43.

Liu, Y., S. Ruan, and A. M. Dean (2006). Construction and analysis of Es^2 efficient supersaturated designs. *Journal of Statistical Planning and Inference* (to appear). (A preprint of this paper is available at http://www.stat.ohio-state.edu/~amd/papers/liu-ruan-dean-rev10.pdf.)

Lopez-Fidalgo, J. and S. A. Garcet-Rodríguez (2004). Optimal experimental designs when some independent variables are not subject to control. *Journal of the American Statistical Association*, **99**, 1190–1199.

Lu, X. and X. Wu (2004). A strategy of searching active factors in supersaturated screening experiments. *Journal of Quality Technology*, **36**, 392–399.

Matthews, J. N. S. and P. W. James (2005). Restricted optimal design in the measurement of cerebral blood flow using the Kety–Schmidt technique. In *Applied Optimal Designs*, Chap. 8 (M. P. F. Berger and W. K. Wong, eds.). Hoboken, NJ: Wiley.

McCullagh, P. and J. A. Nelder (1989). *Generalized Linear Models*, 2nd ed. New York: Chapman and Hall/CRC.

McKay, M. D., R. J. Beckman, and W. J. Conover (1979). A comparison of three methods for selecting values of input variables in the analysis of output from a computer code. *Technometrics*, **21**(2), 239–245.

McDaniel, W. R. and B. E. Ankenman (2000). Comparing experimental design strategies for quality improvement with minimal changes to factor levels. *Quality and Reliability Engineering International*, **16**, 355–362.

Mee, R. (2002). Three-level simplex designs and their role in sequential experimentation. *Journal of Quality Technology*, **34**, 152–164.

Montgomery, D. C. and G. C. Runger (2003). *Applied Statistics and Probability for Engineers*, 3rd ed. New York: Wiley.

Montgomery, D. C., C. M. Borror, and J. D. Stanley (1997). Some cautions in the use of Plackett–Burman designs. *Quality Engineering*, **10**(2), 371–381.

Montgomery, D. C., E. N. Loredo, D. Jearkpaporn, and M. C. Testik (2002). Experimental designs for constrained regions. *Quality Engineering*, **14**(4), 587–601.

Morris, M. D., T. J. Mitchell, and D. Ylvisaker (1993). Bayesian design and analysis of computer experiments: Use of derivatives in surface prediction. *Technometrics*, **35**, 243–255.

Myers, R. H. (1999). Response surface methodology—current status and future directions. *Journal of Quality Technology*, **31**(1), 30–44.

Myers, R. H. and D. C. Montgomery (1995). *Response Surface Methodology: Process and Product Optimization using Designed Experiments*. New York: Wiley.

Nelson, L. S. (1982). Technical aids: Extreme screening designs. *Journal of Quality Technology*, **14**(2), 99–100.

Nelson, P. R. (1993). Additional uses for the analysis of means and extended tables of critical values. *Technometrics*, **35**(1), 61–71.

Nelson, P. R., P. S. Wludyka, and K. A. F. Copeland (2005). *The Analysis of Means: A Graphical Method for Comparing Means, Rates, and Proportions*. Philadelphia: Society for Industrial and Applied Mathematics; Alexandria, VA: American Statistical Association.

Ossenbruggen, P. J., H. Spanjers, and A. Klapwik (1996). Assessment of a two-step nitrification model for activated sludge. *Water Research*, **30**(4), 939–953.

Park, Y., D. C. Montgomery, J. W. Fowler, and C. M. Borror (2006). Cost-constrained G-efficient response surface designs for cuboidal regions. *Quality and Reliability Engineering International*, **22**(2), 121–139.

Piepel, G. F. (1988). Programs for generating extreme vertices and centroids of linearly constrained experimental regions. *Journal of Quality Technology*, **20**(2), 125–139.

Plackett, R. L. and J. P. Burman (1946). The design of optimum multifactorial experiments. *Biometrika*, **33**, 305–325.

Pukelsheim, F. (1993). *Optimal Design of Experiments*. New York: Wiley.

Qu, X. and C. F. J. Wu (2005). One-factor-at-a-time designs of Resolution V. *Journal of Statistical Planning and Inference*, **131**, 407–416.

Reece, J. E. (2003). Software to support manufacturing systems. In *Handbook of Statistics*, Vol. 22, Chap. 9 (R. Khattree and C. R. Rao, eds.). Amsterdam: Elsevier Science B. V.

Reece, J. E. and M. Shenasa (1997). Characterization of a vertical furnace chemical vapor deposition (CVD) silicon nitride process. In *Statistical Case Studies for Industrial Process Improvement*, pp. 429–468. (V. Czitrom and P. D. Spagon, eds.). Philadelphia: Society for Industrial and Applied Mathematics; Alexandria, VA: American Statistical Association.

Roquemore, K. G. (1976). Hybrid designs for quadratic response surfaces. *Technometrics*, **18**, 419–423.

Ryan, T. P. (1997). *Modern Regression Methods*. New York: Wiley.

Ryan, T. P. (2000). *Statistical Methods for Quality Improvement*, 2nd ed. New York: Wiley.

Samset, O. and J. Tyssedal (1998). Repeat and mirror-image patterns and correlation structures for Plackett–Burman designs, their foldovers and half fractions. Technical Report #13, Department of Mathematical Sciences, The Norwegian University of Science and Technology, Norway. (The report is accessible from the list of technical reports at http://www.math.ntnu.no/~sam/reports.html.)

Santner, T., B. Williams, and W. Notz (2003). *The Design and Analysis of Computer Experiments*. New York: Springer-Verlag.

Satterthwaite, F. (1959). Random balance experimentation (with discussion). *Technometrics*, **1**, 111–137.

Scholtens, D., A. Miron, F. M. Merchant, A. Miller, P. L. Miron, J. D. Iglehart, and R. Gentleman (2004). Analyzing factorial designed microarray experiments. *Journal of Multivariate Analysis*, **90**, 19–43.

Simon, R. M., E. L. Korn, L. M. McShane, M. D. Radmacher, G. W. Wright, and Y. Zhao (2004). *Design and Analysis of DNA Microarray Investigations*. New York: Springer.

Simpson, T. W., D. K. J. Lin, and W. Chen (2001). Sampling strategies for computer experiments: Design and analysis. *International Journal of Reliability and Applications*, **2**(3), 209–240.

Silvey, S. D. (1980). *Optimal Design*. London: Chapman and Hall.

Snee, R. D. (1985a). Computer-aided design of experiments—some practical experiences. *Journal of Quality Technology*, **17**, 222–236.

Snee, R. D. (1985b). Experimenting with a large number of variables. In *Experiments in Industry: Design, Analysis, and Interpretation of Results*. Milwaukee, WI: American Society for Quality Control.

Speed, T. P. and T. H. Yang (2002). Direct versus indirect designs for cDNA microarray experiments. *Sankhya, Series A*, **64**, 707–721. (This article is available at http://www.biostat.ucsf.edu/jean/Papers/Sankhya02.pdf.)

Steinberg, D. M. and D. Bursztyn (2001). Discussion of "Factor Screening and Response Surface Exploration" by Cheng and Wu. *Statistica Sinica*, **11**, 596–599.

Tack, L. and M. Vandebroek (2002). Trend resistant and cost-efficient block designs with fixed or random block effects. *Journal of Quality Technology*, **34**, 422–436.

Tang, M., J. Li, L.-Y. Chan, and D. K. J. Lin (2004). Application of uniform design in the formation of cement mixtures. *Quality Engineering*, **16**(3), 461–474.

Tyssedal, J. (1993). Projections in the 12-run Plackett–Burman design. Report No. 106, Center for Quality and Productivity Improvement, University of Wisconsin-Madison. This report is downloadable at http://www.engr.wisc.edu/centers/cqpi/reports/pdfs/r106.pdf.

Voelkel, J. G. (2005). Efficiencies of fractional factorial designs. *Technometrics*, **47**(4), 488–494.

Wang, P. C. and H. W. Jan (1995). Designing 2-level factorial experiments using orthogonal arrays when the run order is important. *Statistician*, **44**(3), 379–388.

Wang, Y. and K.-T. Fang (1981). A note on uniform distribution and experimental design. *KeXue TongBao*, **26**, 485–489.

Waterhouse, T. H. (2005). *Optimal Experimental Design for Nonlinear and Generalised Linear Models*. (Ph. D. Thesis). Brisbane, Australia: School of Physical Sciences, University of Queensland. (This is available at http://www.maths.uq.edu.au/~thw/research/thesis.pdf.)

Wen, Z.-Y. and F. Chen (2001). Application of statistically based experimental designs for the optimization of eicosapentaenoic acid production by the diatom *Nitzschia laevis*. *Biotechnology and Bioengineering*, **75**(2), 159–169.

Wit, E., A. Nobile, and R. Khanin (2005). Near-optimal designs for dual channel microarray studies. *Applied Statistics*, **54**, 817–830.

Wu, C. F. J. and M. Hamada (2000). *Experiments: Planning, Analysis, and Parameter Design Optimization*. New York: Wiley.

Xu, H. and C. F. J. Wu (2005). Construction of optimal multi-level supersaturated designs. *Annals of Statistics*, **33**(6), 2811–2836. (also available at http://www.stat.ucla.edu/~hqxu/pub/XuWu2005.pdf)

Yamada, S. and D. K. J. Lin (1999). Three-level supersaturated designs. *Statistics and Probability Letters*, **45**, 31–39.

Yamada, S., Y. T. Ikebe, H. Hashiguchi, and N. Nicki (1999). Construction of three-level supersaturated design. *Journal of Statistical Planning and Inference*, **81**, 183–194.

Zahran, A. and C. M. Anderson-Cook (2003). A general equation and optimal design for a 2-factor restricted region. *Statistics and Probability Letters*, **64**, 9–16.

Zahran, A , C. M. Anderson-Cook, R. H. Myers, and E. P. Smith (2003). Modifying 2^2 factorial designs to accommodate a restricted design space. *Journal of Quality Technology*, **35**(4), 387–392.

Zelen, M. (1962). Linear estimation and related topics. In *Survey of Numerical Analysis*, pp. 558–577 (J. Todd, ed.) New York: McGraw-Hill.

EXERCISES

13.1 Critique the following statement: "I don't see the need to set a factor to a neutral position and then putting it back where it was before I changed the level to a neutral position. This seems much to-do about nothing."

13.2 Explain how you would design an experiment if you had three factors to examine in the experiment and one of the three was a hard-to-change factor.

13.3 Close-to-orthogonal designs, such as supersaturated designs, are often (appropriately) used. When this is done, it is necessary to consider the correlations between the factors. Explain why these are not true correlations.

13.4 What is one potential danger in using a supersaturated design?

13.5 Construct a standard OFAT design for four factors and compute the correlations between the columns of the design. Do you consider the latter to be unacceptably large? Why, or why not?

13.6 Respond to the following statement: "I don't see any point in using an equi-leverage design because whether or not points are influential will depend partly on the response values, and they of course cannot be controlled."

13.7 Wen and Chen (2001, references) initially used a Plackett-Burman (PB) design to investigate the factors that affected eicosapentaeonic acid (EPA) production by the diatom *Nitzschia laervis* in heterotrophic conditions. A 20-run PB design was used to investigate the effects of 13 factors. Since the number of available degrees of freedom exceeded the number of factors in the study, the experimenters created dummy columns D_1-D_6 "to evaluate the standard errors of the experiment." Are the dummy columns necessary? Why or why not? They stated that "only confidence levels above 90 percent were accepted in the experiment." This is not common statistical jargon. What did they actually mean? Once they identified the factors that they declared significant, they used a central composite design to investigate these factors further, using a 2^{5-1} design with 6 centerpoints and 10 axial points. Does this seem like a reasonable strategy? Comment.

13.8 Using the methodology given at the stated URL, derive the expression for the test item estimate that was given in Section 13.16.

13.9 What will be the leverage values if all the estimable effects in a 2^{5-1} design are estimated and are used in a model?

13.10 Determine using either software or by direct computation the leverages for a 2×4 design when only linear terms are fit, then compute the leverages when the linear, quadratic, and cubic terms in the second factor are fit. Comment.

13.11 Assume that you have used MINITAB or other software to construct a PB design for 15 factors in 16 runs. Since 16 is a power of 2, the design must be a full factorial in a subset of these factors, and the PB design must thus be a fractional factorial. What must be the size of this subset. Show this after first appropriately rearranging the runs, if necessary. Is the fractional factorial that the PB design is equivalent to inferior in any way (e.g., in terms of minimum aberration or maximum number of clear effects) to any other fractional factorial design of this type that could be constructed? Explain.

13.12 State the form of the fractional factorial design to which a PB design with 32 runs is equivalent.

13.13 Assume that a PB design with 24 runs has been constructed to investigate the effect of each of 23 factors. Would it be appropriate to perform a conditional effects analysis, along the lines of what was illustrated in Chapter 5? Why, or why not? If this would not be possible, although a PB design is used because of run size economy, would replicating the design permit a conditional effects analysis? In general, what must be true before such an analysis can be performed?

13.14 Consider the problem posed in Example 13.2. We can label the factors that were involved in the debarred observations as D and E, referring to substrate holder and heat, respectively. If the other half fraction had been used (i.e., with $I = -ABCDE$), would the problem still have existed, and if so, would the number of debarred observations still have been 4? Explain.

13.15 What is the motivation for using a "reverse" PB design.

13.16 Since a PB design is a small-run design, what would be the motivation for using a half-fraction of such a design? Consider a half-fraction of a 12-run PB design and the correlation structure given in Table 82 at www.math.ntnu.no/preprint/statistics/1998/S13-1998.ps . Would you recommend such a design? Why, or why not?

CHAPTER 14

Tying It All Together

In statistics books in general, a method is presented and an illustrative example shortly follows. Simple exercises are given at the end of sections and the student/reader knows that they are to be worked using one of the methods just presented in the section.

The real world is not so accommodating. Specifically, an experimenter has a very large assortment of designs from which to choose and, in the absence of expert systems software, may have difficulty making a good choice. Frequently, the physical scenario is "bent" to make it conformable to published designs, just as an unscrupulous mechanic may try to bend a tailpipe assembly to make it fit a car it is not designed to fit. The bending of scenarios has been done often to allow the use of Taguchi designs, in particular.

Since the analysis of data from a designed experiment is relatively straightforward when the experiment is well designed, the focus in training for experimentation with statistical designs should be on gaining expertise in the selection of good designs. The end-of-chapter exercises provide an opportunity to move in that direction.

14.1 TRAINING FOR EXPERIMENTAL DESIGN USE

Before we look at some scenarios and try to decide how to proceed, it is worth noting that experience in experimentation can be gained "off-line" in various ways. As the late statistician Bill Hunter often said, one should "do statistics," and this certainly applies to engineering students, medical students and researchers, and students in other fields in which experimentation is routinely performed.

An excellent way to learn experimental design, especially some of the nuances that don't come through strongly in textbooks, is to conduct actual, simple experiments that do not require much in the way of materials or time. The primary motivation for students conducting such simple experiments was Hunter (1977), and there have been various other articles on the subject written since then, especially articles appearing in the *Journal of Statistics Education*, an online publication (see http://www.amstat.org/publications/jse/).

One example of a hands-on exercise that I have used in training industrial personnel in experimental design is commonly referred to as the catapult experiment, with a small catapult used to catapult a ping-pong ball, which is measured for distance traveled. Factors that can be varied include the rubber band type, position of the arm, type of ball used, and so on. Data are generated from the experiment, which would be analyzed using the methods given in various chapters such as Chapter 4 if each factor has two levels, as is desirable for the exercise. Of course an important part of the exercise is trying to figure out why certain effects (possibly including interactions) are significant. A good description of an actual experiment of this type is given in the paper "Training for design of experiments using a catapult" by J. Anthony (*Quality and Reliability Engineering International*, **18**, 29–35, 2002). Quoting from that article, "The engineers in the company felt that the experiments were useful in terms of formulating the problem, identifying the key control factors, determining the ranges of factor settings, selecting the experimental layout, assigning the control factors to the design matrix, conducting the experiment, and analysing and interpreting the results of the experiment."

There are variations of the catapult experiment that have been presented in the literature and one variation is to use it to try to determine the factor settings to hit a target value, as in multiresponse optimization that was presented in Chapter 12. An example of this is given in Section 5.4.7.2 of the *NIST/SEMATECH e-Handbook of Statistical Methods* (Croarkin and Tobias, 2002), which can be viewed at http://www.itl.nist.gov/div898/handbook/pri/section4/pri472.htm. In that example, a 2_{IV}^{5-1} design was used with four centerpoints and the objective was to determine the factor settings that should be used to cause the ping-pong ball to reach three different distances—30, 60, and 90 inches. Note that there is only a single response variable but a desirability function approach can still be used, as was used at the end of that section.

REFERENCES

Bjerke, F., A. H. Aastveit, W. W. Stroup, B. Kirkhus, and T. Naes (2004). Design and analysis of storing experiments: A case study. *Quality Engineering*, **16**(4), 591–611.

Box, G. E. P. and S. Jones (1992). Split-plot designs for robust product experimentation. *Journal of Applied Statistics*, **19**, 3–26. (This is available as Report No. 61, Center for Quality and Productivity Improvement, University of Wisconsin-Madison (see http://www.engr.wisc.edu/centers/cqpi.)

Croarkin, C. and P. Tobias, eds. (2002). *NIST/SEMATECH e-Handbook of Statistical Methods* (http://www.itl.nist.gov/div898/handbook), a joint effort of the National Institute of Standards and Technology and International SEMATECH.

Hunter, W. G. (1977). Some ideas about teaching design of experiments with 2^5 examples of experiments conducted by students. *The American Statistician*, **31**(1), 12–17.

EXERCISES

14.1 Assume that an experimenter has a strong need to conduct an experiment but knows that a factor that will affect the response variable cannot be maintained in a state of statistical control, with the latter interpreted to mean fixed parameter values. He has a few factors that he wants to study and will probably use two levels for each factor. What would you advise him to do? (*Note*: A possible design approach was covered in one of the chapters but was not emphasized.)

14.2 An experimenter wants to use a replicated 2^3 design to investigate the three factors of interest. He wants to be able to detect main effects of a certain minimum magnitude and can afford several replicates but suspects that the AB and AC interactions may be significant. He intends to consult tables to determine how many replicates he should use. What advice would you give him for determining the number of replicates to use when these interactions are expected to be significant. Is there any other advice that you would give him?

14.3 An experimenter is considering using a $2^2 \times 3^2$ design because there are three levels of interest for two of the factors. A 2^4 design may be used instead, however, because of a desire to use a simpler design, for which the analysis would also be simpler. A 2^4 design with two replicates and four centerpoints would have the same number of runs as the mixed factorial. Could the replicated 2^4 design be used as a suitable substitute for the mixed factorial? What, if anything, would be lost with this substitution? Explain.

14.4 The case study given by Bjerke, Aastveit, Stroup, Kirkhus, and Naes (2004, references) is a good example of how experimental settings can have various complications, which was discussed in Example 9.1. The authors described an experiment that was performed to investigate the effects of starch concentration on the quality of low-fat mayonnaise during different processing and storing conditions. A face-centered central composite design for three factors was used and there was a split-plot structure with repeated measures. The three factors had been selected in close cooperation with the manufacturer, as well as the levels of those factors.

There were 18 experimental runs and the response variable, which was (Brookfield) viscosity, was measured after 7, 14, and 38 weeks. It is worth noting, as the authors did, that these are not equally spaced time intervals.

At most six runs could be made per day, so the experiment was run over three days, with the axial points run on one day and a half fraction of the factorial points run on each of the other two days, with the four centerpoints scattered among the three days.

All the production samples were split into two halves for storing, with one half stored at room temperature (21°C), and the other half stored in a refrigerator at 4°C. This was labeled factor "D."

The manufacturer decided to use the measurements obtained from the second storage temperature because those most closely resembled the real-life storage of mayonnaise.

The following statement by the authors is worth noting:

> Throughout the production of the samples, some relevant process parameters were monitored to ensure stable process conditions. Observing these parameters during the production stage of the experiment did not reveal anything unusual. Unfortunately, these data were not recorded for further use (p. 599).

Of course such monitoring is important, as was stressed in Section 1.7, although it is undoubtedly not considered in most experiments.

One complication was that the Brookfield viscometer had an upper limit of 100. The viscometer cannot measure the viscosity if the samples are too thick. If the samples are too thick for the viscometer to perform properly, then the viscosity must be greater than 100. During the experiment, some of the mayonnaise samples became so thick that the spindle of the Brookfield meter could not rotate as required. Three of the data values listed in the authors' Table 2 were 100, including two of the three response values when all three factors were set at their highest level. Thus, part of the data was censored, which raises the question of whether the censored (i.e., incorrect) values should be used, should the true values be estimated, or should the censored values simply be treated as missing values.

Another problem is that since the three production factors were difficult to vary, the run order that was used each day "... was partly due to ease rather than randomization."

The authors considered three modeling approaches, which included a mixed model approach and a robustness approach. Address the following questions.

(a) How would you deal with the censored data problem, recognizing that using the recorded values could introduce spurious nonlinear effects?

(b) Are you concerned about the fact that there were hard-to-change factors (apparently all of them) and so randomization was not used? The authors analyzed the data, specifically their Table 5 analysis, as if complete randomization had been performed during the experiment. Would you have analyzed the data in some other way rather than constructing a standard ANOVA table and looking at p-values, which are not valid, strictly speaking, when there is restricted randomization, as was discussed in Section 4.19.

(c) The authors stated, "A more detailed analysis would include modeling of the time data," which they then proceeded to perform. Why would this be better than a repeated measures approach?

(d) Time was actually involved in two different ways as the experimental runs were made over three consecutive days. These would logically constitute three blocks and the blocking factor would be analyzed, but this was not performed in the authors' analyses. Can you think of a way of justifying this, or do you believe that a possible block effect should have been investigated? Explain.

(e) Read the article, if possible, and comment on approaches used with which you either agree and disagree.

14.5 Assume that you have been drawn into a debate about hierarchical versus nonhierarchical models, with one person arguing in favor of the former and another person arguing in favor of the latter. Assume initially that interest is focused only on effect estimates. What would you say to the hierarchical model proponent who forces main effect terms into models when the terms are not statistically significant, and what would you say to the person who favors nonhierarchical models? Now assume that the emphasis is on fitted values rather than effect estimates. Would you adopt a different position in regard to each of these two people? Explain. (In answering the fitted values question, you may wish to consider a configuration similar to Figure 4.2, except that there are two replicates and the replicated values at each design point differ only slightly.)

14.6 Assume that there are at most eight factors that are believed to be related to a response variable. You can make at most 100 runs in one or more experiments with an end objective of identifying the important factors and the levels of those factors that are necessary to hit a target value for the response, or at least come as close as possible to doing so. How will you proceed?

14.7 Would you advocate the use of supersaturated designs? What is one very important thing to guard against in selecting a supersaturated design?

14.8 Assume that you have limited resources and intend to conduct an experiment in a somewhat unusual way: by making runs until you have enough information to identify important effects. Would a one-factor-at-a-time (OFAT) design be a viable alternative to a more conventional design? Explain.

14.9 Assume that you have a scenario for which experimental runs are very inexpensive, as in many computer experiments. If a 2^4 design is to be used, why would it necessarily *not* be a good idea to use, say, 10 replicates?

14.10 What are the advantages of ANOM over ANOVA? Can you think of a design scenario for which an ANOM display could be constructed (such as for a 2^3 design) but it would be better to use ANOVA? More specifically, could ANOM and ANOVA give different results for a 2^3 design for any of the seven estimable effects? If so, give an example in which this occurs. If not, explain why it could not happen.

14.11 Consider a single factor with five levels and an analyst uses both ANOM and ANOVA in analyzing the data. The F-test in ANOVA has a p-value of .031

but all of the plotted points on the ANOM display are within the .05 decision lines (two of the points are barely inside the decision lines in one dataset that can be constructed). If you were handed these data, what general relationship would you expect between the five means? Guided by your answer to this question, construct an example with four observations for each of the five levels that results in the ANOVA p-value being less than .05 but all of the five means falling within the ANOM .05 decision limits.

14.12 Critique the experimenter's following statement: "I routinely use 2_{IV}^{k-p} designs because although I know that two-factor interactions are confounded, this generally doesn't bother me because I usually know which two-factor interactions are likely to be real, so it is simply a matter of selecting a particular design such that each two-factor interaction that is likely to be real is confounded with a two-factor interaction that probably isn't a real effect."

14.13 There was no discussion in any of the chapters about possibly computing conditional effects for a Plackett–Burman design. Could those effects be computed for that type of design? What would be the motivating factor for computing the conditional effects, or would there even be a motivating factor? Explain.

14.14 Assume that you have a need to run an experiment using a simple 2^2 design, but just before the experiment is to begin, a scientist who is part of the experimental team states that, although unlikely, an explosion could occur if the $(1, 1)$ treatment combination were used, because of how extreme the high level was for each of the factors, pointing out that it would be much safer if the $(1, 1)$ design point were replaced by something like $(0.8, 0.7)$. What are you going to do? Do you have all the information that you need to make a decision? If yes, what is your decision? If not, what additional information do you need?

14.15 An experimenter is advised to consider a uniform design rather than selecting a central composite design. The experimenter responds by stating "No, I will never consider using a design that is not an orthogonal design." Respond to that statement.

14.16 Assume that you have encountered an experimenter who has studied several books on experimental design and is convinced that OFAT designs should never be used. The person needs to estimate interactions and each experimental run is quite expensive. You are serving as a consultant to the company and you are working with this person. What will be your recommendation?

14.17 Assume that you have used a factorial design with a mixture of fixed and random factors. Why is it important to label each one correctly when the data are analyzed with statistical software?

14.18 As discussed in Chapter 9, split-plot design configurations that result from hard-to-change factors have often been ignored in the analysis of data, with the data analyzed as if there were no restrictions on randomization. Consider the cake mix data given in Exercise 9.5 in Chapter 9.

   (a) Analyze the data as a completely randomized experiment and compare with the analysis of the data as a split-plot experiment.

   (b) There was no discussion of conditional effects in Chapter 9. However, if you look at the split-plot analysis given by Box and Jones (1992, references), you will see a two-factor interaction estimate that is 70% of one of the main effect estimates and almost eight times the other main effect estimate. Do you believe it would practical, or even possible, to perform a conditional effects analysis for that dataset? If so, perform such an analysis and comment. If it is not possible, explain why.

Answers to Selected Exercises

CHAPTER 1

1.1 **(a)** $z = \frac{-13.5}{4.5\sqrt{\frac{2}{20}}} = -9.49$ The p-value is essentially zero.

(b) $z = \frac{-13.5}{4.5\sqrt{\frac{1}{30} + \frac{1}{10}}} = -8.21$ The p-value is still essentially zero.

The numbers show the importance of having a state of statistical control when experimentation is performed.

1.2 (b)

1.3 In order to maximize the amount of information that results from a designed experiment, we need to know as much about the experimental region as possible and how the response values will vary over that region. We gain that information from a well-designed first experiment. Thus, we could have almost certainly done a better job of designing the first experiment if we had that knowledge. Generally more than one experiment will be performed however, so not having good information about the experimental region before performing the first experiment would be a handicap only when experimentation is extremely expensive.

1.4 No, there is an obvious pairing of the data, as would occur in a taste-testing experiment, for example.

1.6 Repeated readings occur when readings are successively made at a fixed experimental condition, with no resetting of factor levels and no randomness involved. With *replication*, we can speak of replicating an entire experiment (in this chapter) or replicating certain factor-level combinations (in subsequent

Modern Experimental Design   By Thomas P. Ryan
Copyright © 2007 John Wiley & Sons, Inc.

chapters). With replication there is a random ordering of experimental runs and resetting factor levels.

1.7 Because there is no randomness associated with the assignment of the last treatment to the last experimental unit.

1.8 With at most four observations per level, there would be no way to check the assumptions of normality and equal variances.

1.12 The world is nonlinear and experiments with two levels, which are appropriate for linearity, won't always give satisfactory results.

1.14 Such information generally does not exist. There would be no point in experimentation only if all other important factors were at their optimal levels and were held constant and the factor of interest were then varied "from one level to another level." Such a state of affairs will very rarely exist.

1.15 This suggests that a process was probably out of control and affected the values of the response variable. My suggestion would be to check the critical processes, and if a process problem is detected, fix the problem if possible and rerun the experiment.

1.21 The imbalance does not create a problem as this can be analyzed as a one-factor ANOVA problem with unequal sample sizes. There is not a significant operator effect because the value of the F-statistic is 1.76 and the associated p-value is .210.

1.23 An increase in the variance for either or both levels will make it harder to detect a difference between the two means because the estimate of the (assumed) common variance will be inflated.

CHAPTER 2

2.1 $\hat{A}_i = \overline{y}_i - \overline{\overline{y}}$ so $\sum_{i=1}^{k} \hat{A}_i = \sum_{i=1}^{k} (\overline{y}_i - \overline{\overline{y}}) = \sum_{i=1}^{k} \sum_{j=1}^{n_i} y_{ij} / n - k\overline{\overline{y}}$
$$= \sum_{i=1}^{k} \sum_{j=1}^{n_i} y_{ij} / n - k \sum_{i=1}^{k} \sum_{j=1}^{n_i} y_{ij} / nk = 0$$

2.4 (a) There were 31 observations and 4 levels of the factor; $31/4$ is not an integer.
(b) 8, 8, 8, and 7 for levels 1–4, respectively.

2.6 The necessary assumptions are the same (and it was stated that the factor is fixed, so that ANOM can be used). There is no evidence that the population variances would differ (by any significant amount) and the normality assumption could not be tested with such a small amount of data. The averages are 16.4, 17, 16.2, and 20.4. The results differ because three of the averages do not differ greatly, with the last average considerably greater than the other three. Although multiple comparisons are usually not performed unless the null

hypothesis of equal population means is rejected, in this case there is reasonably strong evidence that there is a difference involving the last mean.

2.7 df (error) $= 4 + 6 + 5 = 15$.

2.9 (a) SS(treatments) $= 40$, so MS(treatments) $= 2$. Since $F = 4$, MS(error) $=$
0.5. Then SS(error) $= 0.5(12) = 6$ and SS(total) thus equals 46.
(b) Since $F_{2,12} = 4$ has a p-value of .047, the hypothesis would (barely) be
rejected using $\alpha = .05$.

2.10 6.67

2.13 (a) The smallest possible value is zero; that will occur when the total of the
observations for the two levels is the same. The largest possible value is
infinity.
(b) Level One: 2, 4, 5, 6, 9, and 11
Level Two: 1, 3, 6, 7, 10, and 10

CHAPTER 3

3.1 Statement (b) suggests that a design with blocks should have been used but we
would not expect the treatment totals to be about the same. The motive for the
experimentation with these particular treatments would have to be examined.

3.3 The five factors would be randomly assigned to the experimental units for each
block.

3.4 Consider the 5×5 Latin square:

A	B	C	D	E
B	A	D	E	C
C	E	A	B	D
D	C	E	A	B
E	D	B	C	A

If we delete the last two rows, the design is not balanced because, for example, the (A, B) pair will occur in the first and second columns, but most other
pairs will occur only once. Hence, the design is not balanced, and is therefore
not a balanced incomplete block design and thus not a Youden design.

3.8 (a)

A	B	C	D	E
B	A	D	E	C
C	E	A	B	D
D	C	E	A	B
E	D	B	C	A

(b) $Y_{ijk} = \mu + A_i + R_j + C_k + \epsilon_{ijk}$ $i = 1, 2, \ldots, 5 \quad j = 1, 2, \ldots, 5$
$k = 1, 2, \ldots, 5$

The model assumes no interactions between rows and columns nor between treatments and either rows or columns. These assumptions should be checked with appropriate graphs, such as shown in Figure 3.2. The error term is assumed to have a normal distribution, but there is no way to test that with only one observation per cell.

3.10 It then becomes a two-factor problem with blocking, but it was not designed as such. That is, an appropriate randomization scheme was not used, so a Latin square analysis should still be performed, with the row or column factor (whichever it is) viewed as a factor of interest.

3.16 The numbers suggest that an RCB design should have probably been used instead of a Latin square design since Columns is not significant.

3.20 (a) $F_{4,12} = 4$ gives a p-value of .027.
(b) Neither rows nor columns are significant, so a design with blocking seems unnecessary.

3.21 16.565

CHAPTER 4

4.1 The main effect of A is estimated by $\frac{1}{2}(a + ab - (1) - b)$. The conditional effects of A are $ab - b$ and $a - (1)$, so $ab - b = (1) - a$ since the conditional effects differ only in sign. Then $ab - b + a - (1) = 0$, so the estimate of the A effect is zero.

4.5 The conditional effects will differ very little since the conditional effects are equal to the main effect plus and minus the interaction effect.

4.6 There is no multicollinearity in a 2^3 design.

4.8 The problem with that line of thinking is that the main effect coefficients in the model will not well represent the conditional effects when there are large interactions.

4.11 A useful order would be $A, B, AB, C, AC, BC, D, AD, BD, CD, ABC, ABD, ACD, BCD, ABCD$. The trick would be to construct the display in such a way that it wouldn't look cluttered.

4.12 (a) The B effect estimate is 0.5 but this should not be reported to management because the clear interaction effect causes the conditional effects to be -2 and 3. It would be better to report those numbers, with the caveat that each number is computed as the difference of just two numbers.
(b) 4

4.14 $(250 - 312.5)/62.5 = -1$ and $(375 - 312.5)/62.5 = 1$

4.15 The first is a design for three factors, each at two levels; the second is a design for two factors, each at three levels.

4.18 Anything of higher order than a four-factor interaction almost certainly won't be real, so it would be safe to use such interactions in estimating sigma. In general, however, it is best to estimate sigma with replications, not with interactions.

4.31 The normal probability plot (see below) suggests that the B and C main effects are the significant effects.

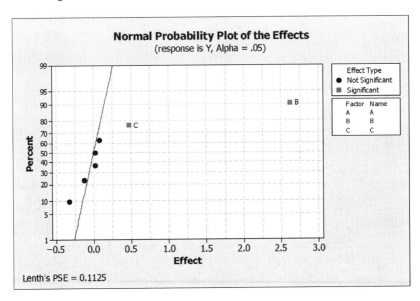

4.32 Homogeneity of variance must also be assumed with ANOM because sigma is estimated the same way with ANOM as it is estimaed with ANOVA.

4.33 Yes, ANOM is a graphical t-test when there are two levels for each factor. The degrees of freedom is 32, assuming that sigma is estimated from the replicates.

CHAPTER 5

5.4 The treatment combination that is incorrectly grouped with the others is *acd*. This has an odd number of letters in common with the interaction in the defining relation, whereas the other four treatment combinations (which would be used with the other 12 treatment combinations) have an even number of letters in common with the treatment combination.

5.5   If $I = ABDE$, 2 two-factor interactions would be confounded in pairs and the design would thus be resolution IV. A resolution V design can be constructed by using the defining relation $I = ABCDE$, which is how the design should be constructed.

5.8   **(a)** The defining relation is $I = ABC$. This result could be obtained using the approach given by Bisgaard (1993) or simply by trial and error for a small design such as this.
  (b) Yes, this is obviously a 2^{4-1} design, which should be constructed using $I = ABCD$, which would confound main effects with three-factor interactions.
  (c) With the design that is given, $A = BC$, $B = AC$, and $C = AB$.

5.10   There would be 32 design points because this would have to be a 2^{6-1} design since E is aliased with only one effect.

5.13   The conditional effects of factor D will differ only slightly because AB is confounded with CD, so when the data are split on factor C, the conditional effects would have to be close, by definition.

5.15   A minimum aberration design won't necessarily be the design that gives the maximum number of clear effects.

5.18   This is a 2^{5-2} design. It was constructed in an unorthodox manner since the generator for factor E is $E = -BC$ and the generator for D was $-ABC$. Letting $E = BC$ would be customary and this would reverse all of the signs in the fifth column of the design. (Alternatively, the generator $E = AC$ could have been used.) It would also be customary to use $D = AB$. If this were used, the first four signs in the fourth column would be the same and the last four would be the reverse of what they are in the problem.

5.19   Highly fractionated designs are usually low-resolution designs, for which there is a risk of one or more factors being declared significant because they are confounded with significant interactions.

5.24   The third column is the column that would be used to generate the fifth factor if the design were constructed as a 2^{5-1} design. We can see that something is awry, however, because the bottom half of the column is the same as the top half of the column. That isn't what happens when a 2^{5-1} design is constructed, as the bottom half should be the mirror image of the top half. Therefore, each of the eight numbers in the top half of the third column should be changed to the other number in order for the design to be resolution V. As it now stands, the design is resolution IV as it can be shown (such as by creating fictitious response values and analyzing the data) that the defining relation is $I = ABCE$.

5.25 We can see by inspection that $E = BCD$, so $I = BCDE$ is the defining relation, which is the same result that would be obtained using the methodology of Section 5.6. Of course this is not the best way to construct the design, however, as using $I = ABCDE$ will maximize the resolution of the design.

5.33 No, a 2^{9-1} design would have 256 design points, which would be extremely wasteful. Estimating the nine main effects and 36 two-factor interactions requires only 46 design points. Any design that has a much larger number of design points is wasteful.

5.40 It is possible to construct a 2^{9-5}_{III} design that has four words of length three in the defining relation. The design used by Hsieh and Goodwin (1986) has considerably more words of that length. Thus, the latter is not the minimum aberration design. A superior design would have the following generators: $E = ABC$, $F = ABD$, $G = ACD$, $H = BCD$, and $J = ABCD$.

CHAPTER 6

6.3 With $I = ABC$, the full alias structure is

$$A = BC = AB^2C^2$$
$$B = AB^2C = AC$$
$$C = ABC^2 = AB$$

6.4 The configuration should be the same, which should be expected because of the X configuration. Furthermore, the middle-level values are the same (and are equal) for each factor, and the set of response values for the high level of each factor is the same, as is the set of response values for the low level of each factor. Thus, the factor designations are interchangeable relative to the interaction plot.

6.11 The designs will be different. For $D = ABC^2$ we have $x_1 + x_2 + 2x_3 - x_4 = 0$ so $x_1 + x_2 + 2x_3 + 2x_4 = 0$ and $I = ABC^2D^2$. For $D = A^2B^2C$ we have $2x_1 + 2x_2 + x_3 + 2x_4 = 0$. Adding $3x_3$ to each side of the equation, and then dividing by 2, we obtain $x_1 + x_2 + 2x_3 + x_4 = 0$ so $I = ABC^2D$. Thus, the designs are different because the defining relations are different. There may not be any reason to prefer one design over the other, unless there are certain combinations of factor levels that cannot be used together for physical reasons.

6.13   No, the fractions are different. This can be verified by constructing the fraction
for which the sum is zero (mod 3) for each design and seeing that the column
for factor D is different in each fraction.

6.15   The full second-order model would have 35 terms (7 linear, 7 quadratic, 21
interaction terms), which would require 98 degrees of freedom. Even a 3^{7-3}
design is only resolution IV and that has 81 points. A 3^{7-2} design would have
243 points and generally be quite impractical, except in computer experiments
when runs are inexpensive.

CHAPTER 7

7.3   We would compute the sum of squares for $D(H)$ as

$$3\sum_{i=1}^{2}\sum_{j=1}^{2}(\overline{H_i D_j} - \overline{H}_i)^2 = 3[(16/3 - 35/6)^2 + (19/3 - 35/6)^2$$

$$+ (23/3 - 50/6)^2 + (27/3 - 50/6)^2]$$

$$= 4.167$$

as given in the table.

7.8   The full degrees of freedom breakdown is:

Source	df
A	4
B (within A)	5
C (within B)	5
D (Error)	5
Total	19

7.13

Source	df
Schools	1
Classroom (School)	6

CHAPTER 8

8.5   The interaction plot suggests that the high level of the factor is better than the
low level, but the line for the high level is not as close to being horizontal as
we would prefer.

8.6   Although slopes close to zero are desirable, if both slopes are close to zero there
is no clear advantage of one level over the other one. This is still preferable,

however, to an interaction profile with the slope of one line close to $+1$ and the slope of the other line close to -1 as this would tell us that neither level of the control factor would be a good choice, relative to the particular noise factor that is plotted on the horizontal axis.

8.7 Since noise factors cannot be controlled during production, we want to find levels of control factors for which the response is virtually the same regardless of the levels of the uncontrollable noise factors.

8.9 The design in Exercise 8.8 does not have the same weakness because it is not a 2^{k-p} design. Since the design has 12 points and is a saturated design, this is also a Plackett–Burman design (see Section 13.4.1) and thus has the same complex alias structure as a Plackett–Burman design. Those designs are resolution III designs. The L_{12} design is suitable for factor screening but not for investigating control $\times$ noise interactions or any other type of interaction.

CHAPTER 9

9.5 Since there are five factors each at two levels and 32 design points, this is analogous to a 2^5 design. Thus, there will be no error terms so two normal probability plots will have to be constructed. The data can be initially analyzed as if a 2^5 design had been used for the purpose of obtaining the effect estimates.

The normal plot for the whole plot effects below shows that the E (egg powder) and F (flour) effects are significant.

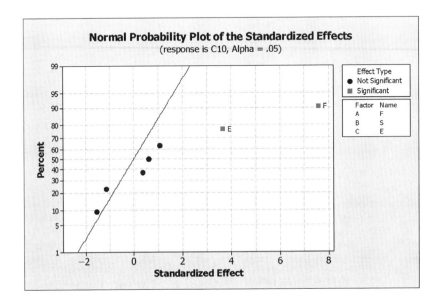

Analyzing the subplot effects is a chore, however, because it requires fitting a nonhierarchical model and producing the normal plot, which most software programs will not allow. We can produce a normal probability plot of effect estimates, however, by using the effect estimates from the 2^5 analysis and then testing manually the points that are well off the line. The two points that stand out in that plot are the A main effect and the SA interaction. The normal plot of Box and Jones (2000) showed these effects to be significant. Manual computations using Lenth's approach (Section 4.10) produce results that are in agreement with those results, with the B effect close to being borderline.

The magnitude of the SA interaction is evidenced by the following graph.

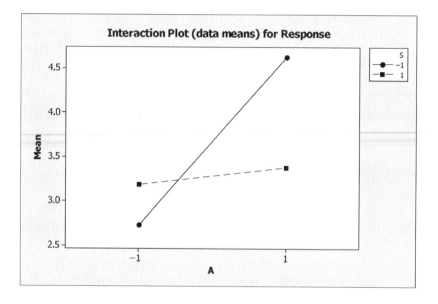

The graph also shows that the high level of factor S is clearly the best from a robust product standpoint because the variability in the response over the two levels of the noise factor A is clearly much less than when S is at the low level.

9.13 In order to have a split-plot design, there must be at least one whole plot treatment in addition to subplot treatments. Here there is only one treatment with two levels, selfing and outcrossing. (This observation was also made by a prominent statistician who responded on the Internet to the experimental setting that was presented. That person felt that the main concern should be a suitable model for the count data. Whether or not that should be a major concern should probably depend on the magnitude and variability of the counts, however.)

9.14 Zero. There would be neither enough observations nor enough factors to permit a normal probability plot estimate.

CHAPTER 10

10.10   The surface plot is a function of the data on the response variable, which is a
random variable. If a second experiment were conducted and another surface
plot displayed, using the same model form as used the first time, the surface
plot will look at least slightly different from the first one.

10.16   The *BC* interaction plot is given below.

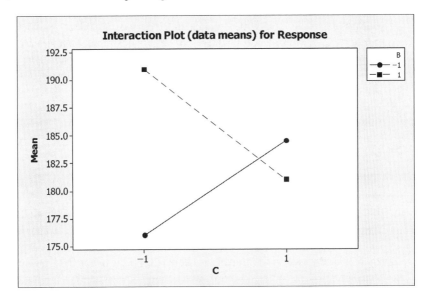

We can see from the plot that the *C* effect estimate will be close to zero
since the average value at high *C* is obviously close to the average value at
low *C*. The conditional effects of *C*, however, are sizable. Therefore, this
should not be overlooked and accordingly the (linear) effect of *C* should
not be overlooked (as the experimenters have apparently done) in terms of
modeling.

10.19   The path of steepest ascent is determined by expressing the amount of change
in each factor as a multiple of a unit change in one selected factor, with the
multipliers computed from the coefficients in the fitted first-order model.
This cannot be done if there is an interaction term in the model because the
contribution to change in the response then results from joint changes in the
two or more factors that comprise the interaction term.

10.25   For the full CCD, all the pairwise correlations between the quadratic effect
columns are $-.124$; for the half CCD the correlations are all $-.067$. Thus,
the half CCD has a slightly better correlation structure, in addition to being a
more economical design.

CHAPTER 11

11.1 No, this Latin square design would be totally unsuitable when there are carry-over effects because each treatment is always preceded by the same treatment when it is not used first in a treatment period (A always precedes B, D always precedes A, and so on.)

11.7 A 3×3 Latin square is not a good design to use when carryover effects may be present. In particular, it is not possible to construct a single Latin square of this size that will be balanced for carryover effects. As discussed in Section 3.3.5, multiple Latin squares are highly desirable, but there are restrictions on the number of Latin squares that can be used and still have balance relative to carryover effects. For example, whereas the following two Latin squares are balanced in regard to possible carryover effects

$$
\begin{array}{ccc} A\,B\,C & \quad A\,C\,B \\ B\,C\,A & \quad B\,A\,C \\ C\,A\,B & \quad C\,B\,A \end{array}
$$

because each treatment is preceded by the same treatment in each period in the first square when the treatment is not used first in a treatment period, and by the other treatment in the second square, the balance would obviously be lost if a third square were used. More specifically, B, for example, obviously has to be preceded by A or C. If we pick one or the other and have B preceded by one of the two in the third Latin square when it is not the first treatment, then the set of three Latin squares will be unbalanced relative to the treatment that precedes B. It is easy to show that we cannot construct the third Latin square in such a way that B is preceded by A in one row and by C in another row, in addition to of course being first in the other row. Therefore, a set of three Latin squares cannot be balanced in regard to possible carryover effects.

CHAPTER 12

12.1 We don't know what desirability function was used because there is no discussion of it. One potential problem with the initial experiment is that some two-factor interactions were stated as being significant, which might be masking important factors that were discarded and not used in the subsequent experimentation.

12.8 There was only a small feasible region when factor C was set at the low level, as was shown in Figure 12.2. When C is set at the high level, there is no feasible region, as shown by the absence of a white (common) area in the contour plot given below.

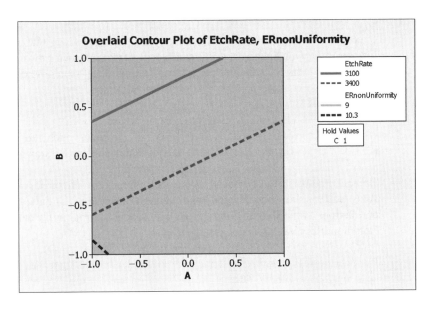

CHAPTER 13

13.2 A split-plot design would be a logical choice, with the hard-to-change factor being the whole plot factor.

13.6 It is certainly true that the response values determine whether or not points are influential, but it would be unwise to create potential problems by using a design that has some high leverage values.

13.12 A 2^{31-26} design, which of course is resolution III.

13.15 The objective is to dealias effects and create a resolution IV design in the process.

CHAPTER 14

14.1 One possibility would be to use a trend-free design. This would protect against an out-of-control state that could be represented by a linear trend, as occurs with tool wear, for example.

14.7 Supersaturated designs can be useful but it is important not to use a supersaturated design that has more than small correlations between the columns of the design.

14.9   Even when data are inexpensive, 160 observations to estimate (at most) 15
           effects can have deleterious effects because the hypothesis tests that are per-
           formed in testing for effect significance will have too much power in the
           sense that effects for which the estimates are quite small could be declared
           real effects.

14.12 Such information generally does not exist. It is hard enough to know which
           main effects can be expected to be real; pretending to know which interactions
           are likely to be real is apt to cause erroneous conclusions to be drawn.

14.15 There are many useful response surface designs that are not orthogonal de-
           signs, such as the Hoke designs, with the larger Hoke designs having small
           correlations. Nonorthogonality is not a major problem as long as the departure
           from nonorthogonality is not great.

14.16 Because the design runs are expensive, an OFAT design of resolution V should
           be strongly considered, despite the person's opposition to OFAT designs. I
           would attempt to be persuasive by using illustrative examples.

Statistical Tables

TABLE A Normal Distribution[a] $[P(0 \leq Z \leq z)$ where $Z \sim N(0, 1)]$

z	0.00	0.01	0.02	0.03	0.04	0.05	0.06	0.07	0.08	0.09
0.0	0.00000	0.00399	0.00798	0.01197	0.01595	0.01994	0.02392	0.02790	0.03188	0.03586
0.1	0.03983	0.04380	0.04776	0.05172	0.05567	0.05962	0.06356	0.06749	0.07142	0.07535
0.2	0.07926	0.08317	0.08706	0.09095	0.09483	0.09871	0.10257	0.10642	0.11026	0.11409
0.3	0.11791	0.12172	0.12552	0.12930	0.13307	0.13683	0.14058	0.14431	0.14803	0.15173
0.4	0.15542	0.15910	0.16276	0.16640	0.17003	0.17364	0.17724	0.18082	0.18439	0.18793
0.5	0.19146	0.19497	0.19847	0.20194	0.20540	0.20884	0.21226	0.21566	0.21904	0.22240
0.6	0.22575	0.22907	0.23237	0.23565	0.23891	0.24215	0.24537	0.24857	0.25175	0.25490
0.7	0.25804	0.26115	0.26424	0.26730	0.27035	0.27337	0.27637	0.27935	0.28230	0.28524
0.8	0.28814	0.29103	0.29389	0.29673	0.29955	0.30234	0.30511	0.30785	0.31057	0.31327
0.9	0.31594	0.31859	0.32121	0.32381	0.32639	0.32894	0.33147	0.33398	0.33646	0.33891
1.0	0.34134	0.34375	0.34614	0.34849	0.35083	0.35314	0.35543	0.35769	0.35993	0.36214
1.1	0.36433	0.36650	0.36864	0.37076	0.37286	0.37493	0.37698	0.37900	0.38100	0.38298
1.2	0.38493	0.38686	0.38877	0.39065	0.39251	0.39435	0.39617	0.39796	0.39973	0.40147
1.3	0.40320	0.40490	0.40658	0.40824	0.40988	0.41149	0.41308	0.41466	0.41621	0.41774
1.4	0.41924	0.42073	0.42220	0.42364	0.42507	0.42647	0.42785	0.42922	0.43056	0.43189
1.5	0.43319	0.43448	0.43574	0.43699	0.43822	0.43943	0.44062	0.44179	0.44295	0.44408
1.6	0.44520	0.44630	0.44738	0.44845	0.44950	0.45053	0.45154	0.45254	0.45352	0.45449
1.7	0.45543	0.45637	0.45728	0.45818	0.45907	0.45994	0.46080	0.46164	0.46246	0.46327
1.8	0.46407	0.46485	0.46562	0.46638	0.46712	0.46784	0.46856	0.46926	0.46995	0.47062
1.9	0.47128	0.47193	0.47257	0.47320	0.47381	0.47441	0.47500	0.47558	0.47615	0.47670

z	0.00	0.01	0.02	0.03	0.04	0.05	0.06	0.07	0.08	0.09
2.0	0.47725	0.47778	0.47831	0.47882	0.47932	0.47982	0.48030	0.48077	0.48124	0.48169
2.1	0.48214	0.48257	0.48300	0.48341	0.48382	0.48422	0.48461	0.48500	0.48537	0.48574
2.2	0.48610	0.48645	0.48679	0.48713	0.48745	0.48778	0.48809	0.48840	0.48870	0.48899
2.3	0.48928	0.48956	0.48983	0.49010	0.49036	0.49061	0.49086	0.49111	0.49134	0.49158
2.4	0.49180	0.49202	0.49224	0.49245	0.49266	0.49286	0.49305	0.49324	0.49343	0.49361
2.5	0.49379	0.49396	0.49413	0.49430	0.49446	0.49461	0.49477	0.49492	0.49506	0.49520
2.6	0.49534	0.49547	0.49560	0.49573	0.49585	0.49598	0.49609	0.49621	0.49632	0.49643
2.7	0.49653	0.49664	0.49674	0.49683	0.49693	0.49702	0.49711	0.49720	0.49728	0.49736
2.8	0.49744	0.49752	0.49760	0.49767	0.49774	0.49781	0.49788	0.49795	0.49801	0.49807
2.9	0.49813	0.49819	0.49825	0.49831	0.49836	0.49841	0.49846	0.49851	0.49856	0.49861
3.0	0.49865	0.49869	0.49874	0.49878	0.49882	0.49886	0.49889	0.49893	0.49896	0.49900
3.1	0.49903	0.49906	0.49910	0.49913	0.49916	0.49918	0.49921	0.49924	0.49926	0.49929
3.2	0.49931	0.49934	0.49936	0.49938	0.49940	0.49942	0.49944	0.49946	0.49948	0.49950
3.3	0.49952	0.49953	0.49955	0.49957	0.49958	0.49960	0.49961	0.49962	0.49964	0.49965
3.4	0.49966	0.49968	0.49969	0.49970	0.49971	0.49972	0.49973	0.49974	0.49975	0.49976
3.5	0.49977	0.49978	0.49978	0.49979	0.49980	0.49981	0.49981	0.49982	0.49983	0.49983
3.6	0.49984	0.49985	0.49985	0.49986	0.49986	0.49987	0.49987	0.49988	0.49988	0.49989
3.7	0.49989	0.49990	0.49990	0.49990	0.49991	0.49991	0.49992	0.49992	0.49992	0.49992
3.8	0.49993	0.49993	0.49993	0.49994	0.49994	0.49994	0.49994	0.49995	0.49995	0.49995
3.9	0.49995	0.49995	0.49996	0.49996	0.49996	0.49996	0.49996	0.49996	0.49997	0.49997

[a] These values were generated using MINITAB.

TABLE B t Distribution[a]

d.f. $(v)/\alpha$	0.40	0.25	0.10	0.05	0.025	0.01	0.005	0.0025	0.001	0.0005
1	0.325	1.000	3.078	6.314	12.706	31.820	63.655	127.315	318.275	636.438
2	0.289	0.816	1.886	2.920	4.303	6.965	9.925	14.089	22.327	31.596
3	0.277	0.765	1.638	2.353	3.182	4.541	5.841	7.453	10.214	12.923
4	0.271	0.741	1.533	2.132	2.776	3.747	4.604	5.597	7.173	8.610
5	0.267	0.727	1.476	2.015	2.571	3.365	4.032	4.773	5.893	6.869
6	0.265	0.718	1.440	1.943	2.447	3.143	3.707	4.317	5.208	5.959
7	0.263	0.711	1.415	1.895	2.365	2.998	3.499	4.029	4.785	5.408
8	0.262	0.706	1.397	1.860	2.306	2.896	3.355	3.833	4.501	5.041
9	0.261	0.703	1.383	1.833	2.262	2.821	3.250	3.690	4.297	4.781
10	0.260	0.700	1.372	1.812	2.228	2.764	3.169	3.581	4.144	4.587
11	0.260	0.697	1.363	1.796	2.201	2.718	3.106	3.497	4.025	4.437
12	0.259	0.695	1.356	1.782	2.179	2.681	3.055	3.428	3.930	4.318
13	0.259	0.694	1.350	1.771	2.160	2.650	3.012	3.372	3.852	4.221
14	0.258	0.692	1.345	1.761	2.145	2.624	2.977	3.326	3.787	4.140
15	0.258	0.691	1.341	1.753	2.131	2.602	2.947	3.286	3.733	4.073
16	0.258	0.690	1.337	1.746	2.120	2.583	2.921	3.252	3.686	4.015
17	0.257	0.689	1.333	1.740	2.110	2.567	2.898	3.222	3.646	3.965
18	0.257	0.688	1.330	1.734	2.101	2.552	2.878	3.197	3.610	3.922
19	0.257	0.688	1.328	1.729	2.093	2.539	2.861	3.174	3.579	3.883
20	0.257	0.687	1.325	1.725	2.086	2.528	2.845	3.153	3.552	3.849
21	0.257	0.686	1.323	1.721	2.080	2.518	2.831	3.135	3.527	3.819
22	0.256	0.686	1.321	1.717	2.074	2.508	2.819	3.119	3.505	3.792
23	0.256	0.685	1.319	1.714	2.069	2.500	2.807	3.104	3.485	3.768
24	0.256	0.685	1.318	1.711	2.064	2.492	2.797	3.091	3.467	3.745
25	0.256	0.684	1.316	1.708	2.060	2.485	2.787	3.078	3.450	3.725
26	0.256	0.684	1.315	1.706	2.056	2.479	2.779	3.067	3.435	3.707
27	0.256	0.684	1.314	1.703	2.052	2.473	2.771	3.057	3.421	3.690
28	0.256	0.683	1.313	1.701	2.048	2.467	2.763	3.047	3.408	3.674
29	0.256	0.683	1.311	1.699	2.045	2.462	2.756	3.038	3.396	3.659
30	0.256	0.683	1.310	1.697	2.042	2.457	2.750	3.030	3.385	3.646
40	0.255	0.681	1.303	1.684	2.021	2.423	2.704	2.971	3.307	3.551
60	0.254	0.679	1.296	1.671	2.000	2.390	2.660	2.915	3.232	3.460
100	0.254	0.677	1.290	1.660	1.984	2.364	2.626	2.871	3.174	3.391
Infinity	0.253	0.674	1.282	1.645	1.960	2.326	2.576	2.807	3.090	3.290

[a] These values were generated using MINITAB.

TABLE C *F* **Distribution**[a,b]

.05

$F\nu_1, \nu_2, .05$

a. $F_{\nu_1, \nu_2}, .05$

ν_2 \ ν_1	1	2	3	4	5	6	7	8	9	10	11	12	13	14	15
1	161.44	199.50	215.69	224.57	230.16	233.98	236.78	238.89	240.55	241.89	242.97	243.91	244.67	245.35	245.97
2	18.51	19.00	19.16	19.25	19.30	19.33	19.35	19.37	19.39	19.40	19.40	19.41	19.42	19.42	19.43
3	10.13	9.55	9.28	9.12	9.01	8.94	8.89	8.85	8.81	8.79	8.76	8.74	8.73	8.71	8.70
4	7.71	6.94	6.59	6.39	6.26	6.16	6.09	6.04	6.00	5.96	5.94	5.91	5.89	5.87	5.86
5	6.61	5.79	5.41	5.19	5.05	4.95	4.88	4.82	4.77	4.74	4.70	4.68	4.66	4.64	4.62
6	5.99	5.14	4.76	4.53	4.39	4.28	4.21	4.15	4.10	4.06	4.03	4.00	3.98	3.96	3.94
7	5.59	4.74	4.35	4.12	3.97	3.87	3.79	3.73	3.68	3.64	3.60	3.57	3.55	3.53	3.51
8	5.32	4.46	4.07	3.84	3.69	3.58	3.50	3.44	3.39	3.35	3.31	3.28	3.26	3.24	3.22
9	5.12	4.26	3.86	3.63	3.48	3.37	3.29	3.23	3.18	3.14	3.10	3.07	3.05	3.03	3.01
10	4.96	4.10	3.71	3.48	3.33	3.22	3.14	3.07	3.02	2.98	2.94	2.91	2.89	2.86	2.85
11	4.84	3.98	3.59	3.36	3.20	3.09	3.01	2.95	2.90	2.85	2.82	2.79	2.76	2.74	2.72
12	4.75	3.89	3.49	3.26	3.11	3.00	2.91	2.85	2.80	2.75	2.72	2.69	2.66	2.64	2.62
13	4.67	3.81	3.41	3.18	3.03	2.92	2.83	2.77	2.71	2.67	2.63	2.60	2.58	2.55	2.53
14	4.60	3.74	3.34	3.11	2.96	2.85	2.76	2.70	2.65	2.60	2.57	2.53	2.51	2.48	2.46
15	4.54	3.68	3.29	3.06	2.90	2.79	2.71	2.64	2.59	2.54	2.51	2.48	2.45	2.42	2.40
16	4.49	3.63	3.24	3.01	2.85	2.74	2.66	2.59	2.54	2.49	2.46	2.42	2.40	2.37	2.35
17	4.45	3.59	3.20	2.96	2.81	2.70	2.61	2.55	2.49	2.45	2.41	2.38	2.35	2.33	2.31
18	4.41	3.55	3.16	2.93	2.77	2.66	2.58	2.51	2.46	2.41	2.37	2.34	2.31	2.29	2.27
19	4.38	3.52	3.13	2.90	2.74	2.63	2.54	2.48	2.42	2.38	2.34	2.31	2.28	2.26	2.23
20	4.35	3.49	3.10	2.87	2.71	2.60	2.51	2.45	2.39	2.35	2.31	2.28	2.25	2.22	2.20
21	4.32	3.47	3.07	2.84	2.68	2.57	2.49	2.42	2.37	2.32	2.28	2.25	2.22	2.20	2.18
22	4.30	3.44	3.05	2.82	2.66	2.55	2.46	2.40	2.34	2.30	2.26	2.23	2.20	2.17	2.15
23	4.28	3.42	3.03	2.80	2.64	2.53	2.44	2.37	2.32	2.27	2.24	2.20	2.18	2.15	2.13
24	4.26	3.40	3.01	2.78	2.62	2.51	2.42	2.36	2.30	2.25	2.22	2.18	2.15	2.13	2.11

TABLE C F Distribution (*Continued*)

ν_1	1	2	3	4	5	6	7	8	9	10	11	12	13	14	15
25	4.24	3.39	2.99	2.76	2.60	2.49	2.40	2.34	2.28	2.24	2.20	2.16	2.14	2.11	2.09
26	4.23	3.37	2.98	2.74	2.59	2.47	2.39	2.32	2.27	2.22	2.18	2.15	2.12	2.09	2.07
27	4.21	3.35	2.96	2.73	2.57	2.46	2.37	2.31	2.25	2.20	2.17	2.13	2.10	2.08	2.06
28	4.20	3.34	2.95	2.71	2.56	2.45	2.36	2.29	2.24	2.19	2.15	2.12	2.09	2.06	2.04
29	4.18	3.33	2.93	2.70	2.55	2.43	2.35	2.28	2.22	2.18	2.14	2.10	2.08	2.05	2.03
30	4.17	3.32	2.92	2.69	2.53	2.42	2.33	2.27	2.21	2.16	2.13	2.09	2.06	2.04	2.01
40	4.08	3.23	2.84	2.61	2.45	2.34	2.25	2.18	2.12	2.08	2.04	2.00	1.97	1.95	1.92

$b.\ F_{\nu_1,\nu_2},\ .01$

ν_2	1	2	3	4	5	6	7	8	9	10	11	12	13	14	15
1	4052.45	4999.42	5402.96	5624.03	5763.93	5858.82	5928.73	5981.06	6021.73	6055.29	6083.22	6106.00	6125.37	6142.48	6157.06
2	98.51	99.00	99.17	99.25	99.30	99.33	99.35	99.38	99.39	99.40	99.41	99.41	99.42	99.42	99.43
3	34.12	30.82	29.46	28.71	28.24	27.91	27.67	27.49	27.35	27.23	27.13	27.05	26.98	26.92	26.87
4	21.20	18.00	16.69	15.98	15.52	15.21	14.98	14.80	14.66	14.55	14.45	14.37	14.31	14.25	14.20
5	16.26	13.27	12.06	11.39	10.97	10.67	10.46	10.29	10.16	10.05	9.96	9.89	9.82	9.77	9.72
6	13.74	10.92	9.78	9.15	8.75	8.47	8.26	8.10	7.98	7.87	7.79	7.72	7.66	7.60	7.56
7	12.25	9.55	8.45	7.85	7.46	7.19	6.99	6.84	6.72	6.62	6.54	6.47	6.41	6.36	6.31
8	11.26	8.65	7.59	7.01	6.63	6.37	6.18	6.03	5.91	5.81	5.73	5.67	5.61	5.56	5.52
9	10.56	8.02	6.99	6.42	6.06	5.80	5.61	5.47	5.35	5.26	5.18	5.11	5.05	5.01	4.96
10	10.04	7.56	6.55	5.99	5.64	5.39	5.20	5.06	4.94	4.85	4.77	4.71	4.65	4.60	4.56
11	9.65	7.21	6.22	5.67	5.32	5.07	4.89	4.74	4.63	4.54	4.46	4.40	4.34	4.29	4.25
12	9.33	6.93	5.95	5.41	5.06	4.82	4.64	4.50	4.39	4.30	4.22	4.16	4.10	4.05	4.01
13	9.07	6.70	5.74	5.21	4.86	4.62	4.44	4.30	4.19	4.10	4.02	3.96	3.91	3.86	3.82
14	8.86	6.51	5.56	5.04	4.69	4.46	4.28	4.14	4.03	3.94	3.86	3.80	3.75	3.70	3.66
15	8.68	6.36	5.42	4.89	4.56	4.32	4.14	4.00	3.89	3.80	3.73	3.67	3.61	3.56	3.52

16	8.53	6.23	5.29	4.77	4.44	4.20	4.03	3.89	3.78	3.69	3.62	3.55	3.50	3.45	3.41
17	8.40	6.11	5.18	4.67	4.34	4.10	3.93	3.79	3.68	3.59	3.52	3.46	3.40	3.35	3.31
18	8.29	6.01	5.09	4.58	4.25	4.01	3.84	3.71	3.60	3.51	3.43	3.37	3.32	3.27	3.23
19	8.18	5.93	5.01	4.50	4.17	3.94	3.77	3.63	3.52	3.43	3.36	3.30	3.24	3.19	3.15
20	8.10	5.85	4.94	4.43	4.10	3.87	3.70	3.56	3.46	3.37	3.29	3.23	3.18	3.13	3.09
21	8.02	5.78	4.87	4.37	4.04	3.81	3.64	3.51	3.40	3.31	3.24	3.17	3.12	3.07	3.03
22	7.95	5.72	4.82	4.31	3.99	3.76	3.59	3.45	3.35	3.26	3.18	3.12	3.07	3.02	2.98
23	7.88	5.66	4.76	4.26	3.94	3.71	3.54	3.41	3.30	3.21	3.14	3.07	3.02	2.97	2.93
24	7.82	5.61	4.72	4.22	3.90	3.67	3.50	3.36	3.26	3.17	3.09	3.03	2.98	2.93	2.89
25	7.77	5.57	4.68	4.18	3.85	3.63	3.46	3.32	3.22	3.13	3.06	2.99	2.94	2.89	2.85
26	7.72	5.53	4.64	4.14	3.82	3.59	3.42	3.29	3.18	3.09	3.02	2.96	2.90	2.86	2.81
27	7.68	5.49	4.60	4.11	3.78	3.56	3.39	3.26	3.15	3.06	2.99	2.93	2.87	2.82	2.78
28	7.64	5.45	4.57	4.07	3.75	3.53	3.36	3.23	3.12	3.03	2.96	2.90	2.84	2.79	2.75
29	7.60	5.42	4.54	4.04	3.73	3.50	3.33	3.20	3.09	3.00	2.93	2.87	2.81	2.77	2.73
30	7.56	5.39	4.51	4.02	3.70	3.47	3.30	3.17	3.07	2.98	2.91	2.84	2.79	2.74	2.70
40	7.31	5.18	4.31	3.83	3.51	3.29	3.12	2.99	2.89	2.80	2.73	2.66	2.61	2.56	2.52

[a] These values were generated using MINITAB.

[b] v_2 = degrees of freedom for the denominator; v_1 = degrees of freedom for the numerator.

TABLE D Analysis of Means Constants[a]

Number of Means, k

a. $h_{0.05}$

d.f.[b](v)	3	4	5	6	7	8	9	10	11	12	13	14	15	16	17	18	19	20
3	4.18																	
4	3.56	3.89																
5	3.25	3.53	3.72															
6	3.07	3.31	3.49	3.62														
7	2.94	3.17	3.33	3.45	3.56													
8	2.86	3.07	3.21	3.33	3.43	3.51												
9	2.79	2.99	3.13	3.24	3.33	3.41	3.48											
10	2.74	2.93	3.07	3.17	3.26	3.33	3.40	3.45										
11	2.70	2.88	3.01	3.12	3.20	3.27	3.33	3.39	3.44									
12	2.67	2.85	2.97	3.07	3.15	3.22	3.28	3.33	3.38	3.42								
13	2.64	2.81	2.94	3.03	3.11	3.18	3.24	3.29	3.34	3.38	3.42							
14	2.62	2.79	2.91	3.00	3.08	3.14	3.20	3.25	3.30	3.34	3.37	3.41						
15	2.60	2.76	2.88	2.97	3.05	3.11	3.17	3.22	3.26	3.30	3.34	3.37	3.40					
16	2.58	2.74	2.86	2.95	3.02	3.09	3.14	3.19	3.23	3.27	3.31	3.34	3.37	3.40				
17	2.57	2.73	2.84	2.93	3.00	3.06	3.12	3.16	3.21	3.25	3.28	3.31	3.34	3.37	3.40			
18	2.55	2.71	2.82	2.91	2.98	3.04	3.10	3.14	3.18	3.22	3.26	3.29	3.32	3.35	3.37	3.40		
19	2.54	2.70	2.81	2.89	2.96	3.02	3.08	3.12	3.16	3.20	3.24	3.27	3.30	3.32	3.35	3.37	3.40	
20	2.53	2.68	2.79	2.88	2.95	3.01	3.06	3.11	3.15	3.18	3.22	3.25	3.28	3.30	3.33	3.35	3.37	3.40
24	2.50	2.65	2.75	2.83	2.90	2.96	3.01	3.05	3.09	3.13	3.16	3.19	3.22	3.24	3.27	3.29	3.31	3.33
30	2.47	2.61	2.71	2.79	2.85	2.91	2.96	3.00	3.04	3.07	3.10	3.13	3.16	3.18	3.20	3.22	3.25	3.27
40	2.43	2.57	2.67	2.75	2.81	2.86	2.91	2.95	2.98	3.01	3.04	3.07	3.10	3.12	3.14	3.16	3.18	3.20
60	2.40	2.54	2.63	2.70	2.76	2.81	2.86	2.90	2.93	2.96	2.99	3.02	3.04	3.06	3.08	3.10	3.12	3.14
120	2.37	2.50	2.59	2.66	2.72	2.77	2.81	2.84	2.88	2.91	2.93	2.96	2.98	3.00	3.02	3.04	3.06	3.08
Infinity	2.34	2.47	2.56	2.62	2.68	2.72	2.76	2.80	2.83	2.86	2.88	2.90	2.93	2.95	2.97	2.98	3.00	3.02

(cont.)

$b.\ h_{0.01}$

3	7.51																		
4	5.74	6.21																	
5	4.93	5.29	5.55																
6	4.48	4.77	4.98	5.16															
7	4.18	4.44	4.63	4.78	4.90														
8	3.98	4.21	4.38	4.52	4.63	4.72													
9	3.84	4.05	4.20	4.33	4.43	4.51	4.59												
10	3.73	3.92	4.07	4.18	4.28	4.36	4.43	4.49											
11	3.64	3.82	3.96	4.07	4.16	4.23	4.30	4.36	4.41										
12	3.57	3.74	3.87	3.98	4.06	4.13	4.20	4.25	4.31	4.35									
13	3.51	3.68	3.80	3.90	3.98	4.05	4.11	4.17	4.22	4.26	4.30								
14	3.46	3.63	3.74	3.84	3.92	3.98	4.04	4.09	4.14	4.18	4.22	4.26							
15	3.42	3.58	3.69	3.79	3.86	3.92	3.98	4.03	4.08	4.12	4.16	4.19	4.22						
16	3.38	3.54	3.65	3.74	3.81	3.87	3.93	3.98	4.02	4.06	4.10	4.14	4.17	4.20					
17	3.35	3.50	3.61	3.70	3.77	3.83	3.89	3.93	3.98	4.02	4.05	4.09	4.12	4.14	4.17				
18	3.33	3.47	3.58	3.66	3.73	3.79	3.85	3.89	3.94	3.97	4.01	4.04	4.07	4.10	4.12	4.15			
19	3.30	3.45	3.55	3.63	3.70	3.76	3.81	3.86	3.90	3.94	3.97	4.00	4.03	4.06	4.08	4.11	4.13		
20	3.28	3.42	3.53	3.61	3.67	3.73	3.78	3.83	3.87	3.90	3.94	3.97	4.00	4.02	4.05	4.07	4.09	4.12
24	3.21	3.35	3.45	3.52	3.58	3.64	3.69	3.73	3.77	3.80	3.83	3.86	3.89	3.91	3.94	3.96	3.98	4.00
30	3.15	3.28	3.37	3.44	3.50	3.55	3.59	3.63	3.67	3.70	3.73	3.76	3.78	3.81	3.83	3.85	3.87	3.89
40	3.09	3.21	3.29	3.36	3.42	3.46	3.50	3.54	3.58	3.60	3.63	3.66	3.68	3.70	3.72	3.74	3.76	3.78
60	3.03	3.14	3.22	3.29	3.34	3.38	3.42	3.46	3.49	3.51	3.54	3.56	3.59	3.61	3.63	3.64	3.66	3.68
120	2.97	3.07	3.15	3.21	3.26	3.30	3.34	3.37	3.40	3.42	3.45	3.47	3.49	3.51	3.55	3.56	3.58	3.58
Infinity	2.91	3.01	3.08	3.14	3.18	3.22	3.26	3.29	3.32	3.34	3.36	3.38	3.40	3.42	3.44	3.45	3.47	3.48

[a] From Tables 2 and 3 of L. S. Nelson, Exact critical values for use with the analysis of means. *Journal of Quality Technology*, **15**(1), January 1983. Reprinted with permission from *Journal of Quality Technology* © 1983 American Society for Quality.
[b] Degrees of freedom for s.

Author Index

Aastveit, A. H., 339, 351, 545, 546
Abraham, B., 499, 533
Addelman, S., 231, 234, 264, 284, 312, 323
Addy, M., 320, 440
Aeschliman, D. S., 411, 418, 419
Afsarinejad, K., 432, 437–439, 441
Ahuja, S. K., 493, 494, 533
Akhtar, M., 405, 409
Akoh, C. C., 361, 412
Albin, S. L. 463, 474
Algina, J., 425, 429, 441
Alldredge, J. R., 437, 443
Allen, J. K., 322, 323, 523, 535
Allen, L., 285
Allen, T., 409, 415, 424, 464, 474, 499, 522, 533
Allison, D. B., 530, 533
Aloko, D. F., 166
Al-Sabah, W. S., 377, 411
Ames, A. E., 464, 472
Anderson, M., 130, 208, 234, 322, 323, 361, 409, 490, 533
Anderson-Cook, C. M., 397, 406, 410, 413, 414, 509–511, 541
Angün, E., 373, 412
Ankenman, B. E., 302, 312, 316, 323, 364, 405, 409, 412, 483, 539
Anthony, J., 495, 496, 498, 533, 545
Antony, J., 325, 328
Arasu, K. T., 528, 533

Armitage, P., 76, 91
Arnau, C., 430, 440
Arviddson, M., 19, 23, 157
Atkinson, A. C., 6, 7, 23, 388, 409, 505, 528, 533, 537

Baczkowski, C., 352, 354
Bailar, J. C., III, 430, 442
Bailey, M., 520, 533
Bailey, R. A., 6, 7, 23, 148, 157, 160, 431, 439, 505, 512, 533, 536
Bainbridge, T. R., 292, 299, 303
Bakir, S. T., 22, 24, 47, 49
Balaam, L. N., 437, 439
Balakrishnan, N., 136, 159, 188, 189, 235
Balusu, R., 494, 533
Bandurek, G., 499, 533
Banerjee, K. S., 524, 526, 528, 533
Barnes, C. G., 285, 290
Barnett, J., 203, 206–208, 234
Barrios, E., 336, 338, 339, 351
Barton, R. R., 14, 22, 24, 523, 535
Bate, S. T., 432, 439
Bates, R. L., 345, 346, 353
Batson, R. G., 292, 303, 308
Baweja, G., 528, 537
Beasley, T. M., 530, 533
Beaumarriage, T., 528, 537
Beckman, R. J., 519, 539
Behnken, D. W., 386, 396, 409, 502, 533

Benet, M., 430, 440

Benjamini, Y., 137, 157, 162

Berger, M. P. F., 507, 533

Berger, R. L., 154, 158

Berk, K. N., 136, 157, 188, 234

Bernshteyn, M., 405, 409, 415, 424, 499, 522, 533

Berry, G., 76, 91

Bertrand, J. C., 393

Beveridge, W. I., 2, 24

Beverly, J. M., 299, 300, 304, 305

Bie, X., 361, 405, 411

Bingham, D., 109, 157, 316, 323, 330, 340–342, 351

Bingham, T. C., 508, 533

Bisgaard, S., 3, 13, 14, 18, 19, 24, 60, 91, 124, 126, 127, 150, 153, 157, 158, 159, 190, 195, 196, 226, 234, 255, 284, 312–314, 316, 322, 323, 330, 336, 338–340, 343, 347, 348, 351, 353, 354, 357, 373, 389, 409, 493, 509, 522, 524, 529, 533, 534, 556

Bishop, T. A., 46, 49

Bjerke, F., 339, 351, 545, 546

Block, R., 194, 228, 234, 236, 406, 409

Bodea, A., 523, 534

Bohrer, R., 107, 158

Bonett, D. G., 107, 158, 162

Bonferroni, C., 37

Boon, P. C., 431, 439

Booth, K. H. V., 499, 534

Borkowski, J. J., 150, 162

Borror, C. M., 322, 325, 405, 406, 410, 412, 413, 490, 507, 522, 523, 529, 534, 537, 539, 540

Bose, M., 432, 439

Bose, R. C., 70, 91

Bowen, B. D., 411, 418, 419

Bowman, D. T., 148, 158

Bowman, K. O., 9, 24

Box, G. E. P., 1, 3, 4, 7, 18, 19, 24, 36, 37, 50, 60, 72, 81, 83, 91, 109, 111, 117, 118, 119, 130, 138, 139, 152, 158, 186, 190, 220, 234, 235, 257, 284, 311, 312, 314, 323, 330, 333, 335, 350, 351, 356, 357, 360, 361, 364, 370–373, 395, 406, 409, 489, 493, 500, 502, 505, 528, 531, 534, 545, 550, 560

Bradley, R. A., 522, 534

Brenneis, F., 108, 162

Brennerman, W. A., 320, 323

Brewster, J. F., 342, 353

Brown, D. R., 107, 425, 439, 444

Brown, M. B., 11, 24, 162

Brownie, C., 345, 352

Brownell, A. K. W., 108, 162

Brownlee, K. A., 124, 158

Buckner, J., 24, 29, 284, 286

Buncick, M. C., 509, 533

Burdick, R. K., 302, 303, 523, 534

Burman, J. P., 489, 539

Bursztyn, D., 314–316, 322, 325, 326, 360, 361, 409, 413, 488, 524, 534, 535, 541

Butler, N. A., 193, 235, 345, 352, 499, 536

Bylund, D., 361, 412

Cahya, S., 403, 404, 410, 413

Cameron, J. M., 525–527, 535

Campbell, C., 13, 25, 148, 160

Campbell, D. T., 24, 26

Campbell, E. D., 404, 409

Carlson, J. E., 93

Carlyle, W. M., 406, 410, 464, 474, 507, 529, 537

Carriére, K. C., 429, 430, 433, 438, 440, 441

Carroll, R. J., 189, 235

Carter, W. H., Jr., 147, 160, 316, 324, 390, 404, 407, 409, 412, 464, 472

Casella, G., 154, 158, 301, 304

Catellier, D., 430, 444

Cawse, J. N., 351, 352

Ceranka, B., 524

Chan, L.-Y., 364, 413, 517, 541

Chananda, B., 138, 140, 160

Chang, J. Y., 438, 440

Chang, Y. C., 406, 411

Chang, Y.-N., 412, 421

Chapman, R. E., 142, 143, 158, 378, 380, 409

Characklis, W. G., 285, 287

Chaturverdi, N., 429, 443

Chen, D., 340, 352

Chen, F., 383, 414, 420, 495, 541, 542

Chen, H., 453, 464

Chen, J., 192, 227, 235, 263, 282, 284, 285

Chen, V. C. P., 523, 535

Chen, W., 322, 323, 524, 538, 540

Cheng, C.-S., 432, 440, 508, 511, 512, 535

Cheng, S.-W., 258, 259, 275–277, 286, 287, 360–363, 410, 414

Cheryan, M., 417

Chevan, A., 277, 285

Chiao, C.-H., 469, 472

Chin, B. L., 24, 29, 284, 286

Chinchilli, V. M., 404, 409, 424, 437, 443, 444

Chipman, H., 109, 158, 499, 533

Cioppa, T. M., 521

Chiu, T.-Y., 412, 421

Chokski, D., 196, 235, 483, 489, 535

Chouinard, P. Y., 93, 100

Chow, W., 107, 158

Chowdhury, S., 312, 325, 326, 328

Churchill, G. A., 529, 538

Cioppa, T. M., 524, 535, 538

Clark, J. B., 254, 285

Clark, L. C., 19, 25, 28

Clatworthy, W. H., 70, 91

Claydon., N. C., 430, 440

Cline, D. B. H., 189, 235

Cobb, G. W., 157, 158

Cobo, E., 430, 440

Cochran, W. G., 68, 70, 77, 86, 91, 179, 235, 430, 433, 434, 440

Coffin, M., 21, 26, 44, 50, 64, 69, 72, 86, 92, 136, 161

Cohen, G. S., 108, 162

Colbourn, C. J., 91

Cole, R. G., 300, 303

Coleman, D. E., 13, 14, 24

Conlon, M., 447, 473

Conover, W. J., 519, 539

Cook, T. D., 24, 26

Cooney, G. A., Jr., 529, 535

Copeland, K. A. F., 26, 44, 47, 50, 64, 69, 72, 86, 91, 92, 136, 161, 228, 229, 235, 452, 472, 523, 540

Cornell, J., 361, 411, 474, 481, 482, 523, 535

Cornfield, J., 43, 50

Cotter, S., 487, 535

Cottrell, J. I. L., 361, 413

Cox, D. R., 14, 23, 364, 410, 499, 522, 534, 535

Cox, G. M., 68, 70, 77, 86, 91, 179, 235, 430, 433, 434, 440

Craigen, R., 528, 535

Creveling, C. M., 322, 324

Croarkin, C., 18, 25, 303, 309, 347, 352, 360, 371, 378, 410, 525–527, 535, 545

Crowder, M. J., 425, 428, 440

Curnow, R. N., 92, 99

Czitrom, V., 7, 17, 18, 25, 33, 50, 148, 158, 159, 203, 206–208, 234, 322, 324, 325, 352, 354, 453, 455, 458, 459, 472, 483, 484, 508, 531, 535

Dallal, G. E., 38, 50, 437, 438, 440

Daniel, C., 4, 25, 106, 116, 118, 120, 121, 128, 130, 136, 151, 159, 172, 177, 188, 190, 203, 209, 221, 235, 330, 335, 352, 484–486, 522, 535, 536

David, O., 431, 441

Davidian, M., 425, 440

Davis, C. S., 425, 439, 440

de Amezua Ayala, J. C., 531, 535

de Palluel, Cretté, 71

de Pinho, A. L. S., 17, 26, 124, 126, 127, 153, 157, 158, 347, 348, 351, 354

De Braekeleer, K., 292, 304

De Beer, J., 292, 304

De Vansay, E., 393, 410

De Vor, R. E., 417

den Hertog, D., 373, 412

Dean, A., 10, 25, 38, 49, 59, 60, 85, 86, 91, 92, 107, 154, 157, 159, 160, 312, 316, 323, 325, 432, 438, 440, 443, 499, 539

Dekker, R., 373, 413

del Castillo, E., 322, 324, 373, 403, 404, 410, 413, 452, 463, 472, 478

Delott, C., 530, 536

Deming, W. E., 500, 536

Deng, J. C., 474, 481

Denton, D. D., 509, 534

Derringer, G., 450, 451, 463, 468, 471, 472, 476, 480

Dershowitz, A. F., 531, 537

Dette, H., 507, 528, 536

Dey, A., 230, 237, 432, 439

Diamond, N., 14, 26, 111, 161, 372, 389, 390, 413

Dickinson, A. W., 522, 536

Ding, R., 406, 410, 452, 472

Ding, Y., 405, 414

Dinitz, J. H., 70, 91

Dobson, A. J., 529, 536

Dobson, D. J. G., 108, 162

Doehlert, D. H., 393, 410
Doganaksoy, N., 452, 473
Donev, A. N., 141, 159, 432, 440
Dong, F., 188, 235
Douglas, P. L., 411, 418
Drabenstadt, D., 245, 474
Drain, D., 406, 410
Draper, N. R., 3, 24, 139, 158, 159, 200, 201,
    235, 311, 323, 360, 361, 371, 372,
    378–380, 403–406, 409–411, 489, 493,
    522, 531, 534, 536, 539
Drouin, H., 411, 418, 419
Du Mochel, W., 504, 536
Dudewicz, E. J., 45–47, 49, 50
Dumencil, G., 393, 410
Duncan, D. B., 38, 50
Dunlop, G., 83, 92
Dunnett, C. W., 38, 50
Dyas, B., 148, 159, 322, 324

Easter, S. M., 405, 412
Eccleston, J. A., 429, 444
Edmondson, R. N., 405, 410
Edwards, J. W., 530, 533
Eibl, S., 149, 159, 222, 235
Ellekjaer, M. R., 453, 472
Emanuel, J. T., 109, 159
Emptage, M. R., 330, 352, 530, 536
Engel, J., 314–315, 322, 324
Erdélyi, J., 464, 473
Eskridge, K. M., 499
Euler, L., 71, 81, 92
Evalgelaras, H., 322, 324
Evans, M. A., 437, 443
Everitt, B. S., 425, 440

Faith, R., 107, 158
Fang, K.-T., 315, 326, 364, 386, 410, 414,
    499, 515, 516, 524, 536, 539, 541
Federer, W. T., 70, 92, 312, 325, 431, 440
Fedorov, V. V., 503, 528, 536
Ferreira, G. M., 493, 494, 533
Ferris, S. J., 431, 441
Filliben, J., 135, 188
Finney, D. J., 169, 235, 430, 440
Fisher, R. A., 38, 50, 84
Fleiss, J. L., 425, 440
Flemming, M., 148, 159, 322, 324
Fletcher, D. J. H., 437, 440

Flick, P. M., 404, 414
Forsythe, A. B., 11, 24
Fowler, J. W., 522, 540
Fowlkes, W. Y., 322, 324
Franklin, M. F., 512, 536
Franks, J., 147, 159
Freeman, G., 71, 92
Freeman, I. C., 537
Fries, A., 192, 235
Fuller, H. T., 150, 157, 159, 336, 338, 339,
    351, 389, 409, 453, 472, 529, 534
Fung, C., 19, 24, 60, 91, 312, 323

Ganju, J., 13, 25, 148, 159, 484, 537
Garcet-Rodriguez, S. A., 507, 539
Garcia, R., 430, 440
Gardner, M. M., 351, 352
Gaudard, M., 464, 473
Gawande, B. N., 245
Gell, M., 312, 324
Gentleman, R., 530, 540
Ghadge, S. V., 361, 410
Gheshlaghi, R., 411, 418
Ghosh, S., 377, 411
Gibb, R. D., 464, 472
Giesbrecht, F. G., 58, 70, 72, 79, 89, 91, 92,
    140, 159, 170, 235, 263, 264, 275, 278,
    280, 285, 332, 346, 352, 364, 370, 383,
    395, 404, 408, 411, 425, 431, 435, 440,
    467, 472, 508, 537
Gilmour, S. G., 403, 405, 406, 411, 499, 536
Giltinan, D. M., 425, 440
Giovannitti-Jensen, A., 406, 411
Girden, E. R., 425, 440
Glasnapp, D. R., 107, 159
Glonek, G. F. V., 530, 537
Goad, C. L., 437, 440
Goh, T. N., 22, 25
Gonzalez-de la Parra, M., 302, 303, 307
Goodwin, D. E., 159, 216, 235, 246, 389,
    411, 529, 537, 557
Goos, P., 149, 159, 330, 335, 339, 352, 353,
    507, 522, 529, 537
Gordon, G., 79
Gorenflo, V. M., 411, 418, 419
Govaerts, B., 460, 473
Grage, H., 414
Graybill, F. A., 302, 303
Greenhouse, S. W., 43, 50

Gregoire, T. G., 107,161
Gregory, W. L., 339, 352
Gremyr, I., 19
Grieve, A. P., 437, 440
Grizzle, J. E., 430, 437, 441
Gruska, G. F., 303, 304
Gumpertz, M. L., 58, 70, 72, 79, 89, 91, 92,
    140, 159, 170, 235, 263, 264, 275, 278,
    280, 285, 331, 332, 345, 346, 352, 364,
    370, 383, 395, 404, 408, 411, 425, 431,
    435, 440, 467, 472, 508, 537
Gunst, R. F., 291, 303
Gupta, A., 241
Gupta, P., 530, 536
Gupta, V. K., 230, 237
Gupte, M., 285, 288
Guseo, R., 349, 352
Guttman, I., 200, 201, 235

Haaland, P. D., 127, 159, 173, 187, 235
Hackl, P., 528, 536, 537
Hahn, G. J., 14, 25, 364, 411, 500, 531, 537
Hahn-Hägerdal, B., 414
Hailes, G. E., 527, 535
Haines, L. M., 528, 533
Hale-Bennett, C., 274, 275, 285
Hall, M. J., 500, 537
Halperin, M., 43, 50
Hamada, M., 81, 93, 107, 109, 121, 136,
    152, 159, 162, 188, 189, 192, 193, 195,
    203, 228, 235, 237, 252, 259, 262–264,
    275, 284, 286, 340, 354, 376, 387, 414,
    469, 472, 490, 493, 497, 498, 537, 541
Hand, D. J., 425, 440
Hanson, R. J., 525, 538
Hardin, R. H., 364, 411, 505
Hare, L. B., 141, 159, 243
Hargit, C., 165
Harrington, E. C., 450, 463, 473
Hartley, H. O., 9, 26, 377, 380, 411
Harwood, C. E., 29
Hashiguchi, H., 499, 541
Hasted, A. M., 92, 99
Haugh, L. D., 453, 455, 458, 459, 472
Hawkins, D. M., 464, 472
Hayford, J. A., 527, 537
Hazel, M. C., 335, 353
Heany, M. D., 285, 290
Heaphy, M. S., 303, 304

Hedayat, A. S., 222, 235, 312, 325, 432, 438,
    441, 453, 472
Helps, R., 322, 324
Henri, J., 23, 29, 284, 286
Heredia-Langner, A., 464–473, 507, 529,
    537
Hess, J. L., 291, 303
Hext, G. R., 364, 413
Hicks, C. R., 140, 159, 303, 308
Hill, H. M., 522, 537
Hill, W. J., 175, 235
Himsworth, F. R., 364, 413
Hinkelmann, K., 65, 69, 70, 91, 92, 93, 107,
    159, 264, 285, 425, 434, 437, 441, 499,
    537
Hochberg, Y., 38, 50, 137, 157
Hoerl, A. E., 403, 411
Hoerl, R. W., 404, 411, 537
Hoke, A. T., 34, 411
Hong, S.-C., 528, 537
Hoogmartens, J., 292, 304
Hooke, R., 450, 473
House, J. D., 92, 99
Hsieh, P. I., 159, 216, 235, 246, 389, 411,
    529, 537, 557
Hsu, J. C., 38, 50
Huang, L. Z., 361, 405, 411
Huang, P., 340, 352
Huang, R., 438, 441
Hubele, N. F., 528, 537
Hudson-Curtis, B., 330, 352, 530
Huele, F. A., 322, 324
Hung, S.-H., 412, 421
Hunter, W. G., 4, 24, 36, 37, 50, 72, 81, 83,
    91, 117, 118, 142, 158, 160, 163, 169,
    186, 192, 235, 237, 330, 351, 357, 493,
    505, 534, 544, 545
Hunter, J. S., 4, 24, 36, 37, 50, 72, 81, 83, 91,
    117, 118, 158, 169, 186, 220, 234, 235,
    330, 351, 357, 395, 407, 409, 493, 505,
    534
Hutchens, C., 322, 324, 326, 327

Iglehart, J. D., 530, 540
Ikebe, Y. T., 499, 541
Illingworth, B. L., 474, 481
Inman, J., 1,19, 25, 28, 148, 160, 266, 285
Inn, K., 188
Ives, S. E., 430, 442

Jackson, R. T., 438, 442
Jacroux, M., 193, 236
Jaech, J. L., 73, 78, 92, 97
Jahanmir, S., 111
Jain, A., 463, 473
James, P. W., 503, 507, 539
Jan, H. W., 484, 541
Jearkpaporn, D., 504, 508, 539
Jeeves, T. A., 450, 473
Jensen, C. R., 302, 303
Jensen, D. R., 392, 411, 502, 537
Jeong, I.-J., 406, 411
Jiang, W., 25, 28
Jillie, D. W., 131, 162
Jin, R., 524, 538
John, J. A., 405, 410, 432, 433, 441
John, P. W. M., 3, 7, 25, 65, 70, 90, 92, 190,
    202, 203, 206–208, 210, 213, 216–219,
    231, 234, 236, 282, 522, 538
Johnson, D. E., 345, 353, 430, 437, 440, 442
Johnson, N. L., 68, 92
Joiner, B. L., 13, 25, 148, 160
Jones, B., 432, 437, 439, 441, 504, 520, 533,
    536
Jones, F. E. E., 524, 538
Jones, S., 335, 350, 351, 356, 545, 550, 560
Jordan, E. H., 312, 324
Jørgensen, T., 244
Joseph, V. R., 149, 160, 322, 324, 524, 538
Joshi, V. M., 107
Ju, H. L., 148, 160

Kahn, W., 352, 354
Kao, L.-J., 107, 160
Karr, A. F., 302, 405, 409
Kasperski, W. J., 188, 237
Kastenbaum, M. A., 9, 23
Kempthorne, O., 69, 70, 74, 81, 91, 92, 93,
    107, 159, 182, 228, 236, 264, 285, 425,
    434, 437, 441, 499, 537
Kempton, R. A., 431, 441
Kennard, R. W., 508, 509, 538
Kenward, M. G., 437, 438, 441
Keppel, G., 430, 441
Kerr, M. K., 530, 538
Kerschner, R. P., 440
Keselman, H. J., 425, 441, 429, 441
Kess, U., 149, 159, 222, 235
Khan, M. A., 389, 413, 424, 473, 477

Khanin, R., 530, 541
Khattree, R., 300, 303
Khuri, A. I., 299, 361, 412, 447, 473, 481,
    482
Kiefer, J., 505, 538
Kim, B. H., 361, 412
Kim, K., 522, 538
Kim, K.-J., 406, 411, 464, 473
Kimmet, S., 82, 92
King, C., 285
Kinzer, G. R., 121, 122, 160
Kirk, R. E., 107, 160
Kirkus, B., 339, 351, 545, 546
Klapwik, A., 507, 540
Klee, V. L., 393, 410
Kleffe, J., 302, 304
Kleijnen, J. P. C., 373, 412, 524, 538
Knowles, G., 325, 328
Kobilinsky, A., 529, 537
Koch, G. G., 430, 444
Koczor, Z., 464, 473
Koksoy, O., 452, 473
Kong, F., 107, 161
Korn, E. L., 530, 540
Koshal, R. S., 393, 412
Koukouvinos, C., 322, 324
Kowalchuk, R. K., 425, 429, 441
Kowalski, S. M., 149, 161, 322, 324,
    333–335, 344, 351–353, 356, 405, 406,
    412–414
Kraber, S. L., 322, 323, 350, 354
Kramer, C. Y., 38, 50
Kros, J. F., 451, 464, 472, 473
Kuehl, R. O., 107, 157, 160, 425, 441
Kuhfeld, W. F., 506, 538
Kuhn, A., 316, 324, 404, 413
Kulahci, M., 336, 340, 343, 353, 354
Kulkarni, P., 285, 288
Kunert, J., 429, 431, 432, 439, 441, 442
Kushner, H. B., 429, 442

Ladstein, K., 453, 472
Laget, M., 393
Lah, C. L., 417
Lamar, J. M., 292, 303
Lambrou, D., 438, 443
Landman, D., 149, 161, 344, 352
Langhans, I., 330, 339, 352, 353
Lasdon, L. S., 463, 473

581

Lasher, W. C., 304, 305
Lavigna, R. J., 47, 51
Lavori, P. W., 430, 442
Lawson, C. L., 525
Lawson, J., 322, 324, 431, 442, 538
Layard, M. W. J., 10, 25
Le Baily de Tilleghem, C., 460, 473
Ledolter, J., 1, 19, 25, 28, 148, 160, 199, 200, 229, 236, 266, 285, 500, 538
Lee, C., 204, 237
Lenth, R., 1, 11, 19, 25, 28, 59, 77, 127, 136, 146, 148, 154, 160, 212, 236, 266, 285, 319, 324, 489
León, R. V., 203, 206–208, 234, 500, 538
Leone, F. C., 68, 92
Leonov, S. L., 503, 536
Lesperance, M. L., 322, 324
Leucuta, S. E., 523, 534
Levene, H., 11, 25, 34
Lewis, D. K., 322, 324, 326, 327
Lewis, S. L., 529
Lewis, S. M., 316, 325, 326, 361, 413, 429, 431, 437, 438, 440, 442, 538
Li, C.-C., 508, 511, 512, 535
Li, J., 364, 413, 517, 541
Li, R., 499, 524, 536, 538
Li, W., 219, 236, 237, 323, 528, 538
Lidy, W. A., 285, 290
Lin, D. K. J., 150, 160, 219, 236, 274, 275, 285, 315, 326, 364, 378, 380, 386, 406, 410, 413, 452, 464, 472–474, 489, 493, 499, 500, 515–517, 524, 536, 538–541
Lin, T., 138, 140, 160
Lindsey, J. K., 425, 442
Liu, H., 302, 405, 409
Liu, M. Q., 499, 524, 536, 539
Liu, S., 292, 303, 308
Liu, Y., 499, 539
Lockyer, J. M., 108, 162
Loeppky, J. L., 200, 236, 340, 353
Loh, W. Y., 188, 236
Lopez-Fidalgo, J., 507, 539
Loredo, E. N., 504, 508, 539
Lorenzen, T. J., 278, 285
Loughin, T. M., 136, 137, 160, 430, 442
Louis, T. A., 430, 442
Love, M. H., 404, 414
Low, J. L., 431, 442

Lowe, C. W., 364, 412
Lü, F., 361, 405, 411
Lu, W.-K., 412, 421
Lu, X., 499, 539
Lu, Y., 361, 405, 411
Lucas, H. L., 528, 534
Lucas, J. M., 13, 25, 123, 148, 149, 150, 159, 160, 162, 335, 353, 484, 537
Lucas, T. W., 524, 538
Lynch, R. O., 11, 25, 120, 160, 176, 187, 221, 236

Ma, C. X., 499, 536
Ma, X., 324
Macdonald, S., 464, 472
MacDougall, D., 2
MacFie, H. J. H., 429, 437, 444
Mack, G. A., 70, 91, 93
MacPherson, K. M., 108, 162
Manly, B. F., 23, 25
Margolin, B. H., 190, 236
Markides, K. E., 361, 412
Marshall, R. J., 438, 442
Martin, A. M., 252, 253, 254, 364, 366, 414
Masinda, K., 378, 380, 409
Mason, R. L., 291, 300, 303
Massart, D. L., 292, 304
Mastrangelo, C. M., 451, 464, 472, 473
Matheson, A. C., 29
Mathews, P., 1, 25
Mattei, G., 393, 410
Matthews, J. N. S., 430, 432, 442, 503, 507, 539
Mattucci, N., 464, 472
Mavris, D. N., 404, 414
Max, M. B., 430, 442
Mays, D. P., 405, 412
McCarville, D. R., 463, 472, 478
McCaskey, S. D., 322, 324
McCullagh, P., 529, 539
McCulloch, C. E., 301, 304
McDaniel, W. R., 364, 412, 483, 539
McGrath, R. N., 150, 160
McKay, M. D., 519, 539
McLeod, R. G., 342, 353
McMahon, L., 79
McShane, L. M., 530, 540
Mead, R., 92, 99, 331, 353, 499, 536

Mee, R. W., 190, 194, 203, 216, 218, 219, 222, 226, 228, 231, 234, 236–237, 243, 246, 345, 353, 364, 405, 406, 409, 412, 500, 538, 539
Meeker, W. Q., 500, 537
Meinander, N. Q., 414
Melas, V. B., 507, 528, 536
Mendicino, M., 347, 353
Merchant, F. M., 530, 540
Meredith, M. P., 108, 161
Meyer, R. D., 109, 138, 158, 314, 323
Michels, K. M., 107, 162, 425, 439, 444
Michelson, D. K., 82, 92
Mielke, P. W., Jr., 430, 442
Miller, A., 188, 219, 236, 322, 324, 335, 346, 347, 348, 353, 530, 540
Milliken, G. A., 345, 347, 353, 428, 442
Miro-Quesada, G., 322, 324
Miron, A., 530, 540
Miron, P. L., 530, 540
Mistree, F., 322, 323
Mitchell, T. J., 520, 539
Moberg, M., 361, 412
Mohammadi, P., 148, 159, 322, 324
Monod, H., 148, 160
Montgomery, D. C., 13, 14, 23, 25, 68, 91, 92, 107, 136, 140, 147, 157, 160, 162, 203, 209, 219, 236, 237, 330, 333, 353, 361, 371, 372, 374, 375, 387, 390, 393, 395, 397, 405–407, 410, 412–414, 430, 442, 452, 463, 464, 471–474, 478, 481, 484, 490, 504, 507, 508, 522, 523, 529, 534, 537–540
Moo-Young, M., 411, 418
Moran, J., 430, 440
Morehead, P. R., 322, 324
Moreira, A. R., 493, 494, 533
Morgan, J. P., 69, 93
Morris, M. D., 405, 412, 520, 539
Mukerjee, R., 364, 410, 432, 443
Mukherjee, B., 432, 439
Muller, W. G., 528, 537
Mustafa, A. F., 93, 100
Muzammil, M., 322, 324
Myers, G. C., 461–463, 473, 474, 475
Myers, R. H., 14, 25, 147, 160, 209, 237, 316, 322, 324, 325, 330, 353, 361, 371, 372, 374, 375, 387, 390, 393, 395, 404, 406, 407, 409, 411–414, 452, 464,

471–474, 504, 505, 509–511, 523, 529, 538, 539, 541
Myers, W. R., 320, 323

Nachtsheim, C. J., 528, 538
Naes, T., 244, 339, 351, 545, 546
Nagaraja, T. G., 430, 442
Naik, D. N., 300, 303
Nair, V., 317, 325
Natrella, M., 61, 63, 70, 86, 92, 98, 161, 164
Nelder, J. A., 529, 539
Nelson, L. S., 43, 44, 50, 140, 161, 229, 237, 301, 303, 493, 539
Nelson, P. R., 21, 26, 44–47, 50, 51, 64, 69, 72, 86, 91, 92, 95, 136, 161, 228, 229, 235, 302, 303, 452, 472, 523, 540
Nester, M. R., 9, 26
Neubauer, D. V., 21, 26
Newcombe, R., 430, 433, 440, 442, 443
Nicki, N., 499, 541
Nicolai, R. P., 373, 413
Niemi, L., 1, 19, 25, 28, 148, 160, 266, 285
Nobile, A., 530, 541
Noble, W., 136, 137, 160
Notz, W. I., 107, 160, 406, 413, 522, 524, 534, 540
Nutan, M. T. H., 389, 413, 424, 473, 477
Nyachoti, C. M., 92, 99

O'Brien, T., 529
O'Carroll, F., 72, 93
O'Connell, M. A., 128, 159, 173, 187, 235
Oehlert, G., 12
Ojima, Y., 300, 303
Oman, S. D., 437, 443
Onifade, K. R., 166
Örnderci, M., 264, 285
Ortiz, F., Jr., 464, 473
Ossenbruggen, P. J., 507, 540
Ott, E. R., 20–22, 26, 43, 50, 134
Ozol-Godfrey, A., 397, 413

Padture, N. P., 312, 324
Paduru, R. M. R., 494, 533
Page, G. R., 530, 533
Pagès, J., 429, 443
Palamakula, A., 388, 413, 424, 473, 477
Palanisamy, M., 109, 159
Palmqvist, E., 414

Park, S.-M., 322, 324
Park, Y., 522, 540
Patkar, A. Y., 245
Patry, M., 237
Patterson, H. D., 70, 93, 433, 437, 438, 443
Peake, R., 59, 93
Pearson, E. S., 9, 26,
Peeler, D. F., 108, 161
Penn, S., 2
Penrod, S. D., 233, 237
Peralta, M., 203, 216, 236, 246
Périnel, E., 429, 443
Peterson, J. J., 403, 404, 413, 464, 473
Peyton, B. M., 285, 287
Phan-Tan-Luu, R., 393, 410
Picard, R. R., 136, 157, 188, 234
Picka, J. D., 302, 405, 409
Piepel, G. F., 505, 523, 540
Piersma, N., 373, 413
Pignatiello, J., 292, 293, 304, 322, 325, 443, 445, 464, 473
Pinho, A., 312, 314, 323
Piret, J. M., 411, 418, 419
Plackett, R. L., 489, 539
Polansky, M., 430, 442
Pontius, J. S., 438, 443
Potcner, K. J., 333–335, 351–353, 356
Prat, A., 149, 161, 210–212, 213, 215, 237
Prescott, P., 405, 409, 431, 442, 443
Pukelsheim, F., 149, 159, 222, 235, 505, 540
Putt, M., 433, 443

Qu, X., 116, 161, 484–486, 540

Radford, R., 188
Radmacher, M. D., 530, 540
Raghavarao, D., 70, 93, 437, 443
Raheman, H., 361, 410
Raktoe, B. L., 312, 325
Ralston, A. R. K., 509, 534
Ramberg, J. S., 322, 325
Ramírez, J. G., 331, 340, 353
Ramsey, P. J., 464, 473
Rao, C. R., 302, 304
Ratkowsky, D. A., 437, 443
Ratner, M., 463, 473
Raulin, F., 393, 410
Raybold, R. C., 525–527, 535
Rechtschaffner, R., 394, 413

Reck, B., 69, 93
Reddy, G., 494, 533
Reece, J. E., 48, 51, 151, 161, 232–233, 237, 252, 285, 291, 304, 321, 325, 345, 347, 350, 353, 469– 471, 473, 508, 531, 532, 540
Reese, A., 41, 51
Reid, N., 364, 410
Reinsel, G. C., 433, 440
Réthy, Z., 464, 473
Ribardo, C., 464, 474
Rightor, E. G., 285, 290
Ritter, J. B., 411, 418, 419
Roach, M., 3, 26
Robbennolt, J. K., 233, 237
Robinson, L. W., 231, 237
Robinson, T. J., 322, 325, 330, 353
Rodriguez-Loaiza, P., 302, 303, 307
Roes, K. C. B., 431, 439
Roets, E., 292, 304
Roof, V., 142, 143, 158
Rosenbaum, P. R., 319, 325, 438, 443
Roquemore, K. G., 390, 391, 413, 502, 540
Rowlands, H., 325, 328
Ruan, S., 499, 539
Runger, G. C., 107, 162, 219, 237, 474, 481, 484, 507, 539
Russell, K. G., 316, 325, 429, 432, 433, 437, 441–443
Ryan, T., 8, 18, 19, 21, 26, 30, 38, 47, 51, 120, 122, 123, 124, 134, 144, 161, 167, 189, 225, 226, 229, 230, 237, 250, 285, 313, 315, 319, 322, 325, 327, 453, 455, 458, 474, 498, 499, 530, 531, 540

Sahai, H., 299, 303
Sahin, K., 264, 285
Sailer, O., 429, 442
Salvia, A. A., 304, 305
Samset, O., 493, 494, 540
Sanchez, A. R., 531
Sanchez, S. M., 322, 325, 524, 535, 538
Santner, T., 524, 540
Sarrazin, P., 93, 100
Satterthwaite, F. E., 271, 286, 498, 540
Sauls, J., 107, 159
Schabenberger, O., 107, 161
Schantz, R. M., 438, 443
Scharer, J. M., 411, 418

Scheffé, H., 38, 51
Schilling, E. G., 21, 26, 43, 51
Schmidt, R. H., 474, 481
Schneider, H., 188, 237
Schoen, E. D., 150, 161, 204, 314, 325, 340, 343, 351
Scholtens, D., 530, 540
Schoonover, R. M., 524, 538
Searle, S., 40, 51, 140, 161, 301, 304
Seddon, I. R., 92, 99
Seenayya, G., 494, 533
Seidden, E., 437
Sen, K., 330, 352, 530
Sen, M., 432, 443
Senn, S., 429, 430, 435, 437, 438, 443
Senturia, J., 312, 325
Sergent, M., 393, 410
Shadish, W. R., 23, 26
Shah, H. K., 464, 474
Shainin, D., 229, 500
Sharma, V., 325
Sheesley, J. H., 83, 92, 267, 268, 286
Shenasa, M., 508, 540
Shewhart, W., 21
Shi, X., 347, 353
Shih, I.-L., 412, 421
Shoemaker, A. C., 318, 319, 325
Shrikhande, S. S., 70, 91
Silvey, S. D., 505, 540
Simon, R. M., 530, 540
Simpson, E., 108, 162
Simpson, J. R., 149, 161, 344, 352, 464, 473
Simpson, T. W., 524, 540
Singh, M., 65, 93
Singh, P. P., 322, 324
Sinha, K., 70, 93
Sinibaldi, F. J., 292, 304
Sitter, R. R., 200, 219, 236, 316, 323, 330, 340–342, 351, 353
Skillings, J. H., 70, 91,93
Sloane, N. J. A., 222, 235, 364, 411, 505
Slominski, B. A., 92, 99
Smith, E. P., 509–511, 541
Smith, J. M., 322, 326
Smith, J. R., 299, 300, 304, 305, 324, 326, 327
Snee, R. D., 187, 220, 237, 238, 292, 304, 306–307, 360, 413, 508, 541
Sniegowski, J., 453, 455, 458, 459, 472

Solomon, P. J., 530, 537
Sotocinal, S. A., 93, 100
Spanjers, H., 507, 540
Speed, T. P., 530, 541
Spendley, W., 364, 413
Srinivasan, M. K., 167
Srivastava, R., 230, 237
Stanley, D. O., 404, 414
Stanley, J. C., 24
Stanley, J. D., 490, 539
Stehman, S. V., 108, 161
Steinberg, D., 109, 137, 158, 162, 169, 237, 314–316, 322, 325, 326, 360, 362, 409, 413, 488, 524, 534, 535, 541
Stephens, M. L., 464, 473
Stephenson, J. V., 245, 474
Sternberg, R., 393, 410
Steudel, H. J., 347, 348, 354
Stevens, L., 429, 443
Stolle, D. P., 237
Stone, L., 508, 509, 538
Stoneman, D. M., 159, 522, 536
Street, D. J., 429, 444
Strigul, N., 507, 536
Stroup, W. W., 339, 351, 545, 546
Stufken, J., 222, 235, 431, 432, 438, 441, 442
Su, C.-T., 464, 474
Subramani, J., 89, 93
Sudjianto, A., 524, 536, 538
Suen, C.-Y., 453, 472
Suich, R., 450, 451, 463, 468, 471, 472, 476
Sun, D. X., 192, 226, 227, 235, 252, 263, 277, 282, 284–286
Sutherland, M., 277, 285, 336, 351
Swersey, A., 229, 236, 500, 538
Szonyi, G., 464, 472
Sztendur, E. M., 109, 161, 372, 389, 390, 413

Taam, W., 339, 352
Tabucanon, M. T., 451, 464, 474
Tack, L., 507, 522, 537, 541
Taguchi, G., 107, 161, 226, 229, 287, 311–313, 322, 325, 328, 346, 354, 357, 452, 474, 500, 524
Taihrt, K. J., 149, 161
Talib, F., 322, 324
Tamhane, A. C., 38, 50
Tang, L. C., 452, 474
Tang, M., 364, 413, 517, 541

585

Testik, M. C., 504, 508, 539
Thaler, J. S., 444, 445
Thayer, J. S., 98
Tippett, L. H. C., 79, 93
Tobias, P., 18, 25, 303, 309, 347, 352, 360, 371, 378, 410, 545
Tobias, R., 340, 353
Tobias, R. D., 506, 538
Toews, J. A., 108, 162
Tomlinson, L. H., 47, 51
Tong, L.-I., 464, 474
Torban, D., 528, 533
Tort, X., 149, 161, 210–212, 213, 215, 237
Travnicek, D. A., 499, 536
Trinca, L. A., 405, 406, 411, 414
Tripolski, M., 137, 162
Trochim, W. M. K., 26
Tsui, K.-L., 318, 319, 322–326, 523, 535
Tu, W., 452, 473
Tuck, M. G., 361, 413
Tudor, G. E., 430, 444
Tukey, J. W., 38, 51, 117, 162
Turnbull, B. W., 19, 25, 28
Turner, K. V., 140, 159
Tyssedal, J., 109, 158, 336, 343, 354, 489, 493, 494, 500, 534, 540, 541

Unal, R., 404, 414
Usher, J. S., 167

van der Linden, W. J., 65, 93
Van Matre, J. G., 14, 26
van Oortmarssen, G. J., 373, 413
Vandebroek, M., 149, 159, 330, 335, 339, 352, 353, 507, 522, 529, 537, 541
Vander Heyden, Y., 292, 304
Vardeman, S. B., 302, 304
Vasudev, P. K., 347, 353
Vázquez, M., 252, 253, 254, 364, 366, 414
Veldkamp, B. P., 65, 93
Verseput, R. P., 529, 535
Vickers, A. J., 444
Vijayan, K., 499, 533
Vine, A. E., 316, 326
Vining, G. G., 147, 160, 390, 405–407, 412–414, 452, 464–469, 474
Vivacqua, C. A., 17, 26, 124, 126, 127, 158, 347–348, 351, 354
Voelkel, J., 340, 352, 483, 541

Vonesh, E. F., 437, 444
Voshi, V. M., 158
Voss, D., 10, 25, 38, 49, 59, 60, 85, 86, 91, 92, 154, 157, 159

Wakeling, I. N., 429, 437, 444
Walls, R. C., 459, 474
Wang, P. C., 204, 237, 267, 286, 484, 541
Wang, Y., 315, 326, 386, 414, 515, 541
Waren, A. D., 463, 473
Wasiloff, E., 165
Waterhouse, T. H., 320, 326, 507, 541
Watts, N., 2
Webb, D. F., 148, 150, 162
Weeks, D. L., 149, 161, 459, 474
Weerahandi, S., 34, 41, 42, 46, 51, 435, 437, 444
Wei, D., 406, 410, 452, 472
Weiss, P., 524, 534
Weissfeld, L., 188, 237
Welch, W. J., 148, 162
Wen, Z.-Y., 383, 414, 420, 495, 541, 542
Wendelberger, J., 302, 304
Westlake, W. J., 378, 414
Wheeler, R. E., 9, 10, 26
Whitaker, D., 432, 433, 441
Whitcomb, P., 12, 130, 157, 208, 350, 354, 361, 409, 490, 533
White, L. V., 148, 162
Wilcoxon, F., 522, 536
Wiles, R. A., 175, 235
Wilk, M. B., 74, 93
Williams, B., 524, 540
Williams, E. J., 433, 434, 437, 438, 444
Williams, E. R., 29, 70, 93
Wilson, K. B., 190, 235, 257, 284, 361, 362, 370, 373, 409, 502, 534
Wilson, W. H., 429, 444
Winer, B. J., 107, 162, 425, 439, 444
Winker, P., 515, 536
Wisnowski, J. W., 107, 162, 474, 481
Wit, E., 530, 541
Wludyka, P. S., 21, 26, 46, 47, 50, 51, 92, 136, 161, 523, 540
Wolfinger, R. D., 429, 441
Wong, W. K., 507, 528, 533, 536
Woodall, W. H., 38, 51, 229, 237
Woodward, A. J., 107, 158, 162

Wright, G. W., 530, 540
Wu, C. F. J., 81, 93, 107, 109, 116, 121, 152,
    158, 161, 162, 192, 193, 195, 203,
    226–228, 235, 237, 252, 258, 259, 262,
    263, 275–277, 282, 284–287, 318, 319,
    322, 324–326, 340, 354, 360–363, 376,
    387, 405, 409, 414, 432, 440, 484–486,
    490, 493, 497–499, 524, 537, 538, 540,
    541
Wu, K. C., 404, 414
Wu, X., 499, 539
Wu, Y., 312, 313, 325, 328
Wurl, R. C., 463, 474

Xu, H., 263, 275, 276, 277, 284, 286, 360,
    414, 499, 541
Xie, L., 312, 324
Xu, K., 452, 474

Yamada, S., 499, 541
Yandell, B., 291, 304
Yang, M., 438, 441

Yang, T. H., 530, 541
Yates, F., 70, 93
Yates, P., 226, 227, 237, 243
Ye, K. Q., 219, 236, 237
Yeh, C. M., 522, 534
Yin, G. Z., 131, 162
Ylvisaker, D., 520, 539
Youden, W. J., 13, 26, 84, 93, 148, 162
Yu, L., 405, 409, 415, 424, 522, 533
Yuan, Y., 361, 405, 411

Zahn, D. A., 188, 237
Zahran, A. R., 406, 414, 509–511,
    541
Zalokar, J., 43, 50
Zelen, M., 525, 542
Zhang, Y., 515, 536
Zhao, Y., 530, 540
Zhu, Y., 292, 304
Zink, P. S., 404, 414
Zirk, W. E., 292, 303
Zubrzycki, S., 393, 410

Subject Index

Analysis of Means (ANOM), 20–22, 42–47, 64, 68–69, 121, 135, 152, 405
  assumptions, 555
  for attributes data, 47
  nonparametric, 47
  with unequal variances, 45
Analysis of Variance (ANOVA), 20–22, 35, 58, 62–64, 75, 81, 83, 116, 117, 123, 134, 152
  identity, 40
  Kruskal–Wallis, 46

Basic concepts
  blocking, 13
  choice of factor levels, 14, 15, 17
  experimental design objectives, 3
  experimental units, 12
  experimentation, 13, 162
    by students, 162
    steps, 13
  hypothesis testing, 9
  power, 38
  randomization, 6, 7, 13, 223
    complete, 31
    restricted, 148
  replication, 8, 123, 551
    built in, 122
  sample size determination, 9–11
  selecting factors, 14

sequential experimentation, 3, 4
  example of, 222–225
training for designing experiments, 544–545
  use of catapult, 545
treatments, 12
unbalanced data, 40
Bayes plot, 138
Box–Cox transformation, 232

Classification and regression trees (CART), 107
Consolidated Standards of Reporting Trials (CONSORT), 430

Designs
  Addelman, 86
  analytic studies, 500
  assessing the capability of a system, 528
  calibration, 525–526
    constrained least squares, 525
    method of Lagrangian multipliers, 525
  catalog of, 284
  categorical response variables, 524
  completely randomized design
    assumptions, 32–33
      checking, 34
    degrees of freedom, 41

Designs (*cont.*)
efficiency relative to RCB design, 61
unequal variances, 41–42
computer experiments, 523
space-filling designs, 523–524
cost-minimizing, 522
Cotter, 487–488
factor correlations, 488
cross-classified (factorial), 292
crossover (changeover), 429
advantages, 430
applications, 429–430
computer analysis, 437
designs for carryover effects, 432–435
Williams squares, 433–434
example, 435–436
disadvantages, 430, 431
examples, 431
optimal, 431, 432
efficiency, 190
df-efficiency, 222
D-efficiency, 320, 330, 486, 487, 506–507
G-efficiency, 283, 330
equileverage, 501–503
factorial, 101
2^2, 101–102
example, 103–106
2^3, 119–120
examples, 120–122
2^5, 136
2^k, 142
blocking, 141
example, 142–144
3^2, 248
decomposing the A*B interaction, 251
example, 252
3^k, 248–257
inference, 252
interaction components, 250, 277, 288
linear effect, 253
quadratic effect, 253
bad data, 127–130
example, 131–134
blocking, 194
missing data, 138–140

mixed factorials, 263–266
constructing, 265
examples, 264, 266–273
need for, 263–264
Graeco-Latin square, 74, 80–84, 91
application, 82
degrees of freedom limitations, 81–82
hyper, 90–91
model, 80
power, 82
sets of, 82, 84
use of ANOM, 83
incomplete block designs, 65–71, 90
α-designs, 70–71, 90
balanced (BIB), 65–69, 84, 85, 526
analysis, 66–68
recovery of interblock information, 68
use of ANOM, 68–69
lattice, 70, 79
nonparametric analysis, 70
partially balanced, 69–70
John's 3/4 designs, 216–219
Latin square, 70–79, 84–85, 228, 562
assumptions, 72–74
efficiency, 77
example, 74–76
missing values, 86
example, 88
model, 74
standard form, 71, 72
use of ANOM, 76, 79
using multiple Latin squares, 77–79
microarray experiments, 529–530
mixture, 522–523
ANOM, 523
optimal, 523
model-robust, 528–529
multiple responses, 452–453
nested (hierarchical), 291
ANOM, 302
applications, 292–293
estimating variance components, 300–301
examples, 294–298
factor, 292
model, 292
factorial, 291
software shortcomings, 295–296
a workaround, 295–296

staggered, 298–300
  with factorial structure, 300
nonlinear models, 528
non-normal responses, 529
nonorthogonal, 212
  inadvertent, 225
nonregular, 170
  defined, 170
one-factor-at-a-time (OFAT) designs, 3,
    115–116, 483–487
  advantages, 486, 487
  nonorthogonality, 486, 487
  OFAT designs versus OFAT
    experimentation, 484
  statistical process control checks, 487
  strict, 485, 486
optimal, 69, 503–507
  applications, 507
  criticisms, 504, 505
  D-optimal, 279, 281, 282, 330, 364,
    504, 507, 508, 512, 513, 532
    Bayesian, 504–505
    CONVERT algorithm, 505
  E-optimal, 505
  G-optimal, 505
  GH-optimal, 511
  incorporating costs, 522
  I-optimal, 364
  L-optimal, 505
  model-robust, 504
  Q-optimal, 511
orthogonal arrays, 102, 138, 170, 229,
    282, 321, 499
  combined, 314, 316, 318
  compound, 318, 319
  inner, 314, 318, 321, 326
  mixed levels, 275–277, 321
  outer, 314, 318, 321, 326
  product, 314–316, 318
orthogonal main effect plans, 278, 489
Plackett–Burman, 212, 215, 230, 231, 336,
    378–379, 381, 383, 405, 489, 532
  applications, 494–498, 542
  foldover, 494
  projective properties, 493, 494
projective properties, 170
randomized complete block (RCB)
    design, 56–64, 77, 195, 344
  assumption, 57–58

  efficiency, 61
  missing values, 86–87
  number of blocks to use, 59
  use of ANOM, 64
repeated measures, 425
  advantages, 425
  carryover effects, 427
  crossover designs, 429
  example, 428
  how many?, 437
  missing data and imputation, 438
response surface, 360
  applications of, 361
  blocking, 394–397
  Box–Behnken, 386–389, 477, 478, 502
    blocking, 396
    rotatability, 387
  central composite (CCD), 361, 363,
      365, 368, 369, 373–377, 385, 389,
      395, 397, 405, 495, 502
    blocking, 394–395
    centerpoints, 373–376
    example, 383–384
    face centered cube, 377, 404
    inscribed (CCI), 377, 388
    uniform precision design, 375
  comparison, 397
  desirable properties, 360–361
    orthogonality, 375
    rotatability, 375–377
  Doehlert (uniform shell) designs, 393
    applications, 393
  Draper–Lin (small composite) designs,
      377–383
    blocking, 396, 397
  eligible projected, 363
  for computer simulations, 404
  Hoke, 394, 397
  hybrid, 390
    311A, 391
  Koshal, 393
  noncentral composite, 405
  number of designs to use, 362–364
  optimal, 405
  row-column, 406
  small factor changes, 364
  split factorial, 405
restricted regions of operability, 508–514
  examples, 508–514

Designs (*cont.*)
  robust, 311
  rotation, 488
  saturated, 105, 138, 489
  screening, 15, 360, 489–500
    p-efficient, 500
  space-filling, 369, 385, 386, 514–521
    Latin hypercube, 369, 519–521
      example, 520–521
    properties, 384–386
    sphere-packing, 369, 385, 518–519
    uniform, 364, 366, 368, 369, 389, 507,
        514–518
      applications, 386
      definition, 515
  split-lot, 345–346, 349
    use of fractional factorials, 345–346
  split-plot, 330–331, 349, 351, 560
    blocking, 342–343
    example, 333
      analysis, 333–335
        versus incorrect complete
        randomization analysis, 335
    in industry, 336
      example, 336–338, 355
    mirror image pairs design, 336
    Plackett–Burman designs, 343
    subplot, 332, 349
      error, 333
        independent of whole plot error,
        339
    whole plot, 331–332, 338, 349
      error, 332
    with fractional factorials, 340–342
      example, 341
    with hard-to-change factors, 343
      examples, 343–345
  split-split-plot, 345
  split-unit, 330–331
  strip-plot (strip-block), 346–349
    applications, 347–349
    example, 346–347
    use of fractional factorials, 346–348
  supersaturated, 489, 498–500, 513, 563
    nonorthogonality, 499
  Taguchi, 312–315, 320–322, 544
    equivalent to suboptimal fractional
        factorials, 313
  trend-free, 521–522

  unreplicated, 114, 116
  weighing, 524–528
  with noise factors, 316–318
  Youden design, 84–86
    lists of, 86
    model, 85
    replicated, 86
Dual response problem, 406

Effects
  conditional main, 107, 109, 114, 115,
        121, 133, 134, 147, 179, 208,
        209, 255–257, 317, 318, 372,
        380, 381, 388, 389–390, 407,
        467, 485, 492
    derivation of, 152
    example, 108, 113
    necessary sample sizes for, 113
    two-split, 181
  confounded, 5, 12
  dispersion, 150, 312
    detecting, 150, 314
  estimates, 114
    precision of, 153
    relationship with regression
        coefficients, 153, 177
  interaction, 102, 106, 134
    control × noise, 315, 319
    generalized, 318
    noise × noise, 316
    transformations, 114
    Tukey test for, 117–118
  location, 312
  main, 102
  partial confounding, 5
  simple, 107
Evolutionary Operation (EVOP), 363–364,
    531
  Box–EVOP, 531
  dealing with interactions, 531
  simplex, 364
Expected mean squares, 144–146, 273
  for replicated 2^2 design, 153–155
  in general, 155–157
  simple method of determining, 146

Factors
  control, 311, 316, 318, 321
  fixed, 32, 101, 146

hard to change, 148–150, 212, 267, 332, 335, 344, 484, 487, 507, 522
  software, 150
noise, 311, 312, 316, 318, 321
not reset, 150
qualitative, 6, 101
quantitative, 6, 101
random, 32, 146
  hypothesis tests, 146–147
False discovery rate (FDR), 137
Fractional factorials, 169
  3/4 fractions, 216–219
  2^{k-p}, 176, 186
    projective properties, 219–220
  2^{k-1}, 170–181
  2^{k-2}, 181–187
    example, 182–184
  2^{3-1}, 171, 176, 178
  2^{4-1}, 175, 180
  2^{5-2}, 191
  2^{6-2}, 202
  3^{k-p}, 257–262, 362
    constructing, 260–262
    linear and quadratic effects, 259
    minimum aberration, 277
    minimum confounded effects, 277
    projective properties, 259
  3^{k-1}, 262–263
    alias structure, 262
  3^{3-1}, 262–263
  4 or more levels
    method of replacement, 278
  4^{3-1}, 279
  16–point designs, 187
  aliases and alias structure, 174, 177–179, 283
    partial aliasing/partial confounding, 174
  alternatives to, 229
  bad data, 230
  blocking, 195
    examples, 196, 199
    size two blocks, 200–201
  confounded effects, 174
  defining relation, 171
    retrieving lost relation, 190–192
  df-efficiency, 222
  foldover, 178, 200–203
    of a 2^{k-1} design, 201
    mirror image, 200, 201

  semi-foldover, 203–216, 233
    of a 2^{k-2} design, 204
    with software, 215
    shortcomings, 203
  for natural subsets of factors, 226–228
  irregular fraction, 216, 220, 221
  minimum aberration, 192–194
  missing data, 230
  mixed level, 274–275
    linear effects, 276
    quadratic effects, 276
  number of clear effects criterion, 192–194
  one fraction better than another?, 179–181
  post-fractionation, 226, 227, 348–349
  pre-fractionation, 226
  projective properties, 170
  relationship with Latin squares, 228–229
  replicated, 223
  resolution, 169, 187, 212, 233
    defined, 170
  small fractions, 220

Gage R&R (reproducibility and repeatability) study, 295
Gantt charts, 13
Generalized F-test, 42, 46

Hadamard matrix, 488

Journal of Statistics Education, 544

Lenth's sample size determination applet, 11, 39, 59, 77
Lenth's PSE method, 124, 126–129, 131, 136–139, 173, 188, 233, 252, 319, 338, 470, 485, 489, 496, 560
Leverage values, 385, 501
  saturated design, 501
Lurking variable, 6

Measurement capability studies, 523
Missing data, 22, 39–40, 48, 230
Modeling variability, 316
Models
  generalized linear, 529
  hierarchical, 147, 390
  mixed, 58
  nonhierarchical, 147, 378, 390, 407
  unrestricted, 156

Modular arithmetic, 251, 279
Multiple comparisons, 36, 37
  Bonferroni intervals, 37, 38
  Scheffé's procedure, 38, 59, 60
Multiple readings, 8, 117, 123
Multiple response optimization, 447
  desirability function, 449
    composite desirability, 450, 457
    example, 450
    exponential, 464
    importance constant, 451, 460, 461,
      468
    maximization, 450, 457
    minimization, 450
    target value, 451
    weight constant, 451, 460, 461
  desirability graph, 456
  dual response optimization, 452
  examples, 453–463
  frequent assumptions, 447
  global optimum, 448
  Hooke–Jeeves method, 450, 463, 464
  local optima, 448, 450
  overlaid contour plots, 447–449
  pitfalls, 447, 455
  variations, 463–464
    genetic algorithm approach, 464
    generalized reduced gradient algorithm,
      463–464
    mean squared error method, 464
    piecewise desirability function, 463
Multi-vari plot, 530–531

NIST/SEMATECH e-Handbook of Statistical
  Methods, 18, 86, 111, 188, 347, 360,
  525, 545
Normal probability plot methods, 136, 187,
  188, 194, 560

Optimum operating conditions, 225, 360,
  404
  methods for determining, 360
Organizations cited
  American Society for Quality (ASQ), 1
  Booth Dispensers, Ltd., 499
  Morton Powder Coatings, 343
  National Institute of Standards and
    Technology (NIST), 111, 135,
    225, 524, 525

Procter and Gamble, 320
Rayovac, 347–348
Rothamsted Experimental Station, 179,
  182

Pareto effects chart/analysis, 126, 180, 213,
  214, 255, 498
Processes in/out of statistical control, 18, 19,
  189, 255, 267
  blocking out-of-control process, 60–61
  checking for, 141, 266, 487
    check runs, 19

Quasi-experimental design, 23

R^2, 208, 255, 266, 271, 273, 373, 497
Region of operability, 364
  irregular design space, 386
  restricted, 387, 404, 508–514
    debarred observations, 387, 508, 509,
      514
Response surface methodology (RSM), 360
  analyzing fitted surface, 398–404
    contours of constant response, 398
    ridge analysis, 403–404
      method of Lagrangian multipliers,
      403
      with noise variables, 404
    rising ridge, 398
    stationary points, 400–403
      confidence regions on, 402–403
  in a three-stage operation, 418
  in the food industry, 417
  method of steepest ascent/descent,
    370–373, 561
    example, 371–372
    modified method, 405
    scale-independent methods, 373

Satterthwaite's procedure, 271, 272
Shainin's variables search approach, 500
Six Sigma, 22
Sliding reference distribution, 36
Software for experimental design, 48, 89–90,
  151, 333, 350, 531–532
  Cornerstone, 471
  Dataplot, 135
  Design-Expert, 1, 11, 48, 89, 90, 151,154,
    157, 175, 176, 177, 192, 201, 215,

216, 217, 230–232, 246, 250, 252,
279–281, 282, 291, 295, 313, 321,
326, 328, 350, 369, 371, 379, 380,
391–393, 396, 397, 407, 438, 452,
457–462, 467–471, 480, 490, 492,
498, 504, 505, 518, 532

D.o.E. Fusion Pro, 48, 151, 230, 232, 283,
295, 321, 350, 356, 392, 407

Echip, 471

Gendex DOE toolkit, 90, 408, 511, 522

GOSSET, 1, 505

JMP, 1, 48, 89, 90, 157, 176, 187, 192,
201, 230–233, 265, 278–279, 281,
282, 289, 295, 297, 321, 366, 370,
375, 392, 407, 408, 427, 438,
455–458, 460, 461, 464, 468–471,
480, 488, 505, 511, 512, 516–519,
532

MathWorks, 520

MAPLE, 403

MINITAB, 1, 48, 64, 67, 87, 89, 90, 109,
124, 127, 131, 135, 143, 146, 149,
150, 154, 157, 176, 180, 186, 192,
201, 230–233, 243, 244, 248–250,
252, 255, 256, 269, 271–273, 275,

280, 287, 295, 300, 301, 321, 350,
365, 366, 374, 386, 392, 395, 396,
398, 401, 407, 408, 427, 438, 445,
452, 470, 471, 475, 490, 491, 493,
495, 501, 532

MIXSOFT, 523

R (CROSSDES), 434, 438

RS/1, 508

RS/Discover, 295, 350, 532

SAS Software, 64, 89–90, 157, 230, 271,
295, 297, 347, 350, 370, 404, 425,
434, 437

SPSS, 438

Stat-ease, Inc., 130, 208, 221, 343

Statgraphics, 90, 233, 408

Statistica, 233

Statistical process control methods, 19

Stepwise regression, 122, 207, 208, 277,
497, 499

Strong heredity assumption, 109

Weak heredity assumption, 109

Yates' algorithm, 172, 174

Yates order, 102–103, 127, 130, 134, 354

BECHHOFER, SANTNER, and GOLDSMAN · Design and Analysis of Experiments for Statistical Selection, Screening, and Multiple Comparisons

BELSLEY · Conditioning Diagnostics: Collinearity and Weak Data in Regression

†BELSLEY, KUH, and WELSCH · Regression Diagnostics: Identifying Influential Data and Sources of Collinearity

BENDAT and PIERSOL · Random Data: Analysis and Measurement Procedures, *Third Edition*

BERRY, CHALONER, and GEWEKE · Bayesian Analysis in Statistics and Econometrics: Essays in Honor of Arnold Zellner

BERNARDO and SMITH · Bayesian Theory

BHAT and MILLER · Elements of Applied Stochastic Processes, *Third Edition*

BHATTACHARYA and JOHNSON · Statistical Concepts and Methods

BHATTACHARYA and WAYMIRE · Stochastic Processes with Applications

BILLINGSLEY · Convergence of Probability Measures, *Second Edition*

BILLINGSLEY · Probability and Measure, *Third Edition*

BIRKES and DODGE · Alternative Methods of Regression

BLISCHKE and MURTHY (editors) · Case Studies in Reliability and Maintenance

BLISCHKE and MURTHY · Reliability: Modeling, Prediction, and Optimization

BLOOMFIELD · Fourier Analysis of Time Series: An Introduction, *Second Edition*

BOLLEN · Structural Equations with Latent Variables

BOLLEN and CURRAN · Latent Curve Models: A Structural Equation Perspective

BOROVKOV · Ergodicity and Stability of Stochastic Processes

BOULEAU · Numerical Methods for Stochastic Processes

BOX · Bayesian Inference in Statistical Analysis

BOX · R. A. Fisher, the Life of a Scientist

BOX and DRAPER · Empirical Model-Building and Response Surfaces

*BOX and DRAPER · Evolutionary Operation: A Statistical Method for Process Improvement

BOX and FRIENDS · Improving Almost Anything, *Revised Edition*

BOX, HUNTER, and HUNTER · Statistics for Experimenters: Design, Innovation and Discovery, *Second Edition*

BOX and LUCEÑO · Statistical Control by Monitoring and Feedback Adjustment

BRANDIMARTE · Numerical Methods in Finance: A MATLAB-Based Introduction

BROWN and HOLLANDER · Statistics: A Biomedical Introduction

BRUNNER, DOMHOF, and LANGER · Nonparametric Analysis of Longitudinal Data in Factorial Experiments

BUCKLEW · Large Deviation Techniques in Decision, Simulation, and Estimation

CAIROLI and DALANG · Sequential Stochastic Optimization

CASTILLO, HADI, BALAKRISHNAN and SARABIA · Extreme Value and Related Models with Applications in Engineering and Science

CHAN · Time Series: Applications to Finance

CHARALAMBIDES · Combinatorial Methods in Discrete Distributions

CHATTERJEE and HADI · Regression Analysis by Example, *Fourth Edition*

CHATTERJEE and HADI · Sensitivity Analysis in Linear Regression

CHERNICK · Bootstrap Methods: A Practitioner's Guide

CHERNICK and FRIIS · Introductory Biostatistics for the Health Sciences

CHILÈS and DELFINER · Geostatistics: Modeling Spatial Uncertainty

CHOW and LIU · Design and Analysis of Clinical Trials: Concepts and Methodologies, *Second Edition*

CLARKE and DISNEY · Probability and Random Processes: A First Course with Applications, *Second Edition*

*COCHRAN and COX · Experimental Designs, *Second Edition*

CONGDON · Applied Bayesian Modelling

---

*Now available in a lower priced paperback edition in the Wiley Classics Library.

†Now available in a lower priced paperback edition in the Wiley–Interscience Paperback Series.

CONGDON · Bayesian Models for Categorical Data
CONGDON · Bayesian Statistical Modelling
CONOVER · Practical Nonparametric Statistics, *Second Edition*
COOK · Regression Graphics
COOK and WEISBERG · Applied Regression Including Computing and Graphics
COOK and WEISBERG · An Introduction to Regression Graphics
CORNELL · Experiments with Mixtures, Designs, Models, and the Analysis of Mixture Data, *Third Edition*
COVER and THOMAS · Elements of Information Theory
COX · A Handbook of Introductory Statistical Methods
*COX · Planning of Experiments
CRESSIE · Statistics for Spatial Data, *Revised Edition*
CSÖRGÖ and HORVÁTH · Limit Theorems in Change Point Analysis
DANIEL · Applications of Statistics to Industrial Experimentation
DANIEL · Biostatistics: A Foundation for Analysis in the Health Sciences, *Eighth Edition*
*DANIEL · Fitting Equations to Data: Computer Analysis of Multifactor Data, *Second Edition*
DASU and JOHNSON · Exploratory Data Mining and Data Cleaning
DAVID and NAGARAJA · Order Statistics, *Third Edition*
*DEGROOT, FIENBERG, and KADANE · Statistics and the Law
DEL CASTILLO · Statistical Process Adjustment for Quality Control
DeEMARIS · Regression with Social Data: Modeling Continuous and Limited Response Variables
DEMIDENKO · Mixed Models: Theory and Applications
DENISON, HOLMES, MALLICK, and SMITH · Bayesian Methods for Nonlinear Classification and Regression
DETTE and STUDDEN · The Theory of Canonical Moments with Applications in Statistics, Probability, and Analysis
DEY and MUKERJEE · Fractional Factorial Plans
DILLON and GOLDSTEIN · Multivariate Analysis: Methods and Applications
DODGE · Alternative Methods of Regression
*DODGE and ROMIG · Sampling Inspection Tables, *Second Edition*
*DOOB · Stochastic Processes
DOWDY, WEARDEN, and CHILKO · Statistics for Research, *Third Edition*
DRAPER and SMITH · Applied Regression Analysis, *Third Edition*
DRYDEN and MARDIA · Statistical Shape Analysis
DUDEWICZ and MISHRA · Modern Mathematical Statistics
DUNN and CLARK · Basic Statistics: A Primer for the Biomedical Sciences, *Third Edition*
DUPUIS and ELLIS · A Weak Convergence Approach to the Theory of Large Deviations
EDLER AND KITSOS · Recent Advances in Quantitative Methods in Cancer and Human Health Risk Assessment
*ELANDT-JOHNSON and JOHNSON · Survival Models and Data Analysis
ENDERS · Applied Econometric Time Series
†ETHIER and KURTZ · Markov Processes: Characterization and Convergence
EVANS, HASTINGS, and PEACOCK · Statistical Distribution, *Third Edition*
FELLER · An Introduction to Probability Theory and Its Applications, Volume I, *Third Edition*, Revised; Volume II, *Second Edition*
FISHER and VAN BELLE · Biostatistics: A Methodology for the Health Sciences
FITZMAURICE, LAIRD, and WARE · Applied Longitudinal Analysis
*FLEISS · The Design and Analysis of Clinical Experiments
FLEISS · Statistical Methods for Rates and Proportions, *Third Edition*
†FLEMING and HARRINGTON · Counting Processes and Survival Analysis

*Now available in a lower priced paperback edition in the Wiley Classics Library.
†Now available in a lower priced paperback edition in the Wiley–Interscience Paperback Series.

FULLER · Introduction to Statistical Time Series, *Second Edition*
†FULLER · Measurement Error Models
GALLANT · Nonlinear Statistical Models
GEISSER · Modes of Parametric Statistical Inference
GELMAN and MENG · Applied Bayesian Modeling and Casual Inference from Incomplete-data Perspectives
GEWEKE · Contemporary Bayesian Econometrics and Statistics
GHOSH, MUKHOPADHYAY, and SEN · Sequential Estimation
GIESBRECHT and GUMPERTZ · Planning, Construction, and Statistical Analysis of Comparative Experiments
GIFI · Nonlinear Multivariate Analysis
GIVENS and HOETING · Computational Statistics
GLASSERMAN and YAO · Monotone Structure in Discrete-Event Systems
GNANADESIKAN · Methods for Statistical Data Analysis of Multivariate Observations, *Second Edition*
GOLDSTEIN and LEWIS · Assessment: Problems, Development, and Statistical Issues
GREENWOOD and NIKULIN · A Guide to Chi-Squared Testing
GROSS and HARRIS · Fundamentals of Queueing Theory, *Third Edition*
*HAHN and SHAPIRO · Statistical Models in Engineering
HAHN and MEEKER · Statistical Intervals: A Guide for Practitioners
HALD · A History of Probability and Statistics and their Applications Before 1750
HALD · A History of Mathematical Statistics from 1750 to 1930
†HAMPEL · Robust Statistics: The Approach Based on Influence Functions
HANNAN and DEISTLER · The Statistical Theory of Linear Systems
HEIBERGER · Computation for the Analysis of Designed Experiments
HEDAYAT and SINHA · Design and Inference in Finite Population Sampling
HEDEKER and GIBBONS · Longitudinal Data Analysis
HELLER · MACSYMA for Statisticians
HINKELMANN and KEMPTHORNE · Design and Analysis of Experiments, Volume 1: Introduction to Experimental Design
HINKELMANN and KEMPTHORNE · Design and analysis of experiments, Volume 2: Advanced Experimental Design
HOAGLIN, MOSTELLER, and TUKEY · Exploratory Approach to Analysis of Variance
*HOAGLIN, MOSTELLER, and TUKEY · Exploring Data Tables, Trends and Shapes
*HOAGLIN, MOSTELLER, and TUKEY · Understanding Robust and Exploratory Data Analysis
HOCHBERG and TAMHANE · Multiple Comparison Procedures
HOCKING · Methods and Applications of Linear Models: Regression and the Analysis of Variance, *Second Edition*
HOEL · Introduction to Mathematical Statistics, *Fifth Edition*
HOGG and KLUGMAN · Loss Distributions
HOLLANDER and WOLFE · Nonparametric Statistical Methods, *Second Edition*
HOSMER and LEMESHOW · Applied Logistic Regression, *Second Edition*
HOSMER and LEMESHOW · Applied Survival Analysis: Regression Modeling of Time to Event Data
†HUBER · Robust Statistics
HUBERTY · Applied Discriminant Analysis
HUBERTY and OLEJNIK · Applied MANOVA and Discriminant Analysis, *Second Edition*
HUNT and KENNEDY · Financial Derivatives in Theory and Practice, *Revised Edition*
HUSKOVA, BERAN, and DUPAC · Collected Works of Jaroslav Hajek—with Commentary
HUZURBAZAR · Flowgraph Models for Multistate Time-to-Event Data
IMAN and CONOVER · A Modern Approach to Statistics
†JACKSON · A User's Guide to Principle Components

*Now available in a lower priced paperback edition in the Wiley Classics Library.
†Now available in a lower priced paperback edition in the Wiley–Interscience Paperback Series.

JOHN · Statistical Methods in Engineering and Quality Assurance

JOHNSON · Multivariate Statistical Simulation

JOHNSON and BALAKRISHNAN · Advances in the Theory and Practice of Statistics: A Volume in Honor of Samuel Kotz

JOHNSON and BHATTACHARYYA · Statistics: Principles and Methods, *Fifth Edition*

JOHNSON and KOTZ · Distributions in Statistics

JOHNSON and KOTZ (editors) · Leading Personalities in Statistical Sciences: From the Seventeenth Century to the Present

JOHNSON, KOTZ, and BALAKRISHNAN · Continuous Univariate Distributions, Volume 1, *Second Edition*

JOHNSON, KOTZ, and BALAKRISHNAN · Continuous Univariate Distributions, Volume 2, *Second Edition*

JOHNSON, KOTZ, and BALAKRISHNAN · Discrete Multivariate Distributions

JOHNSON, KOTZ, and KEMP · Univariate Discrete Distributions, *Third Edition*

JUDGE, GRIFFITHS, HILL, LÜTKEPOHL, and LEE · The Theory and Practice of Econometrics, *Second Edition*

JUREČKOVÁ and SEN · Robust Statistical Procedures: Asymptotics and Interrelations

JUREK and MASON · Operator-Limit Distributions in Probability Theory

KADANE · Bayesian Methods and Ethics in a Clinical Trial Design

KADANE and SCHUM · A Probabilistic Analysis of the Sacco and Vanzetti Evidence

KALBFLEISCH and PRENTICE · The Statistical Analysis of Failure Time Data, *Second Edition*

KARIYA and KURATA · Generalized Least Squares

KASS and VOS · Geometrical Foundations of Asymptotic Inference

†KAUFMAN and ROUSSEEUW · Finding Groups in Data: An Introduction to Cluster Analysis

KEDEM and FOKIANOS · Regression Models for Time Series Analysis

KENDALL, BARDEN, CARNE, and LE · Shape and Shape Theory

KHURI · Advanced Calculus with Applications in Statistics, *Second Edition*

KHURI, MATHEW, and SINHA · Statistical Tests for Mixed Linear Models

KLEIBER and KOTZ · Statistical Size Distributions in Economics and Actuarial Sciences

KLUGMAN, PANJER, and WILLMOT · Loss Models: From Data to Decisions, *Second Edition*

KLUGMAN, PANJER, and WILLMOT · Solutions Manual to Accompany Loss Models: From Data to Decisions, *Second Edition*

KOTZ, BALAKRISHNAN, and JOHNSON · Continuous Multivariate Distributions, Volume 1, *Second Edition*

KOVALENKO, KUZNETZOV, and PEGG · Mathematical Theory of Reliability of Time-Dependent Systems with Practical Applications

LACHIN · Biostatistical Methods: The Assessment of Relative Risks

LAD · Operational Subjective Statistical Methods: A Mathematical, Philosophical, and Historical Introduction

LAMPERTI · Probability: A Survey of the Mathematical Theory, *Second Edition*

LANGE, RYAN, BILLARD, BRILLINGER, CONQUEST, and GREENHOUSE · Case Studies in Biometry

LARSON · Introduction to Probability Theory and Statistical Inference, *Third Edition*

LAWLESS · Statistical Models and Methods for Lifetime Data, *Second Edition*

LAWSON · Statistical Methods in Spatial Epidemiology

LE · Applied Categorical Data Analysis

LE · Applied Survival Analysis

LEE and WANG · Statistical Methods for Survival Data Analysis, *Third Edition*

LEPAGE and BILLARD · Exploring the Limits of Bootstrap

LEYLAND and GOLDSTEIN (editors) · Multilevel Modelling of Health Statistics

LIAO · Statistical Group Comparison

†Now available in a lower priced paperback edition in the Wiley - Interscience Paperback Series.

LINDVALL · Lectures on the Coupling Method

LINHART and ZUCCHINI · Model Selection

LITTLE and RUBIN · Statistical Analysis with Missing Data, *Second Edition*

LLOYD · The Statistical Analysis of Categorical Data

LOWEN and TEICH · Fractal-Based Point Processes

MAGNUS and NEUDECKER · Matrix Differential Calculus with Applications in Statistics and Econometrics, *Revised Edition*

MALLER and ZHOU · Survival Analysis with Long Term Survivors

MALLOWS · Design, Data, and Analysis by Some Friends of Cuthbert Daniel

MANN, SCHAFER, and SINGPURWALLA · Methods for Statistical Analysis of Reliability and Life Data

MANTON, WOODBURY, and TOLLEY · Statistical Applications Using Fuzzy Sets

MARCHETTE · Random Graphs for Statistical Pattern Recognition

MARDIA and JUPP · Directional Statistics

MASON, GUNST, and HESS · Statistical Design and Analysis of Experiments with Applications to Engineering and Science, *Second Edition*

McCULLOCH and SEARLE · Generalized, Linear, and Mixed Models

McFADDEN · Management of Data in Clinical Trials

*McLACHLAN · Discriminant Analysis and Statistical Pattern Recognition

McLACHLAN, DO, and AMBROISE · Analyzing Microarray Gene Expression Data

McLACHLAN and KRISHNAN · The EM Algorithm and Extensions

McLACHLAN and PEEL · Finite Mixture Models

McNEIL · Epidemiological Research Methods

MEEKER and ESCOBAR · Statistical Methods for Reliability Data

MEERSCHAERT and SCHEFFLER · Limit Distributions for Sums of Independent Random Vectors: Heavy Tails in Theory and Practice

MICKEY, DUNN, and CLARK · Applied Statistics: Analysis of Variance and Regression, *Third Edition*

*MILLER · Survival Analysis, *Second Edition*

MONTGOMERY, PECK, and VINING · Introduction to Linear Regression Analysis, *Fourth Edition*

MORGENTHALER and TUKEY · Configural Polysampling: A Route to Practical Robustness

MUIRHEAD · Aspects of Multivariate Statistical Theory

MULLER and STOYAN · Comparison Methods for Stochastic Models and Risks

MURRAY · X-STAT 2.0 Statistical Experimentation, Design Data Analysis, and Nonlinear Optimization

MURTHY, XIE, and JIANG · Weibull Models

MYERS and MONTGOMERY · Response Surface Methodology: Process and Product Optimization Using Designed Experiments, *Second Edition*

MYERS, MONTGOMERY, and VINING · Generalized Linear Models. With Applications in Engineering and the Sciences

†NELSON · Accelerated Testing, Statistical Models, Test Plans, and Data Analysis

†NELSON · Applied Life Data Analysis

NEWMAN · Biostatistical Methods in Epidemiology

OCHI · Applied Probability and Stochastic Processes in Engineering and Physical Sciences

OKABE, BOOTS, SUGIHARA, and CHIU · Spatial Tesselations: Concepts and Applications of Voronoi Diagrams, *Second Edition*

OLIVER and SMITH · Influence Diagrams, Belief Nets and Decision Analysis

PALTA · Quantitative Methods in Population Health: Extentions of Ordinary Regression

PANJER · Operational Risks: Modeling Analytics

PANKRATZ · Forecasting with Dynamic Regression Models

PANKRATZ · Forecasting with Univariate Box-Jenkins Models: Concepts and Cases

---

*Now available in a lower priced paperback edition in the Wiley Classics Library.

†Now available in a lower priced paperback edition in the Wiley - Interscience Paperback Series.

*PARZEN · Modern Probability Theory and Its Applications

PEÑA, TIAO, and TSAY · A Course in Time Series Analysis

PIANTADOSI · Clinical Trials: A Methodologic Perspective

PORT · Theoretical Probability for Applications

POURAHMADI · Foundations of Time Series Analysis and Prediction Theory

PRESS · Bayesian Statistics: Principles, Models, and Applications

PRESS · Subjective and Objective Bayesian Statistics, *Second Edition*

PRESS and TANUR · The Subjectivity of Scientists and the Bayesian Approach

PUKELSHEIM · Optimal Experimental Design

PURI, VILAPLANA, and WERTZ · New Perspectives in Theoretical and Applied Statistics

†PUTERMAN · Markov Decision Processes: Discrete Stochastic Dynamic Programming

QIU · Image Processing and Jump Regression Analysis

*RAO · Linear Statistical Inference and its Applications, *Second Edition*

RAUSAND and HØYLAND · System Reliability Theory: Models, Statistical Methods and Applications, *Second Edition*

RENCHER · Linear Models in Statistics

RENCHER · Methods of Multivariate Analysis, *Second Edition*

RENCHER · Multivariate Statistical Inference with Applications

*RIPLEY · Spatial Statistics

*RIPLEY · Stochastic Simulation

ROBINSON · Practical Strategies for Experimenting

ROHATGI and SALEH · An Introduction to Probability and Statistics, *Second Edition*

ROLSKI, SCHMIDLI, SCHMIDT, and TEUGELS · Stochastic Processes for Insurance and Finance

ROSENBERGER and LACHIN · Randomization in Clinical Trials: Theory and Practice

ROSS · Introduction to Probability and Statistics for Engineers and Scientists

ROSSI, ALLENBY, and MCCULLOCH · Bayesian Statistics and Marketing

†ROUSSEEUW and LEROY · Robust Regression and Outline Detection

*RUBIN · Multiple Imputation for Nonresponse in Surveys

RUBINSTEIN · Simulation and the Monte Carlo Method

RUBINSTEIN and MELAMED · Modern Simulation and Modeling

RYAN · Modern Experimental Design

RYAN · Modern Regression Methods

RYAN · Statistical Methods for Quality Improvement, *Second Edition*

SALEH · Theory of Preliminary Test and Stein-Type Estimation with Applications

*SCHEFFE · The Analysis of Variance

SCHIMEK · Smoothing and Regression: Approaches, Computation, and Application

SCHOTT · Matrix Analysis for Statistics, *Second Edition*

SCHOUTENS · Levy Processes in Finance: Pricing Financial Derivatives

SCHUSS · Theory and Applications of Stochastic Differential Equations

SCOTT · Multivariate Density Estimation: Theory, Practice, and Visualization

†SEARLE · Linear Models for Unbalanced Data

†SEARLE · Matrix Algebra Useful for Statistics

†SEARLE, CASELLA, and MCCULLOCH · Variance Components

SEARLE and WILLETT · Matrix Algebra for Applied Economics

SEBER and LEE · Linear Regression Analysis, *Second Edition*

†SEBER · Multivariate Observations

†SEBER and WILD · Nonlinear Regression

SENNOTT · Stochastic Dynamic Programming and the Control of Queueing Systems

*SERFLING · Approximation Theorems of Mathematical Statistics

SHAFER and VOVK · Probability and Finance: Its Only a Game!

*Now available in a lower priced paperback edition in the Wiley Classics Library.

†Now available in a lower priced paperback edition in the Wiley - Interscience Paperback Series.

SILVAPULLE and SEN · Constrained Statistical Inference: Inequality, Order, and Shape Restrictions

SMALL and McLEISH · Hilbert Space Methods in Probability and Statistical Inference

SRIVASTAVA · Methods of Multivariate Statistics

STAPLETON · Linear Statistical Models

STAUDTE and SHEATHER · Robust Estimation and Testing

STOYAN, KENDALL, and MECKE · Stochastic Geometry and Its Applications, *Second Edition*

STOYAN and STOYAN · Fractals, Random and Point Fields: Methods of Geometrical Statistics

STYAN · The Collected Papers of T. W. Anderson: 1943–1985

SUTTON, ABRAMS, JONES, SHELDON, and SONG · Methods for Meta-Analysis in Medical Research

TANAKA · Time Series Analysis: Nonstationary and Noninvertible Distribution Theory

THOMPSON · Empirical Model Building

THOMPSON · Sampling, *Second Edition*

THOMPSON · Simulation: A Modeler's Approach

THOMPSON and SEBER · Adaptive Sampling

THOMPSON, WILLIAMS, and FINDLAY · Models for Investors in Real World Markets

TIAO, BISGAARD, HILL, PEÑA, and STIGLER (editors) · Box on Quality and Discovery: with Design, Control, and Robustness

TIERNEY · LISP-STAT: An Object-Oriented Environment for Statistical Computing and Dynamic Graphics

TSAY · Analysis of Financial Time Series, *Second Edition*

UPTON and FINGLETON · Spatial Data Analysis by Example, Volume II: Categorical and Directional Data

VAN BELLE · Statistical Rules of Thumb

VAN BELLE, FISHER, HEAGERTY, and LUMLEY · Biostatistics: A Methodology for the Health Sciences, *Second Edition*

VESTRUP · The Theory of Measures and Integration

VIDAKOVIC · Statistical Modeling by Wavelets

VINOD and REAGLE · Preparing for the Worst: Incorporating Downside Risk in Stock Market Investments

WALLER and GOTWAY · Applied Spatial Statistics for Public Health Data

WEERAHANDI · Generalized Inference in Repeated Measures: Exact Methods in MANOVA and Mixed Models

WEISBERG · Applied Linear Regression, *Second Edition*

WELISH · Aspects of Statistical Inference

WESTFALL and YOUNG · Resampling-Based Multiple Testing: Examples and Methods for p-Value Adjustment

WHITTAKER · Graphical Models in Applied Multivariate Statistics

WINKER · Optimization Heuristics in Economics: Applications of Threshold Accepting

WONNACOTT and WONNACOTT · Econometrics, *Second Edition*

WOODING · Planning Pharmaceutical Clinical Trials: Basic Statistical Principles

WOODWORTH · Biostatistics: A Bayesian Introduction

WOOLSON and CLARKE · Statistical Methods for the Analysis of Biomedical Data, *Second Edition*

WU and HAMADA · Experiments: Planning, Analysis, and Parameter Design Optimization

WU and ZHANG · Nonparametric Regression Methods for Longitudinal Data Analysis

YANG · The Construction Theory of Denumerable Markov Processes

YOUNG, VALERO-MORA, and FRIENDLY · Visual Statistics: Seeing Data with Dynamic Interactive Graphics

ZELTERMAN · Discrete Distributions—Applications in the Health Sciences

*ZELLNER · An Introduction to Bayesian Inference in Econometrics

ZHOU, OBUCHOWSKI, and MCCLISH · Statistical Methods in Diagnostic Medicine

*Now available in a lower priced paperback edition in the Wiley Classics Library.